# The Appraisal of Rural Property

# The Appraisal of Rural Property

Second Edition

American Society of Farm Managers and Rural Appraisers®

Appraisal Institute®

| | |
|---|---|
| *Reviewers:* | Frank E. Harrison, MAI, SRA<br>John H. Love, SRA<br>Stephen A. Manning, MAI, SRA<br>Michael S. MaRous, MAI, SRA<br>John A. Schwartz, MAI<br>Janice F. Young, MAI, RM |
| *Vice President, Educational Programs and Publications:* | Larisa Phillips |
| *Director, Content Development & Quality Assurance:* | Margo Wright |
| *Manager, Book Development:* | Stephanie Shea-Joyce |
| *Supervisor, Book Design/Production:* | Michael Landis |
| *Cover Design:* | Amy Reichard |

**For Educational Purposes Only**

The material presented in this text has been reviewed by members of the Appraisal Institute, but the opinions and procedures set forth by the authors are not necessarily endorsed as the only methodology consistent with proper appraisal practice. While a great deal of care has been taken to provide accurate and current information, neither the Appraisal Institute nor its editors and staff assume responsibility for the accuracy of the data contained herein. Further, the general principles and conclusions presented in this text are subject to local, state, and federal laws and regulations, court cases, and any revisions of the same. This publication is sold for educational purposes with the understanding that the publisher is not engaged in rendering legal, accounting, or other professional service.

**Nondiscrimination Policy**

The Appraisal Institute advocates equal opportunity and nondiscrimination in the appraisal profession and conducts its activities in accordance with applicable federal, state, and local laws.



Printed in the United States of America

**Library of Congress Cataloging-in-Publication Data**

The appraisal of rural property.—2nd ed.
p. cm.
Includes bibliographical references and index.
ISBN 0-922154-62-7
1. Farms—Valuation—United States. I. American Society of Farm Managers and Rural Appraisers. II. Appraisal Institute (U.S.)

HD256 .A67 2000
333.33'52—dc21

00-050280

# Contents

## Part IV The Valuation of Selected Rural Properties



## Appendices

# Foreword

Today's rural and agricultural property market is radically different from the market that existed in 1983 when *The Appraisal of Rural Property* was first published. Massive changes have occurred in agriculture over the past 18 years and the size of most agricultural enterprises has increased tremendously. The technology employed in the production of crops and livestock has completely revolutionized the agricultural landscape.

While these trends have changed agriculture, the growing populations of major metropolitan areas across the United States have created extensive demand for rural properties. Properties that were once devoted completely to agriculture are now dominated by recreational and homesite uses, with agriculture as an interim use.

To address these changes in agriculture and in the rural land market, appraisers have implemented new, innovative techniques to value both agricultural and nonagricultural rural land. Special-purpose agricultural properties, which continue to increase in number and complexity, also require valuers to apply new skills and techniques.

This text has been written to promote an understanding of the changes that continue to occur in rural property markets and to provide readers with the most current analytical tools and techniques for valuing all types of rural properties. The methodology demonstrated is based on the three traditional approaches to value as adapted to rural properties. Specific techniques of analysis are presented to ensure that appraisers give proper consideration to the factors involved in virtually all assignments to value rural land. Selected special-purpose properties have been analyzed in detail to assist professionals and students alike.

The development of this text has been a cooperative effort of the American Society of Farm Managers and Rural Appraisers and the

Appraisal Institute. Both organizations endorse and recommend this text to those doing rural appraisal work. From the inception of this project to its conclusion, we have focused on a single goal—publishing a rural appraisal text that is both current and definitive. We trust that readers will find *The Appraisal of Rural Property,* second edition, to be useful in both their daily practice and their academic endeavors.

Tharrell D. Ming, ARA—*1993 President*

John T. Widdoss, ARA, MAI—*1995 President*

James B. Vine, Jr., ARA—*1997 President*

Arthur E. Clapp, ARA—*1999 President*

**American Society of Farm Managers and Rural Appraisers**

Woodward S. Hanson, MAI—*2000 President*
**Appraisal Institute**

# Acknowledgments

The development of this text was organized and conducted by the Rural Textbook Committee of the American Society of Farm Managers and Rural Appraisers composed of Andrew P. Brorsen, ARA, MAI; John Frome, ARA, MAI; Kim Johnson, ARA; Charles Scott Seely, ARA, MAI; and James B. Vine, Jr., ARA. The committee wishes to recognize the major contribution of Stephanie Shea-Joyce, Manager of Book Development for the Appraisal Institute, who performed the technical writing and editing tasks required to produce this text.

The Rural Textbook Committee, which saw the work through to its completion, wishes to acknowledge the contributions of the creative writers and reviewers who gave of their time and talent to produce this work over the development period: J. H. Adkinson, PhD; Paul A. Bierschwale, ARA; William E. Bliss, ARA; James C. Burge, ARA, MAI; James D. Cannon, ARA, MAI; John M. Childears, ARA, AFM; Arthur E. Clapp, ARA; Tony Correia, ARA; Dave Damron, RPRA; J. B. Scott Dilworth, ARA; Glen W. Dittmar; D. Delos Ellsworth, ARA; R. Mark Finley, Jr., ARA, RPRA; Donald A. Fisher, ARA, MAI; Stan Forbes, Farmer Mac; James D. Garvey, ARA; David J. Heine, ARA; Douglas K. Hodge, ARA, MAI; Donald E. Howell, ARA; Jerald A. Hulm, ARA, MAI; Darreld D. Kletke, PhD; Henry A. Long, Jr. ARA, AFM, MAI; Jay D. Luse, ARA, AFM; Sam D. McAnally, ARA; Tharrell D. Ming, ARA; Laurel L. Olsen, MAI; Clifford M. Pettis, ARA; John E. Reynolds, PhD; William R. Schott, ARA, AFM; Lee B. Smith, ARA, MAI; James Connor Smith, ARA, MAI; Dan Smith, PhD.; David W. Snook; David G. Springer, ARA; James Synatzske, ARA; Allen E. Thode, ARA-Retired, MAI; Neill S. Thompson, ARA-Retired, MAI; Tim Vining, MAI; John T. Widdoss, ARA, MAI; Bruce Rowell Willmette; Donald C. Wood, ARA; and Juliana E.Young, ARA, MAI.

PART • I •

# The Foundations of Rural and Agricultural Property Valuation

CHAPTER

• 1 •

# *Property and Property Rights*

> Much can be said about the basic importance of land resources and real estate in the modern world. They provide people with living space, with raw materials necessary for filling material needs, and with opportunities for satisfactions dear to the heart of man. People look to land for their physical environment, for the food they eat, for fibers and the other materials needed to clothe their bodies and to provide housing and manufactured goods, for building sites, for recreation opportunities and for scenery and open space.[1]

Throughout recorded history, people have lived in close association with the soil, fields, forests, and fishing areas that provide food, fiber, shelter, and products for trade. Wars have been fought for control of land and land rights are a key factor in a society's economy and politics.

Over the course of the twentieth century, population has shifted away from rural areas and concentrated in cities. In the same period, farming practices have changed radically and, with the help of modern technology, fewer people can cultivate vast land areas. Although both the historical uses of land and people's relationship with the land have changed, the ownership of land and real estate has remained very important. In fact, investments in land and improvements account for most of the wealth in the United States.

The productivity of agricultural land is measured primarily in terms of its capacity to produce commodities, but rural land also furnishes its owners with amenities in the form of housing and recreation. Real estate accounted for about 71% of the assets in the farm sector in 1991.[2]

---

1. Raleigh Barlow, *Land Resource Economics: The Economics of Real Estate,* 4th ed. (Englewood Cliffs, N.J.: Prentice-Hall, 1986).
2. U.S. Department of Agriculture, "Agricultural Income and Finance: Situation and Outlook," Economic Research Service Publication No. AFO-41, June 1991.

## The Nature of Property

Land is an area identifiable on the earth's surface which, in theory, extends downward to the center of the earth and upward into the atmosphere. Land is defined in space and, like matter, can neither be created nor destroyed. In most cases, land is considered indestructible and nondepreciable. Permanent, natural features of land include lakes, soil, streams, trees, rocks, and subsurface deposits of gravel, oil, gas, and minerals. Air space above the earth's surface is also included in land and may be valuable in certain situations. For example, air rights over existing improvements are bought and sold in many large cities for the construction of buildings and walkways.

The concept of property is a social institution. It provides rules under which man can use the earth's natural resources. Because land is a shared resource, the use of land can lead to competition and conflicts. The nature of property involves the relationship between people and material objects. Some land area is classified as common property because it cannot be exclusively appropriated and possessed by individuals.[3] The right of private property is an exclusive right, giving an individual the right to use and enjoy the property and the right to exclude others from using and enjoying it.

### Real Estate, Real Property, and Personal Property

The terms *real estate, real property* and *personal property* are widely used but often not clearly defined. Real estate refers to land and the appurtenances affixed to the land—i.e., permanent improvements and attachments such as buildings and fences. Improvements and attachments to land are man-made and, unlike land, they are depreciable. Improvements that are generally considered a part of the land, called *site improvements,* include tile drainage, grass waterways, terraces, access roads, and irrigation wells.

Real property refers to the interests, benefits, and rights inherent in the ownership of physical real estate. A right or interest in real estate is also referred to as an *estate.* Specifically, an estate in land is the degree, nature, or extent of the interest held.

Personal property refers to identifiable portable and tangible objects that are considered to be personal and are not permanently affixed to the real estate. Therefore, personal property is not included in real estate ownership. Livestock, machinery, and household furniture are examples of personal property. Personal property may become part of the real estate if it is attached to the land or installed in structure. A fixture is an article that was once personal property, but has been installed or attached in a more or less permanent manner. A fixture is legally part of the real estate, and the courts use the following criteria to decide whether an item is real estate or personal property.

---

3. James C. Hite, *Room and Situation: The Political Economy of Land-Use Policy* (Chicago: Nelson-Hall), 1979.

1. The manner in which the item is affixed. Generally, an item is considered personal property if it can be removed without serious injury to the real estate or to itself. There are exceptions to this rule.
2. The character of the item and its adaptation to the real estate. Items that are specifically constructed for use in a particular building or installed to carry out the purpose for which the building was constructed are generally considered permanent parts of the building.
3. The intention of the party who added the attachment. Frequently, the terms of the lease reveal whether the item is permanent or is to be removed at some future time.

In each assignment the appraiser faces two tasks: 1) identification and description of the real estate or personal property, and 2) determining what property rights are included in the appraisal. Typically farm appraisal poses problems because farm real estate is often difficult to describe and identify. Multiple ownership rights are present and grading standards are imprecise or nonexistent. And, as is true in any appraisal, each tract of land differs from every other tract by virtue of its location.

## Property Rights and Ownership

Property rights play an important role in determining what people can and cannot do with real estate. Property rights are exclusive, but not absolute. When property rights are held by individuals they can exclude all other persons from access to or use of the property. But the exclusive rights that individuals hold are always subject to controls and limitations that are vested in the sovereign power.

Ownership includes the right to use property subject to various restrictions. Exclusive ownership is essential so individuals who invest their labor and money in property can be assured of receiving the returns from their efforts.[4] The rights of ownership must be specified. In case of conflicts, it is important to be able to determine whose rights predominate. For example, a landowner usually has the right to keep others from trespassing. He or she also has the right of transferability, which allows for the sale, trade, or transfer of property rights to some other party. Of course, for a system of property rights to be effective it must be enforced. An unenforced right is effectively no right at all. The sovereign power protects the rights of the owner.

Property ownership involves a bundle of rights, which can be held together or separately. These rights are comparable to a bundle of sticks, with each stick representing a separate right. The property owner has the right to enter and use the property, the right to sell or lease specified

4. Alan Randall, *Resource Economics: An Economic Approach to Natural Resource and Environmental Policy* (Columbus, Ohio: Grid Publishing, Inc., 1981).

ownership rights to the property, and the right to give these rights away. The owner of the bundle of rights can elect to exercise all or none of these rights. The owner can mortgage the property to obtain capital to pay off debt, can grant easements for specified uses, and can subdivide the land (usually subject to restrictions). The owner also may have mineral, water, or air rights. In some situations, the owner may possess development rights which can be separated from the bundle.

A person may own all or only some of the rights in a parcel of real estate. An individual who owns all property rights is said to have fee simple title. A fee simple estate implies absolute ownership unencumbered by any other interest or estate and subject only to the limitations imposed by government.[5] It is the broadest or most complete form of private property rights. Fee simple owners can exercise and enjoy their rights to the exclusion of others as long as they have not disposed of them. In many cases owners do not hold all of these rights. They may have obtained a mortgage to finance the purchase, granted an easement for a power line, or sold the mineral rights to the property.

## Governmental Rights

A fee simple owner holds complete, but not absolute, rights. The owner's rights are always subject to public limitations on private use. Four public limitations are enforced through the powers of government and used to serve the public interest: 1) taxation, 2) eminent domain, 3) police power, and 4) escheat.

1. The government has the right to levy and collect taxes. Real estate taxes have long been a major source of revenue for government. In addition to providing a means to obtain public funds, the power of taxation is an important control measure. For example, a heavy tax on irrigation wells may reduce water use and result in an increase in the water table.
2. Eminent domain is the power of government to take private property for public use without the owner's consent. Many federal, state, and local agencies have the right of eminent domain. In exercising the right of eminent domain, governments must show that they are acting in the public interest, must pay just compensation, and must follow due process of law. The power of eminent domain is used to take land for public roads, schools, and other public facilities.
3. The government uses its police power to perform functions necessary for public safety and the well being of society. With respect to real estate, police power is used to regulate the use and development of property to promote the public's safety, health, morals, and general welfare. Zoning laws, building codes,

5. Appraisal Institute, *The Appraisal of Real Estate,* 11th ed. (Chicago: Appraisal Institute, 1996).

sanitation requirements, and environmental regulations are examples of the use of police power.

4. Escheat is the right of government that gives the state titular ownership of property if an owner dies without a will and known heirs.

The four powers of government which restrict the private use of property are as applicable today as they were when our forefathers delineated them. In fact, as population density has increased and society has become more conscious of environmental and conservation issues, these limitations have become more evident. Many rural property owners feel that their real property rights are being eroded by zoning, environmental regulations, and other restrictions. Others object to the taking of private real estate to create parks and recreation areas. Conflicts between private and public interests in land will continue and one thing is fairly certain: It will become increasingly difficult for owners to do as they please with their land.

## Private Ownership Rights

In addition to public limitations, the property owner's rights may be restricted by private limitations. These limitations may be created by easements, deed reservations, or restrictive covenants. Easements give others the right to use the owner's land for a specific purpose. A utility company may hold an easement that allows it to run its lines across the property and gives its workers access to service these lines. Easements are attached to the property and are transferred to new owners unless revoked.

Property owners are frequently affected by provisions in their deeds that limit their ownership rights. A deed may include restrictions regarding mineral rights or rights-of-way or it may specify that the property remain in agricultural use for a specified period of time. Restrictive covenants may also impose private controls on the use of the land. Many residential subdivisions are subject to private controls which stipulate that the lots must be used for single-family residences with a minimum amount of square footage constructed a specific distance from the lot lines.

In the early history of this country, public policy emphasized private ownership and a laissez-faire economic philosophy. At that time, private property rights were emphasized and the public limitations on private rights were much less significant. However, the mix of private and public rights in land has changed as society's needs have changed over time.[6] Because rights in property are granted by society, these rights evolve over time as the priorities and perceived needs of society change. Private

---

6. Raleigh Barlowe, "Who Owns Our Land?" Southern Rural Development Center Publication No. 126, Mississippi State, March 1990.

property rights depend on the ability and willingness of the sovereign power to recognize and defend them.

During this century we have seen a movement toward more public limitations on private rights in property. Legislative action and court decisions have broadened the use of the public powers of government. No one can be certain how far this movement will go. As the U.S. population continues to grow, there will be increased competition for land and growing conflicts of interest concerning the most appropriate uses of land.

## Fee Simple Ownership and Estates in Property

As mentioned previously, the rights one holds in the ownership, possession, and control of property are referred to as one's *estate*. Each form of estate places certain rights, duties, responsibilities, or limitations on the owner.[7] The type of rights and their duration are important in classifying ownership. The person or party with the greatest number of rights for the longest duration is usually referred to as the owner.

Fee simple ownership is the most complete estate a person can hold in land. Fee simple ownership rights are not limited and there are no restrictions on transferring these rights. Fee simple ownership attaches no conditions on possession or use and endures without limit so long as the use is not illegal. The owner of the fee simple estate can sever certain rights of ownership and sell, lease, or transfer them to others. The disposition of some of the rights in the bundle of rights creates partial interests in real estate.

Generally property rights are divided between the owner and a tenant or between a lessor and a lessee. Other, less common, divisions include life estates created through a will or deed of conveyance and trusts.

## Tenancy

Tenancy is a division of property rights between two or more parties with each having an undivided interest. Properties held by two or more persons in undivided ownerships are usually classified as tenancy in common, joint tenancy, or tenancy by the entirety. Tenancy in common describes an estate in which property is held by two or more persons, each of whom owns an undivided share in the property. Each party may sell his or her undivided interest or dispose of it at will. In a joint tenancy, two or more persons own an undivided share in property, but each has rights of survivorship—i.e., if one of the owners dies, that owner's undivided share of the property goes to the survivor(s) rather than to heirs. Tenancy by the entirety is an estate held by a husband and wife in which neither has a disposable interest in the property during the lifetime of the other. This co-ownership arrangement, like joint tenancy, cannot be broken without the consent of both parties.

---

7. Gene Wunderlich and Russell W. Bierman, "What Do We Mean By Ownership?" in *LAND: The Yearbook of Agriculture, 1958*. (Washington, DC.: U. S. Government Printing Office, 1958), 287–294.

### Lease Interests

A lease is a contract that gives a tenant (the lessee) the right to possess and use property owned by a landlord (the lessor). Property rights are divided upon creation of the lease, which specifies the rights of both parties. The leased fee estate is held by the lessor, or landlord, who has the right to receive rental payments during the lease term and regains the possession and use of the property upon termination of the lease. The leasehold estate is the lessee's, or tenant's, estate and includes the right to use and occupy the property for a specified duration under the terms of the lease.

### Life Estates and Trusts

A life estate is a partial interest created by a will or deed of conveyance. In this legal arrangement, the designated party, known as the *life tenant,* has the rights of use, occupancy and control of the property for the duration of his or her life. The person to whom the property reverts upon the death of the life tenant is known as the *remainderman;* that person has a remainder interest in the life estate. The value of the life estate depends on the life expectancy of the life tenant, which can be estimated using actuarial tables.

Trusts are similar to life estates. A trust is a contractual arrangement in which property is held and administered for the benefit of a specified beneficiary. A living trust is established during the property owner's lifetime and is similar to a life estate. A land trust may be used by a group of owners to operate and manage a property, but the legal title is conveyed to a trustee. Land trusts may be used by owners as a means of concealing property ownership while gaining control of property. Rural property rights and interests will be discussed in more depth in Chapter 4.

## Conclusion

Land is a basic resource that is the subject of a variety of disciplines. The uniqueness, fixity, durability, finiteness, and usefulness of land are acknowledged in the fields of geography, law, sociology, and economics. Geography is concerned with the physical characteristics, demographic patterns, and economic uses of land, while the legal profession is concerned with land ownership. Sociology explores how private and public interests compete for the use and distribution of land. In economics land is regarded as a source of wealth and as one of the four agents of production, which also include labor, capital, and coordination.[8]

The appraiser must be aware of the differences between real estate and the personal property that may be involved in the proposed appraisal and must understand what items are to be included in the real estate to be appraised. Appraisers must also determine the ownership,

8. Appraisal Institute. *The Appraisal of Real Estate,* 11th ed. (Chicago: Appraisal Institute, 1996).

possession, and control of property to define the appraisal problem. It is important to clearly identify the type of estate involved. Several parties may have ownership interests in a given property and the extent of these interests and their duration must be specified. Whether the appraiser is asked to estimate the value of the fee simple estate or the value of a partial interest, the specific rights to be valued must be identified at the outset of the appraisal problem.

CHAPTER

• 2 •

# Concepts of Value

The concept of value pervades every aspect of the real estate industry. Value considerations are of central importance in a broad range of real estate activities. The term *value* is often used imprecisely in common speech, but in economics it has a specific meaning which distinguishes it from the related concepts of price, market, and cost.[1]

## Factors of Value

Value is not inherent in the commodity, good, or service to which it is ascribed; it is created in the minds of people who make up the market for that commodity, good, or service. The relationships that create value are complex and changes in value relate most directly to the four interdependent economic factors that must be present for a commodity to have value: utility, scarcity, desire, and effective purchasing power.

### Utility

Utility is the ability of a product to satisfy a human want, need, or desire. Individual properties possess different degrees of utility in terms of their ability to provide shelter, lifestyle amenities, and income. The characteristics of a property, such as its size, design, and location, influence its usefulness and value.

Residential properties satisfy the human need for shelter and may provide amenities that enhance the quality of life. The value of a residential property is usually related to its desirability and utility to owner-occupants, which may be converted into income in the form of rent.

The utility of income-producing properties is tied to the cash flow they produce. The desirability of an income-producing property may be

1. Most of the definitions and much of the discussion in this chapter has been taken from *The Appraisal of Real Estate,* 11th ed. (Chicago: Appraisal Institute, 1996).

different to different owners. An owner-operator may measure returns in terms of business profit or rental income. Like urban and suburban properties, rural properties are affected by utility factors and may be desired for their residential lifestyle amenities as well as their income-producing capabilities.

The utility and benefits of property ownership are derived from the legal rights that the owner possesses. Restrictions on ownership rights can inhibit the flow of benefits and influence the property's value. Environmental regulations, zoning or land use regulations, deed restrictions, and other limitations on the rights of ownership can enhance or detract from a property's utility and value.

### Scarcity

Scarcity is the present or anticipated supply of an item relative to the demand for it. If demand is constant, the declining supply of a commodity will usually increase its value. Land is generally abundant, but useful, desirable land is limited and can have greater value. No object has value unless scarcity is combined with utility. Air is abundant but, in most cases, it has little definable economic value. Insufficient water or drainage impacts the utility of the land and has a major effect on the value of rural properties.

### Desirability

Desire is a purchaser's wish for an item to satisfy human needs (e.g., shelter, clothing, food, companionship) or individual wants beyond essential life-support needs. Utility, scarcity, and desire are all considered in relation to effective purchasing power.

### Effective Purchasing Power

Effective purchasing power is the ability of an individual or group to participate in a market, i.e., to acquire goods and services with cash or its equivalent. The value of a property is affected by the ability of market participants to pay for that property. Effective purchasing power is the monetary ability to fulfill desire.

Utility, scarcity, desire, and effective purchasing power comprise the four economic factors of value. The interaction of these forces over time determine the value of a property at a specific point in time. An understanding of the past, present, and future interaction of these factors is essential in the valuation of property.

## Principles of Economic Theory

The characteristics of land and real property, which are the focus of appraisal activity, reflect our society's attitudes and beliefs. Real estate has value because the possession and successful management of real estate create opportunities for individuals to achieve economic goals that are generally perceived by society to be desirable. Financial stability, freedom from excessive restrictions, and high returns on investments are a few of the goals that can be achieved by participating in real estate markets.

In determining their level of participation in the market, individuals consider their wants and needs as well as the choices available to them at different times. These choices represent a variety of market characteristics and help support our free market economy, in which both individual and collective decisions contribute to the nation's economic success. Similarly, the production of the goods, services, and income that comprise the consumable portion of our economy depends on the combined effects of the four economic agents of production: labor, capital, coordination, and land. Thus the combination of human effort, tangible wealth, production systems, and natural resources provides the means for society to pursue its economic goals.

To analyze the many dynamic and interactive factors that influence people's attitudes and beliefs about value, a number of fundamental economic concepts must be addressed. The concepts described on the following pages provide the foundation for appraisal theory. It is critical that students of appraisal know and understand these economic concepts.

## Anticipation

Anticipation is the perception that value is created by the expectation of benefits to be derived in the future. Market participants base purchase decisions on their perceptions of the future benefits of acquiring a particular property, not on its historical price or the cost of creation. If a property is owner-occupied, its value depends on both its future income potential and the amenities derived from ownership and occupancy. Income-producing properties are purchased based on the anticipation of receiving future income. In the market for agricultural properties, both owner occupancy and income production contribute to the purchase decision. Although a farm or ranch property may be purchased as an investment with the expectation of future income benefits, it is also likely to provide a home for its owner. For farm and ranch properties, anticipation is tied to productivity, which will determine the income stream that is derived from the operation. This differs from the anticipation of an investor who holds land in expectation of an increase in value resulting from a change in use.

Real property appraisers must be aware of local, regional, and national real estate trends that will affect the perceptions of buyers and sellers and their view of the future. Historical data relating to the subject property or a particular market are relevant only insofar as they help interpret current market anticipations. A property's historical income may reveal favorable or unfavorable factors that will affect the property's ability to produce a certain net income in the future. The assembled data can then be analyzed and appropriate factors can be weighed to assess whether, under typical management and business conditions, the net income stream is expected to continue unchanged, to decline, or to increase.

## Change

Change is the result of the cause and effect relationship among the forces that influence real property value. Change is an inevitable and continuous phenomenon. The process of change may be relatively gradual and not readily discernible or, in an active market, change may occur rapidly with new properties being offered for sale and others being sold daily. The economic conditions created by plant closings, revisions of tax laws, or new construction starts may also cause change.

Other forces of change may include governmental influences, social trends, and environmental forces. During the 1980s all of these factors impacted the market for farmland in the United States and the result was a significant decline in land values. As economic conditions improved and governmental influences changed in the late 1980s, land values once again began to increase. Property values can decrease if commodity prices are adversely affected by environmental restrictions. All of the forces affecting value are interrelated; changes in these forces influence the demand for and supply of realty and, therefore, individual property values. Appraisers try to identify trends and anticipate future market circumstances that can influence current property values. Nevertheless, change is continual and value estimates are valid only for the point in time specified in the appraisal report.

Change is also evident within a particular property. Building improvements change as they age and become subject to physical, functional, and economic impairments. These impairments create depreciation, which is defined as a loss in property value from any cause. Depreciation may be seen as the difference between the cost to reproduce or replace a property and its present value.

Rural appraisers should be aware of changes in land use as well as changes in technology and farming practices that can make land more productive. The appraiser must recognize and weigh observed changes to ascertain their effect on the value of a specific property and on farmland values in general. When trends are correctly identified and interpreted, the impact of change on the market value of property can be estimated effectively.

## Supply and Demand

In economic theory, the principle of supply and demand states that the price of a commodity, good, or service varies directly, but not necessarily proportionately, with demand, and inversely, but not necessarily proportionately, with supply. Similarly, in a real estate context, the price of real property varies directly, but not necessarily proportionately, with demand, and inversely, but not necessarily proportionately, with supply. Thus, increasing the supply of an item or decreasing the demand for an item tends to reduce the price obtainable in the market; the opposite is also true. Although the relationship between supply and demand may not be directly proportional, the interaction of these forces is fundamental to economic theory. The interaction of suppliers and demanders, or sellers and buyers, constitutes a market.

Usually property values vary inversely with changes in supply. If properties for a particular use become more abundant than they were in the past, their values decline; by contrast, if properties become scarcer, their values increase. The supply of and demand for commodities always tend toward equilibrium. At this theoretical point, market value, price, and cost are equal. Most of the other economic principles of real estate valuation revolve around the interplay of supply and demand.

Supply is the amount of a commodity, good, or service made available for sale at different prices. In real estate, supply is the amount of a certain type of real estate available for sale or lease at various prices in a given market at a given period of time. Typically, more of an item will be supplied at a higher price and less at a lower price. Therefore, the supply of an item at a particular price, at a particular time, and in a particular place indicates that item's relative scarcity, which is a basic factor of value.

The supply of real estate is a function of the four agents of production, including the amount of capital needed to put a parcel of land to a given use relative to the current price of existing properties providing that use. An acceleration in price levels for rural land may come from two sources. In the agricultural community, farmers may have a strong demand for more land. At the fringes of urban communities, farmland may be desired for subdivision development. In either case the supply of land will likely increase because the capital necessary to buy the land, either for agricultural development or urban expansion, is available in sufficient quantities to bring the land to market.

Since land is both a commodity and the source of commodities, the supply of rural land refers to the amount of land suitable for a specific use, rather than simply the total number of available acres. As such, the supply of real estate incorporates measures of both quality and quantity. For agricultural properties in particular, the quality of the land may greatly affect property values. Quality is a function of both the tangible attributes of a property (e.g., size, shape, condition) and intangible attributes and amenities such as a desirable lifestyle, a scenic view, and a favorable location.

Property values usually vary inversely with changes in the supply. That is, if properties for a particular use are abundant, their values decline; in contrast, if properties for a desirable use are in short supply, values increase.

Demand is the desire and ability to purchase or lease goods and services. In real estate, demand is the amount of a type of real estate desired for purchase or rent at various prices in a given market at a given period of time. Typically less of an item will be demanded at a higher price, and more will be demanded at a lower price. Property values tend to vary directly with shifts in demand; prices tend to increase if demand increases and other factors remain constant. Usually the relationship between demand and price is not fixed. This changing relationship is described as the elasticity of demand.

Because the supply of real property for specific uses cannot be adjusted immediately, values are most affected by current demand. Demand supported by purchasing power results in effective demand, which is the relevant market consideration. The relationship between demand and price is usually direct, but not proportional. If demand increases and supply remains constant, prices tend to increase; if demand falls, prices tend to fall. Appraisers must interpret market behavior to ascertain the existing relationship between the supply of, and the demand for, the type of property being appraised.

For example, consider the interaction of supply and demand in farmland markets in the 1970s and 1980s. During the 1970s farm commodity prices were high and farmers were in the market for additional land, which created substantial demand. As a result, the prices received for farmland escalated rapidly in most agricultural areas.

Figures 2.1 shows the effect of the increasing demand that was experienced in the late 1970s as commodity prices rose and producers searched out additional land to farm. The equilibrium price $(E_0)$ rose from \$1460 to \$1840 per acre $(E_1)$ as demand increased. Original demand is indicated by curve $D_0$, supply is indicated by curve $S$, and the increased demand is indicated by curve $D_1$.

During the early 1980s, however, economic factors beyond the control of producers caused commodity prices to fall. Falling prices coupled with high interest rates created problems for farmers who had expanded in the late 1970s when land prices were escalating. As a result of the economic downturn, many farmers were forced to place their land on the market. Lending institutions had also obtained land which was subsequently placed on the market and further contributed to the oversupply. This oversupply, combined with the lack of qualified buyers and thus a lack of demand, resulted in the dramatic drop in farmland values. Only

**Figure 2.1** Increasing Demand

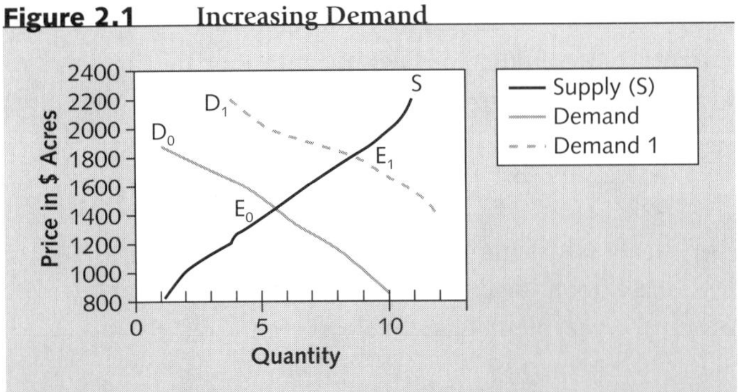

recently has a level of equilibrium returned to the market.

**Figure 2.2** Decreasing Demand

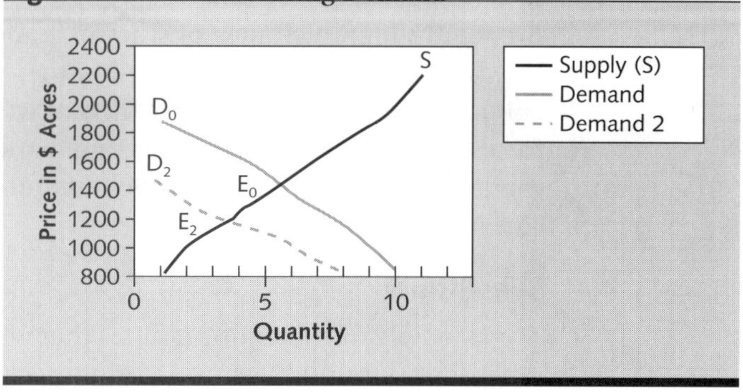


Figure 2.2 shows the effect of the decreasing demand experienced in the early to mid-1980s as commodity prices declined, interest rates rose, and producers had difficulty amortizing the land debt they incurred in the late 1970s. The equilibrium price $(E_0)$ of \$1460 began to fall to \$1190 per acre $(E_2)$ as demand waned. Original demand is indicated by curve $D_0$, supply is indicated by curve *S*, and the decreased demand is indicated by curve $D_2$.

## Competition

The concept of competition is related to the principle of supply and demand. Competition between buyers or tenants represents the interactive efforts of two or more potential purchasers or tenants to make a sale or secure a lease. Between sellers or landlords, competition represents the interactive efforts of two or more potential sellers or landlords to effect a sale or lease. Competition can also be defined as the active demand by two or more market participants for an item in short supply. Competition is fundamental to the dynamics of supply and demand in a free- enterprise, profit-maximizing economic system. Competition causes continual shifts in the levels of commodities available and, therefore, affects the levels of demand for those commodities. Buyers and sellers of real property operate in a competitive market setting. In essence, each property competes with all other properties suitable for the same use.

Competition for real property used for agricultural production is intricately related to national and international markets for agricultural commodities. When demand and prices for commodities increase, an investment in agricultural production is perceived as profitable. Thus, competition for land that can be used for commodity production tends to increase, with a resultant increase in agricultural land values.

Over time, competitive market forces tend to reduce unusually high (excess) profits. While profit encourages competition, excess profits tend to create destructive competition. This is easily observed in agricultural support businesses such as suppliers of production needs and commodity-processing and marketing facilities. The first facility in a new and expanding area generates higher profits than are considered typical for that type of enterprise. The excess profits result from the facility's local monopoly. Soon other, similar businesses move to the area, attempting to earn surplus profits and create more competition in the marketplace. Eventually, if enough new businesses come into the area, there may not be enough business to support all of them. Consequently, one or a few will earn "normal" profits, and the rest will fail. By recognizing the forces of competition and projecting market trends, appraisers can avoid capitalizing temporary, extraordinary profits into inflated estimates of current market value.

### Substitution

The principle of substitution states that when several similar or commensurate commodities, goods, or services are available, the one with the lowest price attracts the greatest demand and widest distribution. This principle assumes rational, prudent market behavior with no undue cost due to delay. According to the principle of substitution, a buyer will not pay more for one property than for another that is equally desirable.

Property values tend to be set by the cost of acquiring an equally desirable substitute property. The principle of substitution recognizes that buyers and sellers of real property have options, i.e., other properties are available for similar uses. Although each real property is unique, an equivalent substitute can usually be found. The substitution of one property for another may be considered in terms of use, structural design, or earnings. The cost of acquisition may be the cost to purchase a similar site and construct a building of equivalent utility, assuming no undue cost due to delay; this is the basis of the cost approach to value. On the other hand, the cost of acquisition may be the price of acquiring an existing property of equal utility, again assuming no undue cost due to delay; this is the basis of the sales comparison approach.

For example, if two farms offer approximately the same advantages, the prospective buyer will select the one with the lower price. Soil, topography, climate, size, and location are some of the factors that are considered in determining if a farm property is equally desirable. In regard to other income-producing properties, substitution applies to alternative investment properties that produce equivalent investment returns with equivalent risk. The limits of property prices, rents, and rates tend to be set by the prevailing prices, rents, and rates of equally desirable substitutes. The principle of substitution is fundamental to all three traditional approaches to value—cost, sales comparison, and income capitalization.

### Opportunity Cost

Opportunity cost is the cost of options foregone or opportunities not chosen. An investor who selects one type of investment forgoes the opportunity to invest in other types of available investments. A well-informed investor will select the investment that seems likely to provide the highest rate of return at the lowest level of risk. The illiquidity the investor experiences over the term of the investment poses a potential opportunity cost if alternative investments at comparable levels of risk outperform the investment chosen. Opportunity costs are especially significant in estimating the rates of return required to attract capital.

The concept of opportunity cost is related to the principle of substitution. By analyzing and comparing the prospective rates of return offered by alternative investment opportunities, an appraiser can estimate the required rate of return indicated by the market for the property being appraised.

### Balance

The principle of balance holds that real property value is created and sustained when contrasting, opposing, or interacting elements are in a state of equilibrium. Balance thus affirms that an appropriate economic mix of types and locations of land uses in an area creates and sustains property value. Three important concepts related to balance are conformity, contribution, and surplus productivity.

The principle of balance also applies to the combination of various property components and to the relationship between the costs of production and the property's productivity. Labor, capital, coordination, and land are the agents of production, but for most real estate investments the critical combination is the land and the improvements. Economic balance is achieved when the combination of land and improvements is optimal—i.e., when no marginal benefit or utility is achieved by adding another unit of capital.

The law of increasing returns holds that larger amounts of the agents of production produce greater net income up to a certain point. At this point, the point of decreasing returns, the maximum value is developed. Any additional expenditures will not produce a return commensurate with the additional investment, according to the law of decreasing returns. This concept is also known as the *principle of diminishing marginal productivity* or *the law of diminishing returns.*

The fertilization of farmland provides a simple example. Applying fertilizer to the land increases crop yields up to a point, beyond which the cost of the fertilizer will exceed the price of the additional crop yield. The optimum amount of fertilization is achieved when the value of the increment in yield resulting from the last unit of fertilizer equals the additional expenditure on fertilizer. This is the point of balance.

The principle of balance and the related concepts of contribution, surplus productivity, and conformity are interdependent and crucial in estimating highest and best use and market value. These concepts form the theoretical foundation for estimating all forms of depreciation in the cost approach, making adjustments in the sales comparison approach, and calculating expected earnings in the income capitalization approach.

### Contribution

The principle of contribution states that the value of a particular component is measured in terms of its contribution to the value of the whole property, or as the amount that its absence would detract from the value of the whole. Basically, the concept states that cost does not equal value. For example, the construction of an expensive residence on a farm or ranch does not necessarily cause the property value to increase in direct proportion to the improvement's cost. The installation of an irrigation system or the development of a specialty crop or orchard may never create value in the market equal to the cost of development. An improvement's contribution to value, which may be higher or lower than its cost, is measured in terms of how valuable its benefit or utility is in the market. As a result, a property's value may not increase even though the real estate has undergone alteration, modification, or rehabilitation.

The value of any part of an enterprise must be economically justified by its effect on productivity and utility. For example, the capital outlay for an irrigation system is justified only if the resulting increased productivity will increase property value by an amount that exceeds the expenditure.

Sometimes the existing improvements do not contribute to a proper balance within the total property. Especially in areas of rapid transition, a property's present use may underutilize the land. Nevertheless, an existing, less optimal use, called an *interim use,* will continue until it is economically feasible for someone to absorb the cost of converting the property to its ultimate highest and best use.

### Surplus Productivity

Surplus productivity is the net income to the land remaining after the costs of the other agents of production have been paid. Classical economists identified this surplus with land rent, suggesting that it tended to fix land value. The concept of surplus productivity has provided the basis for the residual concept of land returns and residual valuation techniques.

Land is fixed and must therefore attract labor, capital, and management to it. The returns generated by the property must be allocated to pay the wages and salaries of the labor necessary for property operation, the interest cost of capital, and a profit for coordination (management). Consequently, land earns rent only if surplus earnings remain after the requirements of labor, capital, and coordination have been met. The residual returns imputed to the land determine the land's value, and properties that are not in economic balance may produce no surplus returns.

### Conformity

Conformity is the appraisal principle that holds that real property value is created and sustained when the characteristics of a property conform to the demands of its market. The styles and uses of the properties in an area may conform for several reasons, including economic pressures and the shared preferences of owners for certain types of structures, amenities, and services. In addition, the government encourages conformity by restricting land use through local zoning ordinances. Standards of conformity set by the market are subject to change, but zoning codes tend to establish conformity in basic property characteristics such as size, style, and design. An individual market also sets standards of conformity, especially in terms of price. Usually, the value of an overimproved property will decline, or regress, toward the value of surrounding, conforming properties; the value of an underimproved property may increase, or progress, toward the prevailing market standard.

### Externalities

The principle of externalities states that factors outside a property can have a positive or negative effect on its value. When external economies affect a great number of people, the product or service is often provided by government. Bridges, highways, police and fire protection, and other essential services can be provided more cheaply through common purchase by the government than through separate acquisition by individuals. External diseconomies result when an individual or firm imposes the costs of inconveniences on others. A person who violates environmental law by dumping hazardous waste, for example, imposes the cleanup costs on others.

Real estate is affected by externalities more than other economic goods, services, or commodities. Because it is immobile, real estate is subject to many types of external influences. Externalities may refer to the uses of properties located near the subject property or the physical attributes of these properties. Externalities may also refer to the economic conditions of the market in which the subject property competes. For example, an increase in the purchasing power of farm owners in the area where a grain elevator or cotton gin is located will likely affect the sales (income-producing) potential of the property.

External influences may be international or national in origin or they may emanate from the region, community, or neighborhood. Externalities may be as broad as international currency and gold prices or as narrow as a neighbor's standard of property maintenance. The appraiser should observe and analyze how external influences affect the parcel of real estate being appraised.

At the international and national levels, trade policy, manufacturing efficiency, interest rates, and socioeconomic priorities can greatly affect real estate values. For example, a combination of these influences caused some U.S. real estate values to stagnate or fall in the early 1980s. Foreign imports depressed some U.S. industries and old plants and equipment tended to make U.S. manufacturing operations less efficient than their foreign counterparts. High interest rates depressed home buying and industrial expansion. Many of the international and national policies of the 1970s and 1980s also affected farmland markets and contributed to the decline in farmland prices in the 1980s. Competition for credit between the private sector and the government caused interest rates to escalate during this period, deepening the economic downturn.

At the regional level, real estate values in the early 1980s fared better in some areas than in others. In general the population migration to the Sunbelt tended to enhance values there at the expense of older communities in the north, where values tended to stabilize or decline. Industrial areas with manufacturing operations that were susceptible to foreign competition suffered more than areas less reliant on such industries. By the mid-1980s, however, the decline in oil prices caused a recession in the oil-producing regions of the Sunbelt. In addition, the large volume of construction activity over the previous decade had resulted in an oversupply of office buildings. Simultaneously, the northern Rustbelt experienced moderate economic recovery. The decline in construction in this section of the country during the early 1980s coupled with pent-up demand led to higher real estate prices and a construction boom. Finally, with many markets overbuilt by the late 1980s, new development slowed nationwide.

At the community and neighborhood levels, property values are affected by local laws, local government policies and administration, property taxes, economic growth, and social attitudes. Different property value trends are frequently noted among communities in the same region and among neighborhoods in the same community. Appraisers should be familiar with external events at all levels and be able to assess their impact on individual property values.

## Highest and Best Use

Highest and best use is a basic premise of value. Like property value, highest and best use is not an absolute fact; it is an appraiser's opinion as to the best use of a property based on an analysis of prevailing market conditions. By identifying and interpreting the market forces that affect a specific property in a local and regional context, the appraiser determines the property's highest and best use. Highest and best use focuses market analysis on the subject property and allows the appraiser to consider the property's optimum use in light of market conditions on a specific date.

Highest and best use is defined as:

> The reasonably probable and legal use of vacant land or an improved property, which is physically possible, appropriately supported, financially feasible, and that results in the highest value. The four criteria the highest and best use must meet are legal permissibility, physical possibility, financial feasibility, and maximum profitability.

Highest and best use reflects a basic assumption about real estate market behavior—that the price a buyer will pay for a property is based on his or her conclusions about the most profitable use of the site or property. Sites and improved properties tend to be put to their highest and best uses, but the highest and best use identified in an appraisal may or may not conform with the property's existing use. The determination of highest and best use must be based on careful consideration of prevailing market conditions, trends affecting market participation and change, and the existing use of the subject property.

Because the use of a parcel of land can be limited by the presence of improvements, highest and best use is determined separately for the land or site as though vacant and available to be put to its highest and best use and for the property as improved. The first determination, the use as though vacant, reflects the fact that land value is derived from potential land use. Land has limited value unless there is a present or anticipated use for it; the amount of value depends on the nature of the land's anticipated use, according to the concept of surplus productivity. Among all reasonable, alternative uses, the use that yields the highest present land value, after payments are made for labor, capital, and coordination, is generally regarded as the highest and best use of the land as though vacant. In other words, the highest and best use of land as though vacant is the use that brings the highest return to the land after the three other agents of production have been compensated.

To determine the highest and best use of the land as though vacant, the appraiser assumes that the parcel of land in question has no improvements. Even a property such as a large dairy farm with extensive building improvements can be made vacant by demolishing the buildings. The question to be answered in this context is: "If the land were vacant, what new improvements should be constructed on the site?"

The second determination, the highest and best use as improved, refers to the optimum use that could be made of the property and all existing structures. Analysis of the highest and best use of a property as improved implies that the existing improvements should be renovated or retained as is so long as they continue to contribute to the total market value of the property, or until the return from new improvements would more than offset the cost of demolishing the existing buildings and constructing new ones. For example, a large equipment storage structure on an orchard property could continue to be used to store equipment or it could be converted for fruit storage or for housing laborers. Even when an existing improvement does not represent the highest and best use of a site as though vacant, it should not necessarily be demolished.

Determining the optimal use of a parcel of land requires careful analysis of all reasonable, alternative uses. Three important points should be noted in estimating a property's highest and best use:

1. Surrounding land uses are not always indicative of optimum use.
2. The existing use may not be optimum.
3. Current zoning may not be consistent with the property's highest and best use.

In appraising agricultural properties, the appraiser must consider whether the agricultural use is long-term or interim. If it is long-term, the type of operation that will provide the greatest return to the land must be identified. For example, conversion of the land use from dry to irrigated cropland or from cropland to vineyard might result in an increase in the net returns to the land.

The existing agricultural use may not be the highest and best use for the property in its entirety, especially in an area undergoing change to a more intensive use such as residential development. Nevertheless, the existing suboptimal use will continue until conversion becomes financially feasible. When a property has an interim use, the related concept of consistent use must also be addressed. The theory of consistent use holds that land cannot be valued on the basis of one use, while the improvements are valued based on another.

Improvements only have value if they contribute to the value of the land. Improvements that do not represent the land's highest and best use, but have substantial remaining physical lives, may have an interim use that adds temporary value or they may have no value at all. They may even have negative value if substantial costs would be incurred to remove them. For example, consider a single operating farm in an area where subdivision development is intense. Use of the property as a farm would seem to be the antithesis of highest and best use. Nevertheless, when all factors are considered, such a use may be the most profitable interim use if the supply of land already devoted to home site use exceeds current demand.

Agricultural use may be an interim use for transitional properties until the ultimate highest and best use is achieved. During this period the land may be valued for its prospective highest and best use. If this is done, however, the value indication must be discounted at an appropriate rate for the remainder of the projected interim holding period.

Determining the highest and best use of land as though vacant is useful for land or site valuation; identifying the highest and best use of an improved property facilitates a decision as to whether the use should continue or the improvements should be demolished. The determination of the property's highest and best use is a critical step in the valuation process and adequate time and effort must be devoted to highest and best use analysis. Highest and best use is the pivotal point upon which the entire valuation process rests.

## Forces That Influence Real Property Values

The value of real property reflects and is influenced by the interaction of basic forces that motivate human activity. These forces are divided into four major categories: social trends, economic circumstances, governmental controls and regulations, and environmental conditions. These forces are dynamic; they exert pressure on human activities and are, in turn, affected by these activities. The interaction of all these forces influences the value of every parcel of real estate in the market.

Property is physically immobile; every property is situated in a particular region, community, and neighborhood. Each geographical level represents a part of the total market within which properties are bought, sold, leased, and exchanged. The appraiser analyzes the forces at each level that influence real property values.

Real estate markets and values are the consequences of ongoing forces. Therefore, the appraiser interprets the ways the market views the property and does not limit the scope of investigation to static, current conditions. At each geographic level, the appraiser is concerned with the underlying trends that influence value. Appraisers analyze trends and assess their direction, speed, duration, strength, and limits.

### Social Forces

Social forces are exerted primarily through population characteristics. The demographic composition of the population reveals the potential, basic demand for real estate; therefore, proper analysis and interpretation of demographic trends are imperative in real estate appraisal. Real property values are affected not only by population changes and characteristics, but also by the entire spectrum of human activity. The total population, the rate of family formation and dissolution, and age distributions strongly influence real property values. A change in use from agriculture to rural residential or recreational use changes the social structure of communities.

### Economic Forces

Appraisers analyze the fundamental relationships between current and anticipated supply and demand and the economic ability of the population to satisfy its wants, needs, and demands through its purchasing power. The specific market characteristics considered in the analysis of economic forces include employment, wage levels, industrial expansion, the economic base of the region and the community, price levels, and the cost and availability of mortgage credit. Most of these are considered demand-side economic characteristics. Supply-side economic characteristics include the stock of available vacant and improved properties, new development under construction or in the planning stage, occupancy rates, the rental and price patterns of existing properties, and construction costs. Other economic trends and considerations may be studied as the appraiser's analysis focuses on successively smaller geographic areas.

In the valuation of rural properties, the appraiser needs to understand the economic forces that affect the specific community in which the subject property is located. Factors such as typical crops, lease rates, cash rents or crop share leasing, and historical trends in commodity prices are all important economic considerations.

### Governmental Forces

Governmental, political, and legal actions have a great impact on property values. The legal climate at a particular time or place may overshadow the natural market forces of supply and demand. The government provides many necessary facilities and services that affect land use patterns and values. Therefore, appraisers must diligently identify and examine the potential value influences of

1. Public services such as fire and police protection, rural water suppliers, and electric cooperatives
2. Local zoning, groundwater regulations, chemical application restrictions, odor abatement rules, and waste disposal regulations
3. Government agricultural programs such as Farm Service Agency commodity programs, lending policies, and other cost-sharing programs including the Conservation Reserve Program
4. Special legislation that influences imports or exports, property taxation, environmental and water quality control, limitations on foreign ownership, and emergency disaster relief programs pertaining to agricultural properties

### Environmental Forces

Both natural and man-made environmental forces influence real property values. Environmental forces that may be analyzed for real estate appraisal purposes include climatic conditions such as snowfall, rainfall, temperature, and humidity; topography and soils; water availability; primary transportation systems including federal and state highways, railroads, airports, and navigable waterways; the nature and desirability of the immediate area surrounding the property; the presence of toxic contaminants such as asbestos, radon, and PCBs; and natural barriers to development such as rivers, mountains, lakes, and oceans.

Environmental forces may best be understood in relation to the location of a property. Location considers the time-distance relationships, or linkages, between a property or neighborhood and all possible origins and destinations of residents going to or coming from the property or neighborhood. Time and distance are measures of relative access. An analysis of locational forces involves identifying linkages and measuring the time required and distances to be traveled by individuals using the most common types of transportation. Depending on the area and the property type, the appraiser may investigate the property's access to all or some of the following: public transportation, schools, stores, service establishments, parks, recreational and cultural facilities, places of worship, sources of employment, product or commodity markets, suppliers of production needs, and processors of raw materials.

An understanding of all value-influencing forces is fundamental to the appraisal of real property. Although the four forces are discussed separately here, they work together and provide the background against which appraisers view every parcel of real property being appraised.

## Types of Value

Valuation is the process of estimating the value of an identified interest or interests in a specific parcel of real estate as of a given date. *Valuation* is often used interchangeably with *appraisal.* The value sought should be clearly defined. To avoid confusion it is essential that both the appraiser and the client understand the nature of the value to be estimated. Valuation analysis typically involves market value, but the appraiser may be directed to estimate value in use, investment value, going-concern value, retrospective value, prospective value, liquidation value, insurable value, or tax or assessed value.

### Market Value

Market value, which is the focus of most real property appraisal assignments, is an objective value created by the collective patterns of the market. It may simply be referred to as what the property can be sold for on the open market. Market value opinions must be based on objective observation of the prevailing market conditions.

Many specific definitions of market value have been promulgated by economists, legislators, the courts, and appraisers. The most often-quoted definitions are derived from court cases concerned with eminent domain and property taxes. The following definition incorporates the concepts that are most widely accepted. This definition is used by agencies that regulate federally insured financial institutions in the United States.

> The most probable price which a property should bring in a competitive and open market under all conditions requisite to a fair sale, the buyer and seller each acting prudently and knowledgeably, and assuming the price is not affected by undue stimulus. Implicit in this definition is the consummation of a sale as of a specified date and the passing of title from seller to buyer under conditions whereby:
>
> 1. Buyer and seller are typically motivated;
> 2. Both parties are well informed or well advised, and acting in what they consider their own best interests;
> 3. A reasonable time is allowed for exposure in the open market;
> 4. Payment is made in terms of cash in U.S. dollars or in terms of financial arrangements comparable thereto; and
> 5. The price represents the normal consideration for the property sold unaffected by special or creative financing or sales concessions granted by anyone associated with the sale.

Market value definitions vary. The definition of market value used in appraisal assignments must be acceptable in the jurisdiction in which the appraisal is being performed. Although the exact wording may differ, all legal-based definitions of market value are based on a "willing buyer, willing seller" concept.

The Uniform Standards of Professional Appraisal Practice (USPAP) require that the following items directly related to the market value definition be included in every appraisal report.

1. Identification of the specific property rights to be appraised, i.e., the fee simple estate, the estate subject to a lease or mortgage, or some other interest in the real estate
2. A statement of the effective date of the value opinion
3. Specification as to whether cash, terms equivalent to cash, or other precisely described financing terms are assumed as the basis of the appraisal

Many properties are leased and most are financed. Because the property rights being appraised and available financing affect the market value, these factors must be taken into consideration. A leased fee interest in a property usually will not have the same market value as the fee simple interest because the tenant holds some of the rights to the property through a contractual agreement.

An opinion of market value is tied to a specific point in time. If the purpose of the appraisal is to develop an opinion of value, USPAP requires the appraiser to develop an opinion of the reasonable exposure time linked to the value opinion. Reasonable exposure time is one of the criteria in most definitions of market value. Exposure time precedes the effective date of value and is defined as: "the estimated length of time the property interest being appraised would have been offered on the market prior to the hypothetical consummation of a sale at market value on the effective date of the appraisal; a retrospective opinion based upon an analysis of past events assuming a competitive and open market."[2]

The market value definition indicated by USPAP does not fill all appraisal needs. This definition assumes a competitive and open market, knowledgeable buyers, typical buyer and seller motivation, and reasonable market exposure. In a given situation, market conditions may not be representative of the conditions assumed in the market value definition. A competitive market is characterized by many individuals acting independently to secure the property from a third party, without restraints, i.e., assuming ample credit is available. In many market segments, however, ample credit has not been available, thus enhancing the position of a few individuals with access to capital. If a

2. The Appraisal Foundation, *Uniform Standards of Professional Appraisal Practice,* 2000 ed., 80.

market is temporarily weak or a quick sale is necessary, the value under current conditions may best be represented as liquidation value.[3]

## Value in Use

Value in use is the value a specific property has to a particular owner for a specific use. Use value is based on the productivity of an economic good to its owner-user.

Use value focuses on the value the real estate contributes to the enterprise of which it is a part, without regard to the property's highest and best use or the monetary amount that might be realized upon its sale. External conditions such as changes in the business or management of the property may cause use value to change. A real property may have a use value as well as a market value. Use value is based on the current productivity of the property, regardless of any functional or external conditions that would limit its market value.

The use value of urban real estate is tied to the improvements and the user's management of the property. The use value of rural properties relates to what can be done with the land. The use value of an agricultural property reflects the income the property generates under a specific use. The land, not the buildings or current farm operation, is the primary consideration in use value appraisals of agricultural properties. Use value appraisal assignments may be performed to value assets for mergers, acquisitions, or security issues. Use value appraisals are most common in assignments to value existing business enterprises that include real property. Farm use value is most often estimated for ad valorem taxation or federal estate tax special use valuation. When farm properties receive special tax breaks, the use value of the property is lower than its market value and taxes will be lower as well.

The market value of limited-market properties may be based on their current use or the most likely alternative use. Limited-market properties are properties that are not commonly exchanged or rented and have relatively few potential buyers. Such properties include large manufacturing plants, research facilities, churches, schools, and public buildings. There may be little evidence to support market value estimates of these properties due to limited market interest and the extended market exposure needed to sell them.

When a use is so specialized that there is no demonstrable market for it, but the use is viable and likely to continue, the appraiser may render an estimate of use value. Such an estimate should not be confused with a market value estimate. If no market can be demonstrated or if data is not available, the appraiser cannot estimate a market value and should state this in the report. Nevertheless, it is sometimes necessary to estimate market value under these conditions for legal purposes. In

3. Peter N. Thompson, MAI, "Market Value: Does One Size Really Fit All?" *The Appraisal Journal* (October 1993).

these cases the appraiser must comply with the legal requirement and rely on his or her judgment rather than direct market evidence.

### Investment Value

Investment value is the value of a specific investment to a particular investor or class of investors based on individual investment requirements. It is distinguished from market value, which is impersonal and detached. Investment value is the price an investor would pay for an investment in light of its perceived capacity to satisfy his or her individual desires, needs, or investment goals. An investment value appraisal may be sought by a potential purchaser of an income-producing property to consider the feasibility of the purchase.

### Going-Concern or Business Value

Going-concern value is the value created by a proven property operation, considered as a separate entity. This value is associated with the business concern, which is distinct from the value of the real estate. Going-concern value includes the intangible value of an operating business enterprise, which is produced by the assemblage of the land, building, labor, equipment, and marketing operation.

Going-concern appraisals are commonly conducted for hotels and motels, restaurants, and shopping centers. In rural applications, going-concern value may be estimated for permanent plantings. The physical real estate is integral to the ongoing business, and it is often difficult to separate the market value of the land and improvements from the total value of the business.

### Retrospective Value

Retrospective value is an estimate of value that is applied as of a specified historical date, i.e., the effective date of the appraisal is prior to the date of the report. Retrospective value estimates may be required in property tax matters, estate or inheritance tax matters, condemnation proceedings, or suits to recover damages. The date of the report is important because it indicates the perspective from which the appraiser is examining the market. In a retrospective analysis, the appraisal should not be impacted by market conditions that occurred after the date of the valuation. Data received subsequent to the effective date of the appraisal may be considered in estimating a retrospective value only as a confirmation of trends that would reasonably be considered by a buyer or seller as of the valuation date. Care must be taken not to go beyond data that reflects the relevant market at the date of valuation. In reporting a retrospective value estimate, appropriate terminology should be used to avoid confusion. The appraiser must establish the date of the value estimate and clearly identify the value as a retrospective market value.[4]

---

4. Statement on Appraisals Standards No. 3, USPAP, 2000 ed.

## Prospective Value

Prospective value is a forecast of the value expected as of a specified future date. A prospective value estimate is most frequently sought in connection with real estate projects that are proposed, under construction, or under conversion to a new use or for projects that have not achieved sellout or a stabilized level of long-term occupancy at the time the appraisal report is written. Prospective value estimates may be sought to estimate the value of permanent plantings that have not yet reached maturity.

Prospective value may also be used to estimate the price a property is likely to command at the end of a typical marketing period. Prospective value is an attempt to reflect the market value at that time as indicated by current market evidence and projected market trends.

In prospective value analysis, the term *market value* must be used with a modifier such as *forecasted* or *prospective*. The appraiser should identify the date when the analysis is made; state whether the prospective value estimate represents value at the time the improvements are completed or when occupancy is stabilized; and estimate the time of completion or stabilization (i.e., the date of the report versus the effective date of the appraisal). Prospective value estimates are intended to reflect the current expectations and perceptions of market participants in light of available factual data; they should include sufficient market research of current and projected economic trends to support the forecasts made.[5]

## Liquidation Value

Liquidation value is the most probable price that a specified interest in real property is likely to bring under all of the following conditions:

1. Consummation of a sale will occur within a severely limited future marketing period specified by the client.
2. Actual market conditions are those currently obtaining for the property interest appraised.
3. The buyer is acting prudently and knowledgeably.
4. The seller is under extreme compulsion to sell.
5. The buyer is typically motivated.
6. The buyer is acting in what he or she considers his or her best interests.
7. A limited marketing effort and time will be allowed for the completion of a sale.
8. Payment will be made in cash in U.S. dollars or in terms of financial arrangements comparable thereto.

5. Statement on Appraisal Standards No. 4, USPAP, 2000 ed.

9. The price represents the normal consideration for the property sold, unaffected by special or creative financing or sales concessions granted by anyone associated with the sale.[6]

Liquidation value is associated with distress sales or forced sales of distressed properties with limited marketing time.

### Insurable Value

Insurable value is based on the replacement and/or reproduction cost of physical items that are subject to loss from hazards. Insurable value may be defined as:

1. The portion of the value of an asset or asset group that is acknowledged or recognized under the provisions of an applicable loss insurance policy.
2. Value used by insurance companies as the basis for insurance. Often considered to be replacement or reproduction cost less deterioration and non-insurable items. Sometimes cash value or market value, but often entirely a cost concept.[7]

Insurable value is often controlled by state law and varies from state to state.

### Tax or Assessed Value

Tax or assessed value is the value of a property according to the tax rolls in ad valorem taxation. This value may be higher or lower than market value and may be based on an assessment ratio that is a percentage of market value.

*Ad valorem tax* is defined as: "A tax levied in proportion to the value of the thing(s) being taxed. Exclusive exemptions, use-value assessment laws, and the like, the property tax is an ad valorem tax."[8]

## Distinctions Between Price, Cost, and Value

Appraisers make careful distinctions among the terms *price, cost,* and *value.* The term *price* usually refers to a sale or transaction price and applies to an exchange; a price is an accomplished fact. A price represents the amount a particular purchaser agrees to pay and a particular seller agrees to accept under the circumstances surrounding their transaction.

Generally the circumstances of a transaction reflect conditions within one or several markets. A market is a set of arrangements in which

---

6. Appraisal Institute, *The Dictionary of Real Estate Appraisal,* 3d ed. (Chicago: Appraisal Institute, 1993).
7. Ibid.
8. International Association of Assessing Officers, *Property Appraisal and Assessment Administration,* Joseph K. Eckert, Ph.D., ed. (Chicago: International Association of Assessing Officers, 1990).

buyers and sellers are brought together through the price mechanism. A market may be defined in terms of geography, products or product features, the number of available buyers and sellers, or some other arrangement.

A real estate market consists of the interaction of individuals who exchange real property rights for other assets, such as money. Specific real estate markets are defined on the basis of property type, location, income-producing potential, typical investor characteristics, typical tenant characteristics, or other attributes recognized by those participating in the exchange of real property. The market for family-operated dairy farms and the market for orchards are examples of specific real estate markets.

The term *cost* is used by appraisers in relation to production, not exchange; cost may be either an accomplished fact or a current estimate. Appraisers distinguish among several types of costs: direct costs, indirect costs, construction costs, and development costs.

Direct costs are expenditures for the labor and materials necessary to construct a new improvement. Direct costs, which are also referred to as *hard costs,* generally include contractor's overhead and profit.

Indirect costs refer to construction expenses for items other than labor and materials. Indirect costs include administrative costs; expenses incurred by the owner for professional fees, financing, taxes, and interest and insurance during construction; and lease-up costs, which are the net expenses of operating the project until it reaches a stable occupancy level. Indirect costs are sometimes referred to as *soft costs.* The construction costs detailed in the contractor's bid price normally include the direct costs of labor and materials plus the contractor's indirect costs.

Development cost is the cost to create a property, including the cost of the land, and to bring it to an efficient operating state, as distinguished from the cost to construct the improvements. Development cost includes the profit to the developer or entrepreneur who brings the project into being.

These real estate-related expenditures are directly linked to the price of goods and services in competitive markets. For example, the costs of roofing materials, masonry, architectural plans, and rented scaffolding are determined by the interaction of supply and demand in specific areas and are subject to the influence of social, economic, governmental, and environmental forces.

Price and cost relationships also incorporate concepts of value. *Value* can have many meanings in real estate appraisal; the applicable definition depends on the context and usage. In the marketplace, value is commonly perceived as the anticipation of benefits to be obtained in the future. Because value exists at a given moment, an appraisal reflects value at a particular point in time. Value as of a given time represents the monetary worth of property, goods, or services to buyers and sellers. To avoid confusion, appraisers do not use the word *value* alone; instead they refer to *market value, use value, investment value, assessed value,* and other specific types of value. Market value is the focus of most real property appraisal assignments and its estimation is the purpose of most appraisals.

## Summary

A number of important principles support society's understanding of value. These principles provide the rationale for market behavior and, therefore, help explain changes in property value. The principle of *anticipation* is the perception that value is created by the expectation of future benefits.

The principle of *change* holds that the social, economic, government, and environmental forces affecting value change continuously. The physical deterioration and functional and external obsolescence of property are other manifestations of the inevitability of change.

The principle of *supply and demand* holds that the price of real property varies directly, but not necessarily proportionately, with supply. Theoretically, the supply of and demand for commodities tend toward equilibrium, where market value, price, and cost are equal. Supply and demand both have qualitative and quantitative dimensions.

*Competition* between buyers or tenants refers to the interactive efforts of two or more potential purchasers or tenants to secure a purchase or lease; competition between sellers or landlords refers to their interactive efforts to effect a sale or lease. Competition is fundamental to the operation of supply and demand.

The principle of *substitution* affirms that when several similar or commensurate commodities, goods, or services are available, the one with the lowest price will attract the greatest demand and the widest distribution. *Opportunity cost,* a related concept, is the cost of options foregone or opportunities not chosen.

The principle of *balance* holds that real property value is created and sustained when contrasting, opposing, or interacting elements are in a state of equilibrium. These elements may include the physical components of the property (i.e., land and improvements), the costs of production or development, and the relationship between the property and its environment. The *laws of increasing and decreasing returns* govern how balance is established. The concepts of *contribution, surplus productivity, and conformity* are related to the principle of balance. Because real estate is physically immobile, its value is subject to the positive and negative effects of *externalities,* economies or diseconomies that exist outside a property and may be local, regional, national, or international.

Four basic forces interact to influence the value of real property: social trends (population characteristics), economic circumstances (demand-side and supply-side economic indicators), government controls and regulations (public services, zoning and building codes, tax policies, and specific real estate laws), and environmental conditions (climate, topography, soil, natural barriers to development, the availability of transportation systems, the suitability of the location in terms of linkages, and the accessibility of the location).

The concept of *highest and best use* is predicated on the assumption that buyers and sellers set prices for properties based on their conclusions about the most profitable use of the site or property. Highest and best use is the reasonable and probable use that supports the highest present value of the vacant land or improved property as of the date of the appraisal; the use must be legally permissible, physically possible, financially feasible, and result in the highest estimate of land value. Highest and best use is determined separately for the land or site as though vacant and available for the anticipated use and for the property as improved. The related concept of *consistent use* states that the land and improvements must be valued on the basis of the same use.

Because the concept of value is central to appraisal, clear distinctions must be drawn among the related terms *price, cost,* and *value.* A *price* represents the amount a particular purchaser agrees to pay and a particular seller agrees to accept under the circumstances surrounding a transaction. A *market* is a set of arrangements in which buyers and sellers are brought together through the price mechanism. Real estate markets involve the exchange of property rights for other assets such as money. *Cost* applies to production rather than exchange, and can be divided into direct, or hard, cost and indirect, or soft, costs as well as construction and development costs. *Value* represents the monetary worth of property, goods, or services to buyers and sellers.

Although *market value* is a relatively simple concept, different beliefs and assumptions about the marketplace and the nature of value render any definition controversial. Most market value definitions share common elements. Today, most appraisals are completed under the definition of market value promulgated by the Uniform Standards of Professional Appraisal Practice with certain jurisdictional exceptions. The USPAP definition is commonly accepted throughout the appraisal profession.

Other types of value are also considered in real estate: use value, investment value, going-concern value, insurable value, and assessed value. *Use value* focuses on the real estate's contributory value to the enterprise of which it is a part, without regard to the property's highest and best use. *Investment value* is the value of an investment to a particular investor, based on his or her investment requirements. *Going-concern value* is the value created by a proven property operation in which the physical real estate assets are an integral part of an ongoing business. *Insurable value* is the portion of value that is covered by loss insurance. *Assessed value* refers to the property's value according to the tax rolls.

Four interdependent factors create value: utility, scarcity, desire, and effective purchasing power. The interaction of these four factors affects the balance of supply and demand.

CHAPTER

• 3 •

# The Valuation Process for Rural and Agricultural Properties

An appraisal involves three major activities:

1. Collection of pertinent data
2. Inspection of the subject property, comparable sales, the area, and the neighborhood
3. Organization and analysis of the data to arrive at a value opinion for the subject property

The valuation process can be thought of as a funnel (see Figure 3.1). The appraiser starts with the defined appraisal problem and sifts through both general and specific data to arrive at a final opinion of value for the subject property.

This chapter will provide an overview of the process and present guidelines for valuing rural property.

**Figure 3.1 The Valuation Process**

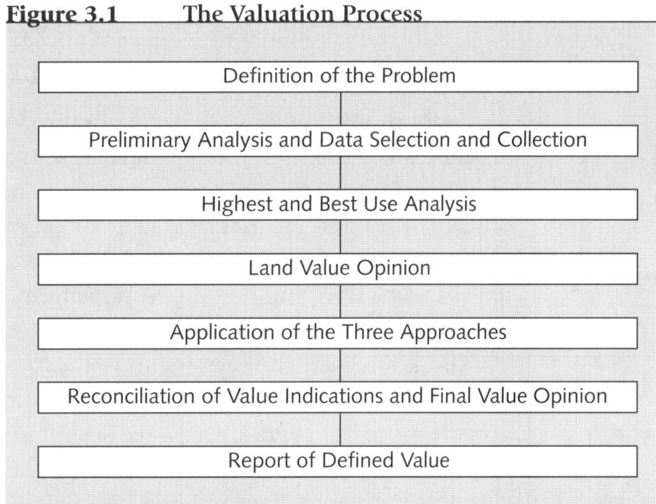


## Definition of the Problem

The valuation process begins with a concise, understandable statement of the appraisal problem. This statement should express the limits of the appraisal and prevent any confusion about the assignment.

The statement of the appraisal problem should include the following items:

1. Identification of the subject property
2. Statement of the property rights to be valued
3. Intended use and users of the appraisal
4. Definition of value used
5. Date of the value opinion
6. Description of the scope of the appraisal
7. Other limiting conditions and assumptions

## Identification of the Subject Property

The subject property should be identified as clearly as possible by road or street address, location from the nearest community, or another description that enables the user to locate the property easily. A complete legal description of the subject property should also be included.

Following is an example of property identification.

| | |
|---|---|
| Reference: | Johnson Ranch |
| Road Address: | 2600 Farm Road<br>Great Falls, Montana. |
| Location: | 6 miles east of Great Falls. |
| Legal Description: | N1/2 Section 26 Township 20 North, Range 4 East<br>Montana Principal Meridian Cascade County, Montana. |

Legal descriptions are derived from land surveys and are typically preserved in public records in a form that meets local or state law requirements. The appraiser must be familiar with the descriptive system used to describe property in the subject area. The appraiser must be sure that the subject legal description is reasonable and accurate. (See Chapter 7 for more information on legal descriptions.)

## Statement of Property Rights To Be Valued

An appraisal is the act or process of estimating value. Identification of the ownership rights (the interest or estate) in the real estate is critical to the final opinion of value. The appraiser may appraise fee simple or partial ownership rights created by easements, right-of-ways, leases, liens, or other encroachments. Typically the property right to be appraised is the fee simple estate, but it is the responsibility of the appraiser to verify the ownership rights to be appraised before accepting the assignment. Without knowing the ownership rights to be valued, the appraiser cannot accurately define the appraisal problem and may produce a value that is irrelevant to the client's problem. Restrictions on ownership generally will complicate the assignment and increase the time and skill necessary to complete it. (See Chapter 4 and Appendix C for more information on undivided fractional interests and lease interests.)

## Intended Use and Users of the Appraisal

The use of an appraisal is the manner in which the client employs the information contained in the appraisal report.[1] The client and the appraiser must agree on the intended use of the appraisal so that the report will contain the information that the client needs to facilitate the decision-making process. Some common uses of an appraisal are value opinions for lending purposes, for purchases or sales, for estate and inheritance planning purposes, for compensation in eminent domain proceedings, and for conservation, power line, and pipeline easements.

## Definition of Value Used

To avoid confusion, a precise written statement of the defined value to be used in the appraisal should be included in the report. The statement should specify whether the value opinion that is reported is in terms of cash, terms equivalent to cash, or other terms. As discussed in Chapter 2, types of appraised value include market value, use value, insurable value, investment value, and going-concern value.

## Date of the Value Opinion

The specific date of the value opinion must be stated. This date is important because the market factors that affect value are constantly changing. For this reason, the defined value is valid only for a specific date.

Typically, appraisal assignments call for a current value opinion, but sometimes a value opinion may be required for a date in the past. A retrospective appraisal is common in appraisals required for inheritance tax (value at date of death), insurance claims (value at date of casualty), federal income tax (date of value loss due to a forest fire or other natural disaster), and condemnation proceedings (value at date of summons).

Since market forces are constantly changing, the date of value is often the date of inspection for current value appraisals.

## Description of the Scope of the Appraisal

Standards Rule 2-2 of the Uniform Standards of Professional Appraisal Practice requires that each appraisal describe the scope of the appraisal—i.e., the extent of the collection and verification of data. Describing the scope of the appraisal protects users of the report by informing them of what process the appraiser followed in completing the valuation. This portion of the appraisal report should provide information that is not stated elsewhere.

> The following is an example of a description of the scope of the appraisal:
>
> The scope of the appraisal includes the necessary research and analysis to prepare an appraisal report that meets the requirements for its intended use, according to the Standards of Professional Practice of the American Society of Farm

---

1. Appraisal Institute, *The Dictionary of Real Estate Appraisal*, 3d ed. (Chicago: Appraisal Institute, 1993).

Managers and Rural Appraisers and the Uniform Standards of Professional Appraisal Practice of The Appraisal Foundation.

For the subject property, the scope involved the following steps:

1. The property was inspected on June 1, 2000. Photographs were taken on the same date and the improvements were inspected and measured. The owner accompanied the appraiser on a tour of the property and provided copies of the farm plan for the gravity flow irrigation system, which was partially completed on the inspection date.
2. Regional, county, and neighborhood information was based on data collected from the local multiple listing service, the State Agricultural Statistics Reporting Service, and other sources identified within the text.
3. In applying the various valuation methods, the appraiser completed a comprehensive search of the community and surrounding competitive market area for comparable cost, rental, and sales data (relating to land and improved properties).
4. After gathering, assembling, and analyzing the data previously mentioned, a final opinion of market value (as defined in the report) was made.

### Other Limiting Conditions and Assumptions

The items described above constitute the definition of the appraisal problem and are the general limiting conditions of the appraisal. If other limiting conditions apply to the appraisal, they should be stated to protect the appraiser, the client, and other users of the report. An example of a limiting conditions would be a statement that maps are provided for illustrative purposes only and no engineering survey has been made. Another example of a limiting condition may be that the appraiser is not an expert in assessing environmental contamination such as hazardous wastes or fuel spills and the client is urged to retain an expert if desired. A statement requiring prior approval from the appraiser before scheduling court appearances could also be a limiting condition.

## Preliminary Analysis and Data Selection and Collection

Once the appraisal problem has been defined, the appraiser should begin the preliminary analysis to determine the nature and scope of the assignment and the amount of work that will be involved in gathering the necessary data. The preliminary analysis and work plan depend on the complexity of the subject property, its market, and the assignment.

Typically, a dry land vacant farm appraisal requires less work than an irrigated, multi-enterprise improved farm. Some appraisals may be finished within a few days while others may take considerably longer. For some assignments, it can take weeks or months to gather the required data to complete the analysis and value opinion.

The appraiser must understand the appraisal problem before a planned work schedule can be completed. The appraiser needs to answer the following questions:

1. Does this assignment require the assistance of experts from other fields—i.e. soil scientists, engineers, etc.?
2. Can this assignment be completed by the appraiser alone or will assistance from other appraisers or staff be needed to complete the project?

Once these questions are answered, a work plan and schedule can be developed and the necessary steps can be taken to ensure that data collection and other required work are completed on schedule. Although others may be involved in the appraisal process, the appraiser must remember that he or she bears the ultimate responsibility for the research and the report.

Three major types of data are collected in an appraisal: general, specific, and competitive market.

- General data refers to the forces that affect property values in the subject market. Included is information on 1) social forces such as trends in the size, makeup, customs, and perceptions of the market; 2) economic forces that contribute to the profitability of agriculture; 3) governmental forces such as changing tax and lending regulations that affect value; and 4) environmental forces such as weather, growing seasons, and the presence of wetlands in the subject market.
- Specific data relates to both the subject property and the comparable properties. This data includes legal descriptions and physical information about the properties such as their size, soils, improvements, views, terrain, and zoning. The appraiser should also research listing prices, financing, potential income, and expenses for the sales and listings as well as factors that may have influenced the participants in the sale transactions. The appraiser should research the presence of potentially hazardous material on the subject property and the terms and conditions of any leases, easements, or other encumbrances affecting the appraised property.
- Competitive market data is important because it helps the appraiser assess how the subject would compete in the market against other properties. Research should also indicate if there is demand for property like the subject. To complete a working farm or ranch appraisal, the appraiser should also research the

neighborhood to determine its financial health, the availability of funds, and the overall economic outlook. This primary research should help the appraiser determine whether there is demand for the subject now or in the future. In the appraisal of a recreational property, however, out-of-town buyers may be influenced more by changes in their area than changes in the subject area. For example, California buyers may decide not to invest in a Wyoming ranch because Congress has passed a new tax law. At the same time, local ranchers may see the ranch as a good investment because they know that calf prices are at an all-time high.

The quality and quantity of data that can be collected and verified set the stage for the appraisal. The data available determines which approaches to value can be used to develop a meaningful value opinion for the subject property. Any data that impacts the final value opinion for the subject property should be reported by the appraiser so that the user of the report can understand how and why conclusions were reached.

## Highest and Best Use Analysis

Highest and best use analysis is an objective economic study conducted by the appraiser to determine which use or uses of the subject property will produce the greatest economic benefit to the subject property owner(s). The study focuses on the four forces that influence value (social, economic, governmental, and environmental) and the impact they have on the subject property.

The appraiser analyzes highest and best use in two ways: as though vacant and as improved. Analyzing the subject as though vacant helps the appraiser identify comparable sales and eliminate from consideration sales that have different highest and best uses. Vacant comparable sales with similar highest and best uses will be used to develop a land value opinion for the subject property in the cost approach. Analyzing the property as improved not only helps the appraiser identify comparable properties, but also indicates whether the improvements contribute to value as they are or if they should be razed or renovated. The improvements should be retained as long as they contribute to value and the return to the existing property is greater than the return produced by a new use less the cost of removing the existing improvements and erecting new ones. The appraiser must be sure to apply the consistent use concept, which states that both the land and improvements must be valued under the same highest and best use. The highest and best use conclusion should state the optimal use of the property on the effective date of the appraisal. (Highest and best use is discussed in detail in Chapter 11.)

## Land Value Opinion

For many rural appraisals, the land component is the primary contributor to the total value of the property. In the income capitalization approach and sometimes in the sales comparison approach, an overall value estimate can be derived without separating the land component from the improvements. In the cost approach, land value is estimated and stated as a separate value.

When rural and agricultural properties are valued, one of three methods is typically used to develop the land value.

1. Allocation of value by land class or type. Sales of similar vacant properties are researched and analyzed. The price paid for each sale is allocated among the various types or uses of land and these values are applied to the corresponding land types or uses in the subject property.
2. Sales comparison. Vacant land sales that have the same or a similar highest and best use as the subject property are compared to the subject property to provide a value opinion.
3. Ground rent capitalization. This method can be applied when good data on land rental rates is available. Capitalization rates are derived and applied to the subject land to estimate the land's contribution to property value.

The first method, allocation by land types using vacant land sales, is the preferred method for land valuation in the cost approach. If vacant land sales are not available, the appraiser may apply the allocation method to improved sales to derive land value but supporting documentation from the ground rent capitalization method is needed.

## Application of the Three Approaches

In appraisal theory there are three classic methods of analyzing data to arrive at an opinion of value. They are the cost approach, the sales comparison approach, and the income capitalization approach. All methods that apply should be used in the appraisal assignment. Sometimes all three approaches are pertinent and provide reliable indications of value. In other cases, only one or two approaches may be applicable.

For example, in the mountainous recreational areas of the Wyoming ranch market, the income capitalization approach may not provide a reliable value indication because the market is being driven by aesthetics and emotion rather than concern over potential earnings. Similarly, the sales comparison approach may not be applicable for appraising a cattle feedlot in a dry cropland market where there have been no feedlot sales.

### Cost Approach

The cost approach is based on the premise that an informed buyer would pay no more for a property than the cost of the land plus the current cost of replacing the improvements less the applicable depreciation from all causes charged to the improvements. This approach can be useful in valuing rural properties that have relatively new, or specialized, improvements.

Current replacement or reproduction costs can be obtained from cost estimation guides; actual subject cost data, if recent; or builders and contractors. The data selected depends on which is most applicable to the appraisal problem. The amount of depreciation to apply to the cost of new improvements is based on market observations. The depreciated value of the improvements is then added to the land value to arrive at a total value opinion. (See Chapter 13 for a complete discussion of the cost approach.)

### Sales Comparison Approach

The sales comparison approach is applicable when a sufficient number of comparable sales have occurred and these sales can be adequately verified. The appraiser must judge the amount of difference between the selected comparable properties and the subject property. Principal elements of comparison, or factors that influence the value of rural properties, are listed below:

Primary factors

1. Land use mix
2. Property rights transferred
3. Terms of sale (financing)
4. Conditions of sale (non-realty)
5. Market conditions (date of sale)

Physical considerations

6. Location
7. Land characteristics
8. Improvement characteristics
9. Other

The appraiser makes either dollar or percentage adjustments to each comparable sale to reflect how the sale differs from the subject in terms of each factor. If, for example, the comparable sale has an inferior land mix, then a positive adjustment to that comparable sale's price would be made. If the comparable sale has a location that is superior to the subject's location, then a negative adjustment would be made to the price of the comparable property.

The real property rights included in the sale is an important factor influencing value. The appraiser needs to study the property rights in each sale and determine if the sale (if encumbered) has the same highest and best use as the subject property. If the sale has a significantly different highest and best use due to a restriction in property rights, it may be best to discard this sale from consideration. (The sales comparison approach is covered in Chapter 14.)

### Income Capitalization Approach

The income capitalization approach converts anticipated income from a property into a value indication based on the following relationships:

$$\text{Income} / \text{Rate} = \text{Value}$$

and

$$\text{Income} \times \text{Factor} = \text{Value}$$

The appraiser derives an appropriate rate or factor to apply by researching comparable data. Selected comparable properties are analyzed to determine the relationship between their estimated incomes and their sales or offering prices. From this relationship, a capitalization rate or an income multiplier can be estimated. Then the subject property is investigated to determine the quality and quantity of the anticipated income stream. Once the appraiser estimates a reasonable income stream for the subject property, the market-derived capitalization rate or income multiplier can be applied to the anticipated income stream to calculate a value indication for the subject property.

## Reconciliation of Value Indications

The critical final step in the valuation process is reconciliation of the different value indications derived in the various approaches to arrive at a final opinion of value. During the reconciliation process, the appraiser examines the strengths and weaknesses of each approach and determines why the approaches produce different value indications. The appraiser must then communicate to the reader why one approach is given primary consideration or why another approach might not be given much consideration.

For example, in appraising a farm with many older improvements, the cost approach might not be given as much weight as the sales comparison approach. Because it is difficult to determine accurate replacement costs for older improvements and the amount of depreciation to apply, the reliability of the cost approach estimate may be weakened. (For a complete discussion of the reconciliation process, see Chapter 16.)

## Final Opinion of Defined Value

The appraisal process is completed when the appraiser derives the final opinion of value subject to the stated assumptions and limiting conditions. The appraisal assignment, however, is not completed until the appraiser prepares and delivers the final opinion of value to the client either orally or in a written report. Typically a formal written report is prepared. This report documents the general and specific data that was used in the valuation process and the reasoning behind the conclusions reached. Under current USPAP requirements, a written report may be presented in a self-contained, summary, or restricted use format. Whichever format is used, the appraiser must keep sufficient evidence on file for a specified period of time to support and defend the opinion of value.

CHAPTER

• 4 •

# *Rural Property Rights and Interests*

A thorough understanding of the property rights to be appraised is essential to proper valuation.[1] In the appraisal of rural and agricultural properties, one or more property rights are usually encumbered or have been severed from the fee simple estate. Rural land is seldom held in fee simple. An estate may be subject to access easements, utility easements, mineral reservations, or other encumbrances. For example, consider a farm that has had the mineral estate severed. In this case the property rights appraised are "fee interests in the surface estate." An analysis of sales in the market area reveals that surface interests in farm properties are selling for $800 per acre and fee mineral interests contribute $200 per acre. If the property were appraised in fee simple, it would be overvalued by $200 per acre.

In defining the estate valued, the appraiser should exercise caution. If there is doubt as to the ownership of the subsurface estate in a mortgaged property, the appraiser may elect to report the value of the surface rights only.

Possession of a title in fee establishes the fee simple estate, i.e., absolute ownership unencumbered by any other interest or estate, subject only to the limitations imposed by the governmental powers of taxation, eminent domain, police power, and escheat. The fee simple estate includes air rights, subsurface rights, and all other property rights associated with real estate ownership. A fee simple estate that is mortgaged, however, does not include the full bundle of rights. Similarly, a property with severed air or mineral rights could not be defined as a fee simple estate. Because the appraiser usually is not valuing the fee simple estate in agricultural properties, a conditional statement explaining the

1. Much of the discussion in this chapter is based on *The Appraisal of Real Estate,* 11th ed. (Chicago: Appraisal Institute, 1996).

estate is often required. In most rural areas, minerals have been partially or totally severed from property rights and many properties are subject to access easements, deed reservations, and other restrictions. The comparable sales used to value a property should reflect the same or similar estates. In some instances, rights may not be included in the estate appraised, but these missing rights may have little or no value, i.e., they do not contribute to value in the context of the highest and best use of the property. The appraiser's job is to determine which rights contribute to value and how much value they add.

An interest in real estate is defined by degrees. A fee simple estate denotes complete, but not absolute ownership. A leased fee estate typically describes a property subject to a lease that has a definite expiration date. A leasehold estate is the interest of the tenant or lessee.

## Life Estate

Life estates transfer ownership while retaining use and control for the life of the grantor. A life estate, by definition, is the total right of use, control, and occupancy invested in the life tenant. The life tenant is typically required to maintain the property in good condition and pay applicable taxes and expenses associated with day-to-day use of the property. Life estates are created by wills or deeds of conveyance. For example, a property can be willed to a son or daughter subject to the surviving parent's life estate. The surviving parent has full use and occupancy of the property until his or her death.

The income stream flowing to the life tenant is normally the basis for valuing a life estate. The duration of the income stream is established from actuarial tables, which indicate the expected remaining life of the tenant. Once the duration of the income stream is set and net operating income from the estate is established, an appropriate market discount rate can be selected and applied to estimate the value of this property interest.

Obviously, a person who has sold a real property interest subject to a life estate would discount that income at a market rate derived from other, similar real estate interests. For example, if actuarial tables indicate a 15-year life expectancy for the life tenant, a 15-year discount period would be applied to the reversionary interest of the estate. Life estates are common in rural areas where parents sell their property to their children and retain a life estate in the buildings and a small acreage around the structures. The value of the whole property depends on when the buildings and acreage are rejoined with the remainder of the property. The contribution of the portion containing the structures and acreage would be discounted at an appropriate market rate and added to the remaining value of the unencumbered property.

## Lease Interests

Lease interests are created when the bundle of rights is divided, usually through a written lease. By definition, a leased fee estate is an ownership interest held by a landlord with the right of use and occupancy conveyed to others through a lease. The rights of the lessor, the leased fee owner, and the leased fee are specified by contract terms contained within the lease.

A leased fee estate

1. Receives income during the holding period
2. Includes the right of the reversion at the expiration of the lease
3. May include a default provision
4. May reserve the right to sell, mortgage, or bequeath the property during the lease period.

The lessor is obligated to surrender possession of the property to the tenant and abide by the leased provisions.

The leasehold estate is the tenant's position. A leasehold is the right to use property for a stated term under the conditions of the lease. When a lease is signed, the tenant acquires the right to possess the property for the leased period and to sublease it subject to the conditions of the lease. In return, the tenant is obligated to abide by the lease terms, pay rent, and surrender possession of the property at the termination of the lease.

In some instances a property is subject to verbal sharecrop arrangements. State laws address the rights of the lessor and lessee with regard to verbal agreements. In some states a verbal sharecrop arrangement must be terminated in writing by the lessor 30 days prior to a specified date set by the state statute. Cash leases that are unwritten may carry the same termination rights, depending on state law. In most areas local custom calls for written agreements or a written lease that is extended from year to year. Appraisers should be aware of local customs and state laws relating to agricultural leases and their implied effect on value.

### Lease Documents

Leases can be verbal or written contracts. Verbal contracts are normally for less than one year and may or may not be enforceable under state law. Both written and verbal leases obligate the tenant to pay rent to the landlord and may include other agreements between the parties. A lease document should include a legal description of the land and any improvements included with the property.

One document similar to a lease is a Conservation Reserve Program (CRP) contract. This is an agreement between the landowner and the federal government to leave cropland idle for a specified period of time—usually 10 years. The program began in the late 1980s and will expire soon unless provisions for its extension are adopted. In most cases CRP contracts are not recorded, but the landowner is subject to very restrictive language during the holding period and a substantial penalty can be imposed if the lease terms are compromised. The appraiser should research and understand the implication of the language in the CRP contract and assess the quality of the income stream during the holding period.

Leases typically include clauses that specify the length of the agreement; the amount of annual or monthly rent; the amount of any deposit; the required insurance on the buildings and liability coverage on the land; the landlord's right of entry; provisions for subleasing, maintenance, and repair on all structures; and which party is liable for taxes. Leases may also include an eminent domain clause, direct payment for a possible taking and/or damages, a default clause, and renewal options. Other elements that may be added include purchase options, rent escalations, alterations, and use restrictions.

It is important that the appraiser understand the effect leases have on the property interest being appraised. Local custom and state laws are of prime importance. A property's value can be influenced positively or negatively by the terms and conditions of the lease agreement.

### Rental Payments

The two most common kinds of leases, gross leases and net leases, delineate different responsibilities for the tenant and landowner. In a gross lease, the landlord receives the stipulated rent and is obligated to pay some or all property expenses including real estate taxes. In a net lease, however, the tenant pays all or most property expenses in addition to the stipulated base rent. Property expenses usually include real estate taxes and the costs of insurance, management, maintenance, and structural repairs. Again, local customs often outline the level of responsibility ascribed to each position.

Rental payments are typically flat or level payments throughout the contract term. A step-up or step-down lease specifies a certain rent for the first period followed by an increase or decrease over the lease term. An index lease provides for periodic rent adjustments based on changes in the Consumer Price Index (CPI) or some other index. A *revaluation lease* is adjusted at periodic intervals based on the current value of the real estate; such a lease recognizes that values change continuously.

Rent payments can be monthly, quarterly, semiannual, or annual, depending on whether the property is subject to a cash lease or a sharecrop arrangement. Cash agricultural leases usually call for annual or semiannual payment in advance or partially in advance. Sharecrop leases are sometimes verbal and typically call for a share of the crop with payment at the time of harvest. Payments can be set on a percentage basis, i.e., one-quarter to one-third of the crop payable to the landowner or a 50–50 split for crop production. The tenant and landlord may also share in any government payments. Under sharecrop leases, the landlord is usually obligated to pay taxes and property maintenance costs; the tenant provides the labor for maintenance and operation of the property. In analyzing sharecrop agreements, the appraiser should look for automatic extensions and consider the perceived risk to the landlord in comparison with a cash lease arrangement.

The most important obligation associated with the rights to use and occupy a leasehold interest is the payment of rent. Contract rent is the actual amount of income specified in a lease. The rent is paid by the tenant to the lessor according to the terms in the lease contract. A leasehold has positive value when contract rent is less than market rent. In turn, the leased fee value may be negatively affected by a positive leasehold. A leased fee encumbered with a long-term, below- market, fixed rental rate may be worth less than an unencumbered estate.

Market rent is estimated based on rentals of similar property in the same locale, i.e., the rental income a property would command in the open market. By contrast, a Conservation Reserve Program (CRP) contract is an example of a positive leased fee, as shown in Figure 4.1. In this example, market rent for comparable land is estimated at $25 per acre. The CRP payment of $40 per acre represents a positive leased fee position of $15 per acre. The example assumes that all expenses associated with the property are equal whether the property receives cash rent or remains in the CRP program.

## Specialized Lease Interests

### Subleasehold Position

A sublease is an agreement in which the lessee (tenant) in a prior lease conveys the rights of use and occupancy to a third party. An assignment is a written transfer of property rights to be held by another legal entity or to be used for the benefit of creditors for the remaining term of the lease agreement.

The lease contract must include a clause allowing the lessee to sublease all or part of the property. Some leases require the lessor's consent and approval of the sublessee. Court decisions over the years have established that subleasing should not be subject to undue constraint by the original lessor. A sublease is sometimes called a *sandwich lease* because the original lessee

**Figure 4.1** Positive Leased Fee

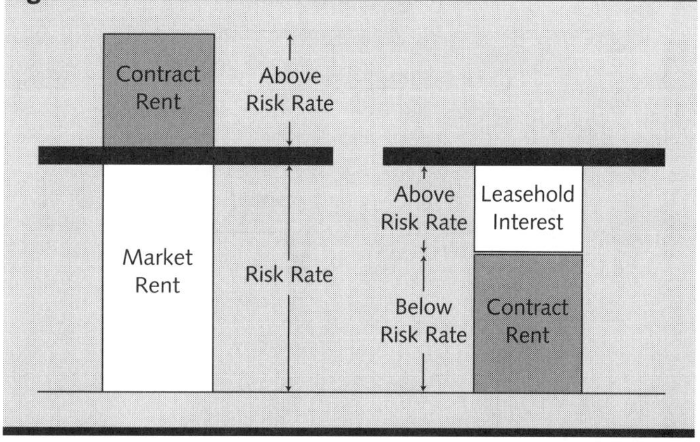

**Figure 4.2** Sandwich Lease

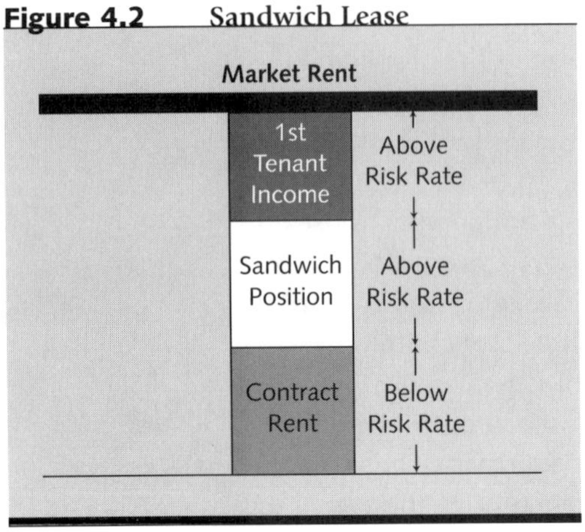


is sandwiched between the landowner and the third party. The original lessee's interest has value if contract rent is less than the amount of rent collected from the sublessee (see Figure 4.2).

The discount rates used in valuing different lease interests will vary and must reflect the risks associated with each position. Normally a lessor's interest, the leased fee, entails less risk than the leasehold. The sublessee is exposed to the greatest amount of risk because either of the remaining positions may change. The sum of the partial interests rarely equals the value of the fee simple interest due to the perceived risk of each position.

A lease contract that prohibits subleases, i.e., the leasehold position cannot be conveyed, will have no market value. If subleasing is not allowed, the only value contribution of the leasehold position is the use value to the tenant.

### Mortgaged Leased Interests

A lease may contain provisions for leasehold mortgaging. Sometimes land is leased on a long-term basis by a lessee who develops the property and must have the lessor's consent to construct improvements. In this type of an arrangement, a long-term ground lease with several renewal options is included in the lease to attract a lender to finance the improvements. If the lessee defaults on the lease, the tenant is subject to eviction because the lessee is subordinating the leasehold position to the mortgage. The leasehold mortgagee is usually protected by a non-disturbance clause, which gives the mortgagee the right to cure any default upon appropriate notice to the lessor. In a lease subordination agreement, the leasehold mortgagee can foreclose on the leasehold interest subject to the lease. Leasehold mortgaging of agricultural properties is not common.

## Other Interests

The bundle of rights in a fee simple estate includes interests other than those created by leases for the use and occupancy of the physical real estate. The ownership of property can be divided by a vertical interest, easements, or transferable development rights. Each property interest can be considered separately in sales, leases, and other realty transactions.

### Vertical Interests

Vertical interests are separated into two broad categories: subsurface rights and air rights. A subsurface right is a right to use and profit from the underground portion of the property. Air rights are associated with the use, control, and regulation of airspace above the property.

The vertical division of property is based on the legal concept of land as a volume of space with boundless height and depth. The government controls the airspace from 500 to 1,000 feet over the land mass within the United States, depending on location. Additional restrictions are imposed over airports, military bases, large cities, testing centers, etc. Generally subsurface rights are described as a cone-shaped area tapering down from the land surface to the center of the earth. Changes in technology have allowed vertical interests to become more useable over time. If subsurface and air rights are being purchased and developed in a particular location, available data can be analyzed to value their contribution or worth. Mineral rights are covered in more detail in Appendix G.

## Easements

An easement is an interest in real property that conveys use, but not ownership, of a portion of an owner's property. Easements permit a specific portion of a property to be used for a public right-of-way, by public utilities, or to gain access to an adjoining property. Surface easements are the most common easements on agricultural property. Overhead easements are used by public utilities for electric and telephone lines; subterranean easements are used by electrical, telephone, natural gas, and water companies. Federal agencies such as the USDA Forest Service and the Bureau of Land Management (BLM) may require underground easements for utilities in an attempt to curtail the development of fee land in areas with a high density of nonfee ownership. An easement is a partial interest that is owned by one party and burdens another. Often easements have measurable value.

In the case of an access easement, a property that acquires an easement is the beneficiary of additional rights, but the property subject to the easement is burdened. The holder of an easement is referred to as the *dominant estate,* while the property burdened is the *servient estate.* Easements can be conveyed in perpetuity or for a stated term. The principle of contribution is used to value the benefit to one property and the diminution to the other.

Conservation easements, which restrict the future use of a property to conservation or a wildlife habitat, are usually donated by a landowner in exchange for income tax benefits. In such cases the easement is ultimately owned by the county, state, or federal government; by a quasi-governmental agency; or by a preservation group such as the Nature Conservancy or the Trust for Public Lands. Open space easements can inhibit the overall use of a property. The basic procedure for valuing these easements is similar to the procedure applied in eminent domain appraisals, i.e., the value of the property before the easement is compared to the value after imposition of the easement.

### Transferable Development Rights

Transferable development rights (TDRs) became popular in the real estate industry during the 1970s. TDRs can be used to preserve property for agricultural production or open space or to protect historical buildings. A TDR is a development right that cannot be used by the landowner, but can be sold to landowners in another location. Under this arrangement a preservation district and a development district are identified. Landowners in the preservation district are assigned development rights which they cannot use to develop their own land but can sell to other landowners in the development district. The recipients can use these transferred rights to build at higher densities than zoning laws in the development district would normally permit. Development rights may also be transferred to land trusts or governmental agencies to curtail development and preserve environmentally sensitive properties.

## Fractional Interests

The term *fractional,* or *partial, interests* refers to co-tenancy rather than the literal division of an interest. Fractional interests include surface rights, subsurface rights, leased fee, leasehold, and various other interests created within a real estate ownership. The legal term for this type of interest is *tenancy in common,* which is an estate held by two or more persons each of whom has an undivided interest. Each tenant has a right to occupy the property held in co-tenancy, but no co-tenant possesses the right to exclusive occupancy of a particular portion of the property. There is an exception to this condition. Exclusive occupancy can be granted through a provision made to the contrary of the instrument creating the tenancy in common or through a subsequent agreement endorsed by all the co-owners.

If a co-owner in a tenancy in common is dissatisfied with the arrangement, several remedies are available: sell the fractional interest, seek voluntary partition, or demand partition of the property by the court.

To sell an undivided interest in property, several negative factors must be overcome:

1. Most buyers prefer fee ownership.
2. Financing is difficult to obtain, except by the seller.
3. Co-owners must jointly approve the decisions required to operate the property.
4. To attract a buyer and meet the requirements of a cash sale, it may be necessary to discount the value of the proportionate share of the property.

A fractional interest in a whole property is typically not as valuable, or as marketable, as an equal percentage of the fee simple value of the property, i.e, a proportionate share, because the co-owner does not have control of the overall operation of the property.

A voluntary partition of property can be time-consuming and expensive. At a minimum, the owners must take the time to facilitate the partition and incur expenses for legal work and surveying. A voluntary partition can be successful if all owners are reasonable and cooperative and the property has similar characteristics throughout. In some cases a partition plan provided by an appraiser may be agreed upon by the co-owners.

Partition of a property by the court places all co-owners in either a plaintiff or defendant classification. If the court concludes that the property cannot be divided equitably, it will order a sale and the proceeds will be distributed among the owners based on their percentages of ownership. The costs associated with a court petition include attorney fees, court costs, appraisal fees, surveying fees, and any other costs ordered by the court's action. Improvements such as fences and irrigation systems may be imputed to the whole or apportioned to the parts to effect an equitable partition. If the court determines that the additional costs are not in the best interest of the parties, it will order the sale of the property and the payment of all costs will be deducted. The remaining proceeds will be distributed to the co-owners based on their ownership interests.

PART
• II •

# *Data Collection and Analysis*

CHAPTER • 5 •

# Data Collection—Sources for Subject and Sales

The expression "Quality in, quality out" is most often associated with the computer world, but it is also applicable to the appraiser's collection, analysis, and reporting of data relevant to the appraisal problem. The information on social, economic, governmental, and environmental forces that the appraiser gathers is fundamental to the determination of property value. Reliable and meaningful data is vital to every appraisal. The selected data provides

1. The background against which the specific property is appraised
2. Information from which to infer possible trends affecting land values
3. Numbers used in the calculations performed within the three approaches to value
4. A basis for judgments about highest and best use and reconciliation of the value indications derived in the approaches into a final opinion of defined value

In essence, the quality of the appraiser's data is directly related to the quality of the final appraisal report. Therefore, the sources from which the appraiser secures data are crucial to the appraisal process.

## Data Sources

### National and USDA Sources

National data of a general nature is seldom a part of the final written report, so its importance to the appraiser's data collection effort is often underestimated. Information is compiled by government agencies that report and monitor national and world economic trends and these trends affect rural properties in many ways. The state of the national economy influences not only the amount of food products purchased, but also the types of products in demand. High demand for one product over another will filter back to the value of the land that produces the preferred product. Despite government programs that attempt to stabilize prices for agricultural commodities, any sustained downtrend in the economy may have decided effects on rural property value.

High inflation tends to accelerate land value and reduce the number of buyers able to acquire or retain property. However, when extended periods of high inflation are accompanied by high interest rates, land values may be adversely affected. Because agricultural production requires extensive capital expenditures, the debt service on property acquisition may become prohibitively high. If the total return on investment (including anticipated property appreciation) is not equal to or greater than the return available from more liquid investment opportunities, owners will tend to sell rural property for other purposes rather that retain it in agricultural production, which may increase the supply of land and lower its value.

The cost of energy is an important consideration. When world petroleum prices increase, the cost of agricultural production increases. The costs to transport and process agricultural commodities are also influenced by an increase in the cost of energy.

World population and the demand for and supply of food and fiber products have an impact on rural land values. Increased demand for agricultural commodities for export or the curtailment of existing demand may be important influences. Increased agricultural production in areas outside the United States can affect commodity prices and lower land values for agricultural properties, as can any economic downswing that affects the purchasing ability of foreign markets. A shift in the monetary exchange rate that makes it more favorable to purchase from other markets can also affect rural land values.

Of course, the effects of national and world economic forces on property value are not immediately evident. They are not directly translatable into adjustments in the sales comparison approach or changes in the gross income estimates for agricultural properties. The effects of economic forces are cumulative and many years of sustained activity may pass before any noticeable results occur. A change in government policy that creates favorable export conditions for agricultural commodities will not cause immediate changes in land values. If these conditions are sustained, however, the gradual increase in agricultural productivity and agricultural income will eventually result in higher land values.

Because government actions change agricultural land values indirectly, these factors are not used by the appraiser in specific value calculations. Rather, knowledge of national and world economic trends alerts the appraiser to specific changes (e.g., an increase in gross income averages or an increase in transportation and/or operational charges due to higher energy costs) that will be useful in explaining, and even in anticipating, changes in land value.

Although the effect of national and international economic trends on rural property value cannot always be quantified, knowledge of and sensitivity to these broad influences can help the appraiser understand and respond to changes in the attitudes and behavior of buyers who are responding to the same trends. Since the anticipation of profit is the essential motivation for purchasing property, any economic trend that affects the buyer's anticipation of profit will ultimately affect property value.

Various agencies within the United States Department of Agriculture compile general national data that the appraiser can use. The Forest Service stores a wealth of timber and minerals information as well as vegetation and grazing fee data.

The Farm Service Agency (FSA), formerly the Agricultural Stabilization and Conservation Service, can provide historical data on national prices, production, and reserves for the country's major agricultural commodities. The National Resource Conservation Service (NRCS), formerly the Soil Conservation Service, is an excellent source of information on soils, water, conservation measures, environmental improvements, and agricultural pollution control.

Current information on the prices, demand, supply, location, quality, and condition of cotton, tobacco, milk, livestock, grains, fruits, vegetables, and other commodities can be obtained from the Agricultural Marketing Service (AMS).

The National Agricultural Statistics Service (NASS) is the statistical and data collection arm of the USDA. Information on acres planted, acres harvested, crop yields and production, and the values of the production is compiled by NASS for every agricultural commodity produced in the United States.

## Regional and State Sources

As the appraiser begins the area analysis, quality regional and state data become very important. Agricultural properties depend on complementary services for economic viability, and the presence of such services within a reasonable distance is vital to the success or failure of the agricultural enterprise. Access to markets, transportation, support businesses, and financial institutions is essential to rural properties and must be considered on a regional or state basis.

The income from agricultural production depends on markets for the commodity produced. The location of markets (including auction markets, terminals, and loading/sorting/buying facilities) has a major impact, especially on the value of large operations. For example, the presence of large commercial confinement feeding operations near rangeland in the West can enhance the value of the rangeland itself.

Winter vegetable-growing regions of Florida, Texas, Arizona, and California have experienced land value increases due to the presence of national cooperative and corporate marketing processing and distribution centers.

Transportation charges from the farm to the market can dictate the profit margin of agricultural enterprises insofar as these charges must be absorbed in the price obtained. A cattle feeding operation, for example, must pass along the transportation costs of moving livestock or face potential losses. Crops grown in areas that are not reasonably close to a market may be not economically feasible to harvest; the same is true for timber located too far from a sawmill. Good facilities for transporting the agricultural commodity from the producing property to a processing or handling property and then on to its final market quickly and effectively are fundamental concerns of agricultural producers. Therefore, purchasers consider transportation in their selection of a property. In the appraisal of rural properties, an analysis of the area to determine the influence of transportation on values is essential.

Support businesses within the region or state greatly impact the desirability, salability, and rentability of rural properties. The importance of seed, feed, fertilizer, and machinery and equipment dealers in rural communities is obvious. However, the importance of businesses such as canneries, vegetable processors, meatpacking plants, livestock commission yards, and grain drying, loading, and storing facilities within the state or region must not be overlooked. In agricultural communities, the suppliers of production, processing, and distribution needs surround the area where the raw product is produced. The accessibility of suppliers is often correlated to the desirability of the area, which ultimately affects land values.

Another factor to be considered is the availability of financing and the type of financing most common in the area. This is the basis of the financing terms adjustment in the sales comparison approach. Current interest rates and terms for both production-type loans and long-term real estate loans must be known because the cost of debt service can figure prominently in the analysis of expenses and property income.

Regional and state data is most often obtained from state agencies or from the state or regional offices of federal agencies. Most of the federal agencies mentioned previously have state offices that can provide information at the state level. The Bureau of Land Management (BLM) and the Forest Service (FS) have excellent timber, minerals, and range data for every state where there are BLM or FS lands to be managed.

Universities, particularly land grant institutions through the Agricultural Extension Services (AES), can provide a wide array of data relating to most aspects of rural property. Although their exact titles vary by state, state agencies of the federal departments of commerce, agriculture, forestry, lands, and environmental services can be excellent data sources.

### Neighborhood and Local Sources

Although general data helps the appraiser gain a broad knowledge of rural properties, rural appraisers must rely on local data to approach the appraisal problem directly. The forces at work within each community define which data will be most useful. An awareness of local forces helps the appraiser understand current market activity and anticipate and identify trends that are affecting or will affect rural properties within a neighborhood.

Every rural property is unique and the appraiser must concentrate his or her information collection and analysis efforts on the data that is most pertinent to the particular appraisal problem. The types of data that the appraiser frequently must procure and analyze and potential sources of this data are discussed in the following section.

## Data Types

Usually small and tightly knit, American rural communities evidence strong independence. Appraisers must be aware of the patterns, customs, and practices that characterize the communities in which they work. One invaluable tool in making accurate value opinions is an understanding that the people of a community, through their opinions and actions, have a strong influence on property value.

The appraiser who works in a given rural community develops a knowledge of that community and its peculiarities. An understanding of the motivations for buying and selling rural land and the extent to which external forces influence property value is often acquired through many years of experience in the community. If an appraiser does not have this type of familiarity with a particular area, extensive efforts must be made to obtain valid information from long-standing community members, particularly those involved in real estate transactions.

Rural communities are not isolated areas on the fringes of urban development. Like any other neighborhood, a rural community is characterized as groups of similar or complementary land uses. In analyzing properties whose basic function is the production of agricultural commodities, the definition of community becomes even more detailed. Farm communities are characterized by size of operation, type of ownership, soils, crops grown, or land use. The boundaries of farm communities may extend over several counties or larger areas. Some rural properties, such as timberland and large cattle ranches, are less susceptible to the influences of the immediate community than those that must depend on a town or neighborhood unit. However, the vast number of agricultural properties must be considered in terms of the community of which they are a part. The data types most frequently considered and sources from which that data can be extracted are discussed at length on the pages that follow.

## Economic Data

The economic stability of a rural community is perhaps the single largest determinant of that community's future. Over the past several years, advances in technology, transportation, and communications have erased many small rural towns and communities from the American road map. Other small rural communities have reaped the benefits of progress and are growing and prospering at an astounding rate.

Economics is most often the underlying force that dictates the destiny of communities. The financial health of a community often can be observed by simply driving around. New construction, consumer activity at local businesses, property maintenance, vacant buildings, and the frequency of "For Sale" signs can often give the appraiser an indication of the current status of a community's prosperity. Discussions with local farmers and ranchers, lenders, community leaders, chamber of commerce officials, business owners, and school officials, among others, can confirm or deny the appraiser's initial impressions. The community's economic direction or stability can be verified with information from a variety of sources. To support or deny the trends noted in the appraiser's preliminary observations, the appraiser looks at information recorded over the past several years, checking 1) building permits issued by the local or county authority, 2) real estate transactions documented in the county recorder's office, and 3) the increase or decrease in real estate sales activity as shown by multiple listing services or classified ads in the local newspaper. Any rumors concerning the relocation of businesses or industries to or from the community must be verified to assess the economy of a rural area accurately.

## Social and Demographic Data

The personal and social needs of the people who make up a rural community are essentially the same as those evident in other communities. The degree to which these needs are met affects community property values. The characteristics of the population often dictate the principal concerns of that population. If many young families live in an area, the number and quality of schools definitely influence property purchases. Accessible health care facilities usually increase property values in a given area. The availability of places of worship, recreational facilities, cultural attractions, and shopping facilities may influence the decision to purchase a property in a rural area. The condition and quality of the roads to and from these facilities also affect the value of properties in the community.

In many rural areas, the availability of off-farm employment is significant. Recent statistics suggest that many farm operators must have income from other sources to meet the debt service on their farms. An established industry in the area can have an impact on property values. If the industry creates housing needs, the development of small parcels could create an increase in both land value and property taxes, which could, in turn, force the sale of agricultural land for the higher values commanded by residential lots. When only seasonal work is available in an area, the unstable nature of the employment opportunities may discourage potential farm purchasers who need additional income to get started in farming.

One of the best sources of social and demographic data for a community is the local chamber of commerce, which generally can supply information such as ethnic composition, number of schools and enrollment, number and denomination of churches, number of doctors and other health care specialists, hospitals and bed capacities, local government structure, unemployment figures, major employers, recreational facilities, and shopping facilities. Some state agencies publish a community profile-type publication which contains much of this data.

## Governmental Data

As mentioned in Chapter 4, the legal right to own land is subject to the powers of government. Part of the appraiser's research must focus on the effects of local, state, and federal government regulations.

An appraiser may look at taxation, including assessment rates, in an area to discern one basic effect of government regulation. Eminent domain takings in proximity to agricultural lands can increase or decrease property values in a community. Pollution and erosion control regulations can increase the cost of running an operation and decrease income and land value.

While most agricultural lands are zoned or designated for agricultural use, local governments have the power to change zoning or land use regulations to accommodate the expansion of a metropolitan area. Urban expansion can make it economically impossible for operators to keep land in agricultural production. Although land values generally increase, the change in zoning to a commercial, residential, or industrial use creates a burdensome increase in real estate taxes. Income tax laws and County Farm Service Agency (CFSA) programs can also impact land values. Governmental forces affecting value will be discussed in depth in Chapter 6.

Appraisers should be familiar with sources of government data. Tax and assessed value information as well as rural area zoning information can be obtained at the county courthouse. City or town planning and zoning authorities control zoning within village limits in many states. State environmental agencies enforce pollution and hazardous waste control laws and can provide information on these regulations. The county CFSA office is the source for data on programs involving bases, yields, and acreages, while the supervisory office of the Bureau of Land Management, Forest Service, or State Land Departments can provide information on the rules governing the leasing of their lands.

## Environmental Data

The appraiser must consider a spectrum of environmental forces that affect land value. Although soils and climate are environmental considerations, they will be discussed separately. The environmental data discussed here relates to detriments and hazards, which are present to some extent on nearly all properties. An appraiser should determine the types of nuisances and detrimental factors present on the property—e.g., odors, undesirable business establishments, dumps, landfills, potential hazardous wastes, noxious plants—as well as neighborhood hazards such as the incidence of hail, flood, and drought and annual variations in yields. In addition, the appraiser must consider the presence of wetlands, highly erodible lands, and endangered species.

Physical inspection of the property will reveal which environmental concerns warrant further investigation. Although data sources vary, the local NRCS office, planning and zoning commission, state environmental agency, and chamber of commerce can generally provide environmental information to satisfy most of the appraiser's needs.

## Climatic Data

Of the many characteristics that determine the agricultural use of land, climate is the most unpredictable. Long-term weather records help farmers establish norms and make predictions, but the climate in any future year cannot be assured. Climatic conditions govern the suitability or unsuitability of an area for producing specific commodities. Agricultural technology has been able to alter some climatic conditions. The installation of irrigation systems in low rainfall areas can ensure adequate water resources for plant growth and heated greenhouses will protect plants from freezing temperatures. However, even a property that has been modified to overcome one extreme may be subject to other climatic influences that can impact production. For example, drought can exhaust irrigation water supplies and hail can destroy a greenhouse.

Appraisers are not expected to be weather forecasters, but they must be well enough acquainted with climatic norms to make sound judgments as to crop suitability, yields, and the supply and dependability of water resources in the area. Climatic information can be obtained from the National Resource Conservation Service, County Extension Service and from the local library and chamber of commerce.

## Soil and Range Site Data

Soil properties and topography combine with climatic conditions to make an area suitable for the production of certain commodities and unsuitable for the production of others. The vegetation of rangeland is dictated by both soils and climate. Soils play a very important role in the appraisal of agricultural properties and they are discussed in detail in later chapters. This discussion will be limited to the sources of soil data.

One extremely valuable tool for rural appraisers are the soil surveys published by the National Resource Conservation Service for individual counties and areas. In cooperation with state agricultural experiment stations and other federal and state agencies, the USDA has been making soil surveys and publishing them since 1899. Soil surveys contain detailed maps and descriptions of soil types in the county or area surveyed. They describe soils and soil characteristics (horizons, consistency, structure, and texture), site characteristics (drainage, topography, climate, and fertility), and natural and land use effects on the soils' productivity. The soil surveys also describe soil properties that provide a scientific basis for comparing one parcel of land with another.

Both the uses for, and the contents of, soil surveys have been expanded over the years. Since 1957 soil surveys have included interpretations for each of the soils found in an area and cited on the area maps. The kinds of interpretations included in recent surveys vary with the needs of the area, but the following interpretations are included in most surveys: estimated yields of the common agricultural crops under defined levels of management, land capability interpretations, soil woodland interpretations, rangeland interpretations, engineering uses of soils, interpretations for community planning, suitability of the soil for drainage and irrigation, and suitability of the soil for wildlife and for recreation.

Information derived from these interpretations is useful in the valuation process. For example, estimated yields can be used in conjunction with expected or average commodity prices to provide a basis for estimating a property's gross annual income. State Department of Agriculture statistical summaries are also excellent sources of this type of information. Rangeland studies indicate the number of animal units that can be maintained on the range without damaging its natural vegetation. Engineering studies of the uses of soils provide information essential to the development of rural land for urban or recreational use. Studies that indicate the suitability of land for irrigation and drainage help the appraiser determine the highest and best use of land.

## Crop and Yield Data

Though of particular importance in the income approach, data on crops and yields can often assist the appraiser in making comparisons in the sales comparison approach. Data for every major commodity produced in the nation is available at national and state levels. However, data sources diminish as the defined area narrows from the state level to the county level to the local level.

At each state headquarters of the National Agricultural Statistics Service, extensive data is compiled for every commodity produced within the state's jurisdiction. The data is published each year in a report called the *(State) Agricultural Statistics.* Crop and yield data for each county and for the state is obtained by surveying the agricultural producers in the state. Other sources an appraiser may find helpful include seed dealers, County Extension Services, farm managers, and marketing companies.

### Livestock Data

In the appraisal of a large western ranch, feed lot operation, dairy farm, poultry operation, swine operation, or other specialty livestock operation, assembling specific data relating to the operation is crucial to the successful completion of the assignment. The breeds of livestock, prices for commodities, stability of markets, location of marketing facilities, prominence of the operation in the neighborhood, potential death losses from disease and predators, feed requirements, and government support programs are a few of the areas that must be researched.

Again, local FSA, NRCS, and County Extension Service offices are excellent sources for livestock data. Information for a specific enterprise can also be obtained from sources such as marketing cooperatives, major poultry companies, creameries, livestock auction yards, meatpacking companies, Agricultural Marketing Service, state agriculture department, universities, and the state office of the National Agricultural Statistics Service.

### Timber and Minerals (Natural Resources) Data

For the appraiser of natural resources, data sources are limited and information from these sources must be scrutinized. The appraiser must have an understanding of the timber or mining operation as well as information about supply and demand, competitive operations, governmental regulations, and expense and income data. A good understanding of all relevant factors is difficult to develop with so few sources available to the appraiser.

Although data sources are few, a wealth of information on timber and minerals is available through the Forest Service and the Bureau of Land Management. These two federal agencies should be the focus of the appraiser's data collection efforts. State agencies associated with these federal agencies, such as the state department of forestry, state bureau of mines, or state land department can provide much meaningful information. Timber and mining companies have much data, but most are reluctant to divulge information. Timber cruisers are excellent suppliers of information on the quantity and quality of standing timber, while geologists or assayers can provide information relating to the supply, price, and concentrations of mineral deposits.

### Water Resource Data

A valuable commodity in the arid regions of the United States, water was the cause of many feuds and deaths in the nineteenth century. Irrigation projects and wells are the lifeblood of agriculture in the arid western states. Many western rivers and streams have been dammed to store water for irrigation and thousands of miles of canals and pipelines distribute that water to farms in the fertile, but arid, valleys. Irrigation projects, districts, companies, and cooperatives govern the use and distribution of the water to their members or subscribers. Most often, water rights are appurtenant to the land and regulated by a state agency, which can provide data on water rights.

Besides individual water districts and companies, the main source of project information is the Bureau of Reclamation, the federal agency that oversees all operations of the dams and reservoirs storing irrigation water. For pump well data, individual well drillers and pump companies can be a good source of information. The NRCS and Extension Service often compile information on the cost of irrigation, which can be invaluable to the appraiser's income approach calculations.

## Lease Data

Collecting lease data is vital to the income approach to value and requires tact and diplomacy because lessees and lessors are usually the only source of lease information. The financial information in a lease is often considered to be of a private nature, so the appraiser needs to earn the trust of the lessee or lessor to secure the necessary data.

When a cropshare lease is being analyzed, details concerning variables such as fertilizer and seeding rates, costs, and crop yields must be obtained to project the landlord's net income accurately. Allowing the appraiser to inspect property records will take time away from the landlord or tenant's schedule. The considerate appraiser will be aware of this demand and arrange the interview for the convenience of the lessor or lessee.

Often lease data can be obtained from government offices such as FSA or private lending institutions such as banks, Farm Credit System lending entities, and life insurance companies. Farm managers can provide lease information in many cases.

## Cost Data

Improvement costs for typical buildings on agricultural properties are readily available, but extensive research and investigation may be needed to acquire accurate information. It may be difficult to estimate the reproduction cost of many older buildings with outmoded designs and structural features without enlisting the assistance of contractors and specialty suppliers.

Most appraisers use cost guide publications such as *Marshall Valuation Service and Boeckh's Agricultural Cost Guide* as sources of building cost information. However, local contractors can provide more accurate cost information in cases where distance may be a factor. For example, a cost guide may not provide the most accurate estimate of the cost to build a dwelling on a ranch 75 miles from the nearest town. A local contractor's estimate will be more accurate because it will include the added travel and transportation costs incurred along with the construction costs.

When atypical improvements are encountered, the appraiser should look to building supply sources to obtain cost estimates. To obtain current cost data on the specialized equipment used in a vegetable cannery or the refrigeration units used in a meatpacking house, the appraiser would contact the equipment manufacturer.

### Market Data

Market data is created by the actions of buyers and sellers in an open and competitive market, thus sources of market data may include market support and market-related businesses as well as the market itself. The reliability of the sales comparison approach to value depends on the appraiser's ability to gather, assemble, analyze, and interpret market data. Poor market data can render the entire approach useless and indefensible, while good market data can provide meaningful and supportable conclusions. It is vital that the market be thoroughly investigated to collect all possible data and that sales be confirmed with the buyer, the seller, or another knowledgeable party. Appraisers can make use of various real estate multiple listing services and sales reporting companies, which periodically provide sales information extracted from county courthouse records. Such a service can be purchased for a reasonable fee, but the data is not necessarily conclusive or accurate. The appraiser must investigate the sales and rely on the reporting service only in the initial stages of investigation. Closing agents such as title insurance companies and attorneys can provide much sales information as can real estate brokers and real estate auctioneers. Lending institutions are excellent sources of information on sales of their acquired properties and sales in which they provided financial assistance to the purchaser. Exchanging sales data with other appraisers is, of course, a popular way to add sales information to the appraiser's data bank. Searching individual county courthouse records remains a very sound method of collecting sales information.

Once a good base of sales has been compiled, the appraiser's analysis can reveal trends in the marketplace and determine if a more thorough investigation of the market is warranted. Discussions with real estate brokers, lenders, and individual buyers and sellers can provide valuable information which the appraiser can use to support trends observed in analyzing the market data.

## Requirements for Data Collection

### Professional

The Uniform Standards of Professional Appraisal Practice (USPAP), as promulgated by the Appraisal Standards Board of The Appraisal Foundation, states the appraiser's professional requirements for data collection. USPAP establishes standards concerning the procedures to be followed in performing an appraisal, review, or consulting service and the manner in which an appraisal, review, or consulting service is communicated. The Preamble to USPAP mandates compliance with the standards and includes this statement, "To maintain a high level of professional practice, appraisers observe these Standards." The professional appraiser's requirements for data collection under USPAP are summarized below.

The Ethics Provision requires the appraiser to maintain written records of all data and statements required by the standards and "all other data, information, and documentation necessary to support the appraiser's opinions and conclusions. The appraiser must meet requirements not only for data collection, but also for data retention.

Standards Rule 1, which governs the development of a real property appraisal, recognizes the importance of data collection. Standards Rule 1-1(b) requires that the appraiser "not commit a substantial error of omission or commission that significantly affects an appraisal." The entire comment to that rule is devoted to gathering and analyzing data. The comment states, "In performing appraisal services, an appraiser must be certain that the gathering of factual information is conducted in a manner that is sufficiently diligent, given the scope of the work as identified according to Standards Rule 1-2(f), to ensure that the data that would have a material or significant effect on the resulting opinions or conclusions are identified and, where necessary, analyzed. Further, an appraiser must use sufficient care in analyzing such data to avoid errors that would significantly affect his or her opinions and conclusions."

Standards Rules 1-2 and 1-3 set forth specific guidelines an appraiser must observe. Standards Rule 1-2(f) lists one of the guidelines as "identify the scope of the work necessary to complete the assignment," while Standards Rule 1-3(a) requires an appraiser to consider the effect on use and value of several factors, all of which require extensive data collection efforts.

Without doubt, Standards Rule 1-4 is the most applicable to the appraiser's professional requirements for collecting data within the three approaches to value. Three relevant sections of the rule are quoted below:

> In developing a real property appraisal, an appraiser must collect, verify, and analyze all information applicable to the appraisal problem given the scope of the work identified in accordance with Standards Rule 1-2(f)
>
> (a) When a sales comparison approach is applicable, an appraiser must analyze such comparable sales data as are available to indicate a value conclusion.
>
> (b) When a cost approach is applicable, an appraiser must:
>
> (i) develop an opinion of site value by an appropriate appraisal method or technique;
>
> (ii) analyze such comparable cost data as are available to estimate the cost new of the improvements (if any); and
>
> (iii) analyze such comparable data as are available to estimate the difference between cost new and the present worth of the improvements (accrued depreciation).

(c) When an income approach is applicable, an appraiser must:
   (i) analyze such comparable rental data as are available to estimate the market rental of the property;
   (ii) analyze such comparable operating expense data as are available to estimate the operating expenses of the property;
   (iii) analyze such comparable data as are available to estimate rates of capitalization and/or rates of discount; and
   (iv) base projections of future rent and expenses on reasonably clear and appropriate evidence.

Standards Rule 1-5 requires that an appraiser "reconcile the quality and quantity of data available and analyzed within the approaches used and the applicability or suitability of the approaches used." It further requires data on prior sales of the subject be analyzed if they occurred within one year for one- to four-family residential property and within three years for all other types of property.

Reporting the results of a real property appraisal is covered under Standard 2. This standard requires the appraiser to describe the extent of the process of collecting, confirming, and reporting data and to summarize the data considered so the reader of the report will understand it. References to the collection of data are made throughout the standards.

In addition to the requirements of the Uniform Standards of Professional Appraisal Practice, appraisers may be subject to detailed data collection requirements imposed by professional organizations to which they belong or by the appraisal firm that employs them. In any case, collecting and retaining data are inherent to appraisal reporting and the importance of these functions are recognized, mandated, and governed by USPAP.

## Governmental

All the governmental requirements for data collection cannot be summarized here. Numerous governmental agencies require appraisals and each has a miscellany of data collection requirements. Each governmental agency writes its own appraisal or appraisal review regulations and the requirements for data collection and reporting vary among the authors of these documents. There is no agreement among the various agencies as to standard data collection and reporting requirements.

Contract solicitation documents and engagement letters usually contain some reference to the agency's appraisal regulations. The reference may indicate that a copy of the regulations can be obtained by request or it may note a specific section of the Code of Federal Regulations (CFR). Unless the appraiser has access to a library that maintains the Code of Federal Regulations, it is imperative that he or she obtain a copy of the appraisal regulations from the agency before the appraisal work commences.

Lenders regulated by government agencies must ensure that the appraisals they obtain comply with the federal regulatory requirements.

For example, a bank regulated by the Federal Deposit Insurance Corporation (FDIC) must ensure that the appraisals they obtain meet the appraisal requirements of the FDIC. A bank mortgage company or savings and loan association may sell loans to several government agencies, all of which have different appraisal regulations with which that lender must comply. Thus, the appraiser may need to consider different data collection and reporting requirements for different appraisal reports prepared for the same client lending institution.

Since the data collection requirements indicated in the different agencies' appraisal regulations vary immensely, a word of caution and bit of advice are warranted. To ensure full compliance with the governmental requirements for data collection, a prudent appraiser will always read carefully the engagement letter, contract, or other information-transmitting document and prepare the appraisal report according to the requirements of the client. Asking for a clarification or explanation when needed will ensure that the appraisal report will meet agency requirements on the first submission.

## Summary

The types of data to be collected, analyzed, and reported by the appraiser are 1) dependent on the specific appraisal assignment, 2) considered in the Uniform Standards of Professional Appraisal Practice, and 3) specific to the client requirements. Data sources can be found at the national, state, and local levels. The quality of data and information an appraiser collects is proportional to the quality of the final appraisal report.

Although we tend to describe data in terms of categories, the forces that affect property value are inseparable and interrelated. National and world economic trends, governmental forces, area climate and land characteristics, land use, area real estate market activity, and community trends are intertwined and each is subject to constant change. The appraiser's opinion of value is based on these changing, complex interrelationships. Therefore, the process of gathering and analyzing data must be ongoing and subject to constant updating and reexamination.

CHAPTER
• 6 •

# Regional and Neighborhood Data and Analysis

Social, economic, governmental, and environmental forces influence property values. The boundaries of the area of influence need to be delineated to analyze the impact of these factors on the value of the subject property. Most of the boundaries are physically observable and areas of influence vary in size.

For rural and agricultural properties, the boundaries generally refer to "regional" and "neighborhood" areas. Regional areas are generally larger than neighborhoods and are characterized by common crop or livestock uses. A neighborhood area is generally smaller and is more closely related to the subject property in land use and community facilities. Properties within the immediate vicinity of the subject should be identified and analyzed in the neighborhood analysis.

The general analysis usually begins with the subject property and extends outward, identifying all relevant actual and potential factors that may influence the value of the subject property. Conclusions regarding the impact of the region or neighborhood on the value of the subject property are significant when the boundaries are properly established. Establishing boundaries will allow the appraiser to identify the area from which comparable sales and other market related information should be obtained and to understand present or future trends or changes in land use.

Regional and neighborhood boundaries are typically established by analyzing a combination of land use, crops, and/or livestock; terrain, topography, soils, and/or range sites; and climatic and/or rainfall patterns.

Land use—crops or livestock—is a significant factor in establishing boundaries since regions and neighborhoods tend to exhibit similar characteristics. Land use, which is observable and easily understood, must be considered in establishing boundaries. Examples of land use would be dry cropland, irrigated cropland, permanent plantings, open improved pasture, native pasture, native rangeland, and timberland. Regions are typically composed of two or more land uses, while neighborhoods may be composed on only one or two uses, although there can be more. Areas are usually characterized by only one or two major crops; in the case of specialty crop areas, the many crops are collectively considered a "specialty crop." Livestock areas are generally thought of as ranch properties, but livestock may be integrated into the crop system in many areas. An example of a livestock area would be the dairy area in California, which is divided into neighborhoods. In ranch country, there may be sheep and goat areas and separate cattle areas. In some areas a region may be referred to as *combination country,* where multiple species are grazed.

Terrain, topography, soils, and/or range sites are factors that are readily identifiable by visual inspection and map analysis. All are physical characteristics that are apparent even to the casual observer. A region may have several types of terrain and topography, while a neighborhood will typically have fewer. Elevations generally play a major role in determining neighborhood boundaries and may or may not be significant in a regional setting. Soils and range sites are typically more homogeneous in neighborhoods than in regions.

Climatic conditions tend to be somewhat related to terrain and topography. In high elevations and coastal regions, diverse microclimates may be found within a single region or neighborhood. Frequently, rainfall patterns in coastal areas dictate the boundaries of irrigated cropland and non-irrigated cropland since the amount of rainfall can decrease rapidly as distance from the coast increases.

Typically, no single factor will circumscribe the boundaries of a region or neighborhood so the appraiser is compelled to consider all of the factors described as well as any other factors that are appropriate in the area of study. Once the boundaries are established, the region or neighborhood can be analyzed in terms of social, economic, governmental, and environmental considerations.

Proper regional analysis is critical to the valuation of rural or agricultural properties since most agricultural properties are used to produce a commodity or service that is marketed outside the neighborhood, but within the region. Residential and commercial properties are generally more affected by neighborhood factors than agricultural properties are. Regional boundaries typically encompass one or more counties and, in some instances, may encompass a significant portion of a state or province. By contrast, a neighborhood typically includes only a portion of a county or counties. The four basic types of considerations (i.e., social, economic, governmental, and environmental) apply equally to the analysis of the region and the neighborhood.

## Social Considerations

The social factors pertinent to the analysis of both regions and neighborhoods are reflected in

- Population and population density, age levels, skill and educational levels, employment, and other demographic trends
- Presence of commercial/agricultural services
- Presence of educational services
- Presence of medical services
- Presence of social, recreational, and community or area civic organizations

The appraiser must recognize that the social preferences of the individuals who create a given market and the effect these preferences have on property value are difficult to measure. Hence, only limited reliance is placed on these factors in reaching a value conclusion. Social factors should only be given significant weight when they can be shown to impact the market and can be accurately measured. Gender, race, religion, ethnicity, and national origin are social considerations, but they have no direct relationship to real property values. Appraisers must keep this in mind to provide unbiased regional and neighborhood analyses.

## Economic Considerations

Economic considerations relate primarily to the financial and business characteristics of an area. The factors of greatest significance to rural and agricultural properties are those that relate to the present or future market mechanisms that will provide a market for production and for the delivery of services to the properties in the area. It is critical that the appraiser consider the typical operating methodology relative to the leasing and sale of properties. Trends in the prices of sales and leases and any evidence of changes in production patterns or changes in property use should also be considered. The following list identifies the most significant items for analysis in most markets, but other considerations may be warranted in specific markets.

- The dominant business character of the market study area, i.e., agricultural, recreational, manufacturing, or service industry
- Presence or absence of markets for grain, feed stuffs, natural fiber, livestock, produce, other cash crops, and game or fish
- Presence or absence of processing facilities for crops and feeding facilities for livestock produced in the area
- Presence or absence of suppliers of inputs such as feed, seed, fertilizer, chemicals, machinery, and building supplies
- Presence or absence of financial services and capital sources oriented to rural, agricultural, and agri-business needs

- Presence or absence of typical retail product suppliers and services not directly related to agricultural production needs
- Typical operating methods prevailing in the area—e.g., cash rent, share rent, owner operation, custom operation, recreational operation
- Typical property size, owned vs. leased properties, vacancy levels, trends in improvements to both land and structures, and changes in any of these factors
- Rent levels, operating income and expense levels and ratios, property tax levels, and anticipated changes in these factors
- Production levels for major crops, livestock numbers, and the capacity of on-farm and off-farm crop storage facilities
- Crop, livestock, grazing, and recreational (hunting and fishing) income levels
- Presence and adequacy of transportation infrastructure including highways and secondary roads, rail service for the movement of commodities, freight services, passenger services by surface and air, and water transportation if applicable
- Presence and adequacy of utilities including electricity, gas, telephone, and domestic water supplies
- Presence or absence of water resources for irrigation or other processing requirements from both ground water and surface water
- Presence or absence of oil, gas and other mineral exploration, production, or processing activities

In analyzing a region or neighborhood's economic characteristics, the appraiser should examine economic trends over a three- to five-year period. Published data on agricultural production is available from the U.S. Department of Agriculture and the state's Agricultural Statistics Services. Demographic information from the U.S. Bureau of the Census and economic data from numerous governmental and private sources is available in print and via the Internet. With this data, the appraiser can decide which economic variables contribute significantly to value differences within the region and how these differences impact the value for the subject property.

## Governmental Considerations

Governmental considerations include factors such as the type and organization of state, county, and local governments, quasi-governmental agencies, and other governing bodies as well as laws, regulations, tax policies, police powers (zoning), environmental regulations, USDA Commodity Programs, and conservation program requirements. These considerations can be quite significant and the appraiser should become familiar with the factors operating in the market study area. The individual factors typically given the most attention in this portion of the analysis are enumerated below.

- USDA regulations relating to commodity subsidy programs administered by the Farm Service Agency
- USDA regulations regarding conservation compliance programs
- Wetlands regulations and requirements imposed by governmental entities
- Regulations regarding grazing rights on federal or state lands and current fee structures.
- Environmental regulations regarding soil and water contamination, agricultural chemicals, air quality, and mineral production activities
- Current real estate tax assessments, rates, and special assessments imposed by various governmental entities, compared with those of competing properties in the area
- Current zoning regulations that impact the subject property
- Building and health codes that impact the subject property
- Quality of police and fire protection.
- Regulations regarding interest rates and the availability of money for mortgages and operations
- Floodplain designations and attendant restrictions on use and development
- Current laws and regulations governing the domestic, livestock, and irrigation uses of water.
- Federal, state, and county game and fish laws that may affect recreational uses
- Eminent domain laws unique to the jurisdiction of the market study
- Private restrictions, easements, or other encumbrances established by condemning authorities, developers, or land owners recorded in official plats or deeds that may restrict property use or development

Government regulations have an impact on the value of property and there appears to be a trend toward increasing governmental regulation in the future. The areas of environmental, wetland, and chemical regulation are growing and changing quickly; zoning and development ordinances and restrictions are also increasing.

The sources of data on governmental regulation are many and varied. Some of the required data is available at the local level from the appropriate regulatory agencies or from local deed records, while other data must be obtained from the state regulatory agencies charged with administering the regulations. Still other data must be obtained at the federal level. Much regulatory requirement data is now available via the Internet at little or no cost. The appraiser must be aware of any current government regulations that could impact the value of the subject property.

## Environmental Considerations

Environmental considerations refer to the physical features of the study area, both man-made and natural. These include the geographic location of the region or neighborhood; topographical and terrain features; rivers, creeks, lakes, and other bodies of water; building improvements and densities; land improvements such as irrigation projects; climatic factors such as temperature, humidity, precipitation, growing season, and storm hazards; and environmental hazards emanating from the environment naturally or as a result of human activities. The environmental considerations most relevent to agricultural and rural regions and neighborhoods are typically terrain, topography, vegetation, and soil or range site types. These factors generally determine the nature of the operations that can be conducted in the area. Thus, the climatic conditions prevailing in the area will determine the nature and intensity of enterprises that can be developed on area properties. Location and roadway linkages often dictate the land uses that will ultimately succeed in a market area. The following factors are most often considered critical to the assessment of environmental considerations:

- Topography considerations, which dictate drainage, accessibility, and the choice of uses. Gentle terrain is best adapted to cropland, while rough terrain is better suited to livestock or recreational uses.
- Types of soil and range sites, which reveal the suitability of the land for crops, grazing, and building types and uses.
- Climate, which is critical to the range of operations and enterprises that prevail in a given region or neighborhood. Numerous micro-climates may exist in a given region or neighborhood and provide production opportunities not available in nearby adjacent areas due to temperature, humidity, and growing season.
- Presence or absence of adequate drainage, either natural or man-made. Drainage systems prescribe limits on most rural properties.
- Precipitation amounts, timing, and type, which must be analyzed to assess crop development potential or grazing potential.
- Roadway, rail, and other transportation routes and services, which determine the adequacy of linkages to and within the neighborhood or region.

- Presence or absence of natural or man-made hazards, which may include odors; waste waters; solid wastes; air-borne contaminants or pollutants; poor quality or contaminated ground or surface water; oil, gas, or hard mineral production not properly operated; and contamination from confinement feeding operations or commodity storage or processing facilities.
- Adequacy of utility services, including public drainage projects.
- Presence or absence of oil, gas, and/or mining operations in the study area.
- Presence or absence of toxic or noxious plants that would limit livestock production or reduce game or fish populations.
- Presence or absence of destructive insect species or plant or animal pathogens which would limit or prevent production.
- Presence or absence of natural predators, which would affect livestock production.
- Presence or absence of endangered or protected species of plants and animals or critical habitat, which would limit property development.
- Presence or absence of wetlands.

The foregoing list includes the considerations commonly associated with rural or agricultural regions and neighborhoods. The professional appraiser must assess each study area individually and identify any environmental considerations unique to the area. Often environmental considerations can make an area adaptable to a unique use or, conversely, prevent a use that would otherwise be appropriate.

## Summary

In conducting regional and neighborhood analysis, the appraiser assesses the four factors that influence the value of property in an area. Defining the study area in terms of region and neighborhood provides the basis for developing the highest and best use analysis and for gathering sales, rental, income, expense, and depreciation data. All of these items are necessary and critical to the appraisal process. The regional and neighborhood analysis in the appraisal report allows the appraiser to describe and present a complete picture of the setting in which the subject property exists. Similarly, it enables the client or reader of the report to visualize the nature of the region and neighborhood. Objectivity is an essential element in the development of the regional and neighborhood analyses. Many users and clients will rely on the appraiser's descriptions to assess the present and future prospects of areas with which they are not familiar.

CHAPTER
• 7 •

# *Legal Descriptions*

Any appraisal assignment begins with identification of the subject real estate. Every parcel of real estate is unique and therefore has a unique legal description. Whether the property is residential, commercial, or agricultural, its legal description will identify it precisely so that it cannot be confused with any other property. To research and analyze the subject property and comparable sales, the appraiser must have a finite definition of the real estate, which is contained in the legal description. In addition to the specific description of the real estate, the legal description may also indicate easements, restrictions, mineral rights, water rights, or partial interests.

History has recorded many court battles relating to the legal description of property. An appraiser has not fully experienced the appraisal profession until he or she has appraised the wrong real estate or included only a portion of the property because of an incomplete legal description. Often the client instructs the appraiser to obtain the legal description from sources such as the local assessor, the register of deeds, or a will. Portions of the original property may have been sold, traded, or altered since the original legal description was created. Ownership of a farm or ranch property is often held by more than one entity.

It is incumbent on the appraiser to demand the exact legal description of the property to be appraised from the client. Obviously, many clients require the appraiser's assistance in obtaining this document. Once obtained, the legal description must be confirmed. Most states require the public recordation of deeds, mortgages, easements, and other documents relating to real estate and most of these records are cross-indexed. This system allows the appraiser to search for a property through the grantor/grantee records or through the land index by legal description. The appraiser should be sure to work with exact legal descriptions since offices such as the assessor often work with abbreviated or laymen's legal descriptions.

Obtaining the legal description often leads to the difficult task of identifying the physical property precisely. A large volume of quasi-public information is available to the appraiser, depending on the political subdivision in which the property is located. Information from county or independent surveyors, state highway authorities, and utility companies may assist the client or appraiser in identifying or correcting imperfections in the written legal description. In addition to these sources, cadastral maps, U.S. Geological Survey maps, FSA/NRCS photos, and numerous other sources of information may be consulted to help the appraiser identify the exact physical property from the written legal description. In extreme situations the client may need to obtain an abstract of title or title insurance commitment to create a correct legal description. A combination of documented official information, support data from unofficial sources, and careful field work can ensure a good legal description of the property for appraisal purposes.[1] Recent federal and state legislation requires that each appraisal comply with USPAP[2] and a legal description is one way to satisfy the USPAP requirement to identify the property.

A property may be difficult to visualize if the field boundaries or land uses are inconsistent with the exact legal description. Federal, state, or private leased lands may be fenced jointly with lands in private ownership. The exact location of irrigation wells, livestock watering facilities, and building improvements are extremely important in determining the legal boundaries of a property. There are often discrepancies between what the owner or client believes are the correct boundaries of the physical property or location of portions of that property and the true legal description. A precise survey may be required prior to the appraiser's completion of the appraisal assignment.

Some of the systems currently in use to identify real estate include: street address, block and parcel number, the U.S. Rectangular Survey System, map-based systems, Dual Independent Map Encoding (DIME), federal surveys based on the coordinate system, and the State Plane Coordinate System (SPSC). Efforts have been made to standardize these systems to accommodate researchers and to make use of modern recordkeeping technology.

The three principal systems used for legal descriptions in the United States are the 1) rectangular survey, 2) metes and bounds, and 3) recorded plat systems. The legal description of a rural or agricultural property often includes portions described by all three systems.

---

1. For a more extensive discussion of the sources for, and construction of, legal descriptions, see Gurdon H. Wattles, *Writing Legal Descriptions* (Orange, CA.: Gurdon H. Wattles Publication, 1976).
2. Uniform Standards of Professional Appraisal Practice, as promulgated by the Appraisal Standards Board of the Appraisal Foundation.

To complete the appraisal assignment, the appraiser must obtain the exact legal description and must be capable of transferring that description to the actual physical property. To accomplish this task, the appraiser must understand the three main systems of legal description.

The rectangular survey system of legal description was created by the United States Congress in the late 1700s. It works well on the open plains but natural phenomenon such as shoreline and rivers or man-made objects such as highways often preclude a complete legal description under the rectangular survey system. Augmentation with the metes and bounds system is then required to complete the legal description. The metes and bounds system is the primary method of real estate description in the 13 original colonies, a few other states east of the Mississippi River, Texas, and Hawaii. This system of legal description is centuries old and was brought by early settlers from the old world to the Americas. Both of these systems are used in conjunction with the recorded plat system for original towns and for modern housing or commercial subdivision developments.

## Rectangular Survey

When the United States acquired the lands of the Louisiana Purchase, the Mexican Territory, and the Oregon Territory, a system of legal descriptions for large parcels of real estate was needed. The U.S. Rectangular Survey System was developed to provide for simple and orderly land description. This system was implemented in May of 1784 through an ordinance passed by the Continental Congress. The ordinance contained the basic elements of the present system, which is sometimes referred to as the *governmental survey,* and included this statement:

> The surveyors shall proceed to divide the said territory into townships of 6 miles square by lines running due north and south and others crossing them at right angles. The plat of the township shall be marked by subdivision into lots 1 mile square of 640 acres in the same direction as the external lines and numbered from 1 to 36.

Although the rectangular survey system appears complicated, it was designed to be mathematically pure, concise, and consistent. The system is based on sets of intersecting lines: principal meridians, which run north and south, and base lines, which run east and west. The intersection of these lines is used as a reference point to locate and describe land. Both meridians and base lines are precisely located in reference to degrees of longitude and latitude. For example, the sixth principal meridian is a north-south line arbitrarily established by the government surveyors west of Lincoln, Nebraska. It intersects an east-west base line, which is the state boundary between Kansas and Nebraska. (See Figure 7.1.)

### Base Lines

Base lines were established by the original government surveys on true parallels of latitude. Several base lines serve as the boundaries between

**Figure 7.1** Principal Meridians and Base Lines

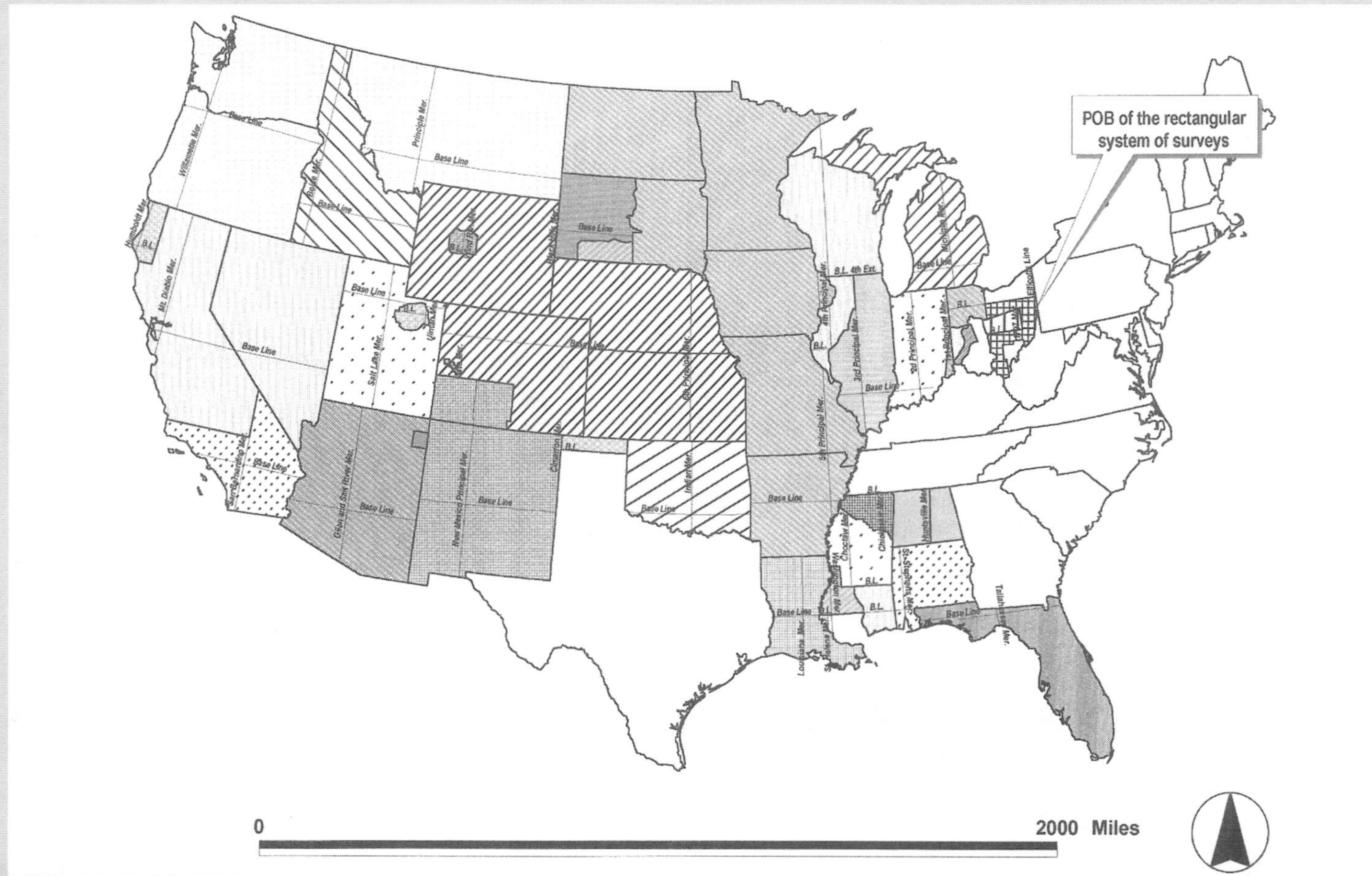

states. Figure 7.1 shows the system of base lines and principal meridians established for the United States.

### Principal Meridians

The second axis for each reference point is a principal meridian, running north and south. Individual meridians were arbitrarily located along the respective base lines, without reference to specific points. Most meridians were originally named for physical or political features within the immediate area. More than 25 meridians are used in the governmental survey of the United States. Some are confined to one state or a portion of a state, while others are used in several states. For example, the fifth principal meridian includes lands in Arkansas, Missouri, Iowa, Minnesota, North Dakota, and eastern South Dakota. Regardless of the specific location, the basis for the rectangular survey is the intersection of two axes, the base line and meridian. From this central point, the rectangular survey lays out blocks of land six miles square.

### Ranges

The next step in the rectangular survey was to draw parallel north-south lines at six-mile intervals east and west of the each principal meridian. The resulting six-mile wide vertical columns are called *ranges* and are numbered consecutively east and west from the respective principal meridian. The first range west is thus referred to as Range One West, and is typically written as R1W. Similarly, the eighth range east of the meridian is called Range Eight East, written R8E. It begins 42 miles east and ends 48 miles east of the principal meridian.

### Townships

In a similar fashion, parallel east-west lines were drawn every six miles north and south of the base lines. These six-mile wide horizontal spaces are called *townships.* They are numbered north and south from the base line in the same fashion, e.g., T4N for Township Four North and T5S for Township Five South. Township Five South would occupy a strip of land starting 24 miles south of the base line and extending to 30 miles south of the base line.

Thus series of parallel lines create north to south columns (ranges) and east to west horizontal spaces (townships). The bisecting of ranges and townships create blocks of real estate six miles by six miles in size that can be referred to with clear numerical definition. The shaded portion in the upper right quadrant of Figure 7.2 identifies T4N-R5E, which begins 18 miles north of the base line and 24 miles east of the meridian. Similarly, T5S-R8W begins 24 miles south of the base line and 42 miles west of the meridian. Under the rectangular survey system, each township is numbered relative to the juncture of the specific base line and meridian (see Figure 7.2).

**Figure 7.2** Numbering of Townships

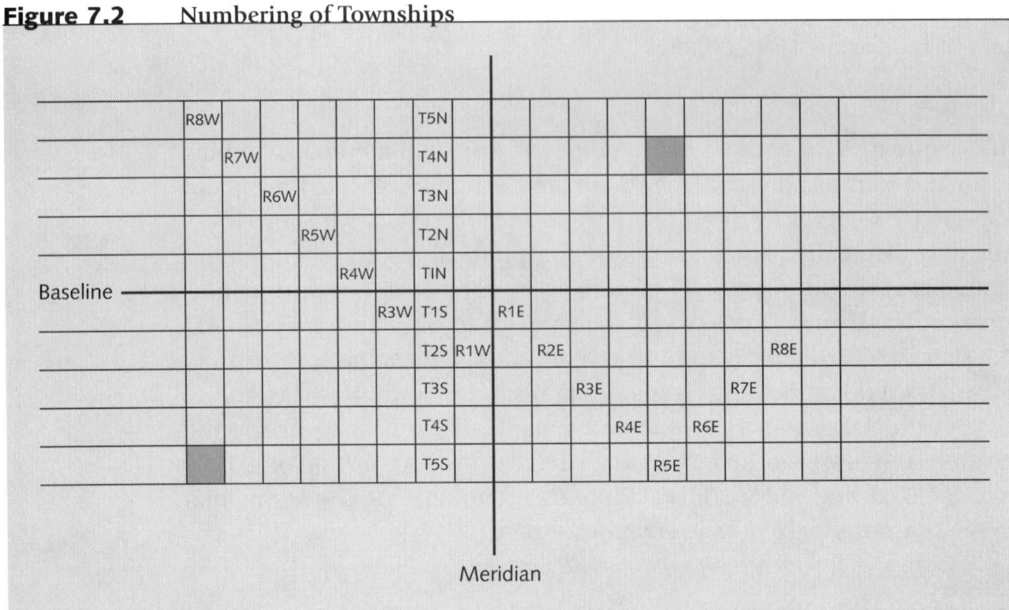


## Sections

Figure 7.3 illustrates the subdivision of each township into 36 parts, called *sections.* Sections within a township are created by parallel lines running both north-south and east-west one mile apart, creating the 36 sections. Within the rectangular survey system, townships are six miles square, or 36 square miles, while sections are one mile square or one square mile. A standard section therefore includes 640 acres. Sections within the township are numbered from 1 to 36, beginning in the northeast corner and proceeding west and east alternately in a southerly manner throughout the township. Figure 7.3 shows the 36 sections of Township 5 South - Range 8 West (T5S-R8W), together with all adjoining sections located in adjoining townships. On a map the section number is written in the center of the section.

**Figure 7.3** Subdivision of a Township

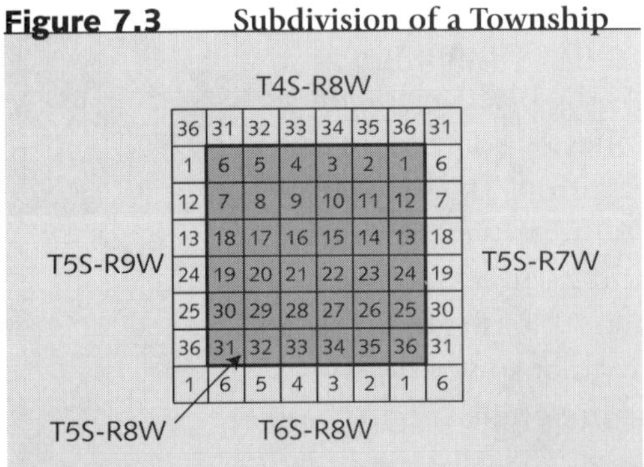


The rectangular survey system further divides each section into quarters containing 160 acres (e.g., NE$^1/_4$ of Section 3); quarter quarters containing 40 acres (e.g., NE$^1/_4$NE$^1/_4$); and quarter quarter quarters containing 10 acres (e.g., NE$^1/_4$NE$^1/_4$NE$^1/_4$). The system can also subdivide sections and portions thereof by one-half (e.g., S$^1/_2$, or N$^1/_2$NE$^1/_4$, or S$^1/_2$N$^1/_2$ of Section 17) as illustrated in Figure 7.4. Figure 7.5 illustrates most of the subdivisions possible within a section. It also provides measurements, acreages, and reference points relative to any individual section within the rectangular survey system.

Since a standard section contains 640 acres, an acre contains 43,560 square feet, regardless of its configuration. Due to the curvature of the earth and human error, however, any section within the rectangular survey might contain more or less acres depending on the original survey. The curvature of the earth creates problems for the square design of the system. An integral part of the system is the process of correction at township lines. The northern and western sides of each township are used for correction of the survey and often contain more or less than 640 acres. Thus, Sections 1 through 6 plus Sections 7, 18, 19, 30 and 31 have special considerations within each township. In general, the S½ of Sections 1 through 5 (see Figure 7.6) are surveyed as 320 acres. Additionally, the S½ of both the NW¼ and the NE¼ are surveyed as 80 acres. Finally, the N½ of the N½ of Sections 1 through 5 are surveyed as lots, typically numbered 1 through 4 from east to west. The surveyor thus created lots along the north boundary of the township containing whatever acres were necessary to correct the survey for the curvature of the earth.

**Figure 7.4** Division of Sections

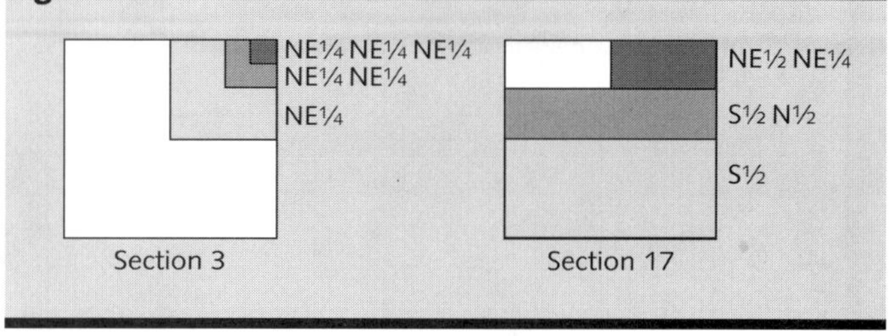


**Figure 7.5** Possible Subdivisions of a Section

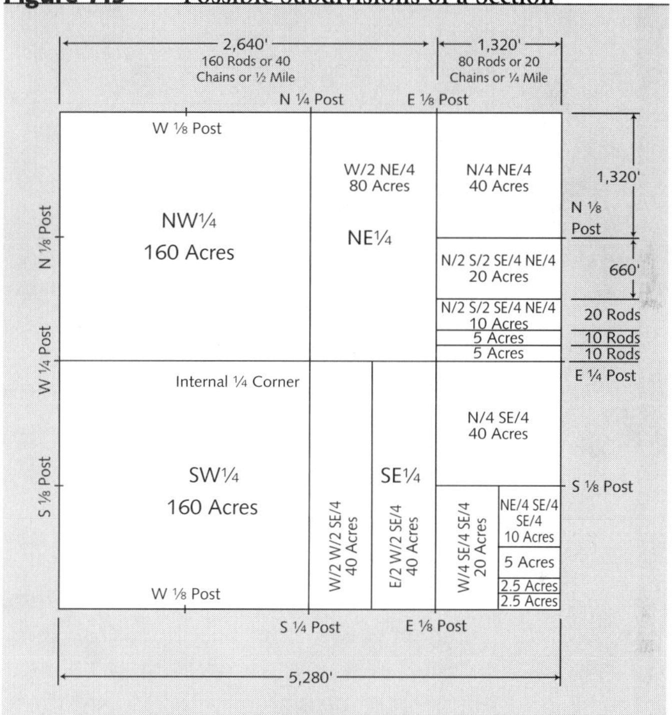


Similarly, the E½ of Sections 7, 18, 19, 30, and 31 were surveyed to contain 320 acres (in two quarters), and the E½ of both the NW¼ and the SW¼ to contain 80 acres each. To finalize these sections, the surveyor used Lots 1 through 4, numbered from north to south along the west side of each section. The required number of acres were included by the surveyor for each section to correct the township line. Section 6 of the rectangular survey is unique in that it requires correction on both

**Figure 7.6** Correction at Township Lines

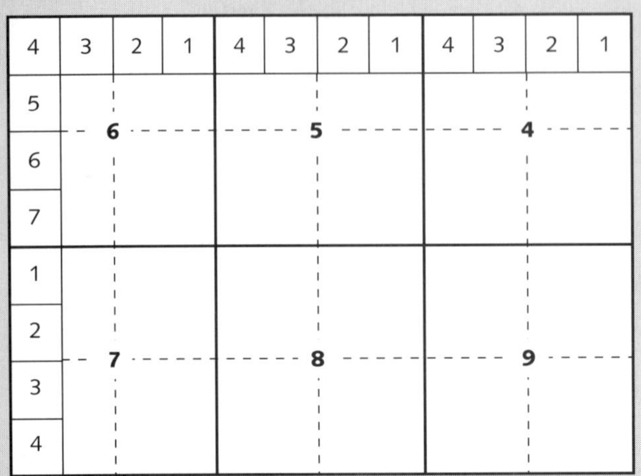


4 Indicates Lot Number
**6** Indicates Center of Section

the north and west boundaries. The survey is completed as described above, starting with the SE¼ as the fixed portion. The S½NE¼, the E½SW¼, and the SE¼NW¼ were then surveyed with 80 or 40 acres respectively. Finally, Section 6 typically includes seven lots numbered from northeast to west and then to south along the north and west boundaries.

Agencies such as the U.S. Geological Survey (USGS) or U.S. Coast and Geodetic Survey (USCGS) have set permanent markers at locations throughout townships within the rectangular survey system. These markers can be located at section corners, half mile lines, etc. Even though the U.S. Rectangular Survey System is semi-precise, any section within the township may contain far more or less than 640 acres.

### Written Description

Several different methods may be used to write a legal description under the U.S. Rectangular Survey System. Although each method is correct, some may be more or less difficult to read or more or less acceptable in the local area. In most cases, the reader can best decipher the description by reading it backwards. To find the exact section and township, the reader must first identify the specific principal meridian. A typical legal description might read:

S¼SW¼NE¼ Section 15-T17N-R21W of the 6th P.M., Custer County, Nebraska.

The reader can most easily understand this legal description by first determining the reference point of the sixth principal meridian (PM). (Note that the state and county are not needed in this type of legal description.) Next the appraiser identifies the township and range, the section, and finally the specific portions of the section. The description is read backwards, first NE¼, then SW¼ of NE¼, and finally S½ of SW¼ of NE¼. The same legal description might also be written:

T17N-R21W of the 6th P.M. Custer County, Nebraska:
S¼SW¼NE¼ Section 15 (or Section 15: S¼SW¼NE¼)

One important factor in writing and understanding legal descriptions is the proper use of punctuation and the words *of* and *and.* A comma, semi-colon, or conjunction *(and)*, denotes a break in the subportions of a section. The above legal description refers to 20 acres total. If the description is written as "S¼, SW¼NE¼", or as "S½ and SW¼NE¼", the land described contains 360 acres. The word *of* denotes a continuation, not a break, in the legal description. Thus "S½ of SW¼ of NE¼" is equivalent to the original tract of 20 acres, and is the proper "reading" of the legal description. Extreme care is required in the writing or reading of legal descriptions prepared according to the rectangular survey system of land description.

## Metes and Bounds

The metes and bounds system was used for legal descriptions of real estate in the United States prior to the development of the rectangular survey system. Under the metes and bounds system, real estate is described using natural or man-made objects as reference points. The starting point for the description is usually an arbitrary point, typically called the *point of beginning* (P.O.B.). Original metes and bounds descriptions often did not relate to the reference points used to describe other properties in the immediate area. Modern metes and bounds descriptions, however, have reference points that are tied to adjoining surveys.

The metes and bounds system is based on distances (metes) and bearings (bounds) from succeeding reference points around the perimeter of the real estate. Distances and directions are thus described relative to naturally occurring or man-made objects, which identify the corners or sides of the parcel. Unfortunately, oak trees, stone heaps, roads, and rivers change over time. Originally there were no consistent general reference points in the area, county, or state from which to describe individual parcels of land.

The following is a portion of an 1812 metes and bounds description from the Hartford, Connecticut probate court:

> Commencing at a heap of stone about a stone's throw from a certain small clump of alders, near a brook running down off from a rather high part of said ridge; thence by a straight line to a certain marked white birch tree, about two or three times as far from a jog in a fence going around a ledge nearby; thence by another straight line in a different direction, around said ledge and the Great Swamp, so called; thence in line of said lot in part and in part by another piece of fence which joins onto said line, and by extension of the general run of said fence to a heap of stone near a surface rock; thence as aforesaid, to the "Horn," so called, and passing around the same as aforesaid, as far as the "Great Bend" so called, and from thence to a squarish sort of jog in another fence, and so on to a marked black oak tree with some stones piled around it; thence by another straight line in about a contrary direction and somewhere about parallel with the line around by the ledge and the Great Swamp, to a stake and stone bounds not far off from the old Indian trail; thence........"

This example points out the potential ambiguity of older metes and bounds descriptions.

The modern metes and bounds system employs permanent man-made monuments and precise descriptions of distance (linear measurement) and bearing (angle measurement). The surveyor installs permanent markers, commonly called *benchmarks,* at some or all reference points around the described real estate. The corner or marker where the survey or legal description actually begins is the point of beginning. A P.O.B. can be used in association with a rectangular survey, recorded plat, or metes and bounds description.

In a metes and bounds description, distances are typically measured in feet, to the nearest tenth or hundredth of a foot. Distances can also be measured in rods, chains, or miles. Bearings or directions are measured in degrees (°) of a 360-degree circle, with further subdivisions of 60 minutes (') per degree and 60 seconds (") per minute. Thus, 29°4'52" would be read as 29 degrees 14 minutes and 52 seconds. It is important to note that the 360 degrees of a metes and bounds description are entirely separate from, and do not relate to, navigational headings.

The 360-degree circle of the metes and bounds system is divided with north-south and east-west lines into 4 quadrants of 90 degrees each. A course is followed to connect monuments or points within the description. Directions for the course are quoted relative to degrees of variation east, west, north, or south from the crossing of the horizontal and vertical lines. New directions are given for each succeeding reference point within the description. Bearings or directions first indicate the direction north or south of the horizontal line, and then the degrees, minutes, and seconds east or west of the vertical line. For example, north 50 degrees east describes a line that runs 50 degrees east of north, or in an easterly direction. Likewise, south 30 degrees west describes a line that runs 30 degrees west of true south. In the diagram in Figure 7.7, minutes and seconds have been omitted for clarity. Note that 90 degrees west can either be north 90 degrees west or south 90 degrees west. Each change in direction within the legal description uses the same quadrants of the 360-degree circle, but the crossing lines are moved to the new point of reference.

**Figure 7.7** Quadrants of a Circle

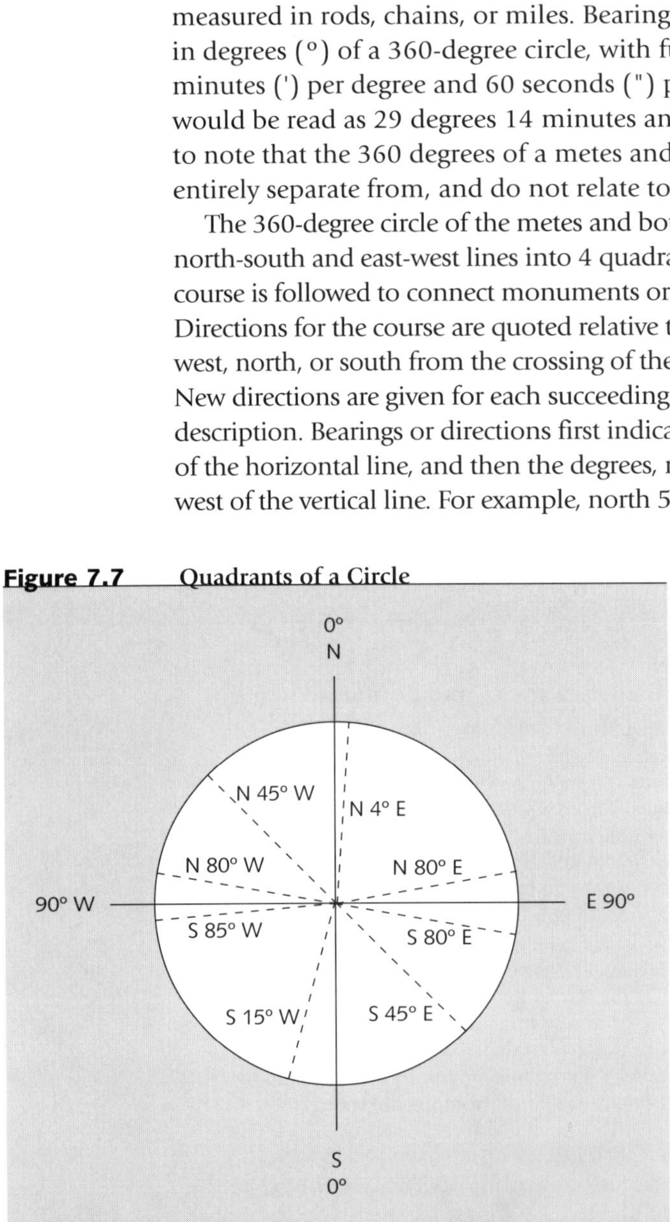


To describe the property in Figure 7.8, start at Point

A—the point of beginning. This P.O.B. is determined by measuring from the Permanent Reference Mark—N80°E, a distance of 152 feet. Imagine that you are in a helicopter looking down on Point A (P.O.B.) with the nose of the chopper pointed north. Then proceed clockwise along the parcel's perimeter. Point B is reached by following the bearing of South 80 degrees East (S80°E) a distance of 180 feet (metes). To get to Point C, travel along the direction (bounds) South 15 degrees West for 160 feet. Point D is reached by traveling from Point C South 85 degrees West for 151 feet. To return to the point of beginning, head North 4 degrees, 10 minutes, 58 seconds East a total of 199.5 feet from Point D.

**Figure 7.8** Metes and Bounds Description

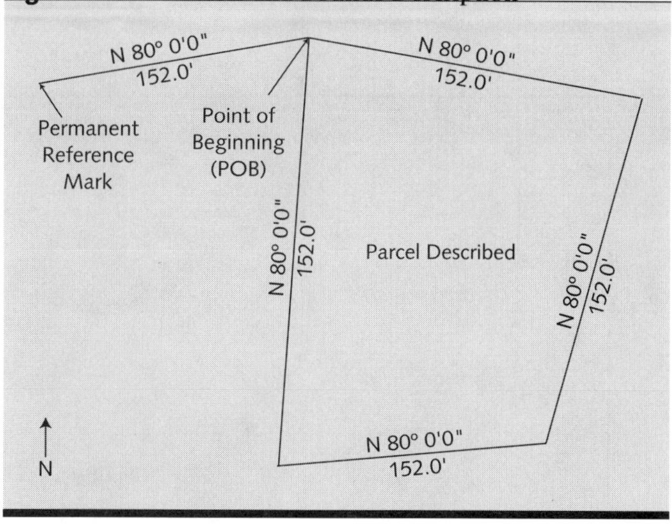


The written legal description for this parcel would first identify the county and state within which it lies and then the permanent marker: "thence N80°0'0"E, 152.0 feet from the point of beginning; thence S80°0'0"E, 180.0 feet, thence S15°0'0"W, 160.0 feet; thence S85°0'0"W, 151.0 feet; thence N4°10'58"E, 199.5 feet to the point of beginning."

Metes and bounds descriptions are used extensively within the rectangular survey system. Because physical features such as rivers, lakes, oceans, highways, railroads, and parcels of excluded land do not conform to the rectangular survey system, a metes and bounds description is required. The description is referenced from the permanent reference mark, which is within the rectangular survey.

## Recorded Plat

When a tract of land is ready for the development of houses or businesses, the recorded plat system is used to divide the tract into blocks and lots. This system is also known as the *lot-block-tract,* the *recorded map,* or the *recorded survey system.* The recorded plat is based on the filing of a surveyor's plat map in the public recorder's office in the jurisdiction in which the land is located. This recording becomes the permanent land description system for that particular subdivision. Referencing properties to a recorded plat is the simplest and most convenient system of land description for blocks, lots, and tracts within a city or subdivision.

Figure 7.9 illustrates a recorded plat map. Note these important features:

**Figure 7.9** Recorded Plat Map

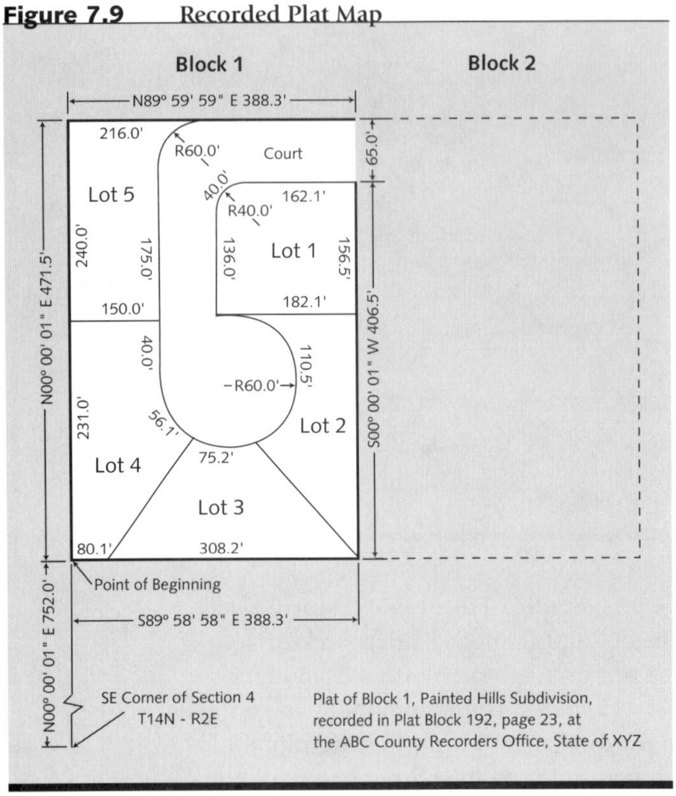


- A metes and bounds survey has been made within a rectangular survey, and this map has been prepared and recorded to show the boundaries of each parcel of land in detail.
- Each parcel is assigned a lot number.
- Each block in the tract is assigned a block number.
- The tract is assigned a name or number prior to being recorded.

Once the requirements of the local political entity have been met, maps showing the entire tract are delivered to the county recorder's office where they are placed in map books or survey books along with the maps (recorded plats) of other subdivisions. Each plat map is given a book and page reference number and all map books are available for public inspection. From that point on, any legal reference to a house or business can be accomplished by identifying the lot and block number, tract name, plat map or book reference, city, county, and state.

With the recorded plat system, a complete, accurate legal description is provided without reference to permanent markers and the lengthy explanation of a metes and bounds description. Note that this legal description is very different from a street address. To find the location and dimensions of a recorded lot, you need only refer to the map book recorded at the county recorder's office.

## Other Methods of Description

Real estate can also be described by referring to a document previously recorded in public records. The document referenced must contain a full legal description of the parcel in question or the parcel to be excluded from the description in question. A previously recorded legal description could be included in a deed, a mortgage, or a similar document. For example, assume that several years ago Mr. Baker received a deed from Miss Adams which contained a long and complicated metes and bounds description. Baker recorded the deed in the public records office, where it became Page 456 of Book 1089. If Baker later wants to deed the land to Mr. Cooper, and Cooper wishes to obtain financing from the bank, the following land description can be used in both the new deed to Cooper and the mortgage to the bank: "All the land described in the deed from Adams to Baker recorded in Book 1089, Page 456, County of ABC, State of XYZ, at the public recorder's office for said county and state." Since the recorder's books are open to the public, anyone could go to Book 1089, Page 456 and find the detailed description of the parcel. Again, this method of legal description can be used to identify the desired parcel or a parcel to be excluded from the description in question.

### Vertical Descriptions

A method of vertical description is required to describe or transfer air rights or subsurface rights. The starting point or surface from which to measure a vertical height or depth is called a *datum*. The most commonly used datum in the United States is mean sea level. A number of political subdivisions have established their own datum surfaces (points) for use in local surveys. From the datum point, benchmarks relative to that point are set by government surveyors. A surveyor need not travel to the datum point to determine an elevation. Rather, the benchmarks, which include the local elevation, are used as metes and bounds reference points. In selling or leasing subsurface drilling (mineral) rights or in restricting air rights, the datum chosen is often the actual surface of the parcel. For example, an oil lease may permit the extraction of oil and gas from a depth greater that 500 feet, but not more than 5,000 feet beneath the surface of the parcel of land.

An air lot or avigation easement, which gives the holder specific rights to a defined space over a given parcel of land, is described by identifying both the parcel of land beneath the air lot and the elevation of the air lot above the parcel. Figure 7.10 illustrates the avigation easement in the following legal description: "An air lot lying between an elevation of 575 feet and 625 feet above sea level over Lot 26, Block 17 of Tanner Subdivision, ABC City in County of DEF, XYZ State."

**Figure 7.10** Avigation Easement

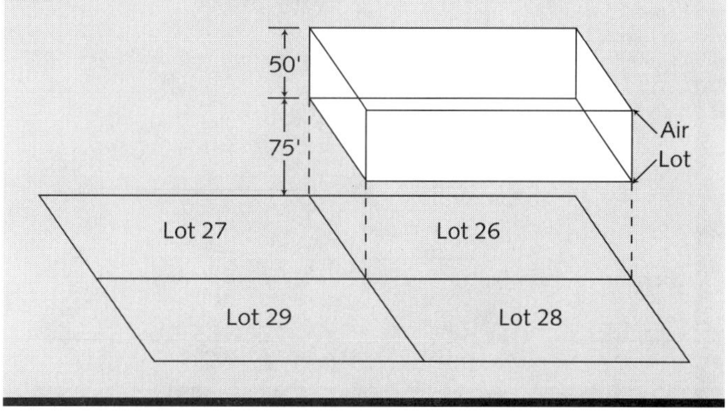


Contour maps, commonly referred to as *topographical maps,* can also be used to describe land.

These maps show contour lines, which connect all points having the same elevation. The purpose of contour maps is to show hills, valleys, slopes and water runoff, etc. A legal description might identify an easement that includes all lands on the subject parcel below the 20-foot contour line. Figure 7.11 illustrates how vertical distances are shown using contour lines.

## Summary

A precise legal description is needed to clearly identify each unique parcel of real property. Proper appraisal technique and USPAP require precise identification of the property being appraised in each appraisal report. The three major systems of land description employed in the United States are the U.S. Rectangular Survey, metes and bounds and the recorded plat systems. Although each system is independent, one system can be used in conjunction with the other two systems.

The U.S. Rectangular Survey System is based on intersecting lines that create an axis as a point of reference; principal meridians run north and south and base lines run east and west. Each principal meridian in the United States has a name or number and is associated with a respective base line. Individual townships are created by the intersection of ranges (strips of land running north to south) and townships (strips of land running east to west). Townships are consecutively numbered north or south and east or west from the axis of the base line and principal meridian. Townships are six miles by six miles, or 36 square miles, in area and include 36 sections. A typical section is one mile square and contains 640 acres.

**Figure 7.11** Contour Lines Show Vertical Distances

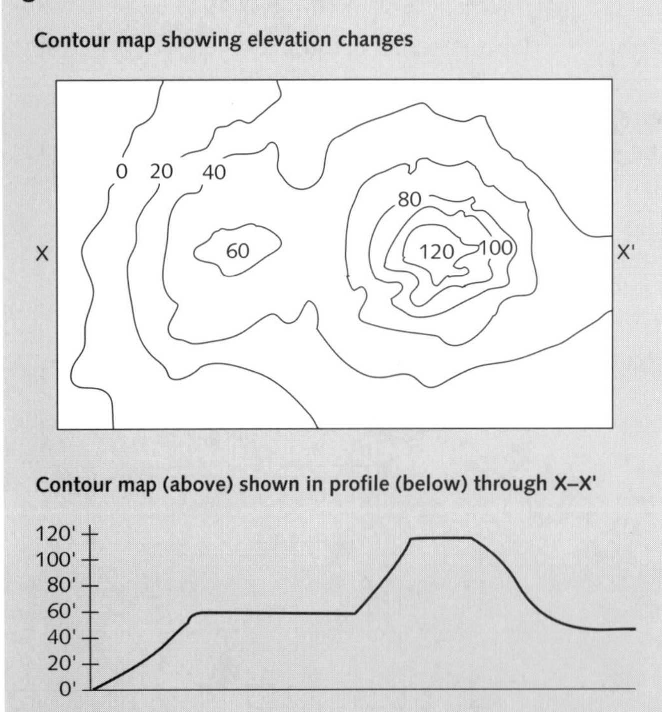


The metes and bounds system of legal description is centuries old. This system uses distances and bearings from succeeding reference points to describe property. Permanent man-made monuments are used with precise descriptions of distance (linear measurement) and bearing (angle measurement) to describe property from the point of beginning. Metes and bounds descriptions are independent for each parcel of real estate. Metes and bounds descriptions are most common in the 13 original colonies, Texas, and Hawaii. This system is often used in conjunction with the rectangular survey to describe real estate along rivers, shores, highways, and other natural and man-made boundaries.

The recorded plat system of legal description relies on a thorough survey of an area of real property that has been divided into platted blocks and lots. Each block and lot is numbered for precise identification. Tract surveys of blocks and lots are recorded for permanent reference at local government offices. The recorded plat system is based on the rectangular survey system but often includes precise surveys developed under the metes and bounds system.

CHAPTER

• 8 •

# *Land Description*

This is the first of three chapters investigating the information the rural appraiser must consider to analyze a rural property properly. This chapter is concerned with the physical aspects of the land that relate to soils and range sites. Site improvements are discussed in Chapter 9 and building improvements are discussed in Chapter 10. All three chapters contain information which must be considered to evaluate a subject property from a physical standpoint.

## Soils and Range Sites

Soil is the basic component of value in a rural property used for the production of agricultural commodities because it is the medium of productive capability. Of course, soil alone will not make a profitable agricultural enterprise; use and management of the soil determine the profit that can be realized from agricultural production. The analysis of soils and their productivity is generally most important in the appraisal of cropland and hayland. In livestock ranching, properties are usually larger and the soil is analyzed less intensively. Rangeland analysis is typically focused on the analysis of the "range sites" found on the ranch. The following discussion is divided into two parts. The first part explores the information to be considered in the soil analysis and the second part concerns the analysis of range sites.

### Soil Characteristics

Every appraisal of a rural property that produces agricultural products should include a full description of the soils and/or range sites on the property and an analysis of their potential uses and productivity. Soil must be evaluated in terms of its potential productivity and desirability, which affect the value of agricultural property. Value may not be directly proportionate to the kinds of soil found on a property, but knowledge of this basic component is essential for all appraisers of rural properties. Although not an expert in soil science or agronomy, the rural appraiser is in contact with experts in these fields. The appraiser gathers this expert knowledge and makes economic sense of it by relating soil information to the earning potential of the rural property. The following discussion explains what soil is and emphasizes the soil characteristics that have the greatest effect on property value.[1]

Soils are natural bodies capable of supporting plants. Soil properties in any one place result from the combined effects of climate and living matter on the parent material (such as rock). The soil development process is influenced by changes in the contours of the land over time. The characteristics of soils and the surrounding cultural environment are important determinants of the uses to which soils are suited. Among the most important soil characteristics are texture, structure, consistence, thickness, and arrangement of the soil horizons, or layers. These characteristics and their interrelationships, together with soil and site features such as depth to bedrock, depth to root-inhibiting layers, available water capacity, drainage, and flooding, give an initial indication of the soil's potential productivity and the type of management that may be required to obtain the greatest productivity from the land.

### Soil Horizons

Knowing the composition of the soil layers can help the appraiser determine how well the soil can support certain plant life. Soil layers, or *horizons,* are three-dimensional areas with certain characteristics produced by soil-forming processes. The horizons are roughly parallel to the soil surface, but they vary in thickness and may have irregular boundaries. For an agricultural appraisal, the top six feet of soil are most significant. The horizons in the top six feet are commonly known as the *surface soil, subsoil,* and *substratum.*

The term *topsoil* has several meanings. It may simply describe fertile soil material, rich in organic matter, that is often used to improve plant growth on road banks, gardens, and lawns. But for the appraiser's purposes, this is referred to as part of the surface soil. The *subsoil* is a layer generally below the surface soil. The surface soil and the subsoil layers together make up the *solum,* the part of the soil where soil formation is most active and most of the soil's plants and animals live. The *substratum* is the part of the soil below the solum. Figure 8.1 illustrates a generalized soil profile.

---

1. The discussion of soil, soil horizons, structure, consistence, and texture in this chapter is based on information from the Soil Conservation Service, *USDA Soil Survey Manual, Soil Survey Staff Handbook No. 18,* (Washington, DC: Government Printing Office, 1951, updated with supplements).

### Texture

An appraisal of agricultural property includes a description of the texture of the surface and subsoil layers. The *texture* of a mineral soil refers to the relative proportions (by weight) of three basic soil particles—sand, silt, and clay. Texture is determined by the weight of the three particles after all coarse fragments (greater than two millimeters in diameter) have been removed. Soil textural class names describe the various kinds of soil textures based on the proportions of clay, silt, and sand. There are 12 basic classes of texture: sand, loamy sand, sandy loam, loam, silt, silt loam, sandy clay loam, clay loam, silty clay loam, sandy clay, silty clay, and clay. Each class has a defined range of clay, silt, and sand content, as shown in the soil triangle in Figure 8.2.

**Figure 8.1** Generalized Soil Profile

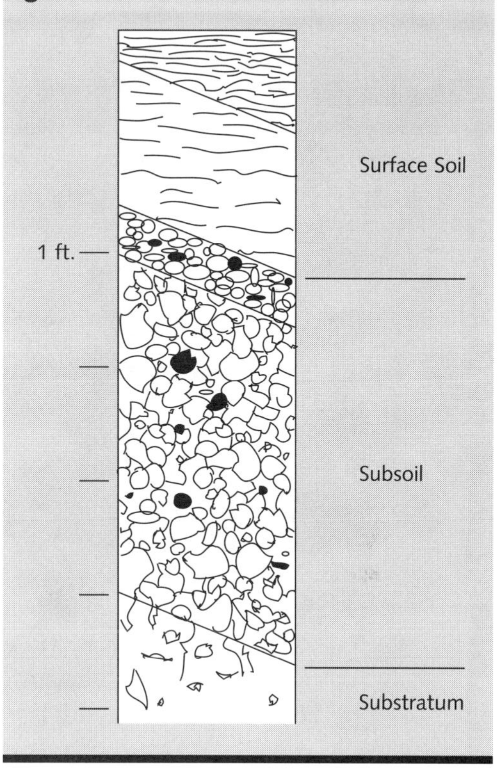


For example, loam is a soil material that has 7%–27% clay, 28%–50% silt, and less than 52% sand. The presence of coarse particles smaller than 10 inches is indicated by modifiers such as *gravelly* sandy loam or *cobbly* loam. A cobbly loam is loam soil that contains fragments 3 to 10 inches in diameter. Larger particles, such as stones or rocks, are defined in terms of the influence they have on soil use. Even if soil is naturally productive, or can be made so through management, stoniness will play a role in determining whether land is used for cropland, rangeland, pasture, or forest.

A soil layer's texture is probably its most stable quality over a long period of time, but there are ways in which texture is altered. For example, heavy and continuous soil blowing during a drought can remove the finer particles from the soil and leave the surface soil coarser. Erosion by wind or water can also change the depth and fertility of the soil. To determine the present fertility and potential of a soil, texture must be considered along with structure, consistence, and other physical and chemical characteristics of each horizon.

### Structure

*Structure,* broadly defined, is the arrangement of primary soil particles into compound particles or aggregates. Very small particles collect together in a mass or cluster known as a *soil aggregate,* which is technically called a *ped.* When scientists discuss soil structure, they are referring to the shape, size, and definition of the natural aggregates. There are four main

**Figure 8.2** Soil Triangle

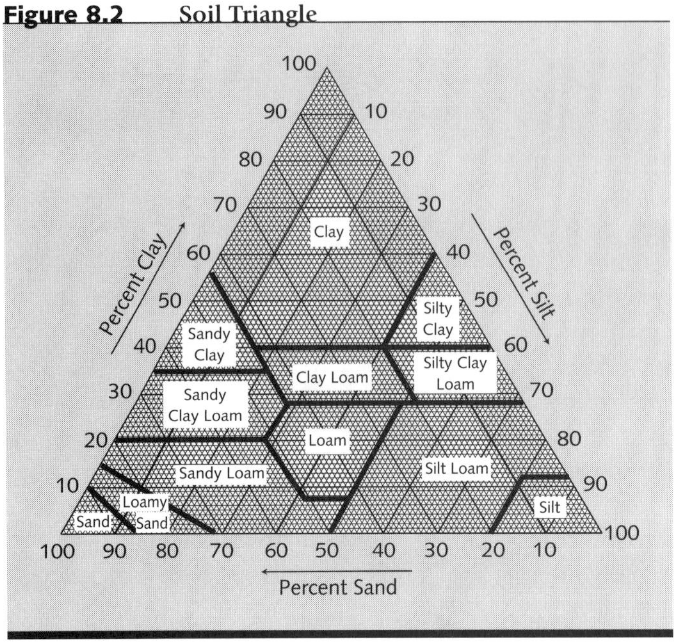


types of structure that describe the shape of the aggregates: platy, prismatic, block-like, and spheroidal. The form of the aggregates has bearing on the productivity of the soil: "Generally, in the United States, soils with aggregates of spheroidal shape have much pore space between aggregates, have more rapid permeability, and are more productive than soils of comparable fertility that are massive or even coarsely blocky or prismatic."[2]

The structure of any soil affects how well plants can grow in it and how well it responds to soil management. A soil's structure can vary by horizon and strongly influences the favorability of the subsoil to root growth. For example, structure affects permeability, i.e., the rate at which water or air can penetrate a soil layer.

### Consistence

Soil *consistence* describes the feel of the soil and how easily a lump of it can be crushed by the fingers. Consistence refers to the nature and strength of the forces that cause soil particles to hold together. Familiar terms used to describe the consistence of a soil include *brittle, crumbly, elastic, fluffy, mealy, mellow, soft, spongy, stiff, tight,* and *tough.* These terms are not used uniformly, however.

The consistence terms used in soil descriptions by Natural Resource Conservation Service soil scientists are defined as follows:

- **Loose-** Noncoherent when dry or moist; does not hold together in a mass.
- **Friable-** When moist, crushes easily under gentle pressure between thumb and forefinger and can be pressed together into a lump.
- **Firm-** When moist, crushes under moderate pressure between thumb and forefinger, but resistance is distinctly noticeable.
- **Plastic-** When wet, readily deformed by moderate pressure but can be pressed into a lump; will form a "wire" when rolled between thumb and forefinger.

2. Ibid., p. 225.

- **Sticky**- When wet, adheres to other material and tends to stretch somewhat and pull apart rather than to pull free from other material. Hard when dry, moderately resistant to pressure; can be broken with difficulty between thumb and forefinger.
- **Soft**- When dry, breaks into powder or individual grains under very slight pressure.
- **Cemented**- Hard; little affected by moistening.[3]

### Fertility

The inherent fertility of a soil depends on the amount of organic matter and plant nutrients that are naturally present in the soil in a form plants can use. Fertility can be altered by past use. For example, a property may have been planted with the same row crop for years with little or no return of plant nutrients and organic matter to the soil, while another property may be virgin soil. Still another property may have been farmed for years, but because farm management has properly rotated crops and carried out other measures, adequate levels of organic matter, nitrogen, phosphorus, potash, and other plant nutrients have been maintained. Both inherent soil characteristics and past land management must be investigated to determine a soil's current fertility level.

### Color

Soil color is another characteristic that the appraiser should not overlook. Typically, the darker the surface soil, the more productive the soil is likely to be; dark soil generally indicates a high level of organic matter. Care must be exercised, however, because in a few soils, soil color simply reflects the color of the parent soil material. While color is an important consideration, conclusions about soil capability cannot be drawn from soil color alone. This single characteristic may suggest a variety of conditions.

### pH

The pH of the soil must also be considered. Strongly alkaline soils with pH levels of 8.5 or higher support limited vegetation; in extreme cases, alkaline can render a soil virtually sterile. At the opposite extreme, high acidity tends to lock up soil nutrients, making them unavailable to plants. Furthermore, some plants are sensitive to high acidity and cannot be successfully grown on acid soils. In most cases, soil alkalinity and acidity can be modified or corrected with the addition of lime (to correct acidity), sulphur (to correct alkalinity), or other soil modifiers. While such soil modification techniques are not typically a direct concern of the appraiser, local practices do affect soil productivity and land value.

---

3. Soil Conservation Service, USDA, in cooperation with Cornell University Agricultural Experiment Station, Soil Survey of Ulster County, New York (Washington, DC: Government Printing Office, 1979).

## Topography, Drainage, and Climate

In addition to the major soil characteristics previously discussed, the appraiser examines several other aspects of the soils and the environment:

### Topography

Topography is an important physical characteristic of the landscape in that it can limit the soil's productive capability or potential uses. Soils that are level, in depressions, or at the base of slopes may require surface drainage and tiling or surface land forming. Sloping areas may be susceptible to erosion, especially if used for row crops, and may require conservation tillage, terracing, a crop rotation program, or other practices, singly or in some combination, to control water erosion and maintain the soil's productivity.

The topography of a property must be seriously considered by the appraiser. In some cases, topography will help identify a soil: slope, an aspect of topography, is a soil characteristic. More importantly, the topography of a property often determines the type of agriculture that the property will accommodate. For example, soils on relatively steep topography in areas of moderate or high rainfall commonly have severe erosion problems. In areas where the soils are used for cultivated crops because of favorable air flow, however, the same soils might accommodate a horticultural enterprise. Sometimes soils on nearly level valley floors have seasonal limitations because of slow air movement or the storage of cold air. Finally, topography suggests the degree of hazard from wind or water erosion. Topography can be described in terms of "percentage of slope," but more frequently appraisers use such terms as *gently rolling, rolling, steep,* and *level.*

### Drainage

Drainage can refer to both the drainage patterns on the landscape and the rate at which water moves through the soil. The relationship between topography and drainage is obvious, but in many areas drainage problems are not revealed by topography alone. Strongly compacted soil layers, which result from natural soil structure, land management, or both, can affect drainage adversely. For example, soil particles can pack together tightly, leaving little room for air or moisture to pass through. Such a subsoil condition, known as *hardpan,* is commonly accompanied by internal drainage limitations that have a serious impact on soil productivity. In areas with no previous history of poor drainage, irrigation combined with certain geologic features can raise the water table to the soil surface and cause drainage problems. In some instances, poor drainage can be corrected through major investments in tiling or land forming. Throughout some parts of the country, drainage districts (much like irrigation districts) have been formed to reclaim or improve land that has poor natural drainage.

Because internal soil drainage can be difficult to measure, the appraiser must be sure to investigate this aspect of the soil very carefully. Soil surveys, which are discussed later in this chapter, physical inspection of the property, and information obtained from the present owner are the best sources of information about the drainage characteristics of soil.

### Climate

Climate and agricultural enterprises are inextricably related. Most successful agricultural enterprises are compatible with the local climate. If they are not, climatic conditions can be mitigated with adaptations such as frost protection, drainage, irrigation, or the growing of tolerant crops. The most important climatic characteristics are rainfall, length of the growing season or frost-free season, temperature, evaporation, sunshine, and winds. The area's susceptibility to weather hazards such as unseasonable frost, hail, early or late snow, hurricane, or drought must also be considered. While total annual precipitation may indicate the desirability of a climate for agriculture, the appraiser should have some knowledge of the rainfall pattern. For example, many agricultural regions in the West experience a warm-season dry spell which necessitates irrigation or the cultivation of early-maturing or drought-resistant crops.

The effects of irrigation cannot be overlooked. For example, drought is generally more serious on sandy soils than on fine-textured soils, but sandy soils often respond well to irrigation. If irrigation water is available or the area receives abundant rainfall, sandy soil will likely have much higher productive capability than it would in an area without irrigation potential or adequate rainfall. Regardless of climate, the rate of internal drainage and the water-holding capacity of a soil have a direct bearing on its productivity.

## Soil Inventory for Appraisal

A soil inventory provides much of the information used in applying the valuation approaches. For example, the soil characteristics of property are considered by the appraiser in making sales comparison adjustments. One kind of soil is judged to be more desirable than another. Consequently, the value the local market ascribes to a given soil can be used in deriving an opinion of value by the sales comparison approach. In the income capitalization approach, the soil of the farm must be considered in relation to potential gross farm production and the operating expenses and level of management required to achieve that potential. The soil of a farm is the principal determinant of its income and expense.

The appraiser makes use of a variety of resources in investigating and inventorying the soils of a rural property. Aerial photographs of the property and the surrounding area are valuable tools as are topographical maps, soil maps, and interpretive maps such as those made to highlight soil productivity. Many kinds of interpretive maps can be prepared to display the suitability or potential of the soils for various uses. Soil surveys and soil productivity maps contain information important to the appraisal of a rural property.

### Soil Mapping

Soil scientists have identified and mapped soils throughout broad areas of the United States. Modern, detailed soil surveys are available for approximately 80% of the United States. The surveys are produced by USDA's Natural Resource Conservation Service in cooperation with the National Cooperative Soil Survey, a joint effort of the USDA and other federal agencies, state agencies including the agricultural experiment stations, and local agencies.

Soil surveys consist of a set of soil maps and a narrative that describes the soils and their suitability for selected uses, including urban and agricultural uses. The text describes soil characteristics such as color, consistence, structure, texture, and thickness of horizons; drainage, topography, climate, and other soil and site characteristics; and the effects of land use and nature on the soils' productivity. Because they describe the properties of major soils found in the area, the soil surveys provide a scientific basis for comparing one property with another. The scale of mapping is large, however, and the maps are not site-specific so the appraiser must still perform an on-site investigation.

In soil surveys, soils are generally identified by a soil series name and a soil map unit designation. The photograph in Figure 8.3 shows a typical page from a USDA soil survey. Note that the soils are identified by a symbol such as DvB, which is keyed in the soil survey to Duval Very Fine Sandy Loam 1% to 3% slope. The appraiser can use soil surveys to ascertain the amount of soil of each type on the subject property and the comparable sale properties. Areas can be measured using a planimeter or various CAD or GIS software programs. An accurate inventory of the soils on the property being valued is essential to the appraisal process.

### Soil Series

Soils that have the same origin and are similar in all major respects except texture of the surface layer are given a name that designates the series to which they belong. The soil series name generally refers to a geographic location or feature and usually indicates the location where the soil was originally classified. Willamette soils, for example, were first identified in the Willamette Valley of Oregon.

The appraiser who is valuing a property that has Willamette silt loam will find in a soil survey a description of the color, structure, consistence, and texture of the horizons; the type of root growth; the pH; and the degree of slope. The survey will also identify what that particular soil series is used for: Willamette soils are used mainly for vegetable crops,

**Figure 8.3** Soil Map

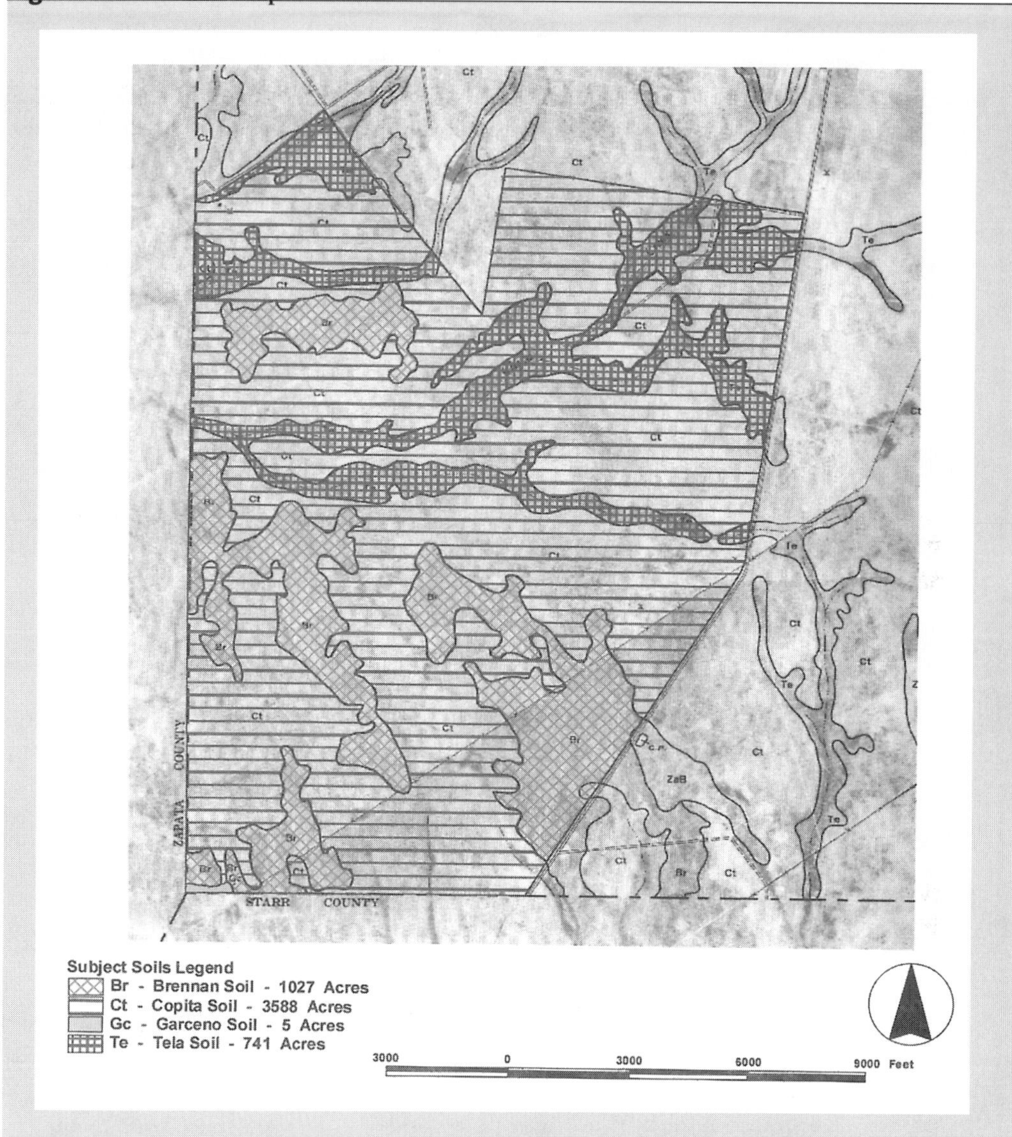


From Page 46, Soil Survey of Frio County, Texas, USDA, Soil Conservation Service, October 1992.

berries, orchards, small grain, hay, and pasture and sometimes for wildlife habitat, recreation, and homesites.

### Productivity Ratings

Productivity ratings are essential to the appraisal of rural properties. Compiled through various methods, productivity ratings summarize the effect of surface soil texture, subsoil characteristics, topography, climate, drainage, soil organic matter, and fertility on the productive capacity of soils. Productivity ratings are used in analyzing the most suitable land use and cropping patterns for a property, and thus are part of the highest and best use analysis of rural properties. Productivity ratings provide a basis for estimating yields and potential income. They are also used as a means for comparing properties and making adjustments for varying productive capabilities.

Soil capability ratings and yield indexes are used to estimate the productivity of all types of rural properties. Yields per acre indicate the productive capabilities of farms, orchards, and vineyards. The productive capability of livestock range is often measured in terms of the animal units (AUs) or animal unit months (AUMs) of forage that each acre can support. Range condition and the corresponding stocking rate can vary considerably, even on the same soil. A range survey showing composition by species and production is needed. The productive capability of timberland may be measured in thousands of board feet (MBF) per acre of usable timber.

Several soil productivity ratings are in use in the United States: the Natural Resource Conservation Service (NRCS) Land Capability Classification System, state soil productivity ratings, Bureau of Reclamation classification, the Storie index, and corn suitability rating. The system most universally applicable to the appraisal of rural property is the NRCS system, which can be found in soil surveys.

### *NRCS Land Capability Classification System*

The most widely accepted land capability classification system is the one developed by the United States Department of Agriculture's Natural Resource Conservation Service. This classification is based on the limitations of the soil rather than its potential yield; it indicates the general suitability of soils for most kinds of field crops. It also indicates the risk of damage to the soil, soil needs, and response to management. NRCS soil surveys indicate the classification of each soil within the system.

The NRCS system was originally designed to help the Natural Resource Conservation Service advise farmers about proper soil management techniques. Although it was not designed specifically for appraisers, it can be extremely useful to them. Because the system is recognized throughout the United States, an appraiser can refer to this system in communicating with a client who is not familiar with the soils in the area where the property being appraised is located.

The NRCS system includes land capability classes, subclasses, and capability units. All soils in one land capability class have similar degrees of limitations and restrictions in use, but they may have different productive capabilities. A subclass designation is used to identify the most important kind of limitation, such as wetness or climate. Finally, the capability unit identifies soils within a subclass that require similar management, have similar risks, and have similar productivity.

**Land Capability Classes.** Capability classes are designated by the roman numerals I through VIII. As the class numbers increase, greater limitations and a narrower choice of practical uses are indicated:

- Class I soils have few limitations that restrict use.
- Class II soils have moderate limitations, which reduce the choice of plants that can be grown or indicate the need for moderate conservation practices.
- Class III soils have severe limitations, which reduce the choice of plants, indicate the need for special conservation practices, or both.
- Class IV soils have very severe limitations, which restrict the choice of plants, indicate the need for very careful management, or both.
- Class V soils generally are not considered suitable for cultivation. These soils are not likely to erode, but they have other limitations that are impractical to remove and limit their use.
- Class VI soils have severe limitations that make them generally unsuited for cultivation.
- Class VII soils have very severe limitations that make them unsuited for cultivation.
- Class VIII soils and land forms have limitations that nearly preclude their use for commercial crop production.

**Capability Subclasses.** Subclasses are soil groups within one class; they are designated by adding a small letter *e, w, s,* or *c* to identify the most important soil limitation.

- Subclass *e*. The letter *e* indicates that the main limitation is the risk of erosion.
- Subclass *w* - The main limitation indicated is excess water in or on the soil due to poor soil drainage, wetness, a high water table, or overflow.
- Subclass *s* - The soil is limited mainly because it is shallow, droughty, or stony or has another soil problem in the rooting zone, such as a low moisture-holding capacity, salinity, or alkalinity.
- Subclass *c* - The chief limitation is a climate that is too cold or too dry.

**Capability Units.** In some local areas, soil capability units are used to identify soils that have similar management requirements, similar risk, and similar productive capability. For example, two soils in subclass IIe may be classed IIe-l and IIe-2 to distinguish the degree of management needed to control the erosion problem.

### *State Soil Productivity Ratings*

Several state agricultural colleges have developed soil ratings that are used in their respective states. Some rating systems attempt to relate soil productivity to the yields of specific crops, while others assign a numerical value, or a fertility index, to soils. In areas where these rating systems are widely used, they can help the appraiser estimate soil productivity.

### *Bureau of Reclamation Classification*

The Bureau of Reclamation land classification system is used to determine the suitability of lands for sustained irrigation farming. It is a systematic classification that divides land into categories based on its characteristics. Suitability is determined by correlating the land's potential productive capacity with the anticipated cost of production and the land's adaptability to a wide range of crops.

### *The Storie Index*

The Storie index is a numerical soil rating system that has been used widely in California. The index expresses the soil's relative suitability for general intensive agriculture at the time of evaluation. The rating is based on soil characteristics only and is derived by evaluating factors such as soil depth, surface texture, subsoil characteristics, drainage, salt and alkali content, and relief. Not considered are factors such as the availability of water for irrigation, climate, and distance to markets, which might determine the desirability of growing certain crops in a given locality. By itself, the index is not a direct indicator of land value. Where economic factors are known to the user, however, the Storie index provides additional, objective information for comparing the value land tracts.

Four general factors are considered in the Storie index rating. These factors are:

A. The characteristics of the soil profile and the depth of the soil

B. The texture of the surface

C. The dominant slope of the map unit

X. Other, less permanent factors that are more readily subject to management or modification, such as drainage, flooding and nutrient level

For some soils, more than one X factor is used. Each of the four general factors is evaluated in relation to a basis of 100%, which expresses the most favorable or ideal condition for general crop production. Lower percentage ratings are assigned for conditions that are less favorable. Factor ratings are selected from tables prepared originally from data and observations that relate soil properties, plant growth, and crop yield.

The Storie index tends to be somewhat rigid and assumes that each of the factors contributes equally to the productivity of the soil, which is seldom the case. Nevertheless, the index can be a useful guide in estimating soil productivity.

### *Corn Suitability Rating or Corn Productivity Rating*

In some Midwestern states, the corn suitability rating (CSR) is used to estimate a soil's capacity to produce primary crops. This system rates each soil on a property and converts the individual ratings into a single index. Where this system is used, it provides a method for making easy and rapid comparisons between farms.

*Applicability of Rating Systems*
Suitability ratings provide a relative ranking of all soils mapped in a county based on their potential for crop production. The CSR is an index that can be used to rate one soil's potential yield production against another's over a period of time. Ratings range from 100, for soils that have no physical limitations, occur on minimal slopes, and can be continuously row cropped, to as low as 5, for soils with severe limitations for row crops.

Appraisers can make excellent use of any soil capability rating available if it is used with a clear understanding of the assumptions and limitations of the rating system. For example, the NRCS land capability classification system assumes improved farm management techniques, while appraisers have traditionally focused on estimated yields under typical management. However, because the market for agricultural properties now seems to be composed primarily of managers with superior management capabilities, the assumptions of the NRCS system are compatible with the rating the appraiser needs to derive.

## Range Site Characteristics

The carrying capacity for a rural property composed partly or wholly of native range land is estimated by analyzing the range sites that exist on the property and then verifying the historical stocking rates and grazing practices through investigation. The appraiser is charged with estimating the carrying capacity of the subject property as well as the comparable sales and rentals which form the basis for valuation. A thorough understanding of range sites is essential to estimate the productivity of rangeland properties properly. A range site is defined as

> a distinctive kind of rangeland that differs from other kinds of rangeland in its ability to produce a characteristic natural plant community. A range site is the product of all the environmental factors responsible for its development. It is capable of supporting a native plant community typified by an association of species that differs from that of other range sites in the kind or proportion of species or in total production. Range sites are named to help users recognize and remember the significant kinds of rangeland in their locality. They are based on such permanent physical features as the kinds of soil, climate, topography, or a combination of these features. [4]

This definition of range site has two parts that are critical to understanding of the range site concept.

- A range site is a distinctive kind of rangeland that differs from other kinds of rangeland in its ability to produce a characteristic natural plant community.

---

4. *National Range Handbook.* USDA, SCS. July 1976

- A range site supports a native plant community typified by an association of species that differs from that of other range sites in the kind or proportion of species or in total production.

The other factors detailed in the definition contribute to the makeup of a range site but, from the perspective of the appraiser, the key is understanding that a range site is unique unto itself and that its uniqueness is due not only to the soils of the site, but also to the climate and topography.

One of the leading research scientists in the field of rangeland management is George Ruyle, Rangeland Management Specialist at the University of Arizona. Ruyle has developed much of the current thinking on the range site concept. The following excerpt from *Rangeland Management, The Range Site Concept - Rangeland Site Potential* by Ruyle provides insight into the development of range site concepts.

> The development of the site concept in range management, range sites, has evolved over the last forty or fifty years. The general concept of range sites is that they are the kinds of rangeland with different soil and vegetation characteristics having different potential productive capacities. Production was originally forage production or grazing capacity but has now been expanded to mean total vegetation production.
>
> Range sites have most often been determined by the study of climax community plant species composition and production. The climax community is assumed to be the plant species assemblage that most nearly achieves long-term steady productivity, structure and composition on a given site. This plant community is primarily an expression of environmental influences and specific plant tolerances of those influences like soil moisture or fertility or competition from other species.
>
> Many studies have quantified plant-soil-climate relationships where minimum human disturbance was thought to have occurred. These studies showed that particular kinds of plant communities reappear wherever equivalent environmental conditions occur. Defining and naming these units is the basis of range site description. General techniques for selection and evaluation of plant associations/range site descriptions include initial reconnaissance, followed by stand selection, detailed analysis, and statistical applications. Areas are selected for sampling based on extent and homogeneity of the soil and vegetation. Interpretations are then made for other areas along these soil-vegetation-climate gradients where studies were not conducted (extrapolation of data).
>
> In practice, the resulting range site delineations are arbitrary landscape divisions used to provide order to a complex system for primarily management purposes and extrapolation of research results to regional areas. The major use of the range site concept has been for estimation of livestock carrying capacities and analysis of range condition; however, the significance of the concept includes its ecological relevance.

With this basic background on the concept of range sites, an appraiser can proceed to develop the data necessary to analyze a property and estimate its carrying capacity in the appraisal process. Approaching the process from a methodology standpoint seems to be the most practical way to demonstrate the analysis of range sites and the development of estimated carrying capacity. In addition to understanding the concept of range sites, the appraiser must possess a general familiarity with the plant community in the region to be analyzed and some understanding of the impact of wildlife on the livestock carrying capacity of a property. The appraiser must be able to identify the major forage species of grasses, forbs, legumes, and woody plants to evaluate range condition with sufficient accuracy to estimate the carrying capacities of both the subject and the sale properties.

In many areas of the West, the estimation of carrying capacity on public lands is determined by the public entity that controls the land asset. However, for land owned in fee simple, the appraiser must assess the range condition and estimate the carrying capacity of the property based on an analysis of the range sites. The accuracy of the estimate should be verified with historical operating data if available.

## Range Site Analysis

Range sites are identified in most jurisdictions through the use of the county soil surveys prepared by the USDA over the last 40 years. Most of the soil surveys for states with rangeland assets include extensive data on the range sites found in the area, including physical descriptions; forage composition by percentage of species; condition indicators; and minimum, average, and maximum forage production in pounds of air-dried forage to be produced annually by a given site. For counties that do not have published soil surveys, the local NRCS office will generally have various range site descriptions applicable to the local area. The information found in the local office's technical guide provides an extensive description of each range site in the area, including the forage species, annual vegetation produced, recommended initial stocking rates, and recommended rates for stocking on improved rangeland as a result of mechanical or other range improvement practices.

A critical part of the property analysis and, ultimately, the estimation of carrying capacity is the ability of the appraiser to assess range condition properly. Range condition is defined as "the state of health or productivity of both soil and forage in a given range, in terms of what productivity could or should be under normal climate and the best practical management. The condition classes generally recognized are excellent, good, fair, and poor. The classification is based on the percentage of original or climax vegetation on the site, as compared to what ought to grow on it if management were good."[5] Thus, range condition is a function of management, not weather. Even if a site is in excellent condition, i.e., with 100% of the climax vegetation present listed as appropriate for the site, its condition can be reduced drastically by poor or inappropriate management. The four classes of range conditions—excellent, good, fair, and poor—are assigned the numerical values shown in Table 8.1.

A layman's assessment of range condition is usually based on whether the range forage produced looks lush, due to normal or above-normal rainfall, or minimal, due to drought or low rainfall. This view is

5. Soil Survey of Jim Hogg County, Texas, 1974, p. 41.

**Table 8.1** Classes of Range Conditions

| Condition | Percent of Climax Vegetation by Species Present on Site as Compared to What Should Be on Site in Climax Condition |
|---|---|
| Excellent | 76% to 100% |
| Good | 51% to 75% |
| Fair | 26% to 50% |
| Poor | 0% to 25% |

in error, however, since range condition is a function of management as opposed to weather. A range site that is appropriately managed can be in excellent condition even in the middle of a drought because the ratio of climax vegetation present to the optimum vegetation is the measure of condition. Certainly, in periods of drought or other adverse climatic conditions, the total annual yield of air-dry vegetation on a site may be very small. Nonetheless, as long as the proper ratios of plant species are maintained, the condition may remain excellent or good. If the ratios are not maintained, the condition is likely to be fair or poor.

Livestock and wildlife tend to graze or browse selectively, repeatedly grazing or browsing those plant species they like most. Continuous, heavy grazing of potential plant communities for extended periods reduces the communities' vigor and ability to reproduce. If such practices are permitted to continue, retrogression occurs and many, if not all, of the most desirable species will be eliminated from the site. The plants first eliminated are usually the tall and midgrasses, most desirable forbs, best legumes, and best browse. These most desirable plants are referred to as *decreasers,* since they tend to decrease as grazing pressure continues. As the population of decreasers begins to decline, the percentage of secondary species that are less palatable tend to increase in the plant community. These less desirable species are called *increasers,* since they tend to increase as the more desirable species decrease.

The terms *decreasers* and *increasers* describe the site's initial response to continuous heavy use. The type of use must be analyzed to identify the source of the grazing pressure and institute appropriate management actions to stabilize the site. As an example, a midgrass or a tall grass that has been decreased from cattle grazing may increase when the site is grazed by sheep and goats.

When improper management techniques are permitted to continue, the population of both *increasers* and *decreasers* will eventually be eliminated. When this occurs, the plants that replace the native climax plants are not native to the site and would not normally be found on the site. These plants are called *invaders.* Invaders usually have special adaptations, such as spines, burrs, toxins, low palatability, or very slow growth, which prevent overgrazing by animals. Common examples of invaders are mesquite, juniper cedar, and cactus.

The process of retrogression described above may occur on a given site or sites and is purely a function of management (grazing practices). All range sites respond to good management; some respond faster than others and all respond faster in favorable years. However, the bottom

**Figure 8.4** Subject Property Layout

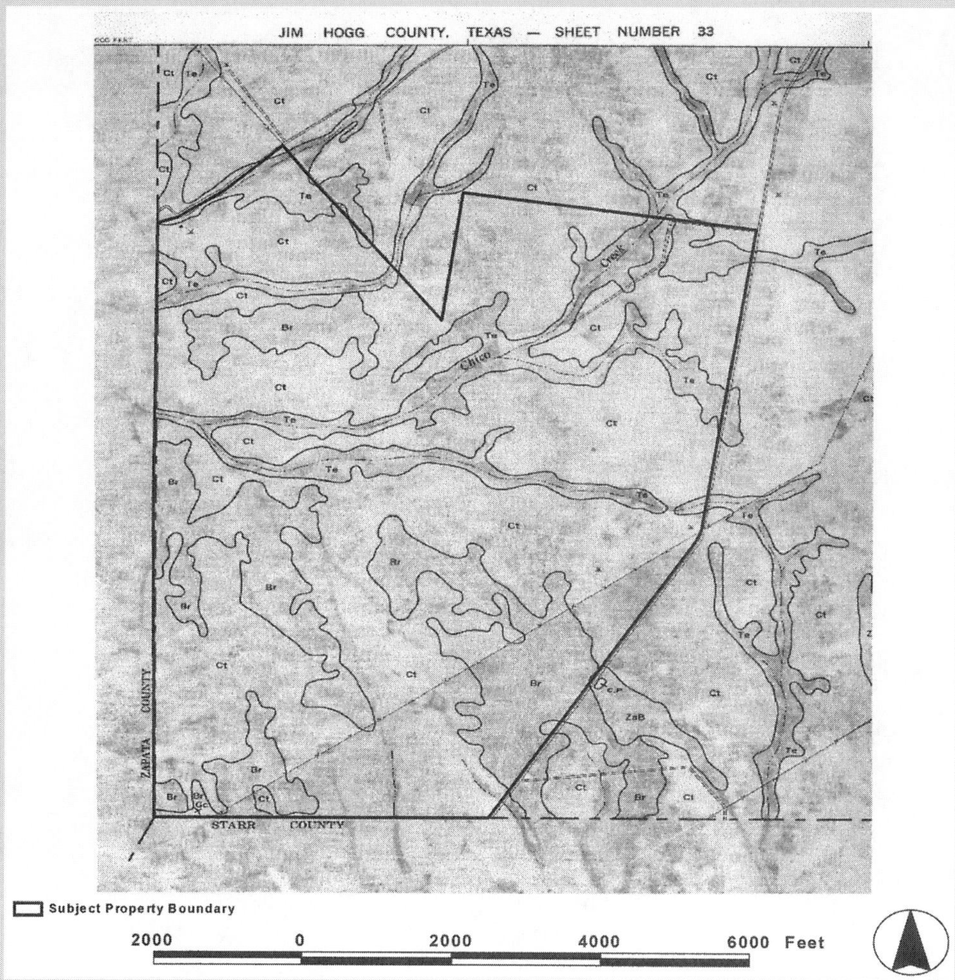


line is that no favorable climatic condition can prevent deterioration in range condition if management practices are inappropriate.

Figure 8.4 shows Sheet 33 of the soil survey of Jim Hogg County, Texas. The boundaries of the subject property are indicated with heavy black lines; the lighter lines are the soils boundaries. The subject property encompasses approximately 5,361 acres. The soils found on the subject property are indicated in Table 8.2.

The soils on the subject property can be measured by planimeter or by digitizing using a GIS or CAD program on a personal computer. The subject soils were mea-

Table 8.2

| Map Symbol | Soil Map Unit Name | Total Acres |
|---|---|---|
| Br | Brennan Soil | 1,027 |
| Ct | Copita Soil | 3,588 |
| Gc | Garceno Soil | 5 |
| Te | Tela Soil | 741 |

**Figure 8.5** Soil Map

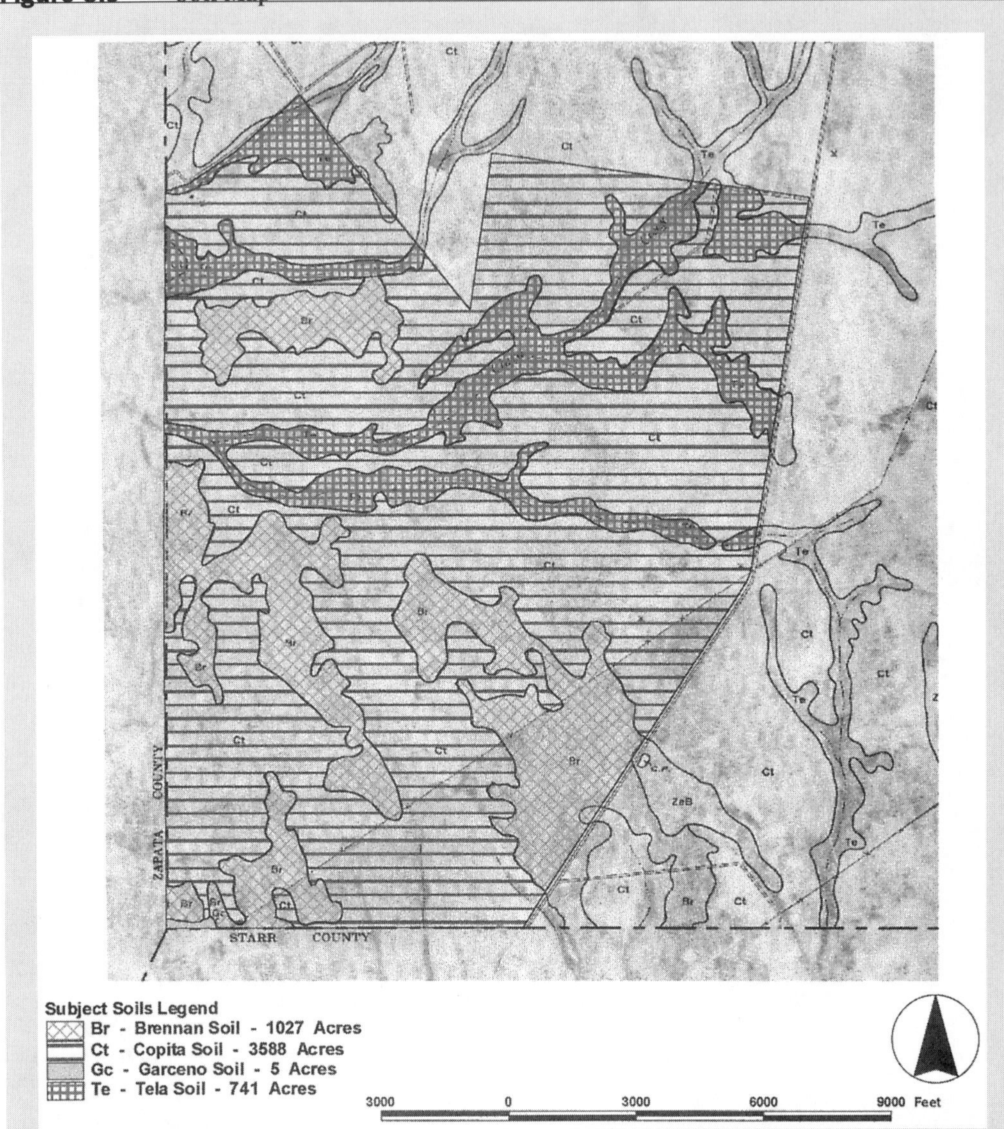


Table 8.3

| Map Symbol | Soil Map Unit Name | Total Acres | Range Site Name |
|---|---|---|---|
| Br | Brennan Soil | 1,027 | Sandy Loam |
| Ct | Copita Soil | 3,588 | Gray Sandy Loam |
| Gc | Garceno Soil | 5 | Clay Loam |
| Te | Tela Soil | 741 | Ramadero |

sured and are displayed on the soils map in Figure 8.5, which indicates the location and dispersion of the soil types on the subject property.

The Jim Hogg County soil survey categorizes the soils on the subject into range sites as indicated in Table 8.3

The individual range site descriptions that follow are from the soil survey of Jim Hogg County, Texas.

**Sandy Loam Range Site**

The soils of this site are Brennan and Hebbronville soils which are nearly level to gently sloping and gently undulating. These soils are well-drained, deep, noncalcareous, and loamy to sandy. They are moderately to moderately rapidly permeable, and their available water capacity is high.

The climax plant community consists of open grassland dotted by ebony trees. Grasses that make up about 65 percent of the climax vegetation are Arizona cottontop, four-flower trichloris, plains bristlegrass, lovegrass tridens, and tanglehead.

Other grasses that account for about 35 percent of the plant community are hooded windmillgrass, Texas bristlegrass, knotroot panicum, slim tridens, fall witchgrass, sand dropseed, pink and whiplash pappusgrass, and slender grama.

This Sandy Loam range site also grows some palatable forbs and woody plants such as bushsunflower, orange zexmenia, daleas, prairie-clover, ephedra, and skeleton-leaf goldeneye. Most of these plants are preferred forage plants for deer. Ebony trees and mottes dot the landscape and add to the attractiveness of the site.

Common invaders on this site are red grama, red lovegrass, whorled dropseed, Halls panicum, and three-awn. As the site deteriorates from its potential, single-stemmed mesquite trees and many kinds of chaparral brush move in and dominate. Among these invaders are spiny hackberry, whitebrush, wolfberry, condalia, amargosa, allthorn, pricklypear and tasajillo, desert yaupon, and others.

Where the site is in fair and poor condition, during years of adequate winter moisture, the site produces a lush crop of annual weeds such as talloweed and bladderpod. These annuals produce palatable and nutritious grazing for short periods in the spring. Growth of more dependable perennial forage plants, however, is retarded following competitive productive periods by annuals.

This site provides a good habitat for deer, javelina, and quail if sufficient cover is available.

Where this site is in excellent condition, the potential acre yield of air-dry herbage ranges from 3,500 pounds in wet years to 2,000 pounds in dry years.

**Gray Sandy Loam Range Site**

The Copita soils are the only soils in this range site. These soils are deep, calcareous, and loamy. They are nearly level to gently sloping, well drained, and moderately permeable. The available water capacity is high.

The climax plant community consists of grasses such as Arizona cottontop, plains bristlegrass, and feathery bluestem. These and grasses such as trichloris, lovegrass tridens, and tanglehead, make up a little more than 50 percent of the vegetation.

Other grasses that make up nearly half of the climax vegetation are hooded windmillgrass, Texas bristlegrass, pink and whiplash pappusgrass, knotroot panic, slim tridens, fall witchgrass, and hairy grama.

The site grows some palatable woody shrubs, such as kidneywood, ratany, guajillo, ephedra, and bumelia. These are preferred forage plants for deer, and this site commonly supports a number of deer. Ebony is a climax tree on the site.

Any deterioration results in a loss of climax plants and an invasion of red grama, whorled dropseed, Halls panicum, red lovegrass, red three-awn and many annual grasses and weeds. Many kinds of woody plants also invade the site. Among these are mesquite, blackbrush, pricklypear, wolfberry, spiny hackberry, condalia, amargosa, and others.

Where this site is in excellent condition, the potential acre yield of air-dry herbage ranges from 3,500 pounds in wet years to 1,000 pounds in dry years.

**Ramadero Range Site**

The Tela soils are the only soils in this range site. These nearly level soils are along small, intermittent drainage ways and are well drained and moderately permeable. They are deep, noncalcareous, and loamy. In some places the site has a mantle of sandy loam or silty clay loam overwash from surrounding soils. The site receives runoff from adjacent higher lying soils, and the available water capacity is high.

The Ramadero site is a preferred grazing site. The potential plant community on this site is an open grassland dominated by four-flower trichloris. Decreaser plants that compose about 70 percent of the climax vegetation are, in addition to trichloris, plains bristlegrass, Arizona cottontop, and lovegrass tridens. Big cenchrus and Wright's sacaton grow on the banks of some of the larger drainage ways. Other grasses that make up about 30 percent of the climax plant community are Nash windmillgrass, pappusgrass, buffalograss, and Texas bristlegrass.

Where this site is in deteriorated condition, invaders such as tumble windmillgrass, whorled dropseed, red grama, Halls Panicum, three-awn, ragweed, and prairie coneflower are common on the site. In the lower condition classes, the site commonly grows a dense stand of large mesquite trees and rank underbrush such as spiny hackberry, white brush, retama, huisache, and condalia.

Where this site is in excellent condition, the potential acre yield of air-dry herbage ranges from 5,000 pounds in wet years to 2,500 pounds in dry years.

**Clay Loam Range Site**

The Garceno soils are the only soils in this range site. These nearly level to gently sloping soils are deep, calcareous, and loamy. They are well drained, are moderately permeable, and have high available water capacity.

Originally this site was open grassland broken by scattered mesquite and chaparral. Among the grasses that make up about 65 percent of the potential plant community on this site are four-flower trichloris, Arizona cottontop, and lovegrass tridens. Other grasses that account for about 35 percent are pink pappusgrass, plains lovegrass, curly mesquite, and slim tridens.

As deterioration of the range site occurs, woody vegetation increases. The principal invading plants are red grama, three-awn, annual forbs and grasses, mesquite, white brush, condalia, spiny hackberry, and pricklypear.

Where this site is in excellent condition, the potential acre yield of air-dry herbage ranges from 3,000 pounds in wet years to 1,500 pounds in dry years.

## Interpretation of the Range Site Descriptions

Note that each range site description includes a brief description of the terrain and topography, a forage inventory guide by percentage, and the annual air-dried forage production when in climax condition. Additionally, a guide is provided to alert the user to signs of deterioration in range condition.

Analysis of the sandy loam range site reveals that the forage on the site should consist of 65% Arizona cottontop, four-flower trichloris, plains bristlegrass, lovegrass tridens, and tanglehead with another 35% of hooded windmillgrass, Texas bristlegrass, knotroot panicum, slim tridens, fall witch-grass, sand dropseed, pink and whiplash pappusgrass, and slender grama.

As range condition declines, common invaders on this site are red grama, red lovegrass, whorled dropseed, Halls panicum, and three-awn. As the site deteriorates from its potential, single-stemmed mesquite trees and many kinds of chaparral brush move in and dominate. Among these invaders are spiny hackberry, whitebrush, wolfberry, condalia, amargosa, allthorn, pricklypear and tasajillo, desert yaupon, and others.

The total annual yield of air-dry forage on this site in excellent condition is 3,500 pounds, 2,000 pounds in dry years, or 2,750 pounds on average.

With these data the appraiser can begin to estimate the indicated carrying capacity of the subject or sale property being analyzed.

### Carrying Capacity Estimation

Developing an estimate of the carrying capacity of a property may be undertaken in a relatively simple series of steps.

1. Identify each range site on the property and estimate the acreage occupied by each.
2. Obtain the appropriate range site description, which specifies the mix of climax vegetation proper for the site.
3. Inspect the site, inventorying the vegetation found to ascertain the extent to which the existing plant community differs from the potential plant community.
4. Estimate the total annual yield of forage on the site based on the range condition observed in the previous step.
5. Calculate the carrying capacity of the property.

Estimating carrying capacity requires a working knowledge of the most common species of forage found on a given site or sites, an understanding of past grazing practices, if available, including the type of livestock; and information on the appropriate management practices applicable to the site. The calculation of carrying capacity is expressed in terms of animal units per month or animal units as described in Chapter 19, Livestock Ranches. Typically, when analyzing a cattle grazing operation, one animal unit is considered equivalent to 1,000 pounds of body weight, the approximate weight of a range cow and calf to weaning. The grazing requirement to maintain an animal unit in a cattle grazing operation ranges from 26 to 30 pounds of air-dried vegetation per day. The precise number of pounds is dependent on local or regional conditions, the nature of the forage produced, and local custom.

Once the first four steps described above have been completed, carrying capacity can be calculated. Assuming 26 pounds of air-dried vegetation is required on a daily basis, the total annual requirement to sustain and allow livestock to reproduce is calculated as 26 pounds daily × 365 days, or 9,490 pounds. Table 8.4 lists the factors used in developing the equation to estimate carrying capacity:

**Table 8.4**

| Symbol | Description | Factor |
|---|---|---|
| AADV | Annual potential air-dry vegetation on site | Pounds |
| GMP | Grazing management practice: take half, leave half of the vegetation on site. | 50% |
| RC | Range condition:<br>76% to 100% = excellent;<br>51% to 75% = good;<br>26% to 50% = fair;<br>0% to 25 % = poor | 0% to 100% |
| PU | Percent of vegetation utilized by livestock | 0% to 100% |
| WD | Vegetation walked down by livestock in the grazing process | 10% to 30% |

These factors are entered into the following equation to estimate the number of acres required to support one animal unit on an annual basis:

Acres per AU annually = 365 days/((AADV × GMP × RC × PU × WD)/26 pounds)

Using the sandy loam range site and the assumptions listed in Table 8.5, the following indicated carrying capacity can be calculated:

Acres per AU annually = 365 days/((AADV × GMP × RC × PU × WD)/26 pounds)
Acres per AU annually = 365 days/((2,750 × .50 × .35 × .95 × .80)/26)
Acres per AU annually = 25.95 acres annually, say 26 acres per animal unit

**Table 8.5**

| Symbol | Sandy Loam Range Site—Jim Hogg County |
|---|---|
| AADV | Total annual forage 3,500 to 2,000 lbs. yearly, average (say) 2,750 |
| GMP | 50%—take half of the forage and leave half for seed stock |
| RC | Estimated range condition 35% by observation and transect |
| PU | 95% utilized by livestock |
| WD | 20% of vegetation walked down by livestock or otherwise not usable—i.e. 80% usable after walk down accounted for on site |

By analyzing each range site on a ranch property, the appraiser can produce estimated stocking rates using the procedures described. In actual practice, it may be necessary to calculate carrying capacity estimates separately for each pasture or similar division when range sites are diverse and productivity varies extensively. The appraiser must also keep the seasonal balance of the property in focus.

In many instances the NRCS may have a cooperation agreement with the property owner and, with the owner's permission, the appraiser may use the calculations recommended in the NRCS plan if the range condition is the same as described in the plan. In any event, the historical carrying capacity of the ranch must be ascertained and compared to the estimated carrying capacity. When the difference is large, it is likely that supplemental feeding or planted grazing was provided to enhance the carrying capacity of the property.

The rural appraiser who engages in the appraisal of livestock ranches needs at least a working knowledge of the foregoing concepts related to range site analysis. Only through the processes described can realistic stocking rates be ascertained and verified for use in the appraisal process.

## Summary

### Soils

Soil is the basic component of value in rural properties used in the production of agricultural commodities because it is the medium of productive capability.

A working knowledge of soils and range sites is required to estimate the value of a rural property properly. Soil must be evaluated for its effects on the potential productivity, desirability, and therefore value of the agricultural property.

Texture, structure, consistence, and the arrangement of the soil horizons are important soil characteristics. Soil layers, or horizons, are three-dimensional areas with certain characteristics produced by soil-forming processes. The composition of the layers of the soil affects the soil's ability to support certain plant life. The texture of a mineral soil refers to the relative proportions of three basic particle sizes—sand, silt, and clay—and soil textural class names are based on proportional combinations of sand, silt, and clay. Soil structure is the arrangement of primary soil particles into compound particles or aggregates. The shape, size, and definition of the natural aggregates are important elements in describing soil structure. Terms for soil consistence describe the feel of soil and how easily a lump of it can be crushed by the fingers.

In addition to these characteristics of soil, the appraiser examines other aspects such as soil fertility, soil color, soil pH, topography, drainage, and climate. The basic fertility of a soil depends on the organic matter and nutrients available in a form plants can use, but fertility can be either enhanced or depleted by past management. Soil color can indicate productivity; generally dark soil contains a high level of organic matter, but other factors must be considered also. The pH of a soil—i.e., its degree of alkalinity or acidity—influences its productivity.

The topography of the physical landscape can limit the soil's use or productive capability. Topography can be described in terms of the percentage of slope, or as gently rolling, rolling, steep, level, etc. Drainage can include the drainage patterns on the landscape and the rate at which water moves through the soil. The most important climatic characteristics are rainfall, length of the growing season or frost-free season, temperature, evaporation, sunshine, and winds.

Aerial photographs, topographical maps, soil maps, and interpretive maps can be useful in appraisal. Soil surveys, which include soil maps and narrative text, provide detailed information about the soils in a particular area. These surveys are available from the Natural Resources Conservation Service of the USDA.

The appraiser makes extensive use of productivity ratings in the appraisal of rural properties. Productivity ratings help in analyzing the most suitable land use and cropping patterns; provide a basis for estimating yields and potential income; and are useful in comparing properties and making adjustments. The primary productivity ratings in use are the NRCS land capability classification system, state soil productivity ratings, Bureau of Reclamation classification, Storie index, and corn suitability rating.

### Range Sites

The carrying capacity of a rural property composed partly or wholly of native rangeland is estimated by analyzing the range sites that exist on the property and then verifying the historical stocking rates and grazing practices through investigation. The appraiser should estimate the carrying capacity of the subject property as well as the comparable sales and rentals used in the valuation.

The definition of a range site has two parts: 1) a range site is a distinctive kind of rangeland that differs from other kinds of rangeland in its ability to produce a characteristic natural plant community, and 2) a range site supports a native plant community typified by an association of species that differs from that of other range sites in the kind or proportion of species or in total production.

Range sites are identified in most jurisdictions through the use of the county soil surveys, which have been prepared by the USDA for 40 years. Most of the soil surveys for states with rangeland assets include extensive data on the range sites found in the area, including physical descriptions; forage composition by percentage of species; condition indicators; and minimum, average, and maximum forage production in pounds of air-dried forage to be produced annually by a given site. In counties that do not have published soil surveys, the local NRCS office will generally have various range site descriptions applicable to the local area. The information found in the local office's technical guide provides an extensive description of each range site in the area, including the forage species, annual vegetation produced, recommended initial stocking rates, and recommended rates for stocking on improved rangeland as a result of mechanical or other range improvement practices.

Each range site description includes a brief description of the terrain and topography, a forage inventory guide by percentage, and the annual air-dried forage production when in climax condition. Additionally, a guide is provided to alert the user to signs of deterioration in range condition.

To develop an estimate of the carrying capacity of a property, the appraiser follows a relatively simple series of steps. Estimating carrying capacity requires a working knowledge of the most common species of forage found on a given site or sites, an understanding of past grazing practices and the type of livestock, and the appropriate management practices applicable to the site.

By analyzing each range site on a ranch property, the appraiser can produce estimated stocking rates using the procedures described. In actual practice, it may be necessary to calculate carrying capacity estimates separately for each pasture or other division of land when range sites are diverse and productivity varies extensively. The appraiser must also consider the seasonal balance of the property.

In many instances, the NRCS has a cooperation agreement with the property owner and, with the owner's permission, the appraiser may use the calculations recommended in the NRCS plan if the range condition is the same as that described in the plan. In any event, the historical carrying capacity of the ranch must be ascertained and compared to the estimated carrying capacity. When the difference is large, it is likely that supplemental feeding or planted grazing was provided to enhance the carrying capacity of the property.

CHAPTER

• 9 •

# Site Improvements

A parcel of land becomes a site when it is improved and ready to be used for a specific purpose. A site may have both on-site and off-site improvements that make it suitable for its intended use or development.[1] Like buildings and other structures, site improvements are subject to physical, functional, and external obsolescence.

On-site improvements on rural properties typically are referred to as *land improvements.* Land improvements commonly include perimeter and interior fencing, interior roads, water and septic systems, irrigation wells, permanent irrigation distribution systems, and livestock handling facilities. Off-site improvements include water supply companies, irrigation and drainage districts, and utilities including electricity, gas, and telephone.

Although the market commonly does not reflect a separate value for rural land improvements, the absence or poor condition of expected improvements may adversely affect the market value of a rural property. Land improvements typically reflect a general level of neighborhood acceptance and use. For example, in a wheat-producing region where cattle provide a secondary income source, perimeter and interior fencing may be important. However, a similar region without a secondary livestock use may have no fences at all. Specific types of rural property may require appropriate land improvements. A cattle ranch needs fences and corrals, a hop farm needs hop trellises, and irrigation and drainage facilities are common land improvements in certain regions.

In inventorying the land improvements, the appraiser considers their age, condition, estimated depreciation, and any annual maintenance charge. The appraiser must assess the adequacy of the improvement and

1. Appraisal Institute. *The Appraisal of Real Estate,* 11th ed. (Chicago: Appraisal Institute, 1996), 219.

its contribution to the operation and value of the property. The appraiser also notes whether the improvements are located advantageously to provide, for example, protection from the elements or access to public roads.

## Definitions

*Site improvements* are land improvements that are necessary to make a property suitable for its intended use or development. Care must be taken to differentiate between site improvements and fixtures and personal property and chattels.[2]

*Personal property* includes moveable items of the property that are not permanently affixed to, or part of, the real estate. Examples of personal property are portable irrigation pipe, livestock squeeze chutes, and portable buildings. A *fixture* is an article that was once personal property, but has since been installed or attached to the land or building in a rather permanent manner; it is legally regarded as part of the real estate. Examples of fixtures include windmills, water troughs, and fences.

*Chattels* or *trade fixtures* are articles that are owned and attached to a rented space or building by a tenant and used in conducting a business. A trade fixture is to be removed by the tenant when the lease expires unless this right has been surrendered in the lease. Examples of chattels include fuel tanks, gearheads and irrigation motors. The courts consider the manner in which the item is affixed, the character of the item, and the intention of the party who attached the item in determining whether an item is a trade fixture, and therefore personal property, or part of the real estate.[3]

The distinction between fixtures (real property) and personal property (chattels) varies significantly from region to region. The appraiser must know how the courts and market participants in the specific jurisdiction or area define these items. Fixtures are part of the realty and must be included in the property value. If personal property is to be included in the value estimate, it should be identified and described as such in the appraisal report.

## Irrigation

Some agricultural land is irrigated to obtain its greatest productive capability. When irrigation is economically feasible and brings the land to its highest and best use, the value of the land is increased. *Irrigation* is the application of supplemental water to soil for any number of the following purposes:

1. To supply the moisture essential for plant growth

---

2. Ibid., 8–9
3. See Chapter 1 for a more complete analysis of the difference between trade fixtures and personal property.

2. To assure a crop during short droughts
3. To moderate the temperature of soil and atmosphere, thereby making more favorable conditions for plant growth
4. To leach or dilute salts in the soil[4]

The value of irrigated property is increased by the smooth integration of all the factors that affect irrigation. These factors include basic aspects of the land such as climate, which establishes the need for irrigation, and the soil and topography, which determine if the land can be irrigated. Soil and topography may also dictate the requirements for an appropriate irrigation system. The effectiveness of irrigation on a particular property depends on the essential element of water—its physical source, the rights governing its use, its quality and conveyance, the methods and rates of applying it, and its cost. In a well-designed and balanced irrigation system, all aspects work together and contribute to increased property value. For example, a better irrigation system produces greater yields and greater net income, thus adding value to the property to the point of diminishing returns. Beyond the maximum point at which the crop can utilize the water provided, the investment in irrigation does not produce a commensurate return.

Typical on-site irrigation improvements include land leveling or shaping, well and casing, underground transmission line, concrete ditches, storage reservoirs, and permanent set sprinkler and drip irrigation systems. Other necessary irrigation equipment such as pumps, gearheads and motors, and side-roll or center-pivot sprinkler systems typically are considered personal property subject to local custom and legal interpretation.

Irrigation and drainage districts service large areas and fund their activities through property taxes and user charges. These service companies have significant improvements, which provide important off-site improvements to rural properties within their jurisdictions. Irrigation districts often hold title to water rights as trustees for the beneficial use of their members. The quality of water rights, financial structure, and charges imposed by individual districts vary dramatically and directly impact the income-producing capability of the properties served. Adjacent properties in different irrigation or drainage districts, which are identical in all other respects, may have significantly different market values due to these factors.

The irrigation system should be designed to meet the needs of a particular soil, its topography, and intended crop production. For example, sandy, undulating soils are better suited to sprinkler irrigation than level soils with a higher silt or clay content, which may require gravity or surface irrigation.

---

4. Orson W. Israelsen and Vaughn E. Hansen, *Irrigation Principles and Practices,* 3d ed. (New York: John Wiley and Sons, 1962), 4.

## Fences

The fencing of agricultural properties serves many purposes including controlling livestock, managing game animals, controlling property access, and providing windbreaks. Fence condition and construction on a specific property should be measured against area or neighborhood norms for properties with the same highest and best use. As is typical for most land improvements, fences that meet area norms often are not assigned a separate value but are treated as part of the land. Common deficiencies requiring attention or adjustment by the appraiser include the absence or poor condition of necessary fencing on a livestock ranch, superadequate construction such as fencing on a row crop farm in a non-livestock area, and inadequacies such as the lack of needed interior or pasture fences.

Fence construction and costs vary with design, materials and intended use. Typical ranch fences may be four or five strands of barbed wire on wood or metal posts. Wood rail snow fences are required in certain intermountain areas. Fences on sheep and goat ranches generally are constructed of net wire on posts. In ranching areas that have experienced a change in highest and best use from livestock grazing to recreational hunting, barbed wire fences are commonly replaced with 8-ft. net wire fences designed to control the movement of deer and other wild game.

Many ranches are located in extremely rough, mountainous terrain where fence construction can be extremely difficult and expensive. Therefore, it is not uncommon to find exterior or perimeter fences that follow natural geographic boundaries or divisions but do not coincide with the property's legal boundary. State law varies, but in some areas title to lands outside such boundary fences may be claimed through adverse possession statutes by the adjacent landowner unless a formal boundary agreement exists.

## Livestock Handling Facilities

Livestock handling facilities, which include corrals, pens, and loading chutes, are important improvements on ranches, stock farms, and farm feedlots. They constitute a necessary improvement because they make such properties suitable for their intended use. Accordingly, the market often views such improvements as a part of the land and accords value consideration only for deficiencies—i.e., when the improvements are absent or deteriorated. Similar improvements on highly improved special-use properties such as commercial feedlots, dairies, and horse farms represent a much more substantial investment and warrant greater attention and detail in site description and value estimation.

Certain livestock facilities such as dipping vats have become obsolete in many ranching areas and increasingly constitute an environmental concern. The current emphasis on environmental issues indicates that such facilities likely will be viewed negatively by market participants and act as a value detriment.

### Livestock Water

Properties used for ranching or livestock grazing must have sufficient livestock water. The amount of water needed varies with the temperature, humidity, and amount of moisture in the feed and with the size, condition, and species of the animals. The availability and distribution of livestock water sources directly affect how efficiently the range is utilized as animals tend to overgraze the forage nearest to water sources.

Typical livestock water sources include wells equipped with a windmill or electric pump, earthen ponds that impound runoff water, creeks, and streams. Water quality for livestock is less critical than it is for domestic use, but animals will tend to congregate to the freshest or least brackish water resulting in excessive grazing pressure in these areas. Pasture arrangement is obviously affected by water sources and all pastures or livestock enclosures should have a water source. Most livestock ranches will have sufficient watering so that animals are not required to walk more than one mile to water. When water sources such as wells or ponds are located at greater distances, water distribution via pipelines and water troughs is necessary.

## Domestic Wells

The availability, quality, quantity, and safety of domestic water sources are important to rural properties. These water sources provide water for rural residences, but they are also important for standard farming and ranching practices, including the on-site preparation of chemical sprays and mixtures. The location of domestic wells in relation to the residence, other headquarters improvements, and septic systems should be noted. Many rural areas now have space, distance, and design regulations affecting the proximity of domestic wells and septic systems. The appraiser should be familiar with such ordinances and assess properties accordingly.

Other factors of importance are the depth and volume of the well(s), pressure, storage, and distribution systems. Local well drilling or service companies can be helpful in assessing the adequacy of existing systems including necessary well volume and storage requirements.

## Sewage

Most rural properties do not have sewer service and must rely on private septic systems. The location and adequacy of the septic tank, absorption fields, or sewage lagoons are important criteria in determining and describing the adequacy of the sewage system. The effectiveness of the system is directly related to soil properties and the USDA Natural Resource Conservation Service (NRCS), formerly the Soil Conservation Service, prepares soil surveys that normally address site suitability for sanitary facilities.

Septic tank absorption fields are subsurface systems of tile or perforated pipe that distribute effluent from a septic tank into the natural soil. Soil permeability, depth to bedrock, and susceptibility to flooding determine a system's effectiveness and the potential impact on groundwater. Sewage lagoons are shallow ponds constructed to hold sewage while aerobic bacteria decompose the solid and liquid wastes. The lagoon floor and sides must be of nearly impervious soil material to minimize seepage and the contamination of groundwater.

Many older septic systems do not have enough capacity to service modern appliances. Moreover, a rural residence originally designed for intermittent recreational use may need septic renovation prior to conversion to a permanent residence.

## Roads

Access or public roads serving rural properties are a location factor. Although they constitute an important off-site improvement, they are not typically discussed with other on-site or off-site improvements. Interior roads are an important site improvement, however, and should be addressed. The suitability or adequacy of interior roads, like many land improvements, must be assessed against area norms for similar properties with the same highest and best use. Construction, quality, location, anticipated maintenance costs, and ability to serve their intended purpose are important review criteria.

Paved interior roads may be important on recreational land or special-use rural properties, but the same improvement on a row crop farm may be superadequate. Similarly, an inadequate headquarters access road may represent a deficiency.

## Drainage

Agricultural drainage may be defined as the removal and disposal of excess water from land. The sources of excess water may be precipitation, snowmelt, irrigation water, overland flow or underground seepage from adjacent areas, artesian flow from deep aquifers, floodwater from channels, or water applied for such special purposes as leaching salts from the soil or for temperature control. Drainage systems are needed to supplement natural drainage in many areas. The amount of water to be removed by such systems depends on the relative effectiveness of any natural and constructed drainage.

Agricultural drainage is divided into two broad classes: surface drainage and subsurface drainage. Some installations serve the dual purpose of surface and subsurface drainage.[5] Adequate drainage can 1) facilitate early plowing and planting, 2) lengthen the effective growing

5. Soil Conservation Service, USDA, *Drainage of Agricultural Land* (Port Washington, New York: Water Information Center, Inc., 1973), 1.

season, 3) increase the depth of the root zone, 4) aerate the soil, 5) minimize erosion, 6) leach excess salts from the root zone, and 7) promote higher soil temperatures.[6]

Surface drainage systems depend on the mechanical forming of land into appropriate terraces, diversion ditches, waterways, open ditches, and dams. Subsurface drainage systems employ drain tiles, which should be strong, dense, and resistant to freezing, thawing, acids, and sulfates. The drainage system should be designed to meet the needs of a particular soil, topography, and irrigation network. For example, soils with a high clay content require closer spacing of underground drainage tile lines than do soils with higher sand content. Drainage water, like irrigation water, can be moved by gravity or by pumping, and the cost of moving it must be considered. Well-maintained drainage systems with clear, unclogged drains have the greatest utility and value.

## Utility Development

The impact of utility development on rural properties is threefold. The appraiser should address

1. The utility services that are available on site on the valuation date
2. Those utilities that impact the intended use or alternative future uses, but are not currently extended to the property
3. Utility easements, such as pipelines and electrical lines, which may be physically present on the property but do not service it

The presence or absence of standard utility services such as water, sewer, electricity, telephone, and natural gas may or may not impact the property's intended use. Water, sewer, and gas service are commonly not available on rural properties necessitating private development of such services via wells, propane or butane systems, and septic systems. The development of broad-based cellular telephone systems has made telephone service more generally available in rural areas.

When the highest and best use or intended use of the property requires utilities that are not present, the appraiser should estimate the cost to cure this deficiency. For example, division of a large rural property may result in the severing of essential utility services to some of the tracts, which may affect the highest and best or intended use. Depending on the highest and best use determination, the appraiser may find it necessary to estimate the cost of providing utility services to the parcels.

The presence of utility easements for electrical transmission lines or oil and gas pipelines may limit alternative uses of the property by preventing the installation of center-pivot irrigation or tile drainage systems.

---

6. Israelsen and Hansen, 351–352.

## Timber

Standing timber is legally considered part of the real estate, not a fixture or personal property. Accordingly, the appraisal of a property with standing timber must consider the contributory value of the timber. Once timber has been severed from the stump, it does become personal property. Felled trees are referred to as *logs*.

*Merchantable timber* is timber of a certain size, quality, and species that has a recognizable market and is ready for harvest. To have value as a commodity, timber must be located in an area accessible for harvesting. This type of timber is referred to as *economically merchantable timber*. *Stumpage value* refers to the contributory value of the economically merchantable timber as it stands in the forest attached to the land.

The market value of timbered properties thus includes the value of the land together with the contributory value of the standing timber. An appraisal of such a property may involve consideration of the timber crop alone, consideration of the cutover land alone, or valuation of both together. Consulting foresters are generally available to perform timber cruises and estimate the contributory value of the timber. Appraisers who lack knowledge of forestry and timber markets should retain the services of such consultants when appropriate.

## Summary

Referring to agricultural land as "raw land" is inappropriate in that it implies a highest and best use other than for agricultural purposes. Such usage also obscures the contribution of site improvements, which may be present or absent. The site improvements discussed in this chapter are not found on all agricultural tracts. Each agricultural area will reflect certain standards for on-site and off-site improvements that make a rural property suitable for its intended use.

Personal property includes moveable items on the property that are not permanently affixed to, or part of, the real estate. Fixtures are articles that once were personal property, but have since been installed or attached to the land or building in a permanent manner; they are regarded as a part of the real estate.

Chattels or trade fixtures are articles that are owned and attached to a rented space or building by a tenant and used in conducting a business. A trade fixture is to be removed by the tenant when the lease expires unless this right has been surrendered in the lease; they are not considered a part of the real estate.

The market does not typically reflect a separate value for rural land improvements, but the absence or poor condition of expected improvements may adversely affect the market value of a rural property. Site improvements should be measured against area or neighborhood norms for properties reflecting the same highest and best use.

CHAPTER
• 10 •

# Building Description

The buildings on rural properties take many forms and vary in importance. Some contribute little to value, while highly specialized buildings may make a major contribution to property value. In transitional areas, single-family residences range from small frame houses to elaborate estates. Appraisals of rural residences are recorded on the same forms used for urban homes and often submitted to the same lenders. Different underwriting problems arise when the home is located in a rural area, but the description portion of the appraisal is very similar.

In ranch areas the buildings located on the ranch may be an insignificant portion of property value, especially if a large acreage is included in the property. Recent purchases of ranch properties by individuals whose main occupation is not ranching have resulted in the construction of elaborate estate homes and weekend retreats on some properties. Typical ranch properties do not include these types of residences.

In farming areas, farm buildings may make a small contribution to the value of the land, but be integral to the operation of the farm. These farms may include a residence, outbuildings for livestock and machinery, shop facilities, and many other structures for the storage of grain or other crops. For building-intensive properties such as grain elevators, cotton gins, feed lots, swine facilities, and dairies, a substantial contribution may be made by the buildings while the land may be an insignificant part of the overall value.

Whatever the type of rural property, the appraiser must consider the building type and describe the buildings included on the property. The construction, use, size, and floor plan of the buildings must be considered as these factors relate to the highest and best use of the property.

## Zoning and Building Codes

Zoning in rural areas varies from nonexistent to extremely restrictive. In areas without zoning, it is common to have a septic system permit requirement for sewage disposal on residential properties. For rural buildings used for livestock or crops, no permits are typically required. For more building- intensive properties, especially those associated with confinement livestock, there are numerous state laws concerning waste disposal and air and water quality. These laws can impact the location and operation of the facilities even in areas without zoning.

In areas that have zoning, the appraiser must consult with the zoning offices to ascertain if proposed improvements are legal and if zoning has any impact on existing improvements. Because environmental, water quality, air quality, floodplain, coastal, and health codes can all impact rural property, the appraiser must be aware of these laws or regulations and understand their impact on existing and proposed structures. Building codes also can affect rural real estate and the appraiser must be aware of any codes that exist in the area.

## Building Size

The first step in the building description is determining the size of the building. For rural residential properties, building size is measured in the same manner as it is for urban residences. Standards for measuring residential property have been developed by several federal agencies, including the Federal Housing Authority (FHA), the Veterans Administration (VA), Fannie Mae, and Freddie Mac. Because there is a close relationship between these agencies and the mortgage market industry, these standards are used in all residential appraisals. The agencies use gross living area to measure single-family residences. *Gross living area* is defined as the total area of finished, above-grade residential space. It is calculated by measuring the outside perimeter of the structure and includes only finished, habitable, above-grade living space. Finished basements and attic areas are not generally included in total gross living area. Local practices, however, may differ.[1]

Gross building area is the standard by which other agricultural buildings are measured. *Gross building area* is the total floor area of the building, excluding unenclosed areas, measured from the exterior of the walls.[2] Sheds attached to the sides of a barn, for instance, are also measured and may be considered separately or as a part of gross building area depending on local custom. Consistency is the key for using measurements to compare buildings. Obviously, when accepting building measurements obtained from another appraiser or from a

1. Appraisal Institute, *The Appraisal of Real Estate,* 11th ed. (Chicago: Appraisal Institute, 1996), 243.
2. Ibid.

public source, the appraiser must know the measurement standards used. Accurate building measurements are very important in the completion of an appraisal. Building measurements are used in the cost approach to estimate total cost new, in the sales comparison approach for comparative purposes, and in the income approach to estimate market rent.

## Building Description Considerations

With the variety of agricultural buildings found in the market, many different factors must be considered depending on the type of building being appraised. The appraiser must consider the complete building description. The following outline lists some of the items to be considered. The appraiser may add or delete from this list depending on the type of building and the area in which it is located.

I. Exterior description
   A. Foundation
      1. Pole
      2. Concrete
      3. Footings
      4. Piers
      5. Beams
      6. Manure pits
   B. Floor structure
      1. None (dirt)
      2. Concrete
      3. Drains
      4. Type (dairy, poultry, swine)
   C. Exterior walls
      1. Solid wall
      2. Curtain wall
      3. Material
      4. Insulation
   D. Interior construction
      1. Clearspan
      2. Walls (type, position)
      3. Insulation
   E. Roof construction
      1. Structure
      2. Cover

3. Drains
4. Insulation

F. Equipment
1. Dairy
2. Poultry
3. Swine
4. Cattle
5. Electrical service
6. Plumbing

G. Utilities
1. Electrical source
2. Water source
3. Gas source
4. Interior distribution
5. Multiple sources (required or not required)

H. Heating systems
1. Type and fuel

## Exterior Description

The exterior building description provides information about the type, quality, and design of the materials that were used or will be used in the construction of the building being appraised. Details concerning the type of foundation, floor, exterior walis, roof structure, roof cover, and insulation should be included. The appraiser should make a determination as to the quality and condition of the structure and, more importantly, its conformity to current standards of construction. As many agricultural buildings are involved with an integrator who supplies the birds or animals that will be housed in these buildings, the integrator's standards are often the measurement of the functional utility of the structure. In addition, the large machinery used in agricultural enterprises today requires larger doors, higher eaves, and less interior partitioning than is typical in older buildings. When a group of buildings makes up a total operation, the functional utility of the group of buildings must be described. Do the buildings provide for an efficient flow of operation so that all buildings are used for maximum efficiency?

## Interior Description

The description of the interior of agricultural buildings should include wall finish, partitioning or lack thereof, wall supports or lack thereof, manure disposal construction, interior height, heating systems, ventilation systems, and lighting. Because agricultural buildings vary tremendously, the appraiser must be familiar with the current interior construction standards of the type of building being appraised. Again, the housing of large machinery is an important consideration in describing the interior of the building. Today's buildings are larger and have higher ceilings than the buildings of 20 years ago. Columns can be a disadvantage insofar as they can cause difficulty in the internal movement of machinery or hamper manure removal in certain buildings. The utility of the building suffers due to these interior supports. Heating can be an important factor, especially if the building houses very young birds or animals. Ventilation is also important to remove dust and odors that can inhibit bird and animal growth.

### Equipment

Equipment can be an integral part of a building and the ability of a building to house modern equipment is often the basis for assessing its functional utility. Dairy, swine, and poultry producers must utilize the most modern equipment to maximize the overall utility of the structure. In many cases, integrators will require the most modern equipment or they will not place birds or animals in the buildings. The refusal of an integrator to place the necessary livestock in the building could indicate obsolescence and totally eliminate the value of the building.

### Utilities

The availability of public utilities is important to many classes of agricultural buildings, especially those that house large quantities of livestock. Water is generally provided by an on-site well because a very large quantity of water is necessary to operate livestock facilities. Electricity is generally provided by local cooperatives. The heat source may be fueled by natural gas, diesel, fuel oil, or electricity depending on availability. In addition to the well, many operations have access to a community water source as a backup. A well is often the primary or only water source. Large electric generators may be in place as a backup electrical source. Many poultry operations require dual sources for water, electricity, and heat if large quantities of birds are housed in the buildings.

The remainder of this chapter provides suggestions for writing building descriptions for various classes of agricultural buildings. The appraiser should be familiar with the standards for the area in which the appraised buildings are located and adapt these general guidelines to the specific area. It is important to understand the dynamics of agricultural product prices and how they can affect the value of farmland and agricultural buildings and plants. Also, technology changes quickly, so external and functional obsolescence can be very significant for many agricultural operations and happen quickly. Many specialized agricultural facilities such as those described below (e.g., swine facilities, poultry facilities) are large-scale, highly mechanized operations. A good understanding of the business operation is essential to accurate valuation of agricultural properties, just as it is for industrial properties.

## Swine Facilities

Commercial swine facilities are located in several parts of the United States and for years have taken the form of farrow-to-finish facilities, but integrators are becoming an increasingly dominant factor in the market. Whether the swine facility is the typical farrow-to-finish facility or one or more segments in an integrator-controlled program, the appraiser must be familiar with the operation of swine facilities and must know what drives the market. In describing a swine facility, the factors to be addressed concern the construction of the facility and the buildings and equipment necessary for a profitable operation.

*Dimensions:* The dimensions of the facility have a direct relationship to its overall size, which is the most relevant factor. The building must be large enough to be financially feasible, but not so large as to be unprofitable. The appraiser must consider the typical size and current standards prevalent in the area and judge the subject property in the light of these standards. Accurate measurements of the facility are needed to ascertain the number of animals that can be housed.

*Foundation:* The foundation will most likely be concrete with concrete slats over a 4- to 8-ft. manure pit.

*Exterior walls:* Corrugated, prefinished steel panels were the standard for several years, but in many areas roll-up curtains are the current standard because they provide more efficient ventilation. The appraiser must consider the standard for the local market and determine which type of exterior walls are preferred in the market.

*Ceilings:* Ceiling types range from corrugated, prefinished steel with blown-in insulation to sloping ceiling with roof vents for additional ventilation. Some barns have aluminum, glass board, or PVC interior ceilings and walls that resist corrosion.

*Heating:* Heating must be sufficient for the climate in which the property is located.

*Lighting:* Lighting and wiring vary with the quality of construction.

*Ventilation*: Depending on the design of the facility, the building ventilation will be in the form of fans, curtain walls, or moveable panels to assist normal air flow. Most ventilation systems will include pit fans to provide manure pit ventilation. Many newer barns have no interior manure pits. In these facilities, an outdoor lagoon is used and there are no fans for pit ventilation.

*Feed/water systems:* These systems are often considered part of the building equipment and typically have a shorter life than the building structure due to wear and deterioration from use.

*Pens/crates:* In the various buildings that make up the facility, different types of pens and crates will be present. These should meet current standards. They will have a shorter life than the building structure.

*Manure storage:* Manure storage will take the form of a lagoon, a storage tank, and/or a manure pit within the building itself. The appraiser must inspect and consider the size of the manure storage area as it relates to the number of animals and the amount of manure that will be produced. The manure facilities will be emptied and injected into the ground for fertilization purposes at various times of the year, depending on local climate and cropping patterns. The facility should have either sufficient acreage or contracts with other producers for disposal of the manure. Local and state regulations concerning manure disposal must be reviewed by the appraiser and the facility must be in compliance or obsolescence may exist.

*Utilities:* Many farms need dual sources of water and electricity, and the lack of dual sources will increase the risk of operations. Most facilities will have a backup generator system and this system must be large enough to supply the farm until normal electrical service is restored. Well water and a community water system can provide dual water sources for a farm.

The building description for a swine facility must include discussion of all physical and functional curable items, the age and condition of short-lived items, the age and condition of the long-lived portion of the facility, and other items of functional and external obsolescence applicable to the facility. Also, the overall integrity of the facility must be addressed. For example, does the facility contain enough finish space to use the farrowing space efficiently? The overall facility must be considered to ascertain functional obsolescence due to either the lack or excess of a particular capacity. External obsolescence in swine facilities is typically the result of lower hog prices.

## Poultry Facilities

Commercial poultry facilities differ considerably depending on the area in which they are located and the type of operation. Poultry facilities are constructed to house broilers, caged laying hens, breeder stock, and pullets; different building construction and equipment are needed to operate each of these facilities.

*Dimensions:* The dimensions of the facility relate to the overall size, which is more relevant. Again, the facility must be large enough to be financially feasible, but not so large as to be unprofitable. The appraiser must consider whether the subject property conforms to the typical size and current standards prevalent in the area. An accurate measurement of the facility is needed to ascertain the number of birds that can be housed.

*Foundation:* Depending on the type of operation, the foundation may vary from a dirt floor to a concrete floor with a manure pit.

*Exterior walls:* Corrugated, prefinished steel panels are used in cold climates, and roll-up curtains are common in warmer climates. The appraiser must consider the standard for the local market and what type of exterior walls controls the market.

*Interior construction:* Clearspan construction, which facilitates manure removal, is predominant in newer buildings.

*Heating:* Heating must be sufficient for the climate.

*Lighting:* Lighting and wiring vary with the quality of construction.

*Ventilation:* Depending on the design of the facility, building ventilation will be provided by fans, curtain walls, or moveable panels that assist normal air flow.

*Feed/water systems:* These systems are considered part of the equipment and will vary greatly with the type of operation. Feed and water systems will have a shorter life than the building structure due to wear and deterioration.

*Cages:* For buildings that house poultry, cages will be a part of the equipment; if broilers are housed, no cages are typically used. If cages are present, they should meet current standards. They will have a shorter life than the building structure.

*Manure storage:* Ground storage is used in many poultry operations. The manure is removed by loader or tractor in batches to manure pits within the building. The appraiser must inspect and consider the size of the manure storage as it relates to the number of birds and amount of manure to be produced. Local and state regulations concerning manure disposal must be reviewed by the appraiser and the facility must be in compliance or obsolescence may result.

*Utilities:* Many farms need dual sources of water and electricity. The lack of dual sources will increase the risk of operations. Most facilities have a backup generator system, which must be able to supply the farm until normal electrical service is restored. Well water and a community water system can provide dual water sources for a farm.

The building description for a poultry facility must include discussion of the physical and functional curable items, the short-lived items and their age and condition, the long-lived components and their age and condition, and the functional obsolescence and external obsolescence applicable to the facility.

## Cattle Feed Yards

Cattle feed yards are designed to accommodate large numbers of cattle prior to slaughter. These facilities are composed of improvements that fall into five general categories.

- Feed pens, feed bunks, shades, working pens, loading and receiving chutes, hospital pens, and horse pens
- Feed mill facility
- Grain storage and handling improvements
- Office and building improvements
- Site improvements

*Pens:* The pens are generally constructed of cable on pipe posts set in concrete, with a pipe top rail and center rub rails. The cables are tied to the end posts by coil springs. Bunk fencing is a pipe rail and an adjustable cable mounted on pipe posts. All pens are normally fitted with welded metal gates. Alleys run along the ends and rear of all pens and are of comparable construction. Each pen has water and possibly shade. The feed bunks are of either concrete or metal construction with a concrete apron and they run the length of each pen. The feed alleys are all weather-surfaced and run along the feed bunk frontage. The hospital pens are of similar construction, but they have metal-sided working and squeeze chutes, all on concrete pads. The shipping and receiving pens are welded pipe with a crowding and working chute on a concrete slab with a sheet metal shade roof. A cattle scale is usually located in these facilities. The loading and unloading chute generally has two levels with concrete ramps and curved metal, steel plate welded on pipe.

*Feed mill:* The feed mill facility is comprised of a mill building and slab area used for batching the feed stuffs. The mill building consists of the mill office and storage area, which is at dock height on a slab foundation with metal or wood frame and walls. A mill truck scale (usually of 100,000-lb. capacity) is located in the mill building under the truck load-out area. The milling equipment includes automatic steamroller mills, which are fed grain from the cleaner or the hopper bins located above the roller mills. The welded steel hopper bins located in the mill building are mechanically operated. There are also concrete bulk material bins located within the commodity barn. Molasses, protein supplement, and fat tanks are located in the mill and supply product to the milling operation. The feed mill equipment includes numerous elevator legs, air lifts, centrifugal pumps, and air compressors. Grain is typically moved by augers. Processed grain and feed are usually moved by airlifts. Elevator legs, augers, airlifts, mixers, shakers or scalpers, steam rolls, and other components are engineered so that all capacities are compatible. Boiler rooms contain 125- to 200-hp boilers with water makeup systems, water softeners, etc.

*Grain storage and handling:* The grain storage and handling improvements consist of either corrugated, bolted metal bins with flat-bottom or concrete-slab sweep augers for unloading or hopper-bottom metal bins that unload into U-trough conveyors. The bins are connected by an elevator system that connects to a concrete dump pit, which receives grain and delivers it to the elevator for movement to the storage bins or directly to the mill.

*Office and building improvements:* The office and building improvements are typically of wood frame or brick veneer construction. The office portion is usually small. A truck platform scale with an electronic printer and a livestock scale with a manual printer are generally located in this building. A shop building of typical construction will also be located on site, with heating depending on the climate. Other small shop buildings, sheds, or employee housing can be found on some sites.

*Site improvements:* Site improvements will include diesel and gasoline storage, LPG storage, water wells and storage system, and a manure storage lagoon.

Again, the appraiser should be familiar with standards for cattle feed lots in the area where the appraised property is located. The run-off lagoon will most likely have to be state licensed and permitted. Some states require a feed yard to include enough land to dispose of the run-off water on site. The appraiser should also consider the availability of grain, feeder cattle, and a slaughterhouse because excess shipping costs can impact the feasibility of the overall operation.

## Grain Elevators

Grain elevators are constructed to handle large quantities of grain and are built of upright concrete, upright steel, or flat steel. Elevators can be of one type or a combination of all three types. Some very old facilities can be made of wood. Grain elevators are found in grain areas predominantly, but they are also located along the major rivers and in port facilities along the coasts of the United States.

The building description must include a detailed description of any of the following components present in the facility to be appraised.

- Grain storage (upright or flat)
- Elevator system
- Warehouse storage
- Driers
- Seed equipment
- Port facilities
- Rail facilities
- Office and scales

*Grain storage:* The description of the grain storage should include a calculation of the total capacity of the facility, usually expressed in bushels. The description needs to include the diameter and height of any upright storage and the exterior measurements and height of any flat storage. The tanks will be either flat-bottom or hopper-bottom and the loading and unloading systems for the tanks must be described. The tanks will usually be loaded by elevator legs from the top and unloaded either by U-trough augers or by a sweep in flat-bottom tanks. Flat storage can be loaded by an auger system in the roof or by portable augers; unloading is generally accomplished with floor augers and portable augers. The description needs to include details such as the size, age, and condition of the electric motors that power the loading and unloading equipment. Any driers or ventilation fans must be described, including the heat source, drying capability, and size of the electric motors.

*Elevator system:* The elevator system includes the dump pit and the associated legs, which move the grain from the dump pit to the various storage areas and to the unloading areas. The length, width, and depth of the dump pit must be measured. The number and length of the elevator legs needs to be determined along with the size of the distributor. Also, the size, age, and condition of all electric motors used in the elevator system needs to be recorded. The bushel capacity per hour of the dump pit and the elevator legs is also essential.

*Warehouse storage:* The warehouse will be used to store fertilizer, seed, or other products sold by the elevator. The floor is typically concrete and the walls and sides are of metal. The amount of insulation and heating required is governed by the climate of the area where the elevator complex is located.

*Driers:* The driers may be small, individual units attached to various bins or a large unit connected to the system with elevator legs that can dry large quantities of grain. They may be flow-through or batch driers.

*Seed equipment:* When seed equipment is present, it adds another dimension to the elevator beyond handling local grain for shipping. Seed equipment includes equipment for cleaning, treating, weighing, sacking, and storing seed. All of this equipment should be itemized. The size, age, and condition of motors and the capacities of the various units should be recorded.

*Port facilities:* Port facilities are located along major rivers to transport grain to the coasts for loading onto ships for export. Port facilities normally have docks for loading grain into barges or ships. The port facility should be described in detail, including the methods for loading and the capacity.

*Rail facilities:* Many grain elevators have rail spurs and elevator legs for loading rail cars. The loading elevators should be described in detail and include capacity. The amount of rail spur is important because additional efficiency results if large numbers of rail cars can be loaded.

*Office and scales:* The office should be described to include size, exterior finish, interior finish, heating and cooling, and other features present. The scales should be described in terms of size, capacity, age, and condition.

For an elevator to be fully efficient, it should be located in an area that produces enough grain to obtain its market share and fully utilize its capacity. There should be an efficient flow of grain from original acquisition, through the drying and storage processes, to the unloading and removal processes. Any deficiencies in the ability to move grain through the process efficiently indicates obsolescence.

## Cotton Gin

The trend is to operate larger gin plants very near the point of production in areas with substantial cotton production. Modern gin plants are being constructed with a 30- to 35-bale-per-hour output. Smaller gins are considered obsolete by current standards. The main components of a cotton gin are:

- Gin plant
- Seed hopper
- Trash bin
- Warehouse

*Gin plant:* A gin plant is typically constructed with a structural steel frame, a concrete floor with the appropriate recesses and tunnels, and pre-engineered steel panel sides and roof. Side walls must be tall enough to house the gin equipment. Due to dusty conditions, adequate windows, doors, and roof ventilators must be present. Attached to the gin building will be a dock-high, covered load-out dock. The plant should include sufficient warehouse area to store ginned cotton until it is ready for shipment.

*Seed hopper:* The seed hopper will be a multiple-hopper, welded-steel overhead bin with metal supports on a concrete foundation. It must have sufficient capacity to store seed from the cotton baled.

*Trash bin:* The trash bin or burr hopper is generally a metal frame building with corrugated metal sides and roof and a concrete floor. An overhead conveyor system is used to move the cotton burs from the gin plant to the trash bin.

*Office:* The office description should include the office size, exterior finish, interior finish, heating and cooling, and any other features present. The scales should be described in terms of size and capacity.

As previously mentioned, cotton gins are typically located in areas where a substantial amount of cotton is grown. Gins located in marginal areas may suffer obsolescence if they cannot gin enough cotton to make the overall operation feasible. New features that impact the feasibility of a gin are the uniform density (U-D) press and module feeder. The U-D press compresses cotton into a smaller bale unit than does the conventional gin press. Module feeders, which are used to feed cotton into the ginning process, are considered almost essential in areas where this equipment is being used. Smaller, less modern gin plants are typically salvaged or parceled out and moved to new locations where the components are used to build larger, more efficient gin plants. Air quality regulations are stringent where gins are concerned and the cost to comply with these regulations may impact the feasibility of a cotton gin.

## Processing, Packing, and Cold Storage Facilities

Facilities for the processing and/or packing of fruits and vegetables typically consist of three major components:

- Receiving area

- Processing area
- Packing and shipping area

Facilities can vary greatly and the physical description of the assets should be consistent with the scale of the facility and its contribution to the value of the total property.

Legal restrictions and requirements differ from region to region. Some areas may have little or no restrictions or requirements, while others may impose a hierarchy of regulations and conditions. Environmental concerns are becoming increasingly onerous. The regulatory environment is a dynamic process, and the appraiser must strive to keep informed of latest changes in zoning and regulatory constraints and requirements. Zoning, permissions, and licensing may involve a hierarchy of government agencies on the local, state, and national levels. A large, modern processing facility may require a plethora of permits and licenses, and the appraiser must research and report on these adequately. Environmental concerns regarding air and water quality, health considerations, traffic flows, noise, dust, etc., may all necessitate a permit or review process. In many cases, existing facilities that are not in compliance with current regulations may be allowed to continue to operate as legal nonconforming uses.

*Receiving:* Receiving areas receive the raw product when it arrives from the field. Depending on the commodity, the facilities may take various forms. The raw product may be delivered to the plant in bins, field boxes, gondolas, or any combination of these. Thus, the receiving area may consist of a simple loading dock or a complex of automated dumping stations. Receiving will typically provide for the quantitative and qualitative measurement of the incoming product. Quantitative measurement is usually accomplished via truck scales, while qualitative analysis may be performed at an inspection station. The inspection station may or may not be certified by a government or industry agency or commission. Receiving may also provide for pre-cooling of the raw field product. Ancillary site improvements may be critical to the receiving process, and they should be described adequately. Incoming truck staging areas should also be assessed.

*Processing:* Main processing and/or packing buildings will typically be of conventional industrial design and construction. Internally framed steel structures are probably the most prevalent, although many packing areas consist of open-walled, covered sheds. Again, the appraiser should describe both the physical structures and the flow of the product through the process. Some consideration may also be given to the possible conversion of the facilities to alternative uses. Many industries have witnessed structural changes in the packing process, with trends moving toward the packing of the product in the field. In such industries, large packing houses and sheds have experienced rapid depreciation and the ability to convert the structures to alternative uses may be a major increment of value.

Utilities are often critical in the sorting and packing process and their description should be consistent with their importance. Modern operations require large electrical inputs, which are typically controlled by centralized computer stations. In some operations, the disposal of culled product is a significant task, and the vehicle for disposal or transfer should be adequately described. Waste water can be an undesirable by-product of packing and processing. The appraiser must explore and report on how waste water generated by the plant's operation is disposed of or processed.

*Packing and shipping:* The final stage in the operation, packing and shipping, typically demands storage areas for packing materials. The actual packing is generally accomplished at the end of the processing line. When wooden or paper packing materials are used, materials storage will require automatic fire sprinklers and alarms and, perhaps, additional safety features. The finished, packed product may either be shipped directly or transferred to storage to await future shipment. With perishable commodities, cold storage is a key component.

## Cold Storage

Building structures for the refrigerated storage of perishable product are typically of conventional industrial design and construction. The basic building shell may be wood framed, internally steel framed, or concrete tilt-up. In most areas, concrete tilt-up construction is popular. The basic shell is improved with refrigeration equipment and fixtures, insulation, receiving and loading docks, and office and dispatch areas. Here again, the appraiser should adequately describe both the physical structure and the flow of the product. The product flow should address temperature changes as the process often involves fairly rapid changes in product temperature.

Some specialized products will demand more precise storage environments, such as the "controlled atmosphere" storage required for apples. Other products may require high or low humidity or other specific atmospheric conditions. The storage facility must be designed and constructed to provide the required environment. Such specialized improvements may suffer significant, and perhaps rapid, obsolescence from changes in market preferences.

Receiving areas may include pre-cooling rooms, with high-volume, forced-air refrigeration, or hydro-coolers, with recirculating refrigerated water cooling. Hydro-coolers may be portable units and even permanent units may be considered personal property; pre-coolers tend to be real property. Entrance doors will typically be sliding or roll-up and should be large enough to handle modern, fully loaded forklifts. Temperature control devices at entrances may be forced-air doors or plastic curtains. Loading dock areas and hallways may also be insulated and refrigerated. They may serve as overflow storage when needed.

The design and layout of main storage areas will be dictated by industry standards. Some products will be stored most efficiently in large, open rooms, while others may demand smaller, isolated bays. Internal racking and shelving may be a desirable asset for some types of storage. Eave or ceiling heights should be measured and described because cold storage is a function of volume in three dimensions, not merely area in two dimensions. Thus, the appropriate unit of measure is often cubic feet, not square feet, which is the more traditional unit of measurement.

Refrigeration equipment and insulation should be matched to the location and operation. Equipment and fixtures should be described commensurate with their importance to the total facility and should be identified as either real or personal property. The appraiser should be conversant with the rules for differentiating between real and personal property in the specific jurisdiction. The refrigeration vehicle (i.e., Freon, ammonia) should also be clearly identified, and any usage restrictions should be discussed. The total refrigeration capacity should be identified and reported, usually in tons of capacity.

Temperature alarms and back-up power systems may be highly sophisticated, involving significant cost. Facilities that store high-value products may employ multiple redundant systems.

## Summary

The appraiser must provide the client and other readers of the appraisal report with a detailed description of the buildings and other improvements located on the subject property. This process is of great importance when the buildings are the predominant feature of the property. The important features of specialized agricultural properties must be addressed. The building description should be tailored to the subject property and the area where it is located. It should include enough detail to allow the reader to visualize the property and all of the major factors that may influence its market value.

CHAPTER
• 11 •

# *Highest and Best Use Analysis*

Previous chapters have introduced principles and concepts that provide the basis for the appraisal of real property. Principles such as supply and demand, substitution, balance, contribution, conformity, and externalities have been discussed and their impact on the appraisal process has been carefully explained. The primary forces influencing real property values have been described including the social, economic, governmental, and environmental issues that are inherent in shaping the decisions made by buyers and sellers in today's agricultural real estate markets. Factors of value are also part of the mix on which real estate appraisal is founded. The value influences of utility, scarcity, desirability, and effective purchasing power have been carefully described.

Buyers and sellers in today's rural real estate markets are aware of the way in which supply and demand influence both the number of offers for a property and the desirability of the products produced by the property. Similarly, buyers and sellers are constantly comparing one offer or sale to another offer or sale to gauge the relative worth of individual parcels in the market.

The factors that are being considered and evaluated by market participants can generally be described in relation to the use of the property in question. The analysis of use draws upon all the concepts, principles, forces, and factors that have been introduced and discussed in previous chapters. Such an analysis is performed to develop a complete understanding of market behavior as it relates to a specific property, a group of properties, or an area.

Analyzing the highest and best use of a rural property may be as simple as understanding what is occurring in a given market and then reporting those observations, or it may require a very specific and detailed study of the market area and all of the forces influencing it. The primary determinant of the scope and depth of the analysis is the nature of the appraisal assignment. A feasibility study of an agricultural tract in a transitional or suburban market may require a significant amount of time and analysis, while estimating the use of a tract of pasture in an area dominated by similar uses may be much simpler.

Highest and best use has been defined and redefined over the years. The most recent definition, found in *The Appraisal of Real Estate,* eleventh edition, describes highest and best use as:

> the reasonably probable and legal use of vacant land or an improved property, which is physically possible, appropriately supported, financially feasible, and that results in the highest value.

The key elements in this definition are consideration of :

- The property as if it were vacant
- The property as it is currently improved (if improved)
- Those uses that are legally permissible
- Those uses that are physically possible
- Those uses that are financially feasible
- Those uses that are appropriately supported, or maximally productive

A distinction must be made between consideration of the property as if it were vacant and consideration of the property as it is currently improved. There are many examples in the marketplace of well-intentioned market participants using rural property for purposes that were not compatible with its highest and best use. Some classic examples include specialized livestock units that are constructed and operated, generally for a short time, in rapidly expanding suburban areas; property used to raise specialty crops long after the market for these specialty crops has declined (e.g., raising tobacco in the Connecticut River Valley); and specialized buildings constructed on a property without regard for the demand, or lack of demand, for the product or commodity to be produced.

If a tract of farmland is improved with structures used for grain production, but the property is located in an area dominated by farming crops such as corn and soybeans, the highest and best use of the property is likely to be use for raising corn and soybeans. If that same property is located in an area dominated by vegetable and small fruit production and the infrastructure required to market small grains is lacking, the highest and best use is not going to be a use that will fully utilize the improvements. However, as long as the value of the improvements located on the property add to the property's value, the total property value will be greater than the value of the land as if vacant and the highest and best use of the property will generally be the highest and best use of the property *as it is presently improved.* Once the value of the land as if it were vacant exceeds the value of the property as improved, the improvements no longer add value to the whole and the highest and best use of the property becomes the use of the land as if it were vacant.

In considering the property as if it were vacant and as currently improved, the appraiser must understand and clearly state in the report what the term *improved* means. In a typical rural appraisal, the improvements to be considered include dwelling units, machinery and equipment storage, feed storage units, and/or livestock housing. In some markets, drainage tile and/or drainage systems, irrigation systems that are permanently affixed to the property, fencing, and corrals are considered to be improvements. It is incumbent on the appraiser to explain clearly in the highest and best use analysis what is, or is not, considered to be an improvement to the property.

Highest and best use is influenced by market participants, but is ultimately shaped by the market and market forces. Competitive economic and social forces within the market where the property is located must be considered to estimate highest and best use. Thus, an analysis of a property's highest and best use is truly a property-specific economic study of market forces.

Data collection was discussed in Chapter 3. Both general and specific data are used in highest and best use analysis. The appraiser must fully analyze the data collected and understand how the data relate to commodity markets, social and economic trends, and property uses prevalent in the market. Trends in factors that influence use range from product prices, expense levels, and demand and supply levels to changing demographics, local land use regulations, and ongoing changes in environmental legislation. The analysis of these factors will help the appraiser arrive at the ultimate use of the property being valued. Thus, if market factors, principles, and forces form the foundation of an appraisal, the highest and best use analysis can be considered the cornerstone.

To consider a property's highest and best use, the motivations of the buyers who are participating in the market must be addressed. In residential markets, real estate is purchased for occupancy and personal use; a buyer's motivation can be described as the desire to acquire the nonmonetary benefits of personal and aesthetic enjoyment. The purchasers of investment property such as multi-family units, apartment complexes, and retail facilities are considered to be motivated by the potential for future monetary benefits (accumulation of net income), for the accumulation of capital, or possibly for tax benefits or advantages.

In rural markets, the same conclusions can be drawn. The buyers of rural residences can be described as being motivated by the desire to benefit from aesthetics and a lifestyle that is not typically measured in monetary terms. Purchasers of income-producing rural land, including farms and ranches, are much like investors in commercial, industrial, and retail properties—they are usually motivated by the anticipation of future monetary benefits. They are primarily interested in the feasibility of the farming or ranching enterprise that is being purchased.

The concepts of feasibility and highest and best use are related to one another and to market analysis. However, the same data may not be required for each type of analysis. Table 11.1 provides a comparison of the tasks involved in these three interrelated processes.

That use that maximizes the value of a particular property is considered its highest and best use. The appraiser needs to distinguish between the highest and best use of the property as if it were vacant and the highest and best use of the property as currently improved. The appraisal report should clearly identify, explain, and justify conclusions for each use.

**Table 11.1** Relationship of Market Analysis, Feasibility Analysis, and Highest and Best Use Analysis

| Market Analysis | Feasibility Analysis | Highest and Best Use Analysis |
|---|---|---|
| To identify demand for alternative uses | To determine respective values based on criterion variables (e.g., residual land value, rate of return, capitalized value of overall property | To determine the use resulting in the maximum value |
| Supply and demand analysis to forecast absorption rate and probable rents for: | Calculation of *NOI*/cash flows and the selection of appropriate cap rate/discount rate to determine property value based on criterion variables for: | Specification in terms of use, timing, and market participants (i.e., user of the property, equity investor, or debt investor) |
| Use No. 1, | Use No. 1, | |
| Use No. 2, | Use No. 2, | |
| Use No. 3, | Use No. 3, | |
| Use No. 4. | Use No. 4. | |

Source: Appraisal Institute, *The Appraisal of Real Estate*, 11th ed. (Chicago: Appraisal Institute, 1996), 299.

## Highest and Best Use of Land as if Vacant

Implicit in the analysis of the highest and best use of the land as if it were vacant is the assumption that the land is vacant or can be made vacant through the demolition of building improvements. Land is generally valued as though it is vacant.[1] Vacant land is a fundamental concept in valuation theory and serves as a basis and preliminary consideration in the cost approach. The cost approach section of an appraisal should include an allocation of value between the land and the improvements affixed to it. To complete the allocation, the appraiser must conduct a highest and best use analysis.

In completing the analysis of land as if vacant, the following questions must be answered. (Remember that any improvements on the property are to be ignored.)

1. The comment to Standards Rule 1-3(b) of USPAP recognizes "...land is appraised as though vacant and available for development to its highest and best use and that the appraisal of improvements is based on their actual contribution to the site."

- What uses of the land are legal?
- What uses of the land are physically possible?
- Of those legal and physically possible uses, which uses are financially feasible and will result in sustained positive net income or will provide nonmonetary benefits?
- Of those legal, physically possible, and financially feasible uses, which use results in the greatest net income or nonmonetary benefits and thus the greatest value for the property.

In analyzing the highest and best use as if vacant, land value is considered based on its potential use rather than its actual use. It is the potential use, not the actual use, that generally governs the price that buyers are willing to pay in the market.

Several appraisal techniques can be used to value land. All of these techniques are predicated on the concept of highest and best use, so this use must be identified to estimate a property's value. The property sales selected for use as comparable sales should have a highest and best use that is the same or similar to that of the subject property. The two primary reasons to identify the highest and best use of land as if vacant in an appraisal are: 1) to estimate a separate land value, and 2) to identify comparable sales of vacant land.

It is not unusual, especially in transitional and suburban markets, to determine that changes are occurring in the market and the uses being considered are temporary in nature or may not be realized until some time in the future. For example, in markets that are adjacent to, or surround, metropolitan areas, land may have a highest and best use for farming purposes in the near term and an ultimate highest and best use for development as a residential subdivision. In this instance, the property has both an *interim* and a *terminal,* or *ultimate,* highest and best use. An interim use is an activity that takes place on a site during a transitional period until the terminal highest and best use is realized. The terminal, or ultimate, use is the likely "end use" of the property, which is expected to be realized at the end of the transitional period.

During the 1970s and 1980s, there was considerable activity in transitional markets based on a highest and best use that could be described as "speculative." Investors were purchasing property with the express intent of retaining the land until development was justified in the market. It was not uncommon for these investors to either leave the land idle or to allow it to be used by tenants or prior owners to minimize carrying costs such as taxes. A speculative use generally occurs in an oversupplied market or in markets where development financing is unavailable. The speculative use of the property is a type of interim use. The land is held in inventory until market forces indicate that further development of the property is warranted or development financing becomes available.

In analyzing highest and best use as if vacant, the appraiser often must consider adding improvements to the property. For example, achieving the highest and best use for a small grain production unit may require that drainage improvements be made to portions of the tillable land base on the property. The appraiser must consider whether open ditches or tile drainage is necessary; if the parcel is physically suited to the type of improvement required; if the improvement is allowed under current local, regional, and federal legislation; and whether the additional cost of drainage is justified by the value added to the land.

Appraisers who are analyzing the highest and best use of transitional or suburban properties must exercise caution. In the comprehensive planning process, it is not unusual for authorities to use zoning regulations to designate areas for specific ultimate uses. Often, portions of communities or municipalities are designated as open-space, residential, industrial, commercial, and/or retail zones. These zones typically are arranged in a manner that is intended to promote controlled growth of the community. The appraiser must not assume that a tract located in an area with preferred zoning—i.e., retail or business versus agricultural or residential—automatically has a higher use and therefore a higher market value.

The use of a property on the date of inspection may or may not be an indication of its highest and best use. The land may be suited to a more or less intense use. The property may be best used as the site of an intensive livestock feeding enterprise while the current use is for pasture. Similarly, although a tract may contain a significant amount of highly erodible land (HEL) and be best suited for pasture, it may be tilled and currently used for cropland.

## Highest and Best Use as Improved

After considering the highest and best use of a property as if it were vacant and arriving at a conclusion, the appraiser must then consider the property as it is improved. The improvements to the property are considered in their current condition. For example, the appraiser might ask, should the old, wood-framed machinery storage area on a quarter section of cropland be renovated, razed, or replaced with a modern structure? The appraiser evaluates the condition, utility, balance, and conformity of any improvements in the context of the market preferences being expressed by the buyers and sellers of real property in the marketplace.

The process followed and the questions asked in this stage of the analysis repeat those considered in the evaluation of the property as if vacant:

- What uses of the land are legal?
- What uses of the land are physically possible?

- Of those legal and physically possible uses, which uses are financially feasible and will result in sustained positive net income or will provide nonmonetary benefits?
- Of those legal, physically possible, financially feasible uses, which use results in the greatest net income or nonmonetary benefits and thus the greatest value for the property.

The concepts of both an interim and a terminal, or ultimate, highest and best use can also apply to the improved property.

There are two reasons to analyze the highest and best use of a property as it is currently improved. First, it is necessary to identify the property use that can be expected to produce the greatest overall return for the investment made. Since most rural property is purchased with the expectation of generating income for the owner, it is reasonable to assume that the potential buyer is economically motivated and informed. If a property is currently being used as a dairy, for example, will this produce the maximum potential return? Would the return increase, or decline, if the improvements were remodeled and converted to horse stalls to conform with the changing demands of the surrounding suburban market? The market value of the property would likely differ under these two scenarios, and the use that results in the higher present value will be the highest and best use as improved.

The second reason to estimate the highest and best use of a property as improved is to assist in the identification and selection of comparable sales, to be used in the sales comparison approach, and comparable rental properties, to be used in the income approach, both of which follow the highest and best use analysis. The highest and best use of the vacant land sales and improved property sales used for comparison should be the same or similar to the highest and best use of the property being valued. It would be illogical to value pasture or rangeland using sales of irrigated cropland; similarly, it would be inappropriate to compare sales of cage-laying poultry farms to confinement swine operations.

## Highest and Best Use Criteria

The four questions asked in the analysis of highest and best use frame the criteria that must be met by any highest and best use. A highest and best use must be legally permissible, physically possible, financially feasible, and maximally productive. The criteria are usually considered sequentially, but they are interrelated and may be considered in concert. Logic dictates that the consideration of legal permissibility and physical possibility precedes consideration of financial feasibility and maximal productivity. The fact that a property use is financially feasible and/or maximally productive is irrelevant if the use is either illegal or not physically possible.

If the property zoning prohibits a potential use, it is pointless to consider that use unless there is a reasonable probability that the zoning can be changed. Physical considerations can be overcome by the addition of capital, but the economic impact of adding the capital must be evaluated and the incremental return from the added investment must be carefully measured.

## Legally Permissible

An appraiser does not need to address all of the legally permitted uses of a property, but it is inappropriate to proceed with a highest and best use analysis if the implied or proposed use is illegal. Greater land use regulation at the local, state, regional, and federal levels increases the need to address legal permissibility. Local zoning and regional comprehensive land use planning regulations are being implemented in most areas. The commissions charged with overseeing these regulations may be unwilling or unable to grant exceptions or variances to the regulations because of past or potential litigation.

Federal, state, and sometimes local regulations control the use and protection of wetlands, highly erodible land, floodplains, historic districts, and other special areas. Strict enforcement of these regulations can have a significant impact on potential uses of land affected by them. Federal and state legislation that is not specifically labeled as land use legislation may also impact property use. Controls that impact the environment such as the Clean Air Act, water quality legislation, and the Endangered Species Act can influence land use in areas surrounding the areas that are the primary focus of the laws. For example, the value of timberland in areas surrounding the Northern California habitat of the spotted owl has been impacted by more restrictive timber-cutting regulations. The appraiser must clearly document the influence of legal factors in the highest and best use analysis. If appraisers feel that they lack the competency to understand or interpret these factors, professional assistance should be obtained.

In some instances, an existing use predates the controlling regulations, and the use is referred to as a *legal nonconforming use.* Also referred to as a *grandfathered use,* a nonconforming use is a use that was established prior to the implementation of a zoning regulation or a change in zoning. Legally nonconforming uses are typically associated with improved property. If the use is transferrable, i.e., it follows the property rather than the ownership of the property, the value "as is," or value as improved, will reflect any value added by the nonconforming use. For example, consider a commercial greenhouse established in a residential zone. The improvements will have value as long as the market recognizes the contribution of the greenhouse improvements over the value of the residentially zoned land. However, if the improvements are demolished and the use ceases, the remaining land will have to conform to the existing residential zoning requirements. Local zoning regulations and the zoning enforcement officer should be consulted to determine what, if any, requirements are in place to deal with legally nonconforming uses.

A title search for deed restrictions and covenants should be made to ensure that there are no legal restrictions on property use in the chain of title. It is not necessary to proceed with the consideration of whether a use is physically possible, financially feasible, or maximally productive if it is prohibited by a deed restriction or covenant. There may be conservation easements placed on the property or restrictions on development; sometimes the development rights may have been severed from the bundle of rights by a previous owner.

Some common restrictions and/or covenants applicable to rural property include:

- Transfer of development rights to a local, state, or quasi-public agency. This type of conveyance may be found when analyzing rural property in an urban region such as Southern New England and New Jersey.
- Life estates. These interests, which typically relate to improved properties such as a dwelling or a farm, are held for a time period that is indefinite on the date the restriction is granted.
- Conservation easements. Easements may be placed on property by a landowner to protect a watershed, wildlife habitat, or portion of a property. They can be conveyed to municipalities or other governmental agencies, to land trusts, or to wildlife agencies such as the Audubon Society.
- Restrictive covenants. Covenants can be used to prohibit subdivision for a specified period of time; limit the size, shape, or configuration of a dwelling; or regulate the use of property –e.g., "for agricultural production not to include livestock raising and/or production."
- Downzoning. This is a change in zoning which results in the conversion of a property from one specified use to a different, less intensive use. For example, a property may have been previously zoned for one-acre residential lots and then be rezoned as an aquifer protection zone with the requirement that all lots subdivided from lots of 25 or more acres average two acres in size. Downzoning is often considered a regulatory taking or inverse condemnation.

## Physically Possible

The utility of real property is impacted by its physical characteristics; this is especially true for rural and agricultural property. Access, drainage, soils, size, shape, terrain, frontage, wetlands acreage, topography, and field size and configuration are characteristics that must be considered in the determination of highest and best use. The property's characteristics will control the selection and consideration of potential uses. Because all the properties in a given market compete for a particular use, in a limited or specialized use assignment the property with the most desirable characteristics, or combination of desirable characteristics, will be the most physically desirable.

Areas of Central California and Southern Arizona have desirable characteristics, including appropriate soils, water, and climate for raising a number of crops ranging from alfalfa to citrus crops to vegetables. Some of these crops can be grown nearly year-round. Other areas of the U.S. have the characteristics necessary for raising other crops or for conducting other agricultural enterprises that are not suitable in the Southwest. For example, the production of cranberries, other fruits and nuts, timber, and specialty crops like sugar cane or tobacco is conducted in various regions in the United States.

The improvements on a property can be considered part of the physical characteristics of a tract. The size, design, utility, and condition of the improvements must be evaluated. If a property is improved with dairy barns that are 30 years old, have stanchion housing, and no pipeline milking system, their contribution to the property is likely to be less than that of a modern, free-stall dairy barn with a milking parlor. At times, however, the social forces operating in the area must be considered. In areas such as Lancaster County, Pennsylvania, where the farming community is dominated by sectarian groups, the preferred improvements may be the older stanchion barn rather than the newer complex.

The location of the improvements on an agricultural property is a physical characteristic that may also influence use. This is especially true in estimating the potential highest and best use of a tract with a dwelling located directly adjacent to livestock housing improvements. The highest and best use of this tract may be different than the use of a similar tract that has the dwelling located some distance away from the livestock housing facility.

Some physical restrictions can be cured. Land that tends to be wet can be drained and hedgerows can be removed with heavy equipment. Laser leveling can improve the use of water and make farming easier on a particular tract. The cost to cure a deficiency needs to be recognized if correction is to be undertaken. An appraiser cannot assume that fields can be leveled, drained, or otherwise improved to attain a higher use without considering the other three criteria of highest and best use—i.e., the cure must be legally permissible, financially feasible, and maximally productive. An economic analysis may be required to evaluate how a physical restriction can be cured.

### Financially Feasible

Considering the criteria of legal permissibility and physical possibility generally results in the elimination of some potential uses. In appraisal theory, the uses that are legally permissible and physically possible are then tested applying the remaining two criteria. All legal and physically possible uses are tested in an economic model to determine which uses will result in positive stabilized cash flows. In the evaluation of commercial and industrial properties, it is generally stated that "...all uses that do, indeed, generate sufficient income to pay in full operating expenses, financial obligations, and capital amortization are deemed to be financially feasible." However, properties such as recreational tracts or single-family residences, which offer nonmonetary benefits, should be analyzed not on an income basis, but in terms of which use provides the highest value.

The financial feasibility of income-producing properties is estimated by developing a pro forma income and expense statement for each use. Anticipated gross income is estimated, allowances for vacancy and/or credit loss are made, operating expenses are calculated and subtracted from the effective gross income, and a likely net operating income *(NOI)* is derived for each use. Then a rate of return for invested capital is estimated. If the *NOI* that results from a potential highest and best use is sufficient to provide the required rate of return for invested capital **and** a positive return to the land, then that use is considered to be financially feasible.

In situations in which the potential revenues from the property do not exceed expenses, the question to be answered is which of the uses is the least costly use. This situation is generally short term. The market participants who are buying and selling such properties anticipate that the gain in land value (i.e., appreciation) will pay for the accumulated losses and provide a positive overall return to the investor.

In conducting property highest and best use analyses, appraisers typically spend more time evaluating the financial feasibility of uses than they spend evaluating any of the other three criteria.

## Maximally Productive

The potential use that satisfies the previous three criteria and results in the highest residual land value is considered to be the property's highest and best use. Residual land value can be estimated by one of two methods. The appraiser can estimate the value of the land and improvements for the selected use; subtract the cost of the labor, capital, and coordination (management) necessary to construct the improvements; and then subtract the overall accrued depreciation attributable to the improvements. These calculations are relatively easy to make when a new or proposed use is contemplated. They become more complex when the depreciation present in existing, older improvements must be estimated.

Alternatively, residual land value can be estimated by measuring the residual income remaining after all expenses have been paid and allocating this income to the land and improvements; then the residual income attributable to the land is capitalized by applying an appropriate capitalization rate. In traditional appraisal methodology, this is referred to as *straight-line, land residual capitalization*. The allocation of income to the improvements and the land is based on market analysis and must be supported by existing market conditions. When a land residual technique is used, the value of the improvements must be multiplied by a building capitalization rate estimated in the market. The income that is attributable to the improvements is subtracted from the *NOI* for the total property and the remaining income is allocated to the land.

The potential highest and best uses considered are generally long-term uses that are expected to be appropriate throughout the economic life of any improvements. The nature of the improvements will tend to dictate the appropriate economic life. For example, specialized livestock housing such as cage-laying poultry buildings typically have a relatively short economic life, which is dictated by the need to replace major components such as cages, feeders, waterers, and/or egg collection and grading equipment periodically. Technological changes and environmental influences both tend to limit the economic lives of poultry facilities. Similarly, most machinery storage buildings, shop buildings, general utility buildings, and dwellings have relatively long economic lives. As will be discussed in Chapter 13, The Cost Approach, economic life can be measured in a number of ways.

## Reconciliation of Highest and Best Use

As is common in appraisal, the appraiser completes a comprehensive analysis of highest and best use as if vacant and as improved and derives multiple solutions to what began as a seemingly simple question. It is not uncommon to find two or more uses that are legally permissible, physically possible, economically feasible, and maximally productive. Thus, the appraiser must, at this point, reconcile the indicated uses and draw a conclusion as to which use truly represents the property's highest and best use based on the market data available.

The conclusion may well be that a single highest and best use exists, especially if the market is "noncomplex"—i.e., largely homogeneous with few, if any, uses that do not support or complement each other. For example, the highest and best use of a grain farm in a grain-producing area, a single-family residence within a large subdivision, or a manufacturing or industrial plant in a planned industrial park may be fairly obvious.

Many rural properties are capable of supporting more than one highest and best use. There often are different uses on different tracts or parcels or complementary uses on the same tract. Depending on the market and market forces, multiple uses can interact as: 1) augmenting uses, 2) complementary uses, or 3) conflicting uses.

- Augmenting uses generally have a synergistic effect on value. The uses seeming to work well together and enhance the value of the property. The property value is greater with the combination of uses than the value supported by each individual use. Some examples of augmenting uses are ranches that have scenic amenities and provide recreational opportunities as well as commercial enterprises and crop farms that serve as hunting plantations subsequent to harvest and prior to planting season.
- Complementary uses are uses that add to value. The uses work together and result in a total property value that equals, or sometimes exceeds, the value supported by each of the uses. Examples of complementary uses are parcels that include cropland, pasture, and timberlands; plantation properties that have various types of cropland; and timberland that can support timber production as well as hunting.
- Conflicting uses can have a negative effect on value. These uses are inherently in conflict with each other and result in a value for the total property, when all uses are considered, that is less than the sum of the parts. Residential use and use as a landfill or another undesirable use are conflicting uses.

The appraiser may conclude that the highest and best use of a property is one that will last for a period of time and then change to another use at some point in the future. The use identified for the short term, generally three to seven years, is an interim use. As social, economic, and demographic forces and factors change over time, land use is also susceptible to change. The areas of change are known as *transitional areas,* and it is in the transitional market that the appraiser will likely encounter interim uses.

In some transitional markets, buyers invest in real estate in anticipation of profiting from increasing real estate prices. This practice was common during the 1970s and early 1980s, but it became less desirable in the economically stressful markets that characterized the late 1980s. In such markets, purchasers and landowners retain their real estate assets in anticipation that real estate price levels will increase significantly. Typically, there is little or no income generated by the property over the holding period. If the real estate has agricultural potential, and a market for rental exists, there may be sufficient income generated to offset the fixed costs of ownership—i.e., real estate taxes and some portion of the interest on borrowed capital. This type of ownership often results in increased risk, especially when the market is comprised of property that has high holding costs (for repairs, maintenance, etc.). This type of property is often described as having a speculative use; the appraiser must recognize that a speculative use is an interim use.

As mentioned previously, a use that was lawfully established and has been maintained over time but does not conform to the zoning and use regulations currently in effect is referred to as a legally nonconforming use. This kind of use frequently results from changes in land use regulations or from the promulgation of land use regulations after some uses have been established. Regulation changes can result in the creation of underimproved or overimproved properties.[2] A farm property that is subsequently zoned for industrial use is an underimproved property. It is likely that any farm-related improvements will be removed from the tract so that it can be improved to its highest and best use. Should immediate conversion of the farm to industrial use not be demanded by the market, the farm and its related improvements may constitute an interim use until the conversion to industrial use is financially feasible.

As an example of a nonconforming, overimproved property, consider a 50,000-sq. ft. greenhouse with 2,500 square feet of retail sales area that was constructed in 1980 and is situated on a three-acre tract. Over the years, the area surrounding the greenhouse has developed into a single-family residential neighborhood. In 1990 zoning was enacted that designated the area as a single-family residential district with a minimum lot size of 40,000 square feet. Since the new zoning, the greenhouse has become an overimproved property.

Zoning and land use regulations vary with the jurisdiction; they may permit a preexisting use to continue, but only without alteration. Or they may require that once the use ceases, it cannot be reestablished.

When valuing land with a legally nonconforming use, the appraiser must recognize the influence on value that results from the current use. The income under the current use may be greater than the income resulting from the legal uses or it may be less. The impact of the income stream on the ultimate underlying property value must be recognized. In most nonconforming use situations, the property is valued in accordance with its nonconforming use, but the land value is based on the legally permissible use of the property as if vacant. The difference between the two values reflects the value contribution of the improvements to the total property and may possibly indicate a bonus for the nonconforming use.

Property that is appropriate for a limited number of uses presents practical problems for the appraiser. The highest and best use of a special-purpose property as improved is probably the continuation of the current use, if that use is economically viable. For example, the highest and best use of a grain elevator is probably continued use as a grain elevator, unless it is not economically feasible to continue that use.

It may be appropriate to analyze and value a special-purpose property on the basis of two highest and best uses—i.e., continuation of the existing, special-purpose use and conversion to an alternative use. This is an analysis of the property's value in use and the potential property value in exchange. For example, a poultry facility may first be analyzed on the basis of its highest and best use. If the demand for similar

2. J. Mark Quinlivan and Vance R. Johnson, "Nonconforming-Use Properties: The Concept of Positive Economic Obsolescence," *The Appraisal Journal* (January 1981).

poultry units is low, an analysis of a highest and best use as warehouse or storage space may also be completed.

## Testing Highest and Best Use

Market forces will generally limit the choice of potential highest and best uses for a property to those that are compatible with a given area and similar to other property uses in the market. The highest and best use analysis must consider all of the potential and possible uses that are reflected in the market. Limitations on the uses are imposed by the four tests that are applied throughout the evaluation process and by the consideration of the property both as if it were vacant and as it is currently improved. The analysis of highest and best use may include a partial feasibility study and involve the use of appraisal tools such as discounted cash flow analysis. The nature of the property, the market area, and the market forces impacting the property will dictate the scope and depth of the highest and best use analysis conducted.

An appraiser should use caution in supporting an estimate of highest and best use. A given site may be well-suited for a particular use, but there may be other sites that are equally or better suited for the same use. An appraiser must test the conclusion of highest and best use to ensure that any potential and existing competition for the use has been adequately recognized and evaluated.

### Sample Highest and Best Use Analyses

The following examples are offered to illustrate the estimation highest and best use for rural property.

#### Highest and Best Use of Property as though Vacant

*Example 1*

Consider a 50-acre property suited to crop production that could also be used as a rural home with excess acreage. Use for crop production would require the construction of a machinery storage building at an estimated cost of $35,000 and would result in an estimated market value of $125,000. If the property is to be used as the site of a rural residence, a dwelling would have to be constructed at an estimated cost of $75,000 and the property would have an estimated market value of $150,000.

The estimated costs of constructing the improvements on the property for the different scenarios and the market values of the two uses can be compared in the analysis of highest and best use. The calculations used in the analysis are shown below:

| | Homesite | Agricultural Production |
|---|---|---|
| Market value | $150,000 | $125,000 |
| Cost to construct improvements | – 75,000 | – 35,000 |
| Residual land value | $75,000 | $90,000 |

Use for agricultural production is suggested by the analysis (if alternative uses have been eliminated based on other criteria). This use results in the highest residual land value.

*Example 2*

In estimating the highest and best use of an income-producing property, the competing uses are generally considered in terms of their ability to generate positive net operating income, the potential rates of return, and the perceived stability of the income. For example, consider an agricultural property consisting of approximately 100 irrigated acres that can be used to raise more than one crop. Analysis has shown that the property is currently used to produce corn and soybeans in rotation, and it generates a stabilized income of $65 per acre per year. It could also be used to raise vegetables, which would produce a stabilized income of $75 per acre per year. Finally, the land is suited to the production of specialized Chinese vegetables and small fruit such as strawberries in rotation; this use would result in a stabilized net income of $100 per acre per year. The overall rates of return for the three alternatives reflect their relative risk: 7.5% for corn and soybeans, 8.0% for vegetables, and 10.5% for Chinese vegetables and strawberries.

The potential net operating incomes, overall rates of return, and indicated land values for the three options are developed as follows:

| | Potential *NOI* | Overall Rate | Capitalized Value |
|---|---|---|---|
| Corn and soybeans | $6,500 | 7.5% | $86,667 |
| Vegetables | $7,500 | 8.0% | $93,750 |
| Chinese vegetables | $10,000 | 10.5% | $95,238 |

Of the three possible uses for the subject property, raising Chinese vegetables in rotation with strawberries results in the highest residual land value for the property. Thus, it is considered the highest and best use.

### Highest and Best Use of a Property as Presently Improved

An analysis of the highest and best use of a property as improved may suggest that little or no capital improvements are required. Alternatively, significant expenditures may be required to convert the property to a different use or to remodel or rehabilitate it for its existing use.

*Example 3*

Consider a free-stall-style dairy barn and its associated land base, which have been used for milking 150 cows a day. The property has been sold for $240,000 subject to mortgage approval. The dairy, including its supporting buildings, has a capacity of 200 milking cows, but to operate the property at capacity levels, a waste water treatment facility would be required. The estimated cost of constructing such a facility is about $150,000. Without the waste water treatment capability, the property cannot be operated at more than 75% of its total capacity—i.e., at a 150-cow level. Analysis of the market reveals that dairy farms similar to the subject sell for about $1,500 per cow without waste water treatment facilities. Farms with waste water treatment facilities sell for about $1,800 per cow. In considering the highest and best use of the dairy farm, the appraiser must assess the financial feasibility of constructing the necessary improvements to expand the property's capacity. The calculations follows:

| | |
|---|---|
| Estimated value of property without waste water treatment facility | |
| 150 cows @ $1,500 per cow | $225,000 |
| | |
| Purchase price of dairy | $240,000 |
| Cost to install treatment facility | 150,000 |
| Total investment to reach capacity | $390,000 |
| Estimated value of property with treatment facility | |
| 200 cows @ $1,800 per cow | $360,000 |

The analysis suggests that the expansion of the dairy to 200-head capacity from the existing 150-head capacity is legally permissible and physically possible, but it is neither financially feasible nor maximally productive.

## Summary and Standards

In the analysis of a property's highest and best use, an appraiser's objective is to identify the most probable use of real property that will result in the highest value. Highest and best use is a market-driven concept. It is generally defined as the probable, reasonable use of vacant land or an improved property that is legally permissible, physically possible, financially feasible, and results in the highest land value.

Highest and best use is an economic study of the market that is focused on a specific property. Market forces are evaluated and competing uses in the market area are observed. The highest and best use analysis rests on the data gathered and presented in the area analysis and property description sections of the appraisal.

When evaluating potential uses for a particular property, the appraiser considers 1) uses that are legally permissible, 2) uses that are physically possible, 3) uses that are financially feasible (result in a positive cash flow), and 4) uses that are maximally productive (result in the highest land value). These considerations are generally applied sequentially, beginning with legal permissibility.

Legal permissibility depends on public restrictions on use enacted through zoning regulations, land use regulations, building codes, preservation and conservation restrictions, and/or environmental regulations. Private restrictions such as deed covenants and the presence of long-term leases also influence the legal use of land.

Soils, climate, terrain, field size and shape, and accessibility must be considered when evaluating physical possibility. Although modern equipment can make many uses possible by leveling hills and improving accessibility, soils and climate can be inhibiting factors for many agricultural uses.

Those uses that are legally permissible and physically possible are next tested for financial feasibility. Simply put, a financially feasible use is one that produces sufficient cash flow to satisfy all expenses and provide a positive return to the property.

Among the uses that satisfy the criteria of legally permissibility, physical possibility, and financial feasibility, that use that results in the highest property price or value is the maximally productive use.

A highest and best use analysis begins with consideration of a property as if it were vacant and available for improvement. This concept assumes that a tract of land is either vacant or can be made vacant by removal of any improvements. It further assumes that buildings can be changed while land and site characteristics cannot. The purpose of determining a property's highest and best use as if vacant is to identify potential uses. The analysis can also help the appraiser identify comparable, or similar, recent land sales for use in the valuation process.

The market limits the number of potential highest and best uses. Competition from other, similar uses and the supply of and demand for the use should be considered. Finally, the use's conformity within the market is evaluated.

All appraisal reports in which an opinion of market value is developed must contain statements of highest and best use. In the Uniform Standards of Professional Appraisal Practice, Standards Rule 1-3(a) and (b) set forth requirements for developing a highest and best use estimate. Standards Rule 2-2 (a)(x), (b)(x), and (c)(x) set forth requirements for reporting highest and best use.

Standards Rule No. 1-3(a) requires consideration of the many factors that can affect the highest and best use of real estate. The comment to Standards Rule No. 1-3(b) says that an appraiser must " . . . recognize that land is appraised as though vacant and available for development to its highest and best use . . . "

Standards Rule 2-2 requires that a written report contain a statement of the appraiser's opinion as to the highest and best use of the real estate. (The accompanying comment recognizes that there are instances when an opinion is unnecessary, e.g., insurance valuation or "value in use" appraisals.) When an opinion is necessary, this standards rule requires that the reasoning in support of the opinion be included in the report.

There are several special considerations relating to highest and best use. The appraiser may have to recognize: 1) augmenting uses (two or more uses with synergistic effects on value), 2) complementary uses (two or more uses that contribute to value; 3) conflicting uses (two or more uses that have a negative effect on value).

Many rural markets are in a state of transition. The changes may be from rural to urban uses or from one rural use to another. Appraisers completing highest and best use studies in these markets should recognize that a use may be temporary in nature. A nonpermanent use is generally called an interim use and will remain in effect for a relatively short period of time (three to seven years). The ultimate highest and best use, i.e., the use realized at the end of an interim period, is called the terminal highest and best use.

Legally nonconforming uses are frequently found in areas that have recently enacted land use or zoning regulations. The nonconforming use predates the implementation of zoning and may represent a highest and best use of the particular property. When evaluating these uses, the appraiser must be familiar with the economics of the use and with any requirements and regulations relating to nonconforming uses.

Property that is considered to have a single, or special-purpose, use may be considered in terms of its ongoing use and in terms of alternative uses. Speculative uses, common in transitional markets, must be recognized as investments by landowners anticipating appreciation in land value.

CHAPTER

• 12 •

# *Sales and Data Analysis*

Market data are studied to give the appraiser a broad overview of the market and to help determine the subject property's highest and best use. Comparable sales data is required to apply the sales comparison approach and solid income data (rentals, leases, crop-share information, profit/loss data, etc.) is needed to develop the income approach. Reliable cost information and data on both improved and unimproved sales are essential to the cost approach to value.

The most important part of any appraisal is the analysis of available data from the market. It is critical that the reader of an appraisal report be given the full benefit of the appraiser's knowledge of the market and analysis of the data from within that market. The appraiser studies comparable sales to derive a market capitalization rate to be used in the income approach, to estimate depreciation in the cost approach, and to make adjustments that are applied in the sales comparison approach.

## Gathering the Market Data

The appraiser should conduct a very thorough review of the real estate market before selecting comparable sales, comparable rentals, and other important data necessary to complete the appraisal report. All information used in a quality appraisal report is derived from the market. A good comparable sale is no more or less important than a good comparable rental when appraising rural properties.

The appraiser collects comparable sales and rentals and carefully verifies all information. Comparable sales should be classified as unimproved and improved. Puritan unimproved sales (which have only one class of land) will lend support for land class ratio analysis. The valuation of the land component of improved sales is facilitated by analyzing information gathered from the unimproved land sales.

Most appraisers will develop their own techniques for analyzing and gathering market data. Suggested sources of comparable sales information include

- Real estate professionals such as brokers
- Title and abstract companies
- Attorneys familiar with real estate transactions
- Local tax appraisal districts and/or assessors
- County deed records
- Local agricultural lenders
- Local farmers and ranchers knowledgeable about the subject area

Possible sources of rental information pertaining to agricultural properties include

- Local county agents
- Local agricultural lenders (Farm Service Agency, Farm Credit Services, and commercial banks)
- Local agricultural suppliers and/or marketing outlets
- Local farmers and ranchers who are known to lease or rent agricultural lands
- Local trust departments and farm management firms

## Data Analysis in the Sales Comparison Approach

The data analysis performed in the sales comparison approach should lead the reader of the report through the development of a land class ratio and land adjustment analysis (if applicable) and through the paired sales analysis, which is performed to ascertain the degree or magnitude of the adjustments required for various elements of comparison.

In applying the sales comparison approach to value, the appraiser directly compares several comparable sales to the subject property to isolate the various elements affecting value. Before the sales are compared to the subject, they must be analyzed and compared to one another to determine the amount of adjustment required for elements of comparison such as land quality, terms of sale, time, location, size, improvements, and other characteristics. When estimating the value of an improved subject property, the appraiser will need to estimate the land value under several improved sales and then estimate how much value the building contributes to each of the sales. The total building contribution estimated for each improved sale must then be allocated to each individual building or other major improvement on the property. Once the appraiser has analyzed the comparable sales and identified the factors that affect market value, the next step is to compare the sales directly to the subject property in a sales comparison grid and eventually arrive at a final indication of value for the subject property.

## Land Class Ratio Analysis and Land Adjustment Process

When analyzing sales to derive a land adjustment, the appraiser must remember that this "land adjustment" may encompass all characteristics of the land, including productivity, topography, open land versus wooded land, and other characteristics. If the appraiser is working in an area where there is typically only one class of land, a productivity or land quality adjustment may adequately reflect the degree of comparability between the land component of a sale and the subject property. An accepted technique for estimating the contributory value of multiple land characteristics is land class ratio analysis.

The first step in developing a land class ratio analysis is to place all unimproved sales in a summary format which reflects their total sale price, date of sale, overall size, and sale price per acre as well as the acreage distribution among the predominant land classes recognized in that market (see Table 12.1). The land classes specified may be Class 1 cropland versus Class 2, 3, and/or 4 cropland; open land versus wooded land; cultivated land versus pasture; good rangeland versus poor rangeland; or any other observable land classifications recognized in the market where the subject property is located. The appraiser is cautioned not to use the eight capability unit classifications established by the USDA Soil Conservation Service. It is very likely that, in some areas, a Class IV capability rating on pasture could be more or less valuable than a Class VI capability rating on similar land. In a recreational land market, the appraiser may encounter situations in which scenic Class V floodplain or river bottom lands are much more valuable than good quality Class II cultivatable cropland.

In areas where the production capabilities of the soil are well-recognized by market participants, productivity indexes may serve to identify the various classes of land in the subject and comparable sale properties. Again, extreme caution should be exercised when relying on productivity indexes. Sometimes the production capabilities of a given soil type or land class do not enhance its value. Productivity indexes should only be used when buyers and sellers in the market appear to readily recognize the production capabilities associated with the farms purchased.

### Analysis of Unimproved Sales

Table 12.1 is an unimproved land sales summary chart, which might be used in an appraisal report to illustrate the development of a market-derived land class ratio. The appraiser can use such a chart to derive the per acre values of cropland and pasture. The completed grid is shown in Table 12.2. Note that this table shows the analysis of unimproved sales only (Sales 1, 2, 3, 4, 5, 6, 11, and 12). The improved sales (Sales 7, 8, 9, 10, and 13) are analyzed in the next section.

**Table 12.1** Unimproved Land Sales Summary For Land Class Ratio Analysis

| | | | | | | Contributory Value | | | |
|---|---|---|---|---|---|---|---|---|---|
| Sale No. | Date | Sale Price | Acres | Location | $/Acre | Crop Acres | Crop $/Acre | Pasture Acres | Pasture $/Acre |
| 1 | 6 months | $225,000 | 450 | Avg. | $500 | 0 | | 450 | |
| 2 | 6 months | $500,000 | 500 | Avg. | $1,000 | 500 | | 0 | |
| 3 | 6 months | $362,500 | 525 | Avg. | $690 | 200 | | 325 | |
| 4 | Current | $320,625 | 375 | Avg. | $855 | 300 | | 75 | |
| 5 | 1 year | $696,100 | 700 | Avg. | $994 | 600 | | 100 | |
| 6 | Current | $517,500 | 690 | Good | $750 | 300 | | 390 | |
| 11 | 6 months | $400,000 | 526 | Good | $760 | 200 | | 326 | |
| 12 | 1 year | $650,000 | 594 | Good | $1,094 | 509 | | 85 | |

One can readily observe that Sales 1 and 2 in Table 12.1 are "puritan" land class sales—i.e., sales having only one class of land. Sale 1 suggests that pasture in an average location had a value of $500 per acre six months ago. Sale 2 reflects a value of $1,000 per acre for cropland in an average location six months ago. The land values by land class for Sale 3 are determined by simple abstraction. Assume that, based on Sale 2, the cropland in Sale 3 has a value of $1,000 per acre. Note that Sale 2 and Sale 3 are similar in size, location, and date of sale.

| Example 1: Analysis of Sale 3 | |
|---|---|
| Total sale price of Sale 3 | $362,500 |
| Less cropland: 200 acres @ $1000 | − 200,000 |
| Residual attributable to pasture value | $162,500 |
| $162,500 / 325 acres of pasture = $500 per acre | |
| Conclude that Sale 3's cropland value is | $1,000/acre |
| Conclude that Sale 3's pasture value is | $500/acre |

The first three sales shown in the table give a strong indication that a 100% to 50% relationship exists between cropland and pasture in this market—i.e., cropland is worth twice as much as pasture. In this first example, the two puritan sales, Sales 1 and 2, were used to indicate the value of cropland and pasture. Then, based on this analysis, it was assumed that the market was indicating a value of $1,000 per acre for the cropland in Sale 3 and a residual value for the pasture of $500 per acre. It should be emphasized that the first three sales all occurred in the same time frame, are approximately the same size, and are similar in location. Sales 4, 5, and 6, and Sale 12 are either current or one-year-old transactions. Sale 11 is similar to Sales 1, 2, and 3 in time (6 months ago) but differs in location.

Analysis of Sales 4 and 5 indicates that one cannot apply the $1,000 per acre for cropland to either of these sales. If a $1,000-per-acre value is assumed for the cropland in Sale 4, then a value of approximately $275 per acre would be indicated for the pasture in Sale 4. Conversely, if the

$1,000-per-acre cropland value is applied to Sale 5, then a residual value of $961 per acre is suggested for the pasture in Sale 5. Both of these value indications for pasture are inconsistent with the information revealed in the first three sales. This suggests that time may be an important factor to consider in the analysis of properties in this market.

To develop a value opinion for the two classes of land in Sales 4 and 5, the 100% to 50% land class ratio derived from Sales 1 through 3 is applied. The following examples illustrate how the appraiser can apply the land class ratio analysis reflected by the first three sales to estimate the value contribution of each class of land in Sales 4 and 5. The same procedure is used to derive land class values for Sales 6, 11, and 12.

| Example 2: Analysis of Sale 4 | | | |
|---|---|---|---|
| Cropland | | 300 acres × 100% = | 300.0 |
| Pasture | | 75 acres × 50% = | 37.5 |
| Total | | 375 acres | 337.5 net crop equivalent acres |
| Total sale price of Unimproved Sale 4 | | | $320,625 |
| Then: $320,625 / 337.5 net crop equivalent acres = | | | $950 |
| $950/ac. × 50% = | | | $475 |
| Conclude | | | |
| Sale 4's cropland value | | | $950 per acre |
| Sale 4's pasture value | | | $475 per acre |
| Proof | | | |
| 300 acres | Cropland | @ $950/ac. = | $285,000 |
| 75 acres | Pasture | @ $475/ac. = | $35,625 |
| 375 acres | Total | @ $855/ac. = | $320,625 |

In this second example, we find that the sale price of $320,625 is simply allocated over what is termed *net crop equivalent acres.* In simple terms, this indicates that if Sale 4 had contained only cropland (and no pasture at all), then the property would have had no more than 337.50 acres. The $320,625 sale price spread over 337.5 crop equivalent acres yields a value indication of $950 per acre for the cropland in Sale 4. Since pasture in this market sells for 50% of the value of cropland, then it stands to reason that the pasture in Sale 4 is valued at 50% of the $950 per acre cropland value indication. This results in a land value of $475 per acre for pasture.

The important part of this example lies in the proof shown. This proof indicates that if the $950 per acre is applied to the 300 acres of cropland, and the $475 per acre is applied to the 75 acres of pasture, then the total sale price should equal $320,625. This confirms that the analysis on Sale 4 is correct.

This same procedure can be applied to Sale 5 to reflect a value indication for the cropland and pasture in that sale. Remember, the proof is the most important part of the analysis, at least until the appraiser becomes totally comfortable with land class ratio analysis.

| Example 3: Analysis of Sale 5 | | |
|---|---|---|
| Cropland | 600 acres × 100% = | 600.0 |
| Pasture | 100 acres × 50% = | 50.0 |
| Total | 700 acres | 650.0 net crop equivalent acres |
| Total sale price of Unimproved Sale 5 | | $696,100 |
| Then: $696,100 / 650.0 net crop equivalent acres = | | $1,070.92 |
| $1,070.92/ac. × 50% = | | $535.46 |
| Conclude | | |
| Sale 5's cropland value | | $1,071 per acre (rounded) |
| Sale 5's pasture value | | $535 per acre (rounded) |
| Proof | | |
| 600 acres | Cropland @ $1,071/ac. = | $642,600 |
| 100 acres | Pasture @ $535/ac. = | $53,500 |
| 700 acres | Total @ $ 994/ac. = | $696,100(R) |

After completing similar land class analyses for the remaining unimproved land sales (Sales 6, 11, and 12), estimated values per acre for each respective class of land can be incorporated into the grid (see Table 12.2).

### Analysis of the Improved Sales: Estimating the Building Contribution

Once the appraiser has calculated the land class values for the unimproved sales in the subject's market, the next step is to estimate the building contribution reflected in the improved sales within the data set. The land values developed for the improved sales are based on the land class values reflected in the unimproved land sales. In analyzing land-intensive properties (i.e., properties in which land value represents the major part of total property value), the appraiser should estimate the land value first. Then the remaining value is attributable to the building improvements.

It has been common practice to simply "estimate" the building's contribution to value using an abbreviated cost approach technique and then deduct this building value from the sale price to arrive at the land

**Table 12.2** Unimproved Land Sales Summary for Land Class Ratio Analysis

| | | | | | | Contributory Value | | | |
|---|---|---|---|---|---|---|---|---|---|
| Sale No. | Date | Sale Price | Acres | Location | $/Acre | Crop Acres | Crop $/Acre | Pasture Acres | Pasture $/Acre |
| 1 | 6 months | $225,000 | 450 | Avg. | $500 | 0 | N/A | 450 | $500 |
| 2 | 6 months | $500,000 | 500 | Avg. | $1,000 | 500 | $1,000 | 0 | N/A |
| 3 | 6 months | $362,500 | 525 | Avg. | $690 | 200 | $1,000 | 325 | $500 |
| 4 | Current | $320,625 | 375 | Avg. | $855 | 300 | $950 | 75 | $475 |
| 5 | 1 year | $696,100 | 700 | Avg. | $994 | 600 | $1,071 | 100 | $535 |
| 6 | Current | $517,500 | 690 | Good | $750 | 300 | $1,045 | 390 | $523 |
| 11 | 6 months | $400,000 | 526 | Good | $760 | 200 | $1,102 | 326 | $551 |
| 12 | 1 year | $650,000 | 594 | Good | $1,094 | 509 | $1,179 | 85 | $589 |

value. **This is a totally unacceptable procedure** since buildings on agricultural properties often have little or no contributory value. The following example is presented to illustrate an appropriate methodology. Note that Sale 7 (shown in Table 12.3) is an improved sale that is similar in location, size, and date of sale to unimproved Sale 3 (shown in Table 12.2).

| Example 4: Analysis of Sale 7 | |
|---|---|
| Total sale price of 300 acres w/buildings | $300,000 |
| Less crop value of 200 acres @ $1,000/acre (Sale 3) | – 200,000 |
| Less pasture value of 100 acres @ $500/acre (Sale 3) | – 50,000 |
| Residual attributable to Sale 7's buildings | $50,000 |
| $50,000 / 300 acres = | $167 per acre building value |

Often less knowledgeable appraisers develop an improper or inappropriate allocation of value. In the following **inappropriate allocation,** it is assumed that Sale 7 has no improvements other than a modest, wood-frame dwelling of approximately 2,500 square feet. This dwelling was built about 10 years ago and has an effective age of 10 years. The building has good utility and is in good condition. Assume that this dwelling would have a reproduction cost of approximately $32 per square foot, indicating a cost new of $80,000. If the value components in this sale had been allocated by simply "estimating" a building contribution using a cost approach technique, one would have arrived at the erroneous residual land value shown below.

| Example 5: Analysis of Sale 7 | | |
|---|---|---|
| ****** INAPPROPRIATE Building Value Allocation ****** | | |
| Sale price | | $300,000 |
| Cost of dwelling 2,500 sq.ft. @ $32/sq.ft. | $80,000 | |
| Less 16.6% depreciation(10 yrs./60 yrs.) | – 13,333 | |
| Building value estimate | | – 66,667 |
| Residual land value | | $233,333 |
| **WRONG !!!!** | | |

It is obvious that this inappropriate allocation has resulted in a low land value. Applying the 100% to 50% land class ratio derived earlier, a value of $933 per acre is indicated for Sale 7's cropland and a value of $467 per acre is indicated for its pasture. Since Sale 7 is very comparable to Sale 3 in all other elements of comparison except the presence of improvements, it is easy to see that, in this particular sale, values have been misallocated by estimating the value of the building incorrectly. When evaluating the building's contribution to the value of an improved rural property, the appraiser should *always* estimate the land value first and let the remainder of the sale price flow to the building.

In Table 12.3, all the improved sales have been analyzed in the same manner shown in the previous example. Note that improved Sale 8's

**Table 12.3** Improved Sale Summary

| | | | | | | Land Class Acres | | | | | |
|---|---|---|---|---|---|---|---|---|---|---|---|
| | | | | | | Crop | Crop | Pasture | Pasture | Bldg. | Bldg. |
| Sale No. | Date | Sale Price | Acres | Location | $/Acre | Acres | $/Acre* | Acres | $/Acre | Acres | $/Acre |
| 7 | 6 months | $300,000 | 300 | Avg. | $1,000 | 200 | $1,000 | 100 | $500 | House | $167 |
| 8 | Current | $391,200 | 446 | Avg. | $877 | 346 | $950 | 100 | $475 | Shed | $34 |
| 9 | 1 year | $362,150 | 330 | Good | $1,097 | 255 | $1,179 | 75 | $589 | sh/brns | $53 |
| 10 | 6 months | $457,500 | 450 | Good | $1,017 | 300 | $1,100 | 150 | $550 | h/s/b | $100 |
| 13 | 1 year | $538,000 | 500 | Avg. | $1,076 | 392 | $1,070 | 108 | $535 | h/s/b | $122 |

h = house; sh = shed; brns = barns

* Values for each land class are derived from those sales in Table 12.2 that are most similar in location and date of sale to each improved sale in this chart. The allocated values have been rounded to the nearest five dollars.

per-acre land class values are the same as those shown for unimproved Sale 4 in Table 12.2. Sale 4 and Sale 8 are very similar in date of sale, location, and size. Also, improved Sale 9's land class values (shown in Table 12.3) are similar to those of unimproved Sale 12 (in Table 12.2). Improved Sale 10's land class values are taken from the values indicated by unimproved Sale 11; the land class values for improved Sale 13, which is reasonably comparable to unimproved Sale 5 in most elements of comparison, are based on Sale 5's land class values.

The key to this land valuation method lies in the appraiser's ability to recognize the similarities in size, location, and date of sale between the improved properties and the unimproved land sales for which values have been previously defined. Once a total land value has been developed for each improved sale, this land value is deducted from the total sale price to reflect the residual value attributable to the buildings or improvements. These residual improvement values are then reduced to a per-acre value contribution and inserted in the last column of Table 12.3.

Table 12.4 is an example of how one might allocate the total building value of each improved sale shown in the last column of Table 12.3 to each of the respective improvements or buildings on the individual properties.

The analysis of Sale 7 in Example 4 indicated that the buildings (in this case, the dwelling) contributed $50,000 to the sale price. Table 12.4 shows the dollars-per-square-foot contributory value of the 2,500-sq.-ft. house on Sale 7, which is in good condition and has good utility.

Sales 8, 9, 10, and 13 are shown to have total building contributions of $15,000, $17,000, $45,000, and $60,780, respectively. To prepare the table, a detailed analysis of these sales was conducted in the same manner shown for Sale 7. All of these residual building or improvement values are "as of the date" of each sale.

As shown in Table 12.4, Sale 8 is improved with only one building—a 4,000-sq.-ft. equipment shed in good condition with good utility. This shed represents all of the $15,000 building contributory value, indicating a value of $3.75 per square foot for a good/good equipment shed. Sales 9, 10, and 13 each have more than one building. The value information

**Table 12.4** Allocation of Buildings on Improved Sales

| Sale No. | 7 | | 8 | | 9 | | 10 | | 13 | |
|---|---|---|---|---|---|---|---|---|---|---|
| Sale date | 6 months | | Current | | 12 months | | 6 months | | 12 months | |
| Total improvement value* | $50,000 | | $15,000 | | $17,000 | | $45,000 | | $60,780 | |
| **Improvements** | **Cond./Utility** | **$** | **Cond./Utility** | **$** | **Cond./Utility** | **$** | **Cond./Utility** | **$** | **Cond./Utility** | **$** |
| | **Size** | **$/Unit** | **Size** | **$/Unit** | **Size** | **$/Unit** | **Size** | **$/Unit** | **Size** | **$/Unit** |
| Dwelling (in sq. ft.) | Good/Good | $50,000 | | None | | None | A-A | $36,000 | G-G | $52,500 |
| | 2,500 | $20 | | | | | 2,400 | $15 | 2,625 | $20 |
| Equip. shed (in sq. ft.) | | None | Good/Good | $15,000 | Good/Good | $12,000 | Fair/Fair | $4,500 | Avg./Avg. | $ 8,280 |
| | | | 4,000 | $3.75 | 3,200 | $3.75 | 4,000 | $1.13(R) | 4,140 | $2.00 |
| Grain bins (in bu. cap.) | | None | | None | Good/Good | $5,000 | Good/Avg. | $4,500 | Poor/Poor | $0.00 |
| | | | | | 5,000 | $1.00 | 6,000 | $0.75 | 5,000 | $0.00 |
| Total improvement contrib | | $50,000 | | $15,000 | | $17,000 | | $45,000 | | $60,780 |

* These total values were derived by applying the market abstraction technique to each improved sale.

revealed by Sales 7 and 8 was used to allocate each sale's total improvement value to the individual components or structures on each property.

Sale 13 took place within six months of Sale 7. Both Sale 7 and Sale 13 have good/good dwellings that are relatively similar in size. Using the $20 per square foot value suggested in Sale 7's analysis, a dwelling value of $52,500 can be estimated for Sale 13. Since the total improvement value of Sale 13 is $60,780, a residual of $8,280 is left for Sale 13's average/average equipment shed and its poor/poor grain bins. It is not unreasonable to assume that Sale 13's poor grain bins have no contributory value. This assumption suggests that the $8,280 residual is attributable to Sale 13's equipment shed. The 4,140-sq.-ft. equipment shed thus has a unit value of $2.00 per square foot. A $2.00 per square foot value for the average/average equipment shed in Sale 13 not only lends support for the good/good shed value of $3.75 in Sale 8, but also supports the $20 per square foot value ascribed to Sale 13's good/good dwelling, which was derived from the Sale 7 analysis.

Since both Sale 8 and Sale 13 give good indications of value for equipment sheds, Sale 9 can be analyzed to suggest a residual value for its good condition/good utility grain bins. Sale 9 is one year old, as is Sale 13. However, Sale 13's equipment shed is judged to be only average in utility and condition, while Sale 9's shed has good utility and is in good condition. Sale 13's analysis reveals that the equipment shed on Sale 9 must be worth more than $2.00 per square foot. The Sale 8 equipment shed is very similar to the Sale 9 equipment shed in utility and condition, and Sale 8 indicates that a good/good shed has a value of $3.75 per square foot.

In the analysis of these sales, no determination has been made about the influence of time on land values or building values. Although Sale 13, which is one year old, sets a $2.00 parameter for average/average shed value, Sale 8 suggests a value of $3.75 per square foot for good/good sheds. Focusing only on condition and utility, it is reasonable to assume that the good/good shed on Sale 9, which is one year old, has a value near $3.75 per square foot.

The total building value for Sale 9, after deducting the land contribution from its actual sale price, was estimated to be $17,000. The $3.75 per square foot equipment shed value estimate reflects a $12,000 contribution for that building ($3.75 × 3,200 sq. ft.). Deducting the shed contribution of $12,000 from the total building contribution of $17,000 leaves a residual of $5,000 for the grain bins on Sale 9. These grain bins have good condition/good utility and a capacity to store 5,000 bushels, thereby reflecting a value of $1.00 per bushel of grain bin capacity.

It should be emphasized that common sense and a working knowledge of new costs for various farm improvements are important in the allocation of building value. For example, if the analysis of Sale 9 had suggested that the equipment shed value should have been $3.00 per square foot rather than $3.75 per square foot, a total equipment shed contribution of only $9,600 ($3.00 × 3,200 sq. ft. ) would have been indicated. With the market-derived residual value of $17,000 for all of Sale 9's buildings, a value of $7,400 would have been indicated for the grain bins on Sale 9. This $7,400 value would have resulted in a per bushel capacity value of $1.48 ($7,400/5,000 bushel capacity). If the cost new of similar grain bins one year ago was $1.20 to $1.30 per bushel of capacity, it would be evident that the equipment shed had been undervalued at $3.00 per square foot. The grain bins cannot contribute more than their cost new at the time the sale occurred.

The analysis of Sale 10 requires significantly more judgment than the other four improved sale allocations. Since most of Sale 10's total building value of $45,000 is likely to be represented by the dwelling, it is reasonable to use the dwelling analyses of Sales 7 and 13 to aid in the valuation of Sale 10's average condition/average utility dwelling. The good/good dwellings on Sales 7 and 13 confirm that an average/average dwelling similar in size should have a value somewhat less than $20 per square foot. Selecting a value of $15 per square foot for Sale 10's dwelling is very reasonable. At $15 per square foot, Sale 10's dwelling has a contributory value of $36,000 ($15 × 2,400 sq. ft.).

It requires slightly less judgment to compare good/good grain bins to good/average grain bins than it does to compare an average/average or good/good equipment shed to a fair/fair equipment shed. Therefore, it is sensible to estimate Sale 10's grain bins based on the results of the Sale 9 analysis. If the good condition/good utility bins on Sale 9 have a value of $1.00 per bushel, it is reasonable to assume that the good/average bins on Sale 10 have a value of about $0.75 per bushel of capacity. This suggests a value contribution of $4,500 for the 6,000-bushel storage facility on Sale 10 ($0.75 × 6,000 bushel capacity).

Once the values of the dwelling and grain bins on Sale 10 have been estimated, the remainder, or residual, of the $45,000 total improvement value can be attributed to Sale 10's fair/fair equipment shed. The following example summarizes the method of abstraction used to estimate Sale 10's shed value:

| Example 6: Analysis of Sale 10 | |
|---|---|
| Total building/improvement value | $45,000 |
| Less value of dwelling | – 36,000 |
| Less value of grain bins | – 4,500 |
| Residual to equipment shed | $4,500 |

The $4,500 residual for Sale 10's equipment shed equates to a unit value of $1.13 per square foot. This finding is well within reason for a fair condition/fair utility shed since average/average sheds and good/good sheds have been shown to contribute value at a rate of $2.00 and $3.75 per square foot, respectively.

The building allocation process can become very complex when the appraiser is not fortunate enough to have at least one or two good puritan (only one building) sales. However, good judgment will be invaluable in allocating the total building value to the individual buildings and improvements. As long as the individual building values are *reasonable* and supported by analyses of other sales, no fault will be found with the appraiser's building allocation methodology.

## Compare Sales to Subject Land and Buildings

Note that up to this point little has been said about the hypothetical subject property. The subject property's characteristics are important in the basic research insofar as the comparable sales to be used in the analysis must have the same highest and best use as the subject and other characteristics of the sales should be "reasonably" comparable to the subject. However, when analyzing sales to derive land class ratios and building value contributions, the subject must almost be left out of consideration. Once the land class ratios and the building allocations have been developed for the improved sales, they should be compared directly to the subject's land and buildings to arrive at an adjustment to reflect each sale's land mix and building contribution.

When the appraiser has arrived at land and building values for each improved sale, several paired sales analyses can be conducted. The improved sales are adjusted for elements of comparison such as terms, time, size (land area), location, and other characteristics affecting property values in the subject's market.

### The Subject Property

For the purposes of this presentation, assume that the subject has the following basic characteristics:

| | | |
|---|---|---|
| Date of appraisal | | Current |
| Location | | Good |
| Acres | 510 | |
| Cropland | 400 | |
| Pasture | 110 | |
| Buildings | | |
| Dwelling: average condition, good utility, 2,250 sq. ft. | | |
| Equipment shed: average condition, average utility, 4,200 sq. ft. | | |
| Grain bins: fair condition, fair utility, 10,000 bushel capacity | | |

When comparing the improved sales to the subject in terms of their land and buildings, all other elements of comparison are ignored or assumed equal.

The land adjustment process is predicated on the following assumption: If all other elements of comparison for the subject and a given sale were equal, then the subject's cropland can be no more or less valuable on a per-acre basis than the cropland value of the sale being compared to the subject. This is true for each and every class of land, regardless of its value contribution in the market being studied.

The building adjustment is accomplished by comparing the individual buildings on each sale property directly to the subject's buildings on a per-building, per-square-foot, or per-bushel basis. Comparisons can also be made based on another unit or on an overall basis. Judgment may be required to compare a good condition/good utility building on a given sale property to a similar subject building with only fair condition and utility. Detailed market research and proper building allocation of the comparable sales will set excellent parameters and inform the appraiser's judgement in this direct comparison process.

The tables that follow will illustrate how to compare the improved sales to the subject for land quality or land mix and for buildings. The sales analyzed are those shown in Table 12.3. Note that the details of each sale are shown on the left side of each table, while the subject characteristics are shown on the right. The per-acre land class values for each improved sale are also taken from Table 12.3, while the sales' building/improvement values are taken from Table 12.4.

From a practical perspective, only sales with a full complement of buildings should be used for direct comparison to a subject property. Although Sales 7, 8, and 9 have fewer buildings than the subject, they have been included to properly "spread" the building values on sales that have more than one building. Good appraisal practice dictates that one should use sales with less than the full complement of buildings found on a subject property only when more similar sales are unavailable.

In Table 12.5 the indicated land adjustment suggested by this comparison is plus (+) $59/acre to Sale 7's $833/ac. land value ($892 – 833 = $59).

The building adjustment required to make Sale 7 more like the subject in buildings is the difference between Sale 7's $167 per acre building value and the $106/acre subject building value ($167 – $106 = $61).

In Table 12.6, the good/good dwelling on Sale 7 is judged to be somewhat superior to the average/good house on the subject. A value estimate of $18 per square foot for the subject's dwelling is within reason, especially if one reviews the overall findings for dwellings shown in Table 12.4. Since Sale 7 has no equipment shed or grain bins, one must again rely on the allocation process completed in Table 12.4 for sales of properties with grain bins and/or sheds. The $2.00 per square foot for the subject's average/average equipment shed is in line with the values for equipment sheds in Table 12.4, and the $0.50 per bushel for the subject's fair/fair grain bins is well supported by the Table 12.4 analyses.

Land and building adjustment analyses for Sale 8 are shown in Tables 12.7 and 12.8.

The subject land value of $848 per acre less the $843 per acre Sale 8 land value indicates that a + $5 per acre adjustment is required to make Sale 8 like the subject in land quality.

**Table 12.5** Land Adjustment Analysis

| | Improved Sale 7 | | | | Subject | | |
|---|---|---|---|---|---|---|---|
| Cropland | 200 ac. @ | $1,000 | $200,000 | Cropland | 400 ac. @ | $1,000 | $400,000 |
| Pasture | 100 ac. @ | $500 | $50,000 | Pasture | 110 ac. @ | $500 | $55,000 |
| Total | 300 ac. @ | $833 | $250,000 | Total | 510 ac. @ | $892 | $455,000 |
| Indicated land adjustment to Sale 7 = + $59/acre | | | | | | | |

**Table 12.6** Building Adjustment Analysis

| | Buildings on Sale 7 | | | | | Buildings on Subject | | | |
|---|---|---|---|---|---|---|---|---|---|
| | Cond./ Utility | Size | $/Unit | $ | | Cond./ Utility | Size | $/Unit | $ |
| Dwelling | G/G* | 2,500 @ | $20 | $50,000 | Dwelling | A/G* | 2,250 @ | $18 | $40,500 |
| Eq. shed | None | | | | Eq. shed | A/A | 4,200 @ | $2 | $8,400 |
| Gr. bin | None | | | | Gr. bin | F/F | 10,000 @ | $0.50 | $5,000 |
| Total | | | | $50,000 | Total | | | | $53,900 |
| Building value per acre | | | | $167 | Building value per acre | | | | $106 |
| Indicated building adjustment to Sale 7 = – $61/acre | | | | | | | | | |

Condition/utility ratings: G = good; A = average; F = fair

The subject's suggested building value of $110 per acre is $76 per acre superior to the $34 per acre building value shown for Sale 8.

In Table 12.7 the subject is shown to be modestly superior in land mix or land quality when compared to Sale 8. Sale 8 has no dwelling or grain bins, so value estimates for these two improvements can be interpolated from the analyses in Table 12.4. If one then assumes that there is no other sale to compare to the subject for equipment shed values, then the good/good shed on Sale 8 might suggest a value of

**Table 12.7** Land Adjustment Analysis

| | Improved Sale 8 | | | | Subject | | |
|---|---|---|---|---|---|---|---|
| Cropland | 346 ac. @ | $950 | $328,700 | Cropland | 400 ac. @ | $950 | $380,000 |
| Pasture | 100 ac. @ | $475 | $47,500 | Pasture | 110 ac. @ | $475 | $52,250 |
| Total | 446 ac. @ | $843 | $376,200 | Total | 510 ac. @ | $848 | $432,250 |
| Indicated land adjustment to Sale 8 = + $5/acre | | | | | | | |

**Table 12.8** Building Adjustment Analysis

| | Buildings on Sale 8 | | | | | Buildings on Subject | | | |
|---|---|---|---|---|---|---|---|---|---|
| | Cond./ Utility | Size | $/Unit | $ | | Cond./ Utility | Size | $/Unit | $ |
| Dwelling | none | | | | Dwelling | A-G | 2,250 @ | $18 | $40,500 |
| Eq. shed | G-G | 4,000 @ | $3.75 | $15,000 | Eq. shed | A-A | 4,200 @ | $2.50 | $10,500 |
| Gr. bin | none | | | | Gr. bin | F-F | 10,000 @ | $0.50 | $5,000 |
| Total | | | | $15,000 | Total | | | | $56,000 |
| Building value per acre | | | | $34 | Building value per acre | | | | $110 |
| Indicated building adjustment to Sale 8 = + $76/acre | | | | | | | | | |

about $2.50 per square foot for the average/average shed on the subject. This Sale 8 building adjustment analysis suggests that Sale 8 requires a positive $76 per acre adjustment to make it more like the subject in the characteristic of buildings.

Analysis of Sale 9 is presented in Tables 12.9 and 12.10. It is assumed in Table 12.10 that no other sales were available for comparison to the subject property. The good/good equipment shed value of $3.75 per square foot for Sale 9 might suggest a value of $2.50 per square foot for the subject's average/average shed. Furthermore, Sale 9's good/good grain bins, at $1.00 per bushel of capacity, strongly support a value of $0.60 per bushel for the subject's fair/fair grain bins. The subject's $18-per-sq.-ft. dwelling value is taken from an overview of the dwelling values in Table 12.4

Sale 10 is analyzed in Tables 12.11 and 12.12. In Table 12.11 the subject is shown to have a significantly superior land mix than Sale 10. A quick review of the percentage of the higher-valued cropland in each property lends support to the need for a land adjustment. Approximately 78.50% of the subject's 510 acres are cropland and Sale 10 is only 66.70% cropland, so a moderate land adjustment in favor of the subject property is appropriate.

**Table 12.9** Land Adjustment Analysis

| | Improved Sale 9 | | | | Subject | | |
|---|---|---|---|---|---|---|---|
| Cropland | 255 ac. @ | $1,180 | $300,900 | Cropland | 400 ac. @ | $1,180 | $472,000 |
| Pasture | 75 ac. @ | $590 | $44,250 | Pasture | 110 ac. @ | $590 | $64,900 |
| Total | 330 ac. @ | $1,046 | $345,150 | Total | 510 ac. @ | $1,053 | $536,900 |
| Indicated land adjustment to Sale 9 = + $7/acre | | | | | | | |

**Table 12.10** Building Adjustment Analysis

| | Buildings on Sale 9 | | | | | Buildings on Subject | | | |
|---|---|---|---|---|---|---|---|---|---|
| | Cond./ Utility | Size | $/Unit | $ | | Cond./ Utility | Size | $/Unit | $ |
| Dwelling | none | | | | Dwelling | A-G | 2,250 @ | $18 | $40,500 |
| Eq. shed | G-G | 3,200 @ | $3.75 | $12,000 | Eq. shed | A-A | 4,200 @ | $ 2.50 | $10,500 |
| Gr. bin | G-G | 5,000 @ | $1.00 | $5,000 | Gr. bin | F-F | 10,000 @ | $ 0.60 | $6,000 |
| Total | | | | $17,000 | Total | | | | $57,000 |
| Building value per acre | | | | $52 | Building value per acre | | | | $112 |
| Indicated building adjustment to Sale 9 = + $60/acre | | | | | | | | | |

**Table 12.11** Land Adjustment Analysis

| | Improved Sale 10 | | | | Subject | | |
|---|---|---|---|---|---|---|---|
| Cropland | 300 ac. @ | $1,100 | $330,000 | Cropland | 400 ac. @ | $1,100 | $440,000 |
| Pasture | 150 ac. @ | $550 | $82,500 | Pasture | 110 ac. @ | $550 | $60,500 |
| Total | 450 ac. @ | $917 | $412,500 | Total | 510 ac. @ | $981 | $500,500 |
| Indicated land adjustment to Sale 10 = + $64/acre | | | | | | | |

**Table 12.12** Building Adjustment Analysis

| | Buildings on Sale 10 | | | | | Buildings on Subject | | | |
|---|---|---|---|---|---|---|---|---|---|
| | Cond./ Utility | Size | $/Unit | $ | | Cond./ Utility | Size | $/Unit | $ |
| Dwelling | A-A | 2,400 @ | $15 | $36,000 | Dwelling | A-G | 2,250 @ | $17 | $38,250 |
| Eq. shed | F-F | 4,000 @ | $1.13 | $4,500 | Eq. shed | A-A | 4,200 @ | $1.75 | $7,350 |
| Gr. bin | G-A | 6,000 @ | $0.75 | $4,500 | Gr. bin | F-F | 10,000 @ | $0.50 | $5,000 |
| Total | | | | $45,000 | Total | | | | $50,600 |
| Building value per acre | | | | $100 | Building value per acre | | | | $99 |
| Indicated building adjustment to Sale 10 = + $1/acre (say $0) | | | | | | | | | |

The building adjustment analysis for Sale 10 in Table 12.12 is predicated on the assumption that this is the only comparable sale available for direct comparison to the subject's buildings or improvements. Sale 10's average/average dwelling, valued at $15 per square foot, would suggest a value of about $17 per square foot for the subject's average/good dwelling. The fair/fair equipment shed on Sale 10 at $1.13 per square foot lends support to a value estimate of $1.75 per square foot for the subject's average/average shed. And finally, the good/average grain bins on Sale 10, which reflect a unit value of $0.75 per bushel of storage capacity, support a $0.50 per bushel value for the fair/fair grain bins on the subject property.

Tables 12.13 and 12.14 present the analysis of Sale 13. The result of Sale 13's land adjustment analysis shown in Table 12.13 reinforces the fact that land class ratio analysis really works. Sale 13 requires virtually no land adjustment because the subject has 78.43% of its 510 total acres in the cropland class, while Sale 13 has a nearly equal 78.40% of its total acreage in the cropland class. This relationship suggests that the properties are nearly equal in land mix, and therefore no land adjustment is warranted.

**Table 12.13** Land Adjustment Analysis

| | Improved Sale 13 | | | | Subject | | |
|---|---|---|---|---|---|---|---|
| Cropland | 392 ac. @ | $1,070 | $419,440 | Cropland | 400 ac. @ | $1,070 | $428,000 |
| Pasture | 108 ac. @ | $535 | $57,780 | Pasture | 110 ac. @ | $535 | $58,850 |
| Total | 500 ac. @ | $954 | $477,220 | Total | 510 ac. @ | $955 | $486,850 |
| Indicated land adjustment to Sale 13 = + $1/acre (say $0) | | | | | | | |

**Table 12.14** Building Adjustment Analysis

| | Buildings on Sale 13 | | | | | Buildings on Subject | | | |
|---|---|---|---|---|---|---|---|---|---|
| | Cond./ Utility | Size | $/Unit | $ | | Cond./ Utility | Size | $/Unit | $ |
| Dwelling | G-G | 2,625 @ | $20 | $52,500 | Dwelling | A-G | 2,250 @ | $18 | $40,500 |
| Eq. shed | A-A | 4,140 @ | $2.00 | $8,280 | Eq. shed | A-A | 4,200 @ | $2.00 | $8,400 |
| Gr. bin | P-P | 5,000 @ | $0.00 | $0 | Gr. bin | F-F | 10,000 @ | $ 0.40 | $4,000 |
| Total | | | | $60,780 | Total | | | | $52,900 |
| Building value per acre | | | | $122 | Building value per acre | | | | $104 |
| Indicated building adjustment to Sale 13 = – $18/acre | | | | | | | | | |

Sale 13 is very comparable to the subject in improvements or buildings too. The good/good dwelling valued at $20 per square foot in Sale 13 supports a value estimate of about $18 per square foot for the subject's slightly inferior average/good dwelling. The equipment sheds on the subject and Sale 13 are both of average condition and average utility, so the $2.00 per square foot value for Sale 13's shed is appropriate for the shed on the subject. The $0 value for Sale 13's poor/poor grain bins and the grain bin results shown in the analyses of the other improved sales in Table 12.4 lend support to a value of approximately $0.40 per bushel of capacity for the subject's fair/fair grain bins.

Once the land and building adjustment analyses shown in Tables 12.5 through 12.14 are completed, all the improved sales can be compared directly to one another and the subject for any other elements of comparison determined to be appropriate, which may include time, location, size, and other characteristics. The elements of comparison found to be significant can be analyzed and individual pairings can be developed and applied in the sales comparison approach. The analysis of improved sales presented here is continued and further developed in Chapter 14, The Sales Comparison Approach.

## Data Analysis in the Income Approach

The income approach analysis should reflect the procedure used to estimate the probable net income for several comparable sales. The appraiser should explain in the report why a cash lease, crop-share, or owner-operator income analysis was developed. Only one methodology should be used in an appraisal, not a combination of two or three different income-estimating methods. Once the net income for each comparable sale is estimated, it should be divided by the sale price of the property to develop an overall capitalization rate. The capitalization rates derived from the sales should then be summarized in a table or grid. The final selection of the capitalization rate to be applied to the subject's estimated net income should be discussed in great detail.

Quality market rental data is just as important as quality sales data. The appraiser should analyze several comparable rentals of properties similar to the subject in terms of highest and best use and building contribution. Whenever possible, cash rentals should be used to avoid the possibility that managerial influences are affecting a given property's ability to generate income. If cash leases are not available, the appraiser should be extremely careful in applying either a cropshare/percentage lease or a direct owner-operator budget analysis to the available comparable sales and the subject property.

Remember that if the subject property is analyzed in the income approach under a cash lease arrangement, then cash lease rentals must be applied to the sales to derive the overall capitalization rate from the market. On the other hand, if cropshare analysis is applied to the subject property, then information gathered from the market regarding cropshare rentals must be used in the analysis of several comparable sales to reflect the capitalization rate from the market. If rental data for a particular property type is unavailable, the appraiser is cautioned to assume a very typical management level in estimating the potential income from comparable sales and the subject property and applying an owner-operator budget analysis in the income approach.

The sales used in the extraction of a market capitalization rate do *not* have to be under some type of rental arrangement to be effective market indicators. A myth has developed in some appraisal circles that a comparable sale that is not being rented at the time of sale cannot be used to extract an overall rate. This is emphatically untrue. As a matter of fact, if one is careful to apply a typical market rent (or typical production levels in the case of cropshare or owner-operator analyses), a much more accurate reading of a comparable sale's ability to generate income can be realized. A sale that is under some type of rental arrangement at the time of sale may or may not be rented at a typical market rent level. The analysis will be skewed if the actual contract rent is either above or below market rent.

This following discussion will address the proper methodology for analyzing cash rental data gathered through detailed market research. Table 12.15 illustrates how an appraiser might summarize the accumulated rental data.

**Table 12.15 Comparable Rental Summary**

| Rental I.D. | Gross Rent | Cult. Acres | Pasture Ac. | Buildings | Cash Rent/Acre | Date |
|---|---|---|---|---|---|---|
| A | $10,000 | 200 | -0- | no | $50.00 | current |
| B | 10,000 | 150 | 100 | no | 40.00 | current |
| C | 14,700 | 250 | 100 | no | 42.00 | current |
| D | 16,000 | 240 | 160 | no | 40.00 | 2 years |
| E | 16,200 | 180 | 300 | no | 33.75 | 1 year |
| F | 8,750 | -0- | 250 | house/barns | 35.00 | 1 year |
| G | 9,800 | 100 | 100 | house/barns | 49.00 | 1 year |
| H | 15,150 | 130 | 250 | house/barns | 39.87 | current |

An analysis of Rentals A and B suggests that cropland currently rents for $50 per acre and pasture rents for $25 per acre. Applying a simple market abstraction technique to Rental B results in the $25 pasture rent conclusion:

| Example 7: Analysis of Rental B | | |
|---|---|---|
| Gross rent on 250 acres @ | $40/acre | $10,000 |
| Less crop rent on 150 acres @ | $50/acre | – 7,500 |
| Residual to 100 acres of pasture | | $2,500 |
| $2,500 / 100 acres = $25 per acre pasture rent | | |

In the analysis of Rental B, the $50 per acre crop rent suggested by Rental A is applied to Rental B's 150 acres of cropland. The remaining $2,500 of Rental B's $10,000 gross rent can be attributable to the rent generated by its pasture. Note that both Rental A and Rental B are current market rental arrangements.

Rental C, which is also a current rental arrangement, includes 250 acres of cropland and 100 acres of pasture. For the purpose of this rental analysis, assume that the pasture in Rental C would rent for $25 per acre based on the Rental B analysis. This allows the balance of the gross rent in Rental C to flow to the cropland.

| Example 8: Analysis of Rental C | | |
|---|---|---|
| Gross rent on 350 acres @ | $42/acre | $14,700 |
| Less pasture rent on 100 acres @ | $25/acre | - 2,500 |
| Residual to 250 acres of cropland | | $12,200 |
| $12,200 / 250 acres = $48.80 per acre cropland rent | | |

The Rental C analysis reflects a suggested current market rent of $25 per acre for pasture and near $50 ($48.80) per acre for cropland. This supports the rental rates shown in the analyses of Rentals A and B.

Rentals D and E, like A through C, are unimproved properties. However, Rentals D and E were initiated two years ago and one year ago, respectively. Without any pure pasture or crop rentals in the same time frame as Rentals D and E, one must analyze these older rentals with the current data to ascertain whether or not time has had an influence on rental rates over the past two years:

| Example 9: Analysis of Rental D | | |
|---|---|---|
| Gross rent on 400 acres @ | $40/acre | $16,000 |
| Less cropland rent on 240 acres @ | $50/acre | - 12,000 |
| Residual to 160 acres of pasture | | $4,000 |
| $4,000 / 160 acres = $25 per acre pasture rent | | |

| Example 10: Analysis of Rental E | | |
|---|---|---|
| Gross rent on 480 acres @ | $33.75/acre | $16,200 |
| Less pasture rent on 300 acres @ | $25/acre | - 7,500 |
| Residual to 180 acres of cropland | | $8,700 |
| $8,700 / 180 acres = $48.33 per acre cropland rent | | |

The analyses of Rentals D and E suggest that rental rates have not changed significantly over the past two years. When the current $50 per acre cropland rent is plugged into the analysis of Rental D, the oldest rental, a residual of $25 per acre for that property's pasture is suggested, which is equal to current pasture rents. Similarly, when the current pasture rent value of $25 is used in the Rental E analysis, a residual of $8,700, or $48.33 per acre, is reflected for that rental property's 180 acres of cropland. This $48.33 approximates the $48.80 to $50.00 per acre cropland rental rates shown in current Rentals A through C.

Once the unimproved rentals have been analyzed to reflect land rental rates in the market, the improved rental properties can be used to suggest appropriate rental rates for farm buildings. Deducting the appropriate land rent from the total rent will indicate a residual attributable to the buildings on each of the properties in Rentals F, G, and H:

| Example 11: Analysis of Rental F | | | |
|---|---|---|---|
| Gross total property rent | | | $8,750 |
| Less cropland rental on 0 acres @ | $50/acre | – | 0 |
| Less pasture rent on 250 acres @ | $25/acre | – | 6,250 |
| Residual to house and barns | | | $2,500 |
| $2,500 / 12 months = $208.33 per month building rent | | | |

| Example 12: Analysis of Rental G | | | |
|---|---|---|---|
| Gross total property rent | | | $9,800 |
| Less cropland rental on 100 acres @ | $50/acre | – | 5,000 |
| Less pasture rent on 100 acres @ | $25/acre | – | 2,500 |
| Residual to house and barns | | | $2,300 |
| $2,300 / 12 months = $191.67 per month building rent | | | |

| Example 13: Analysis of Rental H | | | |
|---|---|---|---|
| Gross total property rent | | | $15,150 |
| Less cropland rental on 130 acres @ | $50/acre | – | 6,500 |
| Less pasture rent on 250 acres @ | $25/acre | – | 6,250 |
| Residual to house and barns | | | $ 2,400 |
| $2,400 / 12 months = $200.00 per month building rent | | | |

The analyses of improved Rentals F, G, and H suggest that farm houses and barns generate a market rent of approximately $200 per month. It is assumed that the quality and quantity of these buildings is not a significant determinant of their rental rates.

Once the appraiser has gathered and analyzed the available market rentals, the findings can be superimposed on the improved comparable sales to ascertain the probable gross market rental potential of each property. Deducting applicable expenses from the gross rent of each improved sale property will indicate the estimated net annual income. The net income of each property can then be divided by the sale price to derive an overall capitalization rate from each improved sale.

In Table 12.16 Improved Sales 7, 10, and 13, which were presented in Table 12.4, are used to demonstrate a method of deriving an overall rate from the market. The estimated expenses for the sales shown in Table 12.16 are based on a management fee allowance equal to 10% of gross income, ad valorem taxes of $2.50 per acre, and building expenses (insurance, maintenance, and repairs) equal to 2% of the building's contributory value. The last column of Table 12.16 shows the estimated income and expenses for the subject property described earlier. The market-derived rental rates and estimated expenses shown above also apply to the subject property:

**Table 12.16** Income Analysis of Improved Sales for Deriving Capitalization Rate

| Property I.D. | Sale 7 | Sale 10 | Sale 13 | Subject |
|---|---|---|---|---|
| Potential gross income: | | | | |
| Cropland | $10,000 | $15,000 | $19,600 | $20,000 |
| Pasture | 2,500 | 3,750 | 2,700 | 2,750 |
| Buildings | 2,400 | 2,400 | 2,400 | 2,400 |
| Total income | $14,900 | $21,150 | 24,700 | $25,150 |
| Expenses: | | | | |
| Management | 1,490 | 2,115 | 2,470 | 2,515 |
| Building expense | 1,000 | 900 | 1,216 | 1,060 |
| Real estate taxes | 750 | 1,125 | 1,250 | 1,275 |
| Total expenses | – 3,240 | – 4,140 | – 4,936 | – 4,850 |
| Net income | $11,660 | $17,010 | $19,764 | $20,300 |
| Sale price | $300,000 | $457,500 | $538,000 | N/A |
| Cap rate | 3.89% | 3.72% | 3.67% | 3.75% |

Sales 7 and 10 are more current than Sale 13 (6 months versus 12 months), but Sale 13 is more comparable to the subject in land quality or land class mix. Sale 13 also has a similar amount of building improvements. Sales 7 and 13 suggest a range of overall rates (capitalization rates) of 3.67% to 3.89%.

This discussion has emphasized sales and data analysis. All three approaches to value must be market driven. This income analysis has been presented to suggest the method(s) appropriate for reading market rental data. Market rental information has been superimposed on improved comparable sales to investigate the relationship between incomes and values in this market.

## Data Analysis in the Cost Approach

The cost approach section of the appraisal report should discuss both the procedure the appraiser used to estimate the value of the subject's land in its unimproved state and the method used to analyze the improved sales and extract a market-derived depreciation estimate.

The appraiser must estimate the reproduction or replacement cost (RC) for the buildings on the subject and the buildings on the improved comparable sales. The cost estimates used in this section of the report must be properly verified with reliable cost services and/or actual construction costs from local contractors. Besides the cost new analysis, the appraiser must address the accrued depreciation inherent in the subject buildings. A depreciation rate can be developed through analysis of several improved sales within the subject's market. Market support for depreciation using the abstraction technique is preferred but other, less precise, procedures such as the age-life method can also be used. See Chapter 13 for a more detailed discussion of these techniques.

To derive a depreciation rate from improved sales, one must first estimate the reproduction or replacement cost for the buildings on each comparable sale. The difference between the costs of the buildings and their actual contributory value is attributable to the overall depreciation they have suffered from all causes (physical deterioration and functional and external obsolescence).

The contributory value of the buildings on each sale property is derived by first estimating land value for the sales. These land value estimates should come from unimproved sales in the same market. Remember that each improved sale's building contribution must be estimated *as of the date of sale.* This will eliminate any confusion that might arise in markets where land is increasing or decreasing in value at a different rate than buildings. Since it is necessary to estimate depreciation as of the date of sale, it is imperative to estimate costs as of the date of sale also.

Once the building contribution "as of the date of sale" is estimated for each improved sale, that contributory value can be subtracted from each sale's respective cost estimate to reflect the total overall depreciation. These overall rates may be refined to reflect a separate depreciation rate for each building based on its effective age and contributory value versus its cost new. Remember, this overall rate encompasses all forms of depreciation that may exist in the buildings.

In Table 12.17, four of the five improved sales discussed earlier in the chapter are analyzed to reflect a market-derived overall rate of depreciation for each building on these properties. The overall rates or percentages of depreciation are then annualized based on the effective age of each building. If a property suffers from either functional or external obsolescence, it is improper to annualize depreciation. Functional and external obsolescence either affect a property or they do not affect a property—they are not functions of age. Therefore, only physical depreciation should be annualized.

In Table 12.17, Sales 7 and 8 are analyzed first to suggest a market-derived overall rate of annual depreciation for houses and sheds in this market. Sale 8's shed analysis has indicated a probable annual overall rate of depreciation of 6%. This 6% rate is then judged to be applicable to the shed on Sale 9. Applying the 6% annual rate of depreciation to the 10-year effective age of Sale 9's shed reflects a total depreciation of 60%, or $18,240. The $12,160 difference between Sale 9's $30,400 shed RC and the property's $18,240 total depreciation is the contributory value of the shed on Sale 9.

Deducting the shed's estimated contributory value of $12,160 from Sale 9's total building value of $17,000 leaves a residual of $4,840 for that

**Table 12.17** Market-Derived Depreciation Rates

| Sale No. | 7 | 8 | 9 | 10 |
|---|---|---|---|---|
| Sale price | $300,000 | $391,200 | $362,150 | $457,500 |
| Land value | 250,000 | 376,200 | 345,150 | 412,500 |
| Bldg. value | 50,000 | 15,000 | 17,000 | 45,000 |
| House RC | $80,000 | n/a | n/a | $81,600 |
| Cont.value | 50,000 | | | 36,450 |
| $ Deprec | 30,000 | | | 45,150 |
| % Deprec. | 37.50% | | | 55.33% |
| Eff. age | 10 years | | | 15 years |
| Deprec./yr. | 3.75% | | | 3.69% |
| Shed RC | n/a | $37,500 | $30,400 | $37,500 |
| Cont.value | | 15,000 | 12,160 | 3,750 |
| $ Deprec. | | 22,500 | 18,240 | 33,750 |
| % Deprec. | | 60% | 60% | 90% |
| Eff. age | | 10 years | 10 years | 15 years |
| Deprec./yr. | | 6% | 6% | 6% |
| Bins RC | n/a | n/a | $6,250 | $7,500 |
| Cont.value | | | 4,840 | 4,800 |
| $ Deprec. | | | 1,410 | 2,700 |
| % Deprec. | | | 22.56% | 36% |
| Eff. age | | | 5 years | 8 years |
| Deprec./yr. | | | 4.51% | 4.50% |

property's grain bins. The difference between Sale 9's $6,250 grain bin RC and the $4,840 contributory value is the depreciation of $1,410. This $1,410 is 22.56% of the $6,250 bin RC and equates to an annual rate of 4.51% when spread over the 5-year effective age of the Sale 9 grain bins.

In the Sale 10 analysis shown in the table, the contributory value of the shed and grain bins were estimated by applying the market-derived annual rates of depreciation derived from the analyses of Sale 8 and Sale 9, respectively, to these improvements. Sale 8 suggested an annual rate of 6% depreciation for sheds in this market. Sale 9 suggests a grain bin depreciation of approximately 4.50% annually. The following example details the steps in applying the market rates of overall depreciation to the shed and grain bins on Sale 10.

| Sale 10 | Shed | Grain Bins |
|---|---|---|
| Cost new | $37,500 | $7,500 |
| % Depreciation | 90%* | 36%** |
| $ Depreciation | -$33,750 | -$2,700 |
| Contributory value | $3,750 | $4,800 |

* 15 years effective age @ 6% per year as derived from Sale 8

** 8 years effective age @ 4.50% per year as derived from Sale 9

Combining the shed and grain bin contributory values and deducting the result from Sale 10's total building value of $45,000 leaves a residual of $36,450 for Sale 10's house [$45,000 – ($3,750 + $4,800) ]. The difference between the house's RC of $81,600 and its $36,450 contributory value reflects a total accrued depreciation of $45,150. This $45,150 is 55.33% of the house's RC and suggests an annual overall depreciation rate of 3.69% when spread over the residence's 15-year effective age.

By allowing the Sale 10 house to fall out as a residual contributory value, a second sale analysis has suggested yet another annual rate of overall depreciation for houses in this market. The 3.69% annual depreciation of Sale 10's residence lends strong support to the findings reached in the analysis of Sale 7. Furthermore, applying the shed and grain bin depreciation rates derived from Sales 8 and 9 to Sale 10 supports the annual rates of depreciation found in those two prior analyses.

This method of extracting depreciation rates for various buildings from the market is much more reliable than "guessing" at depreciation by applying a simple age-life concept. Once the annual overall rates of depreciation for various buildings have been obtained from the market, they are applied directly to the subject's building cost new on a building-to-building basis. The depreciated value of the subject's buildings is then added to the estimated current land value to arrive at an indicated value for the subject property.

## Conclusion

All three approaches to value will not be applicable in all appraisal assignments. A proper appraisal report should contain strong market support. Properly presented, sales and data analyses create a strong foundation for any appraisal report.

PART

• III •

# Valuing Rural Property

CHAPTER
• 13 •

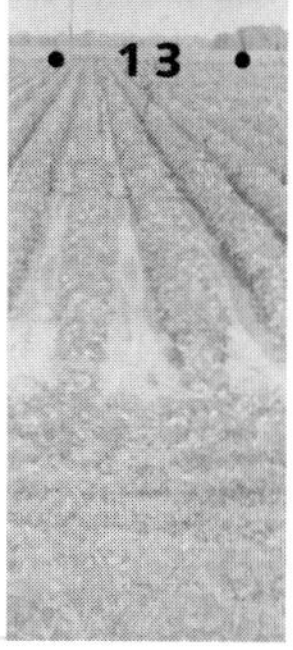

# The Cost Approach

The cost approach is one of the three accepted approaches for valuing real estate. It is based on the premise that the informed purchaser will not pay more for a property than the cost of constructing an equally desirable substitute property minus applicable depreciation and assuming no undue delay. The appraiser compares the subject property to a new property with optimum functional utility. The best way to estimate the total amount of depreciation is from market data. When comparable sales are used to estimate depreciation, the land value is deducted from the sale price of each comparable and the comparable building's contribution to value is compared to the reproduction or replacement cost of the building.

The cost approach employs the technique of summing all the land and building components. The approach is used when data to value the components (land classes or types and buildings) is available, but data to value the property as a whole is limited. For example, the cost approach may be the most reliable indicator of value for a property with minimal improvements and numerous classes of land. Each comparable sale may have one or two land classes, but a group of comparable sales are needed to analyze all the land classes contained in the subject property. In this situation, the appraiser would apply the cost approach and sum the land components since there is little or no data on properties with land components similar to the subject.

The cost approach is most applicable under the following conditions:

- The improvements are new and represent the highest and best use.
- The subject property has characteristics that are typical for the area.

- The subject is a special-use property.
- Adequate data are available to value property components, but limited data are available to value the whole property.

The cost approach is least applicable under these conditions:

- No vacant land sales are available.
- Construction costs are difficult to obtain.
- Depreciation is difficult to measure.
- The improvements are very old.

## Cost Approach Procedure

Application of the cost approach includes the following steps:

1. Develop a land value opinion. Land is valued as if vacant and available for its highest and best use, regardless of its present use. The land valuation source data should be based on comparable vacant land sales with a similar highest and best use.
2. Estimate the reproduction or replacement cost of the improvements. The reproduction or replacement cost is estimated as of the effective date of the appraisal and should include all direct and indirect costs and entrepreneurial profit.
3. Estimate the amount of depreciation. Depreciation is divided into three primary categories: physical deterioration, functional obsolescence, and external obsolescence. The primary techniques used to estimate depreciation are market abstraction, cost to cure, straight-line estimates, capitalized rent losses, capitalized added ownership cost, and costs in excess of an equally desirable substitute.
4. Deduct depreciation from reproduction or replacement cost. Depreciation from all causes is deducted from the new cost of each structure to calculate the contributory value of each building.
5. Total the land and building components. The contributory value of each building is added to the land value opinion to arrive at the indicated value from the cost approach.

The format can be summarized as follows:

1. Cost new of improvements
2. *Less* depreciation
3. *Equals* depreciated value of improvements
4. *Plus* land value estimate
5. *Equals* indicated value from the cost approach

In applying the cost approach the appraiser must follow some basic procedures to extract and apply rates consistently and accurately. The basic guidelines are

- Reproduction or replacement cost is estimated as of the effective date of the appraisal for the subject property
- Market abstraction measures depreciation as of the sale date of the comparable sale. The new cost estimates and land values analyzed are also based on the sale date of the comparable.
- Depreciation is always estimated in this order: 1) physical deterioration, 2) functional obsolescence, and 3) external obsolescence.

## Land Valuation

The cost approach to value has two parts—the development of an opinion of land value and the estimation of improvement value. Although more emphasis is usually placed on valuing the improvements, valuing the land is an equally important part of the cost approach, particularly in appraising land-intensive properties. The opinion of land value must be based on the subject's land value as though vacant, or unimproved, and available for its highest and best use. Unimproved land sales within the subject market area should be used to derive land value.

Chapter 12, Sales and Data Analysis, introduced the reader to the techniques used to analyze both improved and unimproved sales. In this chapter, these analytical methods will be applied to demonstrate the development of a land value opinion for a hypothetical subject property. A "laundry list" of land values will be used to arrive at a land value for the subject property. This method is not the only acceptable way of developing a land value opinion, but it is widely used in rural appraisal. After this method is presented, a sales comparison example will be shown and five alternative methods will be discussed.

A search for sales in the market area of the subject property must be conducted to provide data. The sales should be as much like the subject as possible, particularly with regard to highest and best use. The hypothetical subject property has a good location and 500 acres. Using the categories established by the Soil Conservation Service, it has 300 acres of Class I land and 200 acres of Class IV land. The yield index on the Class I land for both the sales and the subject is 45, with normal production at 80 to 90 bushels per acre for small grains.

The sales search of the market area reveals eight unimproved sales that are similar to the subject. The sales have various mixtures of Class I and IV land, as summarized in Table 13.1.

The summary grid shows the land value-per-acre calculations for Sales 1 and 2. However, the remaining sales have two types of land and

**Table 13.1** Sales Summary

| Sale | Date | Sale Price | Acres | Terms | Location | $/Acre | Class I Acres | Class I $/Acre | Class IV Acres | Class IV $/Acre |
|---|---|---|---|---|---|---|---|---|---|---|
| 1 | 6 mos. | $125,000 | 250 | Cash | Average | 500 | n/a | | 250 | $500 |
| 2 | 6 mos. | $500,000 | 500 | Cash | Average | 1,000 | 500 | $1000 | n/a | |
| 3 | 6 mos. | $362,500 | 525 | Cash | Average | 690 | 200 | | 325 | |
| 4 | Curr. | $320,625 | 375 | Cash | Average | 855 | 300 | | 75 | |
| 5 | 1 year | $696,100 | 700 | Cash | Average | 994 | 600 | | 100 | |
| 6 | Curr. | $517,500 | 690 | Cash | Good | 750 | 300 | | 390 | |
| 7 | 6 mos. | $400,000 | 526 | Cash | Good | 760 | 200 | | 326 | |
| 8 | 1 year | $650,000 | 594 | Seller finan. | Average | 1,094 | 509 | | 85 | |

the per-acre contributory values for these sales cannot be completed without further analysis. It is apparent that there are a number of differences between the sales and the subject. These differences include: time, ranging from current to one year old; favorable seller financing; good to average locations; and various mixtures of land. To provide a reliable indication of the value of the subject land, these differences need to be reconciled.

The first difference to be considered is land mix. In this market the productivity and earnings capacity of Class I and Class IV land differ. This difference is reflected in sale prices. Sale 1, with all Class IV land, sold for $500 per acre; Sale 2, which is all Class I land, sold at the same time for $1,000 per acre.

In identifying land types, the appraiser must take care to reflect the market. Different market areas will indicate different types of land categories. Some markets are very sensitive to Soil Conservation Service land classifications, while others are not. For example, the difference between Class I land and Class III land in a corn- or soybean-producing area would most likely be very great. However, the market may perceive the difference between Class IV and VI pasture to be very slight or nonexistent. In some areas productivity indexes such as the crop equivalency rating or corn suitability rating are the recognized measures of land categories. Some areas are so homogenous as to provide no indications of differing land types. In locations with wetlands, the per-acre sale prices of land may differ greatly based on the amount of tillable land. In other areas much broader land types are recognized, such as simply cropland and pasture or irrigated land and dry cropland. The appraiser must identify how the local market differentiates between various types of land. An adjustment should be made only if there is a market-derived basis for the adjustment.

In this demonstration, Sales 1 and 2 are considered puritans—i.e., they each have only one type of land. These puritans form the basis for value allocations by land class for the sales with a mix of land classes. Sale 3 is the mixed land class sale that is most similar to Sales 1 and 2 in other respects, so this sale is the first to be analyzed. If the $1,000 per-acre value for Class I land indicated by Sale 2 is applied to Sale 3, the remaining $500 per acre can be allocated to the Class IV land in Sale 3. For example:

| | |
|---|---|
| Sales price | $362,500 |
| Less: 200 acres Class I × $1,000 per acre = | 200,000 |
| Residual to Class IV land | $162,500 |
| $162,500 / 325 acres Class IV = $500 per acre | |

The application of data from Sales 1 and 2 to Sale 3 supports the similarities between the puritans and the mixed properties. The ratio analysis method is applied to the remaining sales. Sales 1 through 3 have basically only one difference—land mix. However, it is unlikely that a given group of sales would not have other differences in addition to land mix. As more differences are introduced, the results produced by applying the puritan land values directly to the sales become less reliable. For example, Sale 8 is one year old and had favorable seller financing—two additional characteristics that differ from Sales 1 through 3. Applying the indication of $1,000 per acre for Class I land directly to Sale 8 produces the erroneous results shown below.

| | |
|---|---|
| Sale price | $650,000 |
| Less: 509 acres Class I × $1,000 per acre = | 509,000 |
| Residual to Class IV land | $141,000 |
| $141,000 / 85 acres Class IV = $1,659 per acre **(WRONG)** | |

With Class I land at $1,000 per acre and Class IV at $1,659 per acre, it is obvious that direct application is not the proper technique to use. However, this analysis does provide the appraiser with an indication that time, or terms, or both factors are very likely to be significant in this market.

The comparisons of Sales 1, 2 and 3 indicate that Class I land sells for $1,000 per acre and Class IV land sells for $500 per acre. Thus Class IV land has a value of 50% of Class I land, calculated as $500 / $1,000. This ratio is used to derive the value contribution of Class I and IV land on the rest of the sales. Consider Sale 8 once again. The procedure for allocating the sales price between the Class I and IV land can be set up as follows:

| | |
|---|---|
| Class I land: 509 acres × 100% (1.00) = | 509.00 |
| Class IV land: 85 acres × 50% (0.50) = | 42.50 |
| | 551.50 |

This calculation converts all the sale property's land to the equivalent of Class I land. Sale 8 has 551.50 net Class I-equivalent acres. With a ratio of 50%, the logic behind the procedure can be stated another way. If the Class IV land is worth half as much as the Class I land, then it would take two acres of Class IV land to equal one acre of Class I land. The 85 acres of Class IV land divided by two is 42.50 acres. In other words, 85 acres of Class IV land is the same as 42.5 acres of Class I land. To complete the ratio analysis calculations, the net Class I-equivalent acreage is divided into the total sale price to provide the indication of the Class I contributory value per acre. The Class IV land allocation is based on the ratio of Class IV to Class I land multiplied by the Class I value. The analysis of Sale 8 is completed as follows:

| | |
|---|---|
| $650,000 / 551.50 = | $1,178.60 per acre (value of the Class I land) |
| $1,178.60 × 50% (0.50) = | $589.30 per acre (value of Class IV land) |

Algebraically, the calculation to find the net Class I-equivalency acres is:

| | |
|---|---|
| 509 (1x) + 85 (0.5x) = | $650,000 |
| 509x + 42.5x = | $650,000 |
| 551.5x = | $650,000 |
| x = | $1,178.60 per acre for Class I land |
| $1,178.60 × 50% = | $589.30 per acre for Class IV land |

The proof of the ratio calculation is summarized as follows:

| | |
|---|---|
| 509 acres Class I × $1,178.60 per acre = | $599,907.40 |
| 85 acres Class IV × $589.30 per acre = | $50,090.50 |
| Total | $649,997.90 |

The total sale price indicated by the "proof" varies only slightly from the actual sale price of $650,000. It is common to round the per-acre allocations to the nearest dollar per acre. The $1,178.60 could be rounded to $1,179 per acre. If the ratio of 50% is applied to the rounded figure, the allocation for Class IV land in Sale 8 is $590 per acre. Rounding the indications will cause the proof to vary somewhat more from the actual sale price, but will make little difference in the total valuation process.

The contributory values of the land in each class for the other sales are calculated in the same manner, using the 50% ratio with the different acreage of Class IV land. The results are shown in Table 13.2.

After the land allocations are made, indications of the effects of favorable seller financing and time can be developed. Sale 8 sold with favorable seller financing. Sale 5 occurred at the same time and is similar except that it is a cash sale.

**Table 13.2** Land Sale Analysis

| Sale | Date | Sale Price | Acres | Terms | Location | $/Acre | Class I Acres | Class I $/Acre | Class IV Acres | Class IV $/Acre |
|---|---|---|---|---|---|---|---|---|---|---|
| 1 | 6 mos. | $125,000 | 250 | cash | average | 500 | n/a | | 250 | $500 |
| 2 | 6 mos. | $500,000 | 500 | cash | average | 1,000 | 500 | $1,000 | n/a | |
| 3 | 6 mos. | $362,500 | 525 | cash | average | 690 | 200 | $1,000 | 325 | $500 |
| 4 | Curr. | $320,625 | 375 | cash | average | 855 | 300 | $950 | 75 | $475 |
| 5 | 1 yr. | $696,100 | 700 | cash | average | 994 | 600 | $1,071 | 100 | $535 |
| 6 | Curr. | $517,500 | 690 | cash | good | 750 | 300 | $1,045 | 390 | $523 |
| 7 | 6 mos. | $400,000 | 526 | cash | good | 760 | 200 | $1,102 | 326 | $551 |
| 8 | 1 year | $650,000 | 594 | seller finan. | average | 1,094 | 509 | $1,179 | 85 | $590 |

| Sale | Date | Financing | Class I | Class IV |
|---|---|---|---|---|
| 5 | 1 yr | Cash | $1,071/ac. | $535/ac. |
| 8 | 1 yr | Owner | $1,179/ac. | $590 |

The differences between allocated values within the respective classes provide indications of the effect of favorable terms. The Class I comparisons indicate a difference of $108 per acre, or -9.16% ($108 / $1,179). The Class IV comparisons indicate a difference of $55 per acre, or -9.32% ($55 / $590). From this information it is reasonable to conclude an adjustment for terms of 9%.

Sales 4 and 6 are current sales. Sales 3 and 5 are older sales with locations similar to Sale 4. Sale 7 is a six-month-old sale with a location similar to Sale 6.

Comparisons between these sales are shown below.

| Sale | Date | Class I | Class IV | Compound Change |
|---|---|---|---|---|
| 4 | Current | $950/ac. | $475/ac. | |
| 3 | 6 months | $1,000/ac. | $500/ac. | 0.85%/mo. |
| 5 | 1 year | $1,071/ac. | $535/ac. | -0.99%/mo. |
| 6 | Current | $1,045/ac. | $523/ac. | |
| 7 | 6 months | $1,102/ac. | $551/ac | -0.88%/mo. |

In this text time adjustments are measured and applied on a compound basis. Using a financial calculator, the keystrokes for the comparison of Sale 3 and Sale 4, using Class I land are: 1000 CHS PV; 950 FV; 6 n; solve for i.

The change can be measured using the Class I or Class IV land. Rounded to the nearest 1/100th of 1%, the indications are the same. With the range from -0.85% per month to -0.99% per month, a time adjustment of -0.90% per month is appropriate.

The indicated adjustments for favorable financing and time are applied to the selected sales. The adjustments can be applied by one of two methods—the total sale price can be adjusted for favorable financing and time, after which each sale is reallocated, or the adjustments can be applied directly to the allocated values of each land class on the eight sales. The indications produced using the methods are essentially the same, although there may be variances of a few dollars per acre due to rounding.

Applying financing and time adjustments was covered in Chapter 12 and will be briefly reviewed here. For example, Sale 7 is a six-month-old sale with a total sale price of $400,000, cash. If this sale is adjusted to the current time period using a time adjustment of -0.90% per month, the time-adjusted price is $378,880. Sale 7 has 363 net Class I-equivalent acres. Dividing 363 into the adjusted price of $378,880 provides an allocation of $1,044 per acre for the Class I land. The Class IV land, at 50% of the value of Class I land, is allocated at $522 per acre. Alternatively, the time adjustment of -0.90% per month can be applied to the Class I allocated price of $1,102 per acre, yielding a time-adjusted price of $1,044 per acre. When the time adjustment is applied to the Class IV allocation, the indicated time-adjusted price is $522 per acre.

Sale 8 requires both a time and a favorable financing adjustment. The adjustment for favorable financing is applied first and then the time adjustment is applied to the sale price adjusted for financing.

| | |
|---|---|
| Sale price | $650,000 |
| Less 9% | – 58,500 |
| Cash equivalent sale price | $591,500 |

The HP-12C calculator keystrokes are

$591,500 CHS PV; 12 n; 0.9 CHS i; solve for FV = $530,687
$530,687/551.5 net acres = $962 per acre.

Alternatively,

| | |
|---|---|
| Sale price Class I | $1,179/ acre |
| Less 9% | – 106/ acre |
| Cash equivalent price Class I land | $1,073/ acre |

The HP-12C calculator keystrokes are

$1,073 CHS PV; 12 n; 0.9 CHS i; solve for FV = $963/acre.

Table 13.3 shows that the value indications for Class I land vary from $947 to $1,045 per acre. Class IV land ranges from $474 to $522 per acre. Sales 6 and 7 have good locations like the subject, while the other sales have average locations. The Class I indications from Sales 6 and 7 are $1,045 and $1,044 per acre. Class IV indications from these sales are $523 and $522 per acre. An opinion of land value for the subject is developed from these similar sales.

**Table 13.3** Sales Adjusted for Favorable Financing and Time

| Sale | Date | Adjusted Sale Price | Acres | Terms | Location | $/Acre | Class I Acres | Class I $/Acre | Class IV Acres | Class IV $/Acre |
|---|---|---|---|---|---|---|---|---|---|---|
| 1 | 6 mos. | $118,400 | 250 | Cash | Average | 474 | n/a | | 250 | $474 |
| 2 | 6 mos. | $473,600 | 500 | Cash | Average | 947 | 500 | $947 | n/a | |
| 3 | 6 mos. | $343,360 | 525 | Cash | Average | 654 | 200 | $947 | 325 | $474 |
| 4 | Curr. | $320,625 | 375 | Cash | Average | 855 | 300 | $950 | 75 | $475 |
| 5 | 1 yr. | $624,533 | 700 | Cash | Average | 892 | 600 | $961 | 100 | $480 |
| 6 | Curr. | $517,500 | 690 | Cash | Good | 750 | 300 | $1,045 | 390 | $523 |
| 7 | 6 mos. | $378,880 | 526 | Cash | Good | 720 | 200 | $1,044 | 326 | $522 |
| 8 | 1 yr. | $530,687 | 594 | Seller finan. | Average | 893 | 509 | $963 | 85 | $482 |

| | |
|---|---|
| 300 acres Class I land × $1,045 = | $313,500 |
| 200 acres Class IV land × $523 = | $104,600 |
| Land value | $418,100 |

In some market areas, sales will exhibit many differences beyond land mix, terms, and time. In the example it is apparent that location is also a factor, but focusing on those sales most similar to the subject eliminated the need to adjust for location. When there are additional characteristics that affect value, such as size, shape, access, and crop allotments, it may be necessary to complete a more detailed land value analysis that includes adjustments for these important market characteristics. A direct comparison technique like the one shown in Chapter 14 can be used to develop an opinion of land value for the cost approach. Using the techniques explained in Chapter 14, a direct comparison grid for the hypothetical subject property can be created (see Table 13.4).

**Table 13.4** Direct Comparison Grid

| | Subject | Sale 3 | Sale 4 | Sale 5 | Sale 6 | Sale 7 | Sale 8 |
|---|---|---|---|---|---|---|---|
| Date | Curr. | 6 mos. | Curr. | 1 yr. | Curr. | 6 mos. | 1 yr. |
| Acres | 500 | 525 | 375 | 700 | 690 | 526 | 594 |
| Financing | Cash | Cash | Cash | Cash | Cash | Cash | Terms |
| Location | Good | Avg. | Avg. | Avg. | Good | Good | Avg. |
| $/Acre | | $690 | $855 | $994 | $750 | $760 | $1,094 |
| Land adj. | | + 110 | − 95 | − 137 | + 86 | + 121 | − 152 |
| Adj. $/acre | | 800 | 760 | 857 | 836 | 881 | 942 |

For terms adjustment, compare Sale 5 to Sale 8, indicating $85, or 9% (rounded)

| | | | | | | |
|---|---|---|---|---|---|---|
| Adjustment | 0 | 0 | 0 | 0 | 0 | − 85 |
| Subtotal | 800 | 760 | 857 | 836 | 881 | 857 |

For time adjustment, compare Sale 3 to Sale 4, indicating $40, or -0.851% per month
compare Sale 4 to Sale 5, indicating $97, or -0.996% per month
compare Sale 6 to Sale 7, indicating $45, or -0.870% per month

Final adjustment = -0.90% per month

| | | | | | | |
|---|---|---|---|---|---|---|
| Adjustment | – 42 | 0 | – 88 | 0 | – 47 | – 88 |
| Subtotal | 758 | 760 | 769 | 836 | 834 | 769 |

For location adjustment, compare Sale 5 to Sale 6, indicating $67
compare Sale 3 to Sale 7, indicating $76
compare Sale 7 to Sale 8, indicating $65

Final adjustment = $70 per acre

| | | | | | | |
|---|---|---|---|---|---|---|
| Adjustment | + 70 | + 70 | + 70 | 0 | 0 | + 70 |
| Adj.$/Acre | 828 | 830 | 839 | 836 | 834 | 839 |

Sales 6 and 7 are most similar to the subject and provide the most reliable indications of value. An appropriate land value for the subject property is:

500 ac. × $835 = $417,500

## Alternative Land Valuation Techniques

The most reliable way to develop a land value opinion is through comparison of unimproved sales. However, when few sales are available or additional support is desired, other techniques can be used. Since the alternative techniques are not widely used in rural appraisals, they are described briefly.

### Land Residual

The land residual method is based on the principles of balance and contribution, which are concerned with the relationship among the four agents of production—land, labor, capital, and coordination or management. The technique is used most effectively when building value is known or can be estimated, net operating income is known or can be estimated, and both land and building capitalization rates can be extracted from the market.

To apply this technique, the net operating income of the property is estimated. Then the appraiser estimates how much of the total income is attributable to the improvements. The remainder of the income is residual to the land. This income to the land is then capitalized to provide an opinion of land value. The capitalization rate used should be a market-derived rate for similar land.

The land residual technique is generally most applicable to highly improved properties such as extensive feeding facilities for hogs or poultry, processing plants, or storage facilities in which a substantial portion of the property value lies in the improvements.

### Allocation

When the allocation method is applied, sales of improved properties are analyzed and the sale prices are allocated between the land and building components. This method is based on the principles of balance and contribution. In this case *balance* refers to a proper economic mix of property components—land and buildings. The principle of contribution states that the value of a particular component is measured by its contribution to the value of the total property.

### Extraction

Extraction is a variation of the allocation method. The land value of a sale is extracted by subtracting the contributory value of the improvements from the sale price. The contributory value of the improvements is the reproduction or replacement cost less depreciation from all causes. This procedure is most applicable when improvement values are low or easily identified, as is the case in appraising a large ranch or farm with few improvements.

### Subdivision Analysis

This subdivision analysis approach is used to value land when subdivision is the highest and best use or when subdivision is imminent. In rural appraisals it is most often used when the property is in a transitional stage from agricultural uses to urban uses.

### Ground Rent Capitalization

Simply stated, ground rent is the amount paid for the right to use and occupy the land according to the terms of the ground lease. Ground rent can be capitalized to provide an indication of the market value of the land. The indicated value is generally equivalent to the value of the leased fee interest.

The capitalization rate used should be derived from analysis of similar ground leases. In choosing an appropriate rate, the appraiser should pay particular attention to the terms and conditions of the leases analyzed and the reversion to the owner. In rural appraising this technique may be used to value land for airports, garbage transfer sites, and power or telephone substations.

In summary, direct sales comparison of various property components, as illustrated in Table 13.3, is the preferred and most widely used technique for developing an opinion of land value in the cost approach. Direct comparison analysis is generally most appropriate when there are significant differences among sales other than financing and time. The alternative methods discussed have very limited practicality in appraising most rural properties.

## Improvement Value

### Reproduction or Replacement Cost

*Reproduction cost* is the estimated cost to construct, at current prices as of the effective appraisal date, an exact duplicate or replica of the building being appraised, using the same materials, construction standards, design, layout, and quality of workmanship, and embodying all the deficiencies, superadequacies, and obsolescence of the subject building.

*Replacement cost* is the estimated cost to construct, at current prices as of the effective appraisal date, a building with utility equivalent to the building being appraised, using modern materials and current standards, design, and layout.[1]

The use of replacement cost eliminates the need to estimate some forms of functional obsolescence, but other forms of functional obsolescence, physical deterioration, and external obsolescence must still be measured. If an appraiser chooses to use replacement cost, the added costs of removing items of curable functional obsolescence and any excess operating costs associated with superadequate construction must still be considered.

Reproduction cost is sometimes difficult to estimate because identical materials may be unavailable and construction standards may have changed. Nevertheless, reproduction cost usually provides a basis for measuring depreciation from all causes. Additional information on this topic can be found in the discussion of functional obsolescence later in this chapter.

An estimate of reproduction or replacement cost includes expenditures for direct and indirect costs plus entrepreneurial profit. Direct costs include all capital expenditures directly related to the physical components of the structure. Indirect costs, such as interest, taxes, and insurance, are capital costs that are indirectly associated with construction of the structure.

### Sources of Cost Data

#### Local Builders

Local builders are excellent sources of current construction costs for the various buildings typically located on rural properties. These builders can supply local costs, which do not need to be adjusted with local or current cost multipliers. A builder's detailed cost estimate is the most reliable indicator of building cost but, due to the time and expense involved, this method is typically used only to value the most difficult

1. Appraisal Institute. *The Appraisal of Real Estate,* 11th ed. (Chicago: Appraisal Institute, 1996), 345.

properties. In assignments to appraise a proposed building, a builder's cost estimate is generally provided. If the appraiser maintains a file of proposed costs from past assignments, cost estimates for future assignments can be developed based on local costs.

### Market Abstraction

Market abstraction is based on the sale of a new building. After land value is deducted from the sale price, the total cost of construction and profit can be obtained. This method is most meaningful if the improvement does not include any obsolescence and if there is adequate support for land value at the time of sale. This method works well in appraising residential subdivisions, but has little application in rural appraisal.

### Cost Services

A cost service summarizes the construction costs of various types of buildings and converts them into average unit costs either on a square-foot or cubic-foot basis. Some services provide more detail than others, breaking down costs by the various building components. In each case, the service provides instructions for modifying the basic, average cost to fit the characteristics of buildings that vary from the prototypes that were used to compute the benchmark costs. In addition, these services usually supply cost indexes and local cost multipliers for different geographic areas. These features allow the appraiser update the cost figures. The appraiser should keep a file of current local costs to check the costs indicated by cost manuals. In some instances, the appraiser may need to alter the modifier supplied to equate the cost service's figures with actual local costs. Cost manuals are widely used by appraisers and, with some in-house quality control, they can provide reliable indicators of new costs. The cost guides available include *Boeckh Building Valuation Manual, Marshall Valuation Service,* and guides published by local or state agencies and organizations.

## Cost Estimating Methods

The techniques used to estimate reproduction or replacement cost include: 1) quantity survey, 2) unit in place, and 3) comparative unit methods.

### Quantity Survey Method

The quantity survey is the most detailed, complex, and costly method of estimating costs. It requires expert analysis and is used in rural appraising when exact figures are needed. The quantity survey method involves adding together all types of labor and materials, subcontractor fees, and equipment required to reproduce or replace an improvement. Each item of cost is expressed in terms of current local prices and wages and calculated by the hour, pound, or other unit of measurement. Overhead, profit, and other indirect costs may be added as a lump sum or as a percentage. This method requires a detailed knowledge of all the components of the construction process and precise measurement and calculation of the quantities used. The quantity survey method is used to estimate the cost of highly customized properties that have unique features and when other information is generally not available. A quantity survey is shown in Table 13.5.

### Unit-in-Place Method

In applying the unit-in-place method, the appraiser estimates the unit costs of materials or components of the structure as installed. The measurement includes both the cost of materials and the labor necessary to put each unit in place. The structure is generally broken down into large units: exterior walls, roof, foundation, floors, etc. A unit cost expressed in dollars per square foot or lineal foot for installation is added to larger unit costs. Equipment and fixtures are usually added as a lump sum, which includes expenses for installation. Other indirect costs and overhead are then added either as lump sums or percentages. Accurate measurements and descriptions are essential for the successful use of this method. A unit-in-place cost estimate is shown in Table 13.6.

### Comparative-Unit Method

In the comparative-unit method, which is often referred to as the *square foot method,* all components of the structure are combined and their costs are estimated based on the cost per unit (e.g., square foot) of building area or cost per volume (e.g., bushels) of building capacity. The costs used are usually completed construction costs, including all installation

**Table 13.5** Reproduction Cost—Quantity Survey Method

| Description | Quantity | Cost |
|---|---|---|
| 42,500-bushel galvanized steel grain bins with ladders | 2 @ $27,625 | $55,250 |
| 1,500-bushel-per-hour, 50-ft. elevator leg with 9-in. x 16-in. grain buckets | 1 | $8,500 |
| 100-ft., 6-in. steel discharge pipe | 1 | $2,250 |
| 75-ft. discharge augers and steel pipe from bins to elevator leg | 2 @ $3,187.50 | $6,375 |
| 10-ft. discharge auger and steel pipe from dump pit to elevator leg | 1 | $850 |
| Steel dump grate spouts | 1 | $450 |
| Concrete work – cone-bottom floors and dump pit | | $22,950 |
| Subtotal – materials | | $96,625 |
| Labor for installation | | $71,000 |
| Electrical subcontract (motors, controls, installation) | | $2,425 |
| Fees and permits | | $1,575 |
| Ground preparation and excavation | | $1,850 |
| Subtotal - materials and installation | | $173,475 |
| Sales tax 5% | | $8,674 |
| Total new cost | | $182,149 |

Note. Although entrepreneurial profit is a valid cost for many types of improvements, including agricultural improvements, on-farm grain storage is generally owner-constructed and there is no intention to sell it to the end user. Therefore, entrepreneurial profit is not added as a cost item.

**Table 13.6** Reproduction Cost—Unit-In-Place Method

| Description | Quantity | Cost Per Quantity | | Cost |
|---|---|---|---|---|
| Two cylindrical steel grain bins with 34-ft. vertical height | 68 ft. | $1,375.00/ft. | | $93,500 |
| Concrete cone-bottom floors | 2,512 sq.ft. | $13.50 | | $33,912 |
| Discharge augers | 75 ft. | $127.00 | | $9,525 |
| Grain elevator leg with cages, ladders, and distributor | 50 ft. | $285.00 | | $14,250 |
| Dump pit construction | | | | $8,500 |
| 6-in. downspouts | 100 ft. | $25.00 | | $2,500 |
| Electrical motors and controls | | | | $3,500 |
| Site excavation and preparation | 20 hours | $75.00 | | $1,500 |
| Subtotal | | | | $167,187 |
| Current cost multiplier | | | × | 1.03 |
| Subtotal | | | | $172,203 |
| Local cost multiplier | | | × | .95 |
| Total direct costs | | | | $163,592 |
| Indirect costs of architecture, engineering, appraisal, interest cost during construction, insurance, and permits | | | × | 1.10 |
| Total cost new | | | | $179,952 |

Note: Again, entrepreneurial profit is not added as a cost item because on-farm grain storage is generally owner-constructed and there is no intention to sell it to the end user.

expenses, builder's overhead, and profit. Most cost estimating services provide information on a square-foot or cubic-foot basis. This method may be the least accurate of the three methods, but it is the simplest to apply, the least time-consuming, and the least expensive. If the appraiser applies it carefully, the comparative-unit method provides a cost estimate that is sufficiently detailed and accurate for estimating the cost of new rural improvements. Some in-office quality control should be exercised to ensure that the cost manual provides a cost indication similar to the costs quoted by local contractors. The comparative-unit method is illustrated in Table 13.7.

## Depreciation

In its simplest form, depreciation is defined as a loss in property value from any cause. To appraisers, depreciation is the difference between the cost to reproduce or replace the property and its contributory value as of the date of appraisal. Depreciation can be caused by physical factors such as breakage, deterioration, and simple aging. A property may have additional depreciation due to functional obsolescence, which is a market resistance caused by an internal characteristic such as poor floor plan, outmoded or inadequate equipment, or an inadequacy or superadequacy due to size or some other characteristic. External obsolescence is caused by conditions outside the property, such as an inharmonious user, lack of demand, changing uses in the area, and unfavorable economic conditions.

The three principal forms of depreciation—physical deterioration, functional obsolescence, and external obsolescence—are defined and their various components are discussed in the following sections.

**Table 13.7** Reproduction Cost by the Comparative-Unit Method

| Description | | Cost Per Bushel |
|---|---|---|
| Base cost for two, 42,500-bu., galvanized steel grain bins with concrete cone bottoms | | $1.50 |
| Unloading augers – 12-ft., U-trough, covered | | $0.13 |
| 50-ft., 1,500-bph.* elevator leg with boot pit | | $0.20 |
| Concrete, steel-lined dump pit | | $0.12 |
| Spouting with 4-hole distributor | | $0.05 |
| Subtotal | | $2.00 |
| Current cost multiplier | × | 1.03 |
| Subtotal | | $2.06 |
| Local cost multiplier | × | 0.95 |
| Total direct costs | | $1.96 |
| Architect's fee @ 8% | | $0.16 |
| Total cost new per sq. ft. | | $2.12 |
| Total cost for 85,000-bu. facility | | $180,200.00 |

* bph= bushels per hour

Note: Again, because on-farm grain storage is generally owner-constructed with no intention to sell it to the end user, entrepreneurial profit is not added as a cost item.

## Physical Deterioration

Physical deterioration is the adverse effect on value caused by deterioration or impairment of the condition of the improvement as a result of wear and tear and disintegration. Physical deterioration is divided into two major categories: curable and incurable. Some causes of physical deterioration include:

- Action of the elements
- Wear and tear
- Use
- Damage or breakage
- Pest or insect infestation
- Deferred maintenance

### Curable Physical Deterioration

Curable physical deterioration is applicable to items with physical deterioration that it is economically feasible to cure. These items are customarily repaired or replaced. Examples of curable physical deterioration include broken equipment, components in need of new paint, damaged roof or siding, and missing fixtures. This depreciation is usually computed as the dollar amount of the cost to restore the item to new or reasonably new condition. Curable physical deterioration is also referred to as *deferred maintenance.* In estimating curable physical deterioration, the appraiser should not include any excess cost to repair or prepare the surface for the item which is included in the reproduction or replacement cost (RC) estimate. For example, if panels on a grain bin need replacing, the cost to remove the old panels and prepare the remaining surface to accept the new panels will exceed the RC of the new panels if they were installed during original construction. Therefore, the curable physical deterioration will include the cost of the panels as if installed during original construction plus an excess cost for preparation. For future depreciation estimates, only the cost actually included in RC should be deducted from RC to obtain the base for estimating incurable physical deterioration in long-lived items.

### Incurable Physical Deterioration

This type of physical deterioration is applicable to items whose physical deterioration either cannot be corrected or can be corrected only at a cost in excess of the amount the item contributes to the value of the property. Incurable physical deterioration is divided into short-lived and long-lived components.

***Incurable physical deterioration in short-lived items.*** A short-lived item is a component that has an economic life shorter than the economic life of the structure. Short-lived items such as floor covering, roof covering, appliances, and heat and cooling systems will typically be replaced several times during the life of the improvement. These items are not economically feasible to replace as of the date of appraisal, but will most likely move into the physical curable category at some future date. This form of depreciation is typically estimated with an age-life analysis. Any items that are considered short-lived and physically curable would not be depreciated at this step in this analysis. The procedure is shown in Table 13.8.

**Table 13.8** Estimate of Incurable Physical Deterioration, Short-Lived

| Item | Cost New | Actual Age | Physical Life | Ratio | Depreciation |
|---|---|---|---|---|---|
| Floor covering | $3,000 | 10 | 20 | 50% | $1,500 |
| Appliances | $2,000 | 6 | 10 | 60% | $1,200 |
| Roof covering | $4,000 | 6 | 20 | 30% | $1,200 |
| Total incurable physical deterioration in short-lived items | | | | | $3,900 |

***Incurable physical deterioration in long-lived items.*** Improvement components that are not and will not be economically feasible to replace in the future are categorized as long-lived items of incurable physical deterioration. These items, which are expected to last throughout the entire economic life of the improvement, include the structure itself, foundation walls, roof structure, etc. This form of depreciation is typically measured with an age-life analysis, with a modified age-life analysis, or through application of a market-derived depreciation rate. The reproduction or replacement cost of the items depreciated in the curable and incurable short-lived categories must be deducted from RC to avoid double depreciation. The procedure is illustrated in Table 13.9.

**Table 13.9** Estimate of Incurable Physical Long-Lived Depreciation

| | | |
|---|---|---|
| **Reproduction Cost** | | **$125,000** |
| Less cost new of curable items included in RC | – | $1,000 |
| Less cost new of incurable short-lived items | – | $11,000 |
| Remaining cost new to be depreciated | | $113,000 |
| Long-lived percentage (by method estimated) | | 40% |
| Dollar amount ($113,000 × 40%) | | $45,200 |
| Depreciated RC | | $67,800 |

## Functional Obsolescence

Functional obsolescence is a loss in value resulting from defects in design. It can also be caused by changes that, over time, have made some aspect of the structure, such as its materials or design, obsolete by current standards. The defect may be curable or incurable.

Some causes of functional obsolescence (curable and incurable) include:

- Functional inadequacies and/or superadequacies
- Improper or inappropriate improvement that does not conform to the highest and best use of the site.
- Poor layout or design
- Outmoded equipment
- Changing public tastes or standards
- Changing technology

### Curable Functional Obsolescence

The major test as to whether an item of functional obsolescence is curable or incurable is the relationship between the cost to cure and the increased income or benefits to be derived by curing the obsolescence. Additionally, some superadequacies, such as an excess foundation or an excessive grain handling system, may be classified as incurable because it is not physically practical to cure them. In the appraisal of residential or commercial property, where a defined rental market exists, the test of financial feasibility is easier to apply than it is for farm properties. Most farm improvements are not rented separately from the farm operation and it is difficult to actually compare economic benefits with cost to cure. In valuing specialized agricultural buildings where rental markets do exist, the relationship between cost and increased rental income can be measured. For the rural appraiser, the question is often, "Would a prudent operator spend the money necessary to cure the obsolescence to receive the additional benefit from the improvement?" Curable functional obsolescence may be divided into three subcategories: deficiency requiring addition, deficiency requiring substitution or modernization, and superadequacy.

A *deficiency requiring addition* is measured by how much the cost of the addition exceeds the cost of the item if it were installed new during construction. The example of curable functional obsolescence shown in Table 13.10 concerns a barn that was built with a dirt floor, while current building design requires a concrete floor.

A *deficiency requiring substitution or modernization* is measured as the cost of the existing component included in the cost estimate minus the physical deterioration already charged against the existing component and salvage value, plus the cost to remove the existing component and the added cost of installation. *Salvage value* is defined as the value attributed to an item that is to be removed from the property, usually for use elsewhere.

The example of curable functional obsolescence shown in Table 13.11 concerns equipment in a commercial broiler facility that is outdated due to changes required by the integrator. There are no problems with the building structure. The question as to whether or not the depreciation is curable is easily answered because the integrator will not place chickens in the house until this modernization is completed.

A *superadequacy* is measured as the current reproduction cost of the item minus any physical deterioration already charged, plus the cost of removal, minus the salvage value, if any. A superadequacy is curable if correcting it on the date of the appraisal is economically feasible; otherwise it is considered incurable.[2]

Table 13.12 concerns a farm that is deficient in hay storage and includes an old barn which has numerous mangers and stalls that could be converted into open space storage by the removal of these structural components. The test of feasibility is obviously met because a prudent operator would most likely convert the old barn, not build a new barn.

**Table 13.10** Estimate of Curable Functional Obsolescence—Deficiency Requiring Addition

| | |
|---|---|
| **Cost to install a concrete floor in existing barn** | **$5,500** |
| Less cost to install a concrete floor in existing barn if the structure were being built on the date of the appraisal | – $5,000 |
| Loss in value | $500 |

**Table 13.11** Estimate of Curable Functional Obsolescence—Deficiency Requiring Substitution or Modernization

| | |
|---|---|
| **Cost new of existing equipment already included in cost new estimate** | **$24,000** |
| Less physical deterioration previously charged | – $21,200 |
| Less salvage value | $1,000 |
| Plus removal cost | $1,500 |
| Plus cost to install new equipment | $26,500 |
| Depreciated RC | $29,800 |

**Table 13.12** Estimate of Curable Functional Obsolescence—Superadequacy

| | |
|---|---|
| **Current reproduction cost of the mangers and stalls** | **$16,000** |
| Less physical deterioration already charged | – $12,650 |
| Plus cost of removal | $2,200 |
| Loss in value | $5,550 |

## Incurable Functional Obsolescence

Incurable functional obsolescence may be caused by a deficiency or a superadequacy. An item not included in the cost new that should have been causes obsolescence due to a deficiency; excess items

2. Ibid., 387

included in construction which are included in reproduction cost cause obsolescence due to superadequacy. The test of curability or incurability can be whether it is physically possible to add or remove the item that causes the obsolescence or whether it is financially feasible to remove or add the item.

*Incurable functional obsolescence due to a deficiency* occurs when a typical building of the same type includes a feature or component that is not present in the subject building. The addition of this feature to the subject would enhance the market value, but not to a level that equals the cost to cure the deficiency. If a defined rental market exists for the type of building being appraised, the determination of feasibility is easier than if no market rental structure exists, which is the case with most farm buildings.

An example of incurable functional obsolescence caused by a deficiency would be a hay barn constructed for use with square hay bales (see Table 13.13). In many areas, square bales have been replaced with round bales and the type of barn required has also changed. The barn in question has a 14-ft. interior height and three round bales stack 17 feet high. The barn can be used for round bale storage, but it is not as efficient as a barn with a 17-ft. interior height that allows the stacking of three bales. The deficiency in question here is height, and it is measured by comparing barns with sufficient height to barns with insufficient height. The obsolescence is incurable, at least from a feasibility standpoint, although it may not be physically impossible to correct. The market in this area indicates similar physical deterioration for both types of barns, but a barn with deficient height sells for 25% less than the physically depreciated value of the higher barn.

*Incurable functional obsolescence due to a superadequacy* relates to features within the building that are in excess of current standards. They may be physically impossible to remove, as is the case with a superadequate foundation, or they may be removable only at a cost that exceeds the added value. Incurable functional obsolescence due to a superadequacy is measured differently depending on whether a reproduction or replacement cost estimate is used. If reproduction cost is the base, the appraiser must consider the cost to construct a building that contains the same superadequacy. The superadequacy is measured as the reproduction cost of the superadequate component, minus any physical deterioration already charged, plus the present value of any added cost of ownership due to the presence of the superadequacy, less any

**Table 13.13** Estimate of Incurable Obsolescence—Deficiency

| | |
|---|---|
| **Current reproduction cost of the barn** | **$50,000** |
| Less physical deterioration previously charged | – $25,000 |
| Physically depreciated RC | $25,000 |
| Functional obsolescence @ 25% | $6,250 |
| Depreciated RC | $18,750 |

value added such as increased rent. The charge for increased operating expense is based on items such as additional taxes, insurance, maintenance, and utility charges. Replacement cost does not include the cost to construct the superadequacy, so no loss in the value of a superadequacy due to physical deterioration is considered when replacement cost is used. However, a charge is required if the superadequacy causes a monetary burden of ownership reflected in additional taxes, insurance, maintenance, and utility charges. The capitalized value of these additional expenses, over and above any increase in market rent resulting from the superadequacy, represents the loss in value caused by incurable functional obsolescence due to the superadequacy. This type of functional obsolescence can also be measured by market abstraction, as illustrated in the case study later in this chapter.

Assume that a farm has a superadequate grain handling system (see Table 13.14). A grain handling system is necessary, but not one as elaborate as the system included in the subject property. Comparing typical systems to the elaborate system indicates that the functional obsolescence can be estimated at 40% of the physically depreciated value.

## External Obsolescence

External obsolescence is the adverse effect on value from influences outside the property itself. External obsolescence affects both land and improvements. The result of the adverse influence on the subject's land is considered in the opinion of land value; only the loss imputable to the improvements is considered depreciation. External obsolescence results in some degree of market rejection of a property, and the appraiser must measure the market rejection to estimate this type of depreciation. If rental data is available, capitalization of the rental loss is one way to measure external obsolescence. Because there is usually no defined rental market for agricultural buildings, another technique must be used. One reliable technique for estimating external obsolescence is to compare sales that have external obsolescence to sales that do not. External obsolescence is almost always incurable.

Causes of external obsolescence include:

- Intrusion of inharmonious uses or users
- Heavy traffic
- Unpleasant odors
- Physical hazards
- A declining market or economy, purchasing power, or demand— all of which result in declining values

**Table 13.14** Estimate of Curable Functional Obsolescence—Superadequacy

| | |
|---|---|
| **Current reproduction cost of the facility including the superadequacy** | **$150,000** |
| Less physical deterioration previously charged | – $42,500 |
| Physically depreciated RC | $107,500 |
| Functional obsolescence @ 40% | – $43,000 |
| Depreciated RC | $64,500 |

Assume that, due to an inharmonious use, buildings on farms in this neighborhood contribute less value than do similar buildings in areas farther away from the use. The decline attributable to the building has been abstracted from market data, which indicates the value decrease to be 30% of the physically and functionally depreciated value (see Table 13.15).

## Methods of Estimating Depreciation

Depreciation is a loss in value from all causes. It is measured from market data and categorized as physical, functional, and external. The separate components of all categories of depreciation are deducted from the new cost of the improvements to estimate their contributory value as of the date of the appraisal. As stated previously, either reproduction or replacement cost can be used, but the handling of certain categories of depreciation may differ depending on the base used. Some forms of functional obsolescence are automatically eliminated when replacement cost is used. Removal costs and added operating costs must be accounted for in the estimation of curable and incurable functional obsolescence, regardless of whether reproduction or replacement cost is used.

In rural appraisal, depreciation is generally estimated using one of four methods: the economic age-life method, the modified economic age-life method, the breakdown method, and market abstraction of depreciation rates. The techniques applied within each method also provide the basis for extracting depreciation from market data.

### Economic Age-Life Method

In the economic age-life method, a ratio of effective age to economic life is applied to the new cost of the improvements. This method of estimating depreciation lumps all forms of depreciation into a single amount, as summarized in the formula:

$$\text{Depreciation} = \frac{\text{Effective age}}{\text{Economic life}} \times \text{reproduction or replacement cost}$$

To understand the economic age-life method of estimating depreciation, the appraiser must be familiar with the terms defined below.

*Actual age* is the number of years that have elapsed since construction of an improvement was completed. It is equivalent to historical or chronological age. This age is not part of the depreciation formula, but it does give the appraiser some indication of the likely effective age, especially with relatively new improvements. For very old improvements that have been remodeled extensively, actual age has little meaning.

**Table 13.15** Estimate of External Obsolescence

| **Reproduction Cost** | | **$150,000** |
|---|---|---|
| Less physical deterioration | – | 42,500 |
| Physically depreciated RC | | $107,500 |
| Functional obsolescence @ 40% | – | 43,000 |
| Physically/functionally depreciated RC | | 64,500 |
| External obsolescence @ 30% | – | 19,350 |
| Depreciated RC | | $45,150 |

*Effective age* is the age indicated by the condition and utility of a structure. This age is estimated as of the date of the appraisal and is based on the physical condition, functional utility, and market acceptance of the structure. Effective age will approximate actual age for relatively new buildings with little physical deterioration. If the building has diminished utility and poor market acceptance, the effective age could exceed the actual age. The effective age of a structure is estimated as of the date of the appraisal. The effective age of a property can change with remodeling, with changes in the market's perception of the building's utility, and with market acceptance of the area. For example, a three-year-old single-family residence may have an effective age of three years until the local airport authority changes flight patterns and directs heavy air traffic across the neighborhood. This change may result in diminished market acceptance of the property, which will increase its effective age. Conversely, a 50-year-old residence in poor condition could have an effective age of considerably less than 50 years after $100,000 is spent in remodeling the property.

*Economic life* is the period of time over which improvements to real estate contribute to property value. As with effective age, economic life can be extended or shortened by market conditions or extensive remodeling. For example, there are many structures that have been in existence for hundreds of years and at some point had little economic life remaining but now, due to remodeling, the economic life has been extended. Economic life is also estimated as of the date of appraisal, although it may change in the future. A change in the highest and best use of a site can end the economic life of a perfectly good structure. Due to increased land value, the structure may be torn down and replaced with a structure that represents the highest and best use of the site.

*Remaining economic life* is the estimated period during which improvements should continue to contribute to property value. It is also the period from the date of appraisal to the end of the improvement's economic life. Remaining economic life can also be changed by remodeling the property or by changes in highest and best use.

Based on these definitions, a relationship between effective age, economic life, and remaining economic life is established. The relationship is summarized in the following formulas:

Economic life = effective age + remaining economic life
Effective age = economic life – remaining economic life
Remaining economic life = economic life – effective age

A major weakness of the economic age-life method is that it obscures many types of depreciation by grouping them together. The curable items are not treated separately, and the short-lived items typically depreciate at a rate much greater than the structure as a whole. This method estimates overall depreciation, but it does not consider functional and external obsolescence and physical deterioration separately, nor does it differentiate between the causes. The economic age-life method is illustrated in Table 13.16.

### Modified Economic Age-Life Method

The effect of curable items of depreciation is recognized in the modified economic age-life method. To apply this method, the appraiser first estimates the cost to cure of all curable items of physical deterioration and functional obsolescence. This sum is deducted from the reproduction or replacement cost as of the date of appraisal. The appraiser then estimates the ratio of effective age to economic life and applies this ratio to the remaining portion of the building cost to be depreciated. The appraiser must recognize that when curable items are recognized separately, the effective age may decrease and the remaining economic life may be extended. As with the age-life method, the appraiser assumes that the ratio can be applied to the entire remaining portion of the structure. If there are many items that depreciate more rapidly than the basic structure (i.e., short-lived items), this method may not be appropriate. For an illustration of the modified economic age-life method, see Table 13.17.

**Table 13.16** Estimating Depreciation by the Economic Age-Life Method

| | |
|---|---|
| **Reproduction Cost** | **$100,000** |
| Effective age | 10 yrs. |
| Economic life | 50 yrs. |
| Remaining economic life | 40 yrs. |
| Ratio applied to cost 10/50 | 20% |
| Total depreciation | – $20,000 |
| Depreciated value of improvements | $80,000 |
| Land value | $250,000 |
| Indicated value by the cost approach | $330,000 |

**Table 13.17** Estimating Depreciation by the Modified Economic Age-Life Method

| | |
|---|---|
| **Reproduction Cost** | **$100,000** |
| Less physical and functional curable items | – $3,000 |
| Effective age | 9 yrs. |
| Economic life | 50 yrs. |
| Remaining economic life | 41 yrs. |
| Ratio applied to cost less physical and functional curable items: 9/50 | 18% |
| RC less curable items | $97,000 |
| Other depreciation (all causes) 18% | – $17,460 |
| Total depreciation ($3,000 curable + $17,460 incurable) | – $20,460 |
| Depreciated value of improvements | $79,540 |
| Land value | $250,000 |
| Indicated value by the cost approach | $329,540 |
| Rounded to | $329,500 |

### Breakdown Method

To apply the breakdown method, the appraiser analyzes each cause of depreciation separately, measures the amount of each, and then totals the estimates to derive a lump sum that is deducted from the replacement or reproduction cost.

The types of depreciation analyzed and estimated are

1. Curable physical deterioration
2. Incurable physical deterioration
    - Short-lived
    - Long-lived

3. Curable functional obsolescence
    - Deficiency
    - Superadequacy
4. Incurable functional obsolescence
    - Deficiency
    - Superadequacy
5. External obsolescence

The various types of depreciation are listed above in the order in which they are typically treated in the appraisal process. The appraiser may find it more convenient to analyze the curable physical and functional items first and then analyze the incurable physical, functional, and external items in the proper order.

Valuations of many types of farm buildings and most single-family residences simply do not require the detail afforded by the breakdown method. An estimate derived using an age-life method or an overall market-abstracted rate is sufficient for many of these properties. However, as the complexity of the property to be appraised increases, the breakdown method becomes more applicable. If a property has many depreciable short-lived items, substantial functional or external obsolescence, or both, the breakdown method may be mandated to furnish the client with the detail the assignment requires. Appraisers should be familiar with this exacting and detailed method of estimating depreciation (see Table 13.18).

**Market Abstraction of Depreciation**

The market abstraction method of depreciation estimation provides the appraiser with the ability to "prove" the depreciation estimate. To extract depreciation from the market, comparable sales similar to the subject must be available and the appraiser must have sufficient data to develop an opinion of the value of the land under the comparable sales. Land value is subtracted from the sale price and the residual is the contributory value of the comparable improvements as of the date of sale. Once the reproduction or replacement cost is derived, the contributory value is deducted from the RC to derive a dollar amount of depreciation. This depreciation amount is then divided by the RC to yield an overall rate of depreciation (see Table 13.19).

This very simple example indicates that the improvements exhibit depreciation of 30% from all causes. This method is sufficient to abstract an overall rate of depreciation for many types of farm improvements, especially when improvements in the area are homogeneous. The improvements on the subject must be very similar to the sale improvements in terms of age, utility, and market acceptance.

Now assume that the improvements described in Table 13.19 are similar to the subject improvements except that the sale improvements

**Table 13.18** Cost Approach – Breakdown Method

| | | | | | | |
|---|---|---|---|---|---|---|
| **Reproduction Cost** | | | | | | $100,000 |
| Depreciation | | | | | | |
| Physical deterioration | | | | | | |
| Curable physical | | | | | $2,000 | |
| Incurable physical short-lived items | | | | | | |
| **Item** | **Cost** | **Age** | **Life** | **Depr.** | | |
| Roof cover | $3,000 | 10 | 20 | $1,500 | | |
| Appliances | $2,000 | 2 | 10 | $400 | | |
| Floor cover | $3,000 | 2 | 10 | $600 | | |
| Total | $8,000 | | | $2,500 | $2,500 | |
| Incurable physical long-lived items (10 yr. effective age/50 yr. economic life) ($100,000 – $2,000 – $8,000 = $90,000 × 20%) | | | | | $18,000 | |
| Functional obsolescence | | | | | | |
| Curable excess ($2,000 – 20% physical charged) | | | | | $1,600 | |
| Curable deficiency | | | | | $3,000 | |
| Incurable excess ($10,000 – 20% physical charged) | | | | | $8,000 | |
| Incurable deficiency | | | | | $2,000 | |
| External obsolescence | | | | | | |
| $62,900 × 30% (Physically and functionally depreciated RC × appropriate percentage) | | | | | $18,870 | |
| Total depreciation (all causes) | | | | | | – $55,970 |
| Depreciated reproduction cost | | | | | | $44,030 |
| Land value | | | | | | $400,000 |
| Value indicated by the cost approach | | | | | | $444,030 |
| Rounded to | | | | | | $444,000 |

**Table 13.19** Market Abstraction of Overall Depreciation Rate

| | |
|---|---|
| Sale price | $720,000 |
| Less land value | – $650,000 |
| Contributory value of improvements | $70,000 |
| Reproduction cost of improvements | $100,000 |
| Less contributory value | – $70,000 |
| Depreciation | $30,000 |
| Overall depreciation percentage ($30,000/$100,000) | 30% |

have an effective age of 10 years and the subject improvements have an effective age of 8 years. The overall rate of depreciation can be further reduced to an annual percentage by dividing 30% by the 10 years of effective age to yield a rate of 3% per year. This rate is applied to the subject's effective age of 8 years to indicate the subject improvements should be depreciated 24%. This additional step eliminates the need for the improvements to be similar to the subject in terms of age, but does not eliminate the need for the improvements to be similar in terms of short-lived items, utility, and market acceptance. If the farms in the area exhibit differences in these characteristics, application of the breakdown method is more appropriate.

Most farms have multiple improvements and often these improvements exhibit varying rates of depreciation. The depreciation rate for individual improvements can be measured by first analyzing a farm that has only one improvement, such as a residence. After abstracting an annual rate of depreciation for the single building, a farm with two buildings can be measured by using the rate for the single building and the residual for the second building. Often groups of buildings such as barns can be lumped together and there are always some buildings that can be eliminated from consideration due to condition, age, etc. The process is illustrated in Tables 13.20, 13.21, and 13.22.

**Table 13.20** Abstracting Depreciation From a Single-Improvement Farm

| | |
|---|---|
| Sale price | $350,000 |
| Less land value | – $300,000 |
| Contributory value of improvements | $50,000 |
| Reproduction cost of improvements | $75,000 |
| Less contributory value | – $50,000 |
| Depreciation | $25,000 |
| Overall depreciation percentage ($25,000/$75,000) | 33% |
| Assume effective age of 11 years | |
| Annual depreciation rate | 3% |

Assuming the sale in Table 13.20 contained only a residence and that other sales will support this indication, it can be concluded that residences in this market exhibit depreciation at the rate of 3% per year. This depreciation rate should only be applied to physical depreciation unless all sales in the market have similar functional problems or external obsolescence is present in all improvements in the market. If functional and/or external obsolescence is present in some improvements but not all, the process requires additional steps to break out the obsolescence.

The process illustrated in Table 13.21 uses an annual depreciation rate from one sale to analyze an additional sale and provides an indication that barns depreciate at an annual rate of 5%. This process can be continued for other improvements and will substantially mature with additional data. If the appraiser keeps this type of information on file, substantial support for market-derived depreciation rates will be available. Table 13.22 summarizes numerous depreciation rates from sales derived over a period of time.

As the appraiser develops additional data from additional sales, these indications can be added to the summary and provide additional support for depreciation rates for different types of buildings. If the rates change substantially, the appraiser will be alerted to changes in the market acceptance of certain types of buildings.

**Table 13.21** Abstracting Depreciation From a Multiple-Improvement Farm

| | |
|---|---|
| Sale price | $360,000 |
| Less land value | – $300,000 |
| Contributory value of improvements | $60,000 |
| Reproduction cost (residence) | $75,000 |
| Less market depreciation for residence (3%/yr. for 15-yr. effective age × $75,000) | – $33,750 |
| Contributory value of residence | $41,250 |
| Contributory value of improvements | $60,000 |
| Less contributory value of residence | – $41,250 |
| Contributory value (barn) | $18,750 |
| Reproduction cost (barn) | $25,000 |
| Less contributory value (barn) | – $18,750 |
| Depreciation (barn) | $6,250 |
| Overall depreciation percentage (barn) ($6,250 depreciation/$25,000 RC) | 25% |
| Assume effective age = 5 years | |
| Annual depreciation rate | 5% |

**Table 13.22** Depreciation Rate Summary (Annual Depreciation Rates)

| Sale | House | Barn | Grain Bins | Farm Shop |
|---|---|---|---|---|
| 1 | 3.0% | N/A | N/A | N/A |
| 2 | 3.0% | 5.0% | N/A | N/A |
| 3 | 2.8% | N/A | N/A | N/A |
| 4 | 3.2% | N/A | N/A | N/A |
| 5 | N/A | N/A | 5.7% | N/A |
| 6 | N/A | 5.2% | 5.7% | N/A |
| 7 | 2.8% | N/A | N/A | 4.4% |
| 8 | N/A | N/A | 5.5% | N/A |
| 9 | 3.0% | N/A | N/A | 4.0% |
| 10 | N/A | 5.0% | 5.6% | 4.0% |

## Depreciation Case Study

In the appraisal of a rural property with substantial improvements that exhibit depreciation from a variety of sources, the market abstraction process can be used to estimate depreciation with the breakdown method. The appraiser can allocate depreciation from the comparable sale properties into the categories appropriate to the individual buildings on the subject property and the residual process can be applied to estimate the remaining types of depreciation. The following case study illustrates the methods employed to abstract depreciation from sales via the breakdown method.

### Area

The subject property and the comparable sales are located in an area where agriculture is the prime use and the motivating factor in all purchases. There is little market pressure for other uses in the area, so agricultural production and improvements are the most significant factors in all sales. Land in the area is typically planted to small grain and typical yields range from 80 to 90 bushels per acre. Approximately two years ago there was a change in government programs which has affected the desirability of on-farm grain storage. Prior to the change, it was common for farms to have the capacity to store two years' of grain production. Currently, one year's storage is sufficient and, due to the design of the storage bins, it is impossible to lease out the excess storage capacity. Grain storage in the area is of two different designs. The most prevalent design is an upright circular steel bin that has a flat-bottom concrete floor with a metal grate for the unloading auger and a portable sweep auger for cleanout. These bins are loaded by a portable auger powered by either an electric or gasoline motor. Each bin has one or two air fans depending on size. The more elaborate systems in the area are cone-bottom, upright steel bins with a 1,500-bushel-per-hour elevator leg with ladder and cage assembly, dump pit, a conveyor to the leg, and electric motors and controls. The cost of such systems is substantially higher and current buyers are unwilling to pay the full, physically depreciated value of these systems.

**Sales**

The following sales are located in the subject area and are similar to the property being appraised.

- Sale 1 sold for $523,000 one year ago. The land contribution at the time of sale was $472,350. The only improvement on the property is a 38,000-bushel grain storage bin which has an effective age of three years and is of flat-bottom, upright design. There is no physical curable deterioration and the RC of the short-lived items is $6,600. The short-lived items are three years old and have a projected life of 10 years. The total reproduction cost as of the date of sale is $58,900. The storage is adequate for one year's production on the 450-acre farm.
- Sale 2 sold for $700,000 one year ago. The land contribution at the time of sale was $639,150. The only improvement is a 52,000-bushel grain storage bin which has an effective age of five years and is of flat-bottom, upright design. The physical curable deterioration is estimated at $2,000, of which $1,750 is included in RC and $250 is excess cost. The RC of the short-lived items is $6,600. These items are five years old and have a projected life of 10 years. The total reproduction cost as of the date of sale is $80,600. The storage is adequate for one year's production on the 610-acre farm.
- Sale 3 sold for $547,775 one year ago. The land contribution at the time of sale was $500,000. The only improvement is a 42,000-bushel grain storage bin which has an effective age of five

years and is of cone-bottom, upright design with a 55-ft. elevator leg with dump pit, conveyor, and electric motors and controls. The physical curable deterioration is $1,000, all of which is included in RC. The RC of the short-lived items is $7,800. These items are five years old and have a projected life of 10 years. The total reproduction cost as of the date of sale is $89,250. The storage is adequate for one year's production on the 575-acre farm.

- Sale 4 sold for $716,400 three years ago. The land contribution at the time of sale was $635,000. The only improvement is a 110,500-bushel grain storage bin with an effective age of 10 years, which is of cone-bottom, upright design with a 55-ft. elevator leg with dump pit, conveyor, and electric motors and controls. The physical curable deterioration is $20,000, of which $17,500 is included in RC and $2,500 is excess cost. The curable deterioration includes some damage to the bin and short-lived items. These short-lived items are fully depreciated and in need of replacing, so they are included as physical curable deterioration. The total reproduction cost as of the date of sale is $215,475. The storage is adequate for two years' production on the 650-acre farm.
- Sale 5 sold for $683,000 and the sale date is current. This is a resale of Sale 4. Land values in the area have not changed, so the land contribution at the time of sale is $635,000. The current effective age of the grain storage bin is 13 years. There are no curable items and the RC of the short-lived items is $12,750. These items are three years old and have a projected life of 10 years. The total reproduction cost as of the date of sale is $232,050. The storage is adequate for two years' production on the 650-acre farm.
- Sale 6 sold for $532,900 and the sale date is current. The land contribution at the time of sale is $465,000. The only improvement is an 80,000-bushel grain storage bin, which has an effective age of two years and is of flat-bottom, upright design. There is no physical curable deterioration. The RC of the short-lived items is $7,400 The short-lived items are two years old and have a projected life of 10 years. The total reproduction cost as of the date of sale is $124,000. The storage is adequate for two year's production on the 650-acre farm.

### Analysis of the Sales

The sales are analyzed to estimate the market depreciation of grain bins with various characteristics. The curable items are considered first, followed by the incurable short-lived items. If there is no functional or external obsolescence, the remainder is incurable physical deterioration, long-lived. After an annual percentage for this type of depreciation is developed, the remainder can be classified as either functional or external obsolescence depending on the situation.

***Sale 1.*** The analysis of Sale 1 includes an age-life analysis of the short-lived components and abstraction of an annual rate of depreciation for the incurable physical deterioration in long-lived items. The grain bin is functional and does not represent an over-improvement. It is of sufficient size to hold one year's production but not two, so no external obsolescence is present. There are no items of curable physical deterioration, therefore only incurable short-lived and long-lived depreciation is considered.

| | | |
|---|---|---|
| | Sale price | $523,000 |
| *Less* | Land contribution | – $472,350 |
| *Equals* | Contributory value of the improvements | $50,650 |
| | Reproduction cost | $58,900 |
| *Less* | Contributory value of the improvements | – $50,650 |
| *Equals* | Depreciation | $8,250 |
| **Depreciation Analysis** | | |
| | Curable physical | $0 |
| | Incurable physical, short-lived—motors & fans [$6,600 cost × (3-yr. age/10-yr. life)] | – $1,980 |
| | Balance of depreciation attributed to long-lived items | $6,270 |

The incurable physical deterioration, long-lived, divided by the reproduction cost of the long-lived items is the percentage of long-lived depreciation. The total percentage of long-lived depreciation divided by the effective age equals the annual rate of long-lived depreciation. The new cost of the short-lived items previously depreciated is deducted from total RC to avoid double counting of depreciation. The calculations are summarized below.

| | |
|---|---|
| Long-lived depreciation | $6,270 |
| Long-lived cost ($58,900 RC – $6,600 short-lived cost) | $52,300 |
| $6,270 divided by $52,300 is equal to .119885, or 12% | |
| 12% divided by the effective age (3) is 4% per year | |

***Sale 2.*** The analysis of Sale 2 requires consideration of the curable deterioration resulting from damage to the side of the grain bin and analysis of short-lived items with the remaining deterioration attributable to incurable physical long-lived items. The deferred maintenance totals $2,000, of which $1,750 is included in RC and $250 is the excess cost needed to remove the old material prior to replacing the damaged panels. This grain bin is functional and does not represent an overimprovement. It is sufficient for one year's storage, therefore external obsolescence is not present.

| | | |
|---|---|---|
| | Sale price | $700,000 |
| *Less* | Land contribution | – $639,150 |
| *Equals* | Contributory value of the improvements | $60,850 |
| | Reproduction cost | $80,600 |
| *Less* | Contributory value of the improvements | – $60,850 |
| *Equals* | Depreciation | $19,750 |
| **Depreciation Analysis** | | |
| | Curable physical (in RC) | – $1,750 |
| | Curable physical excess (not in RC) | – $250 |
| | Incurable physical short-lived—motors & fans [$6,600 cost × (5 yr. age/10 yr. life)] | – $3,300 |
| | Balance of depreciation attributed to long-lived items | $14,450 |

As with Sale 1, the depreciation balance (after deductions are made for deferred maintenance and incurable short-lived items) is divided by the reproduction cost of the long-lived items (after deductions for those portions of RC that have been depreciated as curable and incurable short-lived) to arrive at the percentage of long-lived depreciation. The calculations are summarized as follows:

| | |
|---|---|
| Long-lived depreciation | $14,450 |
| Long-lived cost ($80,600 RC – $1,750 curable cost – $6,600 short-lived cost) | $72,250 |
| $14,450 divided by $72,250 equals .20000, or 20% | |
| 20% divided by the effective age (5) is 4% per year | |

***Sale 3.*** The analysis of Sale 3 requires consideration of curable physical deterioration due to damage to the elevator leg and analysis of the short-lived items. The appraiser must also apply the incurable physical long-lived rate derived from Sales 1 and 2 to arrive at a residual applicable to the incurable functional obsolescence due to excess. The deferred maintenance totals $1,000, all of which is included in the RC estimate; there is no curable excess. The grain bin on this property exhibits incurable functional obsolescence due to the excess handling system. The grain bin itself is functional, but the market will not pay for the excess cost represented by the elevator leg, dump pit, boot pit, and U-trough unloading augers. Nevertheless, the bin is sufficient for one year's storage, so external obsolescence is not present.

| | | |
|---|---|---|
| | Sale price | $547,775 |
| *Less* | Land contribution | – $500,000 |
| *Equals* | Contributory value of the improvements | $47,775 |
| | Reproduction cost | $89,250 |
| *Less* | Contributory value of the improvements | – $47,775 |
| *Equals* | Depreciation | $41,475 |
| **Depreciation Analysis** | | |
| | Curable physical (in RC) | – $1,000 |
| | Curable physical excess (not in RC) | $0 |
| | Incurable physical short-lived—motors & fans [$7,800 cost × (5 yr. age/10 yr. life)] | – $3,900 |
| | Incurable physical long-lived [$80,450 (long-lived cost) × 0.20 (4%/yr. for 5 years)] | – $16,090 |
| | Balance of depreciation attributed to curable functional items | $20,485 |
| | Physically depreciated RC $89,250 – $20,990 total physical depreciation | $68,260 |
| | Incurable functional obsolescence, superadequacy | $20,485 |

Incurable functional obsolescence due to a superadequacy is reported as a percentage of the reproduction cost after curable and incurable functional obsolescence due to a deficiency has been deducted. In this sale, there is no curable or incurable functional obsolescence due to a deficiency, so the RC base is the same as the physically depreciated RC. The estimated incurable functional obsolescence due to the superadequacy ($20,485) is divided by the physically depreciated RC ($68,260), indicating 30% depreciation.

***Sale 4.*** Due to the age and condition of Sale 4, the short-lived items are fully depreciated and in need of replacing. Therefore, these items are classified as curable physical deterioration. There is additional curable deterioration due to deferred maintenance on the tanks. The total curable deterioration (both short-lived and damage) is $20,000, of which $17,500 is included in RC and $2,500 is excess cost to remove the old motors and fans and damaged panels prior to replacement. There is no incurable physical deterioration in short-lived items because all of the short-lived items were handled in the curable category. The grain bin on this property exhibits incurable functional obsolescence due to the excess handling system. The bin is sufficient for two years' storage. Because this sale occurred prior to the change in the government program, however, no external obsolescence is present.

| | | |
|---|---|---|
| | Sale price | $716,400 |
| *Less* | Land contribution | – $635,000 |
| *Equals* | Contributory value of the improvements | $81,400 |
| | Reproduction cost | $215,475 |
| *Less* | Contributory value of the improvements | – $81,400 |
| *Equals* | Depreciation | $134,075 |
| **Depreciation Analysis** | | |
| | Curable physical (in RC) | – $17,500 |
| | Curable physical excess (not in RC) | – $2,500 |
| | Incurable physical short-lived—motors & fans (Motors & fans included as curable) | $0 |
| | Incurable physical, long-lived [$197,975 (long-lived cost) × 0.40 (4%/yr. for 10 years)] | – $79,190 |
| | Balance of depreciation attributable to curable functional obsolescence $116,285 | |
| | Physically depreciated RC [$215,475 (RC) – $99,190 total physical depreciation] | $116,285 |
| | Incurable functional obsolescence, superadequacy | $34,885 |

The estimate of incurable functional obsolescence due to the superadequacy ($34,885) is divided by the physically depreciated RC ($116,285, indicating 30% depreciation As with Sale 3, this sale is indicating 30% incurable functional obsolescence due to the superadequate grain handling facility. Because this property sold prior to the change in the government program that reduced the need for on-farm storage, it exhibits no external obsolescence.

***Sale 5.*** This is a resale of Sale 4. The property has been well-maintained since the repair and replacement of short-lived items three years ago. Again, the grain bin exhibits functional obsolescence due to the excess handling system. The bin is sufficient for two years' storage and, since the change in government programs, it exhibits external obsolescence. Note that the RC has changed due to an increase in cost over the past three years.

| | | |
|---|---|---|
| | Sale price | $683,000 |
| *Less* | Land contribution | - $635,000 |
| *Equals* | Contributory value of the improvements | $48,000 |
| | Reproduction cost | $232,050 |
| *Less* | Contributory value of the improvements | - $48,000 |
| *Equals* | Depreciation | $184,050 |
| | Depreciation Analysis | |
| | Curable physical (in RC) | $0 |
| | Curable physical, excess (not in RC) | $0 |
| | Incurable physical, short-lived—motors & fans [$12,750 cost × (3-yr. age/10-yr. life)] | - $3,825 |
| | Incurable physical, long-lived [$219,300 (long-lived cost) × .52 (4%/yr. for 13 years)] | - $114,036 |
| | Functional obsolescence [$114,189 (phys. dep. RC) × 30%] | - $34,257 |
| | Balance of depreciation attributed to external obsolescence | $31,932 |
| | Physically and functionally depreciated RC [$232,050 (RC) - $117,861 total phys. depr. - $34,257 (funct. obs.)] | $79,932 |
| | Incurable functional obsolescence, superadequacy | $31,932 |

The external obsolescence is reported as a percentage of the physically and functionally depreciated RC. The RC base is the same as the physically and functionally depreciated RC. The external obsolescence divided by the physically and functionally depreciated RC is $31,932 divided by $79,932, or 40%.

*Sale 6.* There are no curable items, so the appraiser analyzes the short-lived items and applies the incurable physical long-lived rate derived from Sales 1 and 2 to arrive at a residual applicable to external obsolescence. The grain bin does not exhibit incurable functional obsolescence due to an excess handling system. It is sufficient for two years' storage so external obsolescence is present.

| | | |
|---|---|---|
| | Sale price | $532,900 |
| *Less* | Land contribution | – $465,000 |
| *Equals* | Contributory value of the improvements | $67,900 |
| | Reproduction cost | $124,000 |
| *Less* | Contributory value of the improvements | – $67,900 |
| *Equals* | Depreciation | $56,100 |
| **Depreciation Analysis** | | |
| | Curable physical (in RC) | $0 |
| | Curable physical, excess (not in RC) | $0 |
| | Incurable physical, short-lived—motors & fans [$7,400 cost × (2 yr. age/10 yr. life)] | – $1,480 |
| | Incurable physical, long-lived [$116,600 (long-lived cost) × .08 (4%/yr. for 2 years)] | – $9,328 |
| | Functional obsolescence | $0 |
| | Balance of depreciation attributed to external obsolescence | $45,292 |
| | Physically and functionally depreciated RC [$124,000 (RC) – $10,808 total phys. depr. – $0 (funct. obs.] | $113,192 |
| | External obsolescence | $45,292 |

The external obsolescence is reported as a percentage of the physically and functionally depreciated RC. The RC base is the same as the physically and functionally depreciated RC. The external obsolescence divided by the physically and functionally depreciated RC is $45,292 divided by $113,192, or 40%. This sale indicates that the external obsolescence is applicable to both flat- and cone-bottom storage.

### Application of Depreciation Data to Subject

The subject property includes an 85,000-bushel grain bin. The bin has a concrete, hopper-bottom floor with a discharge auger. There is a 1,500-bushel-per-hour (bph) elevator leg with 9-in.-by-16-in. buckets. The elevator leg has a 50-ft. discharge height and two downspouts, one to the grain bin and one for loading trucks. The comparative-unit method is used to estimate the reproduction cost of the subject improvements. This method of estimating reproduction cost was used on the sales and consistency in the analysis and application of depreciation rates dictates that the same method be used to estimate reproduction cost for the subject. The reproduction cost of the subject's grain bin is $180,200.

The subject's grain bin was constructed six years ago. Curable physical deterioration is present due to the four damaged metal panels. The estimated cost to replace these panels is $5,500, of which $5,000 is included in RC and $500 is the excess cost incurred in removing the old panels to replace them with new ones. There is curable functional obsolescence due to a deficiency because the electrical motor that powers the elevator leg is undersized and needs to be updated. The cost if included at the time of construction is $500; the cost to remove and update the motor now is $800, indicating curable functional obsolescence of $300. There is no salvage value for the motor. The new cost of the short-lived items is $7,800 and the projected life is 10 years. Thus the incurable physical depreciation is $7,800 cost × 6-yr. age/10-yr. life = $4,680.

Sales 1 and 2 provide the basis for estimating incurable physical long-lived deterioration of 4% per year over a six-year effective age for total incurable physical long-lived deterioration of 24%. This percentage will be applied to the new cost of the long-lived items to avoid double depreciation of the items previously depreciated. The RC to be depreciated as incurable physical long-lived deterioration is calculated as follows:

| **Long-lived Cost Calculation** | | |
|---|---|---|
| RC | | $180,200 |
| Curable cost in RC | – | 5,000 |
| Cost of short-lived items | – | 7,800 |
| Cost of long-lived items | | $167,400 |
| | × | 24% |
| Incurable physical long-lived deterioration | | $40,176 |

Sales 3 and 4 provide the basis for estimating incurable functional obsolescence due to an excess. This depreciation reflects market resistance to the elaborate grain handling system and the depreciation estimate is 30%. This percentage is applied to the physically depreciated cost new to avoid double depreciation of the items previously considered. The physically depreciated value is calculated as follows:

| | | |
|---|---|---|
| Reproduction cost | | $180,200 |
| Total physical deterioration ($5,000 + 500 + 4,680 + 40,176) | – | 50,356 |
| Physically depreciated cost | | $129,844 |
| Functional incurable obsolescence ($129,844 × .30) | | $38,953 |

Sales 5 and 6 provide the basis for estimating external obsolescence due to an excess amount of storage. This depreciation is the result of a change in the government programs that govern farm grain storage. The percentage of depreciation indicated by these sales is 40%. This percentage is applied to the physically and functionally depreciated value, again to avoid double depreciation of the items previously considered.

| | | |
|---|---|---|
| Reproduction cost | | $180,200 |
| Total physical deterioration ($5,000 + 500 + 4,680 + 40,176) | – | 50,356 |
| Functional obsolescence | – | 38,953 |
| Physically and functionally depreciated cost | | $90,891 |
| Functional incurable obsolescence ($90,891 × .40) | | $36,236 |

### Depreciation Estimate for the Subject Property

The case study concludes with the final estimate of depreciation in the subject property (see Table 13.23).

**Table 13.23** Depreciated Cost of Subject

| | | |
|---|---|---|
| **Reproduction Cost** | | **$180,200** |
| Depreciation | | |
| Physical deterioration | | |
| Curable (in RC) | $5,000 | |
| Curable (excess) | 500 | |
| Incurable short-lived | 4,680 | |
| Incurable long-lived | 40,176 | |
| Total physical deterioration | | $50,356 |
| Physically depreciated RC | | $129,844 |
| Functional obsolescence | | |
| Curable | 300 | |
| Incurable | 38,953 | |
| Total functional obsolescence | | $39,253 |
| Physically and functionally depreciated RC | | $90,591 |
| External obsolescence @ 40% | | $36,236 |
| Depreciated reproduction cost | | $54,355 |

## Summary

This chapter has described the process for developing land value and improvement value opinions via the cost approach. Land value is the most important part of most rural appraisals, but many types of rural property are building-intensive.

In this chapter land value is developed by valuing the individual components that make up the subject property. This method is preferred because it is very difficult to find two farms that are alike in terms of land mix. Usually, a simple pairing process cannot be used to derive adjustments. For farms with very similar land types (in terms of value), the process is not as difficult as it is for farms with several different soils, use types, and productive capabilities, which may be selling for varying prices in the market.

Local builders will be an excellent source of cost data for most agricultural improvements. This is not to say that a cost source cannot be used. If local builders are available, however, they will have cost information that is up-to-date and applicable to the local area.

Several methods of estimating depreciation are presented in this chapter. All are sometimes applicable to the appraisal of rural property. If buildings are not a significant part of the subject property, abstracting an overall rate of depreciation is sufficient. An age-life comparison may also be performed, using market information to develop the effective age and economic life estimates. For more building-intensive properties, the breakdown method with market-derived rates is the most applicable and will yield the most defendable estimate of value.

All of the methods and techniques described here depend on sales data from which to derive estimated rates of depreciation as well as values for the individual land components. As with all of the approaches to value, the cost approach is only as good as the data available and the appraiser's ability to analyze the data.

CHAPTER

• 14 •

# The Sales Comparison Approach

The sales comparison approach is a set of procedures in which the appraiser derives a value by comparing the property being appraised to similar properties that have recently been sold, applying appropriate units of comparison, and making adjustments to the sale prices of the comparable sales.

## Steps in the Sales Comparison Approach

The sales comparison approach is applied in five steps:

1. Research the market to obtain information about transactions, listings, and other offerings of property similar to the subject.
2. Verify the information obtained in Step 1. The appraiser verifies that the data obtained are factually accurate and that the transactions were arm's-length market considerations. Verification is accomplished by obtaining information from a knowledgeable source, usually one of the participants in the transaction.
3. Identify relevant units of comparison (e.g., acres, square feet, animal unit, bushel per acre) and develop a comparative analysis for each relevant unit.
4. Adjust the sale price of each comparable property, as appropriate, or eliminate the property as a comparable. Compare the subject and comparable sales on the basis of the seven or more elements of comparison, including property rights conveyed, financing terms, conditions of sale, time, location, physical characteristics, and income characteristics.
5. Reconcile multiple value indications derived from the comparables into a single value indication.

### Research the Market for Comparable Sales

The major consideration in comparable sale selection is the degree of comparability between sale properties and the subject property. The appraiser seeks sales that are representative of the probable market for the subject property. There must be a significant degree of comparability between the sale properties and the subject property in terms of their physical and economic characteristics. A good question to ask is, "Could the circumstances surrounding the sale be duplicated on the subject property?

In choosing comparable sales from the available data inventory, the appraiser is particularly concerned with the broad elements of comparison—property rights conveyed, financing terms, conditions of sale, market conditions (time or date of sale), location, physical characteristics, and income characteristics. These elements of comparison are used in analyzing data and typically include all the variations that must be addressed in the comparison process. Sources of sales data include real estate brokers and agents, sales services, other appraisers, courthouse records, buyers and sellers, and people in industries related to the property type being appraised, such as grain elevator managers, cotton gin managers, and field men.

In selecting comparable sales, the appraiser should choose data from the market area in which the property to be appraised is located. Market areas vary depending on property types. The market for a grain farm may be a few adjoining townships, while the market for a processing facility may be multi-state or national. The number of sales necessary to complete the appraisal is dictated by the market area and the type of property being appraised. No set number of sales is necessary or required to apply the sales comparison approach. It is important that the appraiser select a representative number of sales with the greatest similarity to the subject property. The appraiser may select recent sales, but the data search is not limited to a particular time period. Adjustments for the change in value over time can be made when older sales are used. In compiling an inventory of sales, the appraiser wants to find and verify truly comparable sales, even if their number is limited.

### Verify Sale Information

To develop valid market value conclusions, the process of comparison and adjustment must be based on sales that represent typical interactions within the market in which the subject property would be offered and sold. Each transaction must be verified with one of the participants involved.

In the interview process, the goal of the appraiser is to obtain the information that the parties used in arriving at the final sale price. The appraiser seeks to have the best possible understanding of the thought processes that were involved in the transactions. In the interview process, the appraiser also asks questions that relate to the items of consideration included in the transaction that influenced the sale price. Crops, equipment, livestock, growing costs, and other items of consideration are often included in the purchase price.

Financing terms are investigated to identify what terms were typical in this particular market at the time the transactions took place. The appraiser should determine if a given property was sold with any special financial considerations that may have affected the price paid. For example, if the terms were cash, was the price lower than it would have been if an owner-financed transaction had been negotiated? If the financing was favorable—e.g., a low down payment, a reduced interest rate, or a longer payoff period—was the price adjusted upward to reflect these differences?

Another item to be considered is the production capability of the property, such as the yield per acre or the carrying capacity. If the property is a special-purpose property, operating history, yield data, and other information relating to the special use should be obtained.

Often the appraiser is not given an opportunity to inspect and measure the buildings, so information about the buildings is only obtained in the verification process. The interview process should include questions about the size, age, and condition of the buildings and elicit the buyer's opinion as to how much the improvements contributed to the sale price. The appraiser will also ask these questions:

- Was any duress involved in the transaction?
- Were the parties related?
- Was the property advertised for sale on the open market?
- How long was the property available for sale?
- Were any special considerations made to facilitate the sale?
- Was the transaction a like-kind exchange?

If the information obtained renders the sale unusable for the particular assignment, the data should not be discarded, but retained in the appraiser's files to be used at another time.

To ensure valid market value conclusions, the process of comparison and adjustment must be based on sales that reflect typical behavior of participants within the subject market. Sales of properties that are comparable to the subject are the basic measure of value, but the appraiser often finds it desirable or necessary to also analyze other sales that illustrate market actions or measure the influence of other factors. Consideration of such sales can provide the appraiser with supplemental information that supports the comparison of the subject property with the sales that are most comparable in character and condition.

### Identify Units of Comparison and Develop a Comparative Analysis

In making comparisons, the appraiser identifies similarities and dissimilarities between the subject property and the comparable sales selected for analysis. The most dependable conclusions are based on comparisons of the most similar factors and conditions. The appraiser considers the extent of the dissimilarities and estimates how much they add to, or detract from, the known price of the sale property. The sale price of the comparable is then adjusted to reflect the value of the property being appraised.

### Units of Comparison

Theoretically, comparisons are made between whole properties. However, appraisers of rural property typically must compare properties with physical characteristics and land and building components that vary significantly. Consequently, it is common practice in rural appraising to reduce the total property to a single unit which is representative of the whole. Various units of comparison are used in analyzing rural properties, but the most common unit of comparison applied to rural property, particularly farms, is price per acre. Applying this unit allows the total property to be represented by a single acre. In analyzing comparable sales, the total price is converted into the price per acre by dividing the total sale price by the number of acres that comprise the property. A similar procedure is used to derive a unit price for livestock operations, but the value is usually expressed in terms of carrying capacity or feed equivalents. The unit of comparison used in this case is the AU (animal unit), AUM (animal unit month) or CYL (cow year long). For dairy farms, the unit of comparison may be pounds of butterfat produced from the operation or produced per cow. Other units of comparison used for dairy operations may be per stall or per square foot of pen area. For timber properties, a common unit of comparison is thousand board feet (MBF). The unit of comparison applied to properties that store or process agricultural commodities could be based on the holding capacity, the amount of the commodity that can be stored or processed, or the number of livestock that can be fed within the facility.

After determining the appropriate unit of comparison, the appraiser reviews the tabulated data to identify which characteristics of the properties and sale transactions to use in applying the sales comparison approach. No matter which unit of comparison is selected, the total property is represented by a single unit. The unit of comparison is used in two ways: to derive a price per unit for all comparable sales and to convert the indicated unit value of the subject property into an estimated total property value.

Unit sale prices that have not been adjusted are generally not applied directly to the subject property. Once a unit of comparison has been applied to all the comparable sales, the appraiser has a range of unit prices that provides a basis for calculating an appropriate unit value for the subject property. To obtain an appropriate unit of value, however, each sale must be adjusted to reflect differences between it and the subject property.

## Adjust Prices Using Elements of Comparison

Application of the sales comparison approach involves identifying the variations among properties, measuring the value impact of the variations, and applying appropriate adjustment techniques. The sale price is adjusted using an appropriate unit of comparison—i.e., the sale price per unit is increased or decreased for each significant variation between the comparable and the subject property. These variations are expressed in terms of the elements of comparison.

By defining as precisely as possible the truly significant variations between the subject property and the comparable properties, the appraiser can measure the dollar value for each variation to arrive at a value indication. The number of potential differences is nearly infinite. Although the appraiser notes all the characteristics of the subject and sale properties, to apply the sales comparison approach the differences that have the greatest significance in the market must be identified and measured. The differences found in rural properties can usually be classified under the basic elements of comparison: property rights conveyed, financing terms, conditions of sale, market conditions, location, physical characteristics, and income characteristics.

### Property Rights Conveyed

The appraiser must clearly define the particular rights or interests in the subject property and compare those interests with the interests conveyed in the sales used in the valuation. This is particularly important in assignments to appraise a partial interest in property, limited rights such as surface or mineral rights, the value of the fee subject to a long-term lease, or a leasehold interest. Any encumbrances such as easements or special occupancy or use restrictions should also be identified and explained in analyzing the sales, income, and cost data used in the assignment.

### Financing Terms

The transaction price of one property may differ from that of an identical property because the financing arrangements in the transactions vary. If the purchasers of a comparable property assume an existing mortgage at a favorable, below-market interest rate, they may have paid a higher price than the purchasers of the subject property under current market financing.

Farms are often sold on land contract with a nominal down payment and the seller providing the financing for the balance of the sale price. The appraiser will need to consider this characteristic of the transaction and determine whether the financial arrangements are, in fact, a reasonable representation of the market and reflect a significant enough equity investment to warrant concluding that the sale is a bona fide transaction. In areas where seller financing is the norm, this type of financing reflects a typical market transaction and no variation may exist. Occasionally, however, because of insufficient data, it may be necessary to include a transaction that includes unusual financial arrangements such as a low down payment or repayment terms favorable to the buyer. Such transactions can be used effectively in the sales comparison approach, but the appraiser must recognize that an additional variable has been introduced.

When unusual financing terms are present in a sale, the appraiser must adjust the sale price to a current price indication based on the typical financing available at the time of sale. This procedure involves the use of paired sales analysis in which a property that was sold for cash is compared to a similar property that was sold with favorable financing. If sales are not available for paired sales analysis, present worth factors can be used to establish the present value of the future mortgage payments at the market interest rates prevailing as of the date of sale compared to the interest rates applicable to of the actual transaction. The difference between the actual transaction and the typical market transaction would indicate the discount applicable to the sale for financing.

### Conditions of Sale

When the conditions of sale are not typical, the result may be a higher or lower price than would occur in a normal market transaction. For example, one family member may sell a rural property to another family member at a reduced price. Or someone may pay a higher-than-market price for a property because it was constructed or owned by the buyer's grandparents. Special income tax situations can also cause the price to differ from the price that would have been paid by typical buyers in the market.

When non-market conditions of sale are detected in a transaction, the sale should be compared to the subject property only with great care. Market-based adjustments are very difficult to make in such situations and, if at all possible, the sale should not be used. Occasionally, such a transaction may be used because of a lack of other comparable sales. However, an adjustment should be made only after the appraiser has carefully researched the non-market motivations of buyers and sellers and the probable amount of premium or penalty paid for the property.

### Market Conditions

This adjustment is sometimes called a *time adjustment,* but it actually refers to the trend in market conditions between the date of sale and the date of valuation. The market price level may have remained stable, decreased, or increased in that time interval. In analyzing the data, the appraiser must analyze whether prices are stable or changing and, if change is evident, the appraiser ascertains the reasons for, and amount of, the change.

The best indicator of a change in market conditions is a sale and subsequent resale of a property. If the appraiser can confirm that no changes occurred in the property from the time of the original sale through the time of the resale, these two transactions can indicate any change that has occurred in the specific market over time. If a sale and resale cannot be found, the next best option is to compare two properties with different sale dates that had very few, if any, differences except the date of sale. Again, the analysis will produce an indication of the change in the market due to time.

The change in price is almost always expressed as a percentage of change over the time interval specified. The rate of change is normally measured on a compound basis. For agricultural properties, the compound rate of change is generally expressed on a monthly or annual basis.

The time adjustment is a combination of two factors: the change over time in the desirability of the property being analyzed and the change in the purchasing power of money. These two elements can change property value in the same direction (up or down) or they can move value in opposite directions. By identifying and measuring changes in market conditions, the appraiser can adjust the price of each comparable sale for any change between the comparable sale date and the date of valuation.

### Location

When properties are located within the same rural region, location does not usually have a major impact on value. There are, however, location-related factors that may differ among properties and make the consideration of location more complex.

Distances from community services, markets, and conveniences are locational features that can affect the marketability and, therefore, the value of a rural property. For example, the distance that a livestock ranch is located from schools and shopping may affect its value. The availability and types of roads that service the area, zoning, and the proximity of utilities may also be considered by potential market buyers. Each of these factors alone might not necessitate a location adjustment but, as in the case of other adjustments, the need for an adjustment is indicated by the market data that are collected and analyzed in the valuation.

An adjustment for location, if needed, is obtained by pairing two of the sales that are going to be used in the valuation. The two sales chosen should be similar in all aspects or should have been adjusted to eliminate any dissimilarities other than location. When the sale prices of the two properties are compared, the difference, if any, can be attributable to the difference in location.

### Physical Characteristics

Physical characteristics require the most extensive comparison among rural properties and produce the greatest number of variations to consider. There are two kinds of physical characteristics:

1. Land features and improvements
2. Building improvements.

In the appraisal of a farm, land features and improvements may be more significant than building improvements, but all differences between the improvements on the sales and the subject property should be considered. Land mix and building adjustments are described in detail in Chapter 12 and reviewed in the example presented later in the chapter.

Depending on the significance and degree of variation in each of the physical characteristics of the properties and their influence on value, the appraiser may choose to deal with each factor separately or in aggregate classifications. Soil, water, topography, yields, drainage, elevation, field size and shape, and hazards, for instance, could be considered individually or together. Some might be considered individually and some together. In determining how to proceed, the appraiser ascertains the significance of each individual factor and the degree to which the data supports the separate consideration of each.

As the size of a property increases, its per-unit value usually decreases. It is generally believed that when fewer total dollars are required to purchase property in any given market, more people will be able to participate in purchases. Smaller properties tend to attract more buyers because fewer dollars are needed to purchase such a property. As the number of potential buyers for a property increases, the competition increases and, as a result of the additional competition, the price generally increases. The reverse is also true—i.e., as property size increases, potential buyers who lack sufficient funds to pursue such properties are eliminated from the market. Thus there is less competition among buyers and the price per unit is generally lower than for a smaller property of the same type. As a result, the size of agricultural properties is a major consideration in the adjustment of comparable sales.

A size adjustment is estimated by comparing two sale properties that are similar in all elements of comparison except size. If the sales differ in other aspects, these adjustments must be made to the sales first so that the only difference than remains for comparison purposes is size. The two sales usually will reflect different unit values. This value difference is attributable to the difference in the sizes of the two properties and becomes the size adjustment. This size adjustment is applicable to properties that are in the same general size range as the properties from which the adjustment was derived. It cannot necessarily be applied to properties outside this range.

Building improvements usually require separate valuation and adjustment if their existence is significant to the value of property. While adjustments for these improvements are made in the sales comparison process, if data is lacking the figures used may sometimes be derived using cost approach techniques, but only if the cost approach indications are derived from the market. Typically, the property is analyzed to estimate how much value the buildings contribute to the sale price. In this analysis, the land value underlying the improved sale is developed based on sales of properties that have no building improvements. These sales are sales that have been used in the sales comparison approach and have been adjusted with the methods outlined above. The value of the underlying land in the improved sales is estimated and this value is subtracted from the total sale price to yield the contributory value of the improvements.

### Income Characteristics

Because agricultural properties produce income, comparisons and adjustments may be made to reflect their differing income characteristics. Gross income, operating expenses, and potential net income are used for these adjustments. Analysis of these characteristics provides indications of expected net returns on investment, which may also be used as a basis for comparison and adjustment in the sales comparison approach.

## Identification and Measurement of Adjustments

After the significant differences between the subject property and all comparable sales have been analyzed, the appraiser measures these differences in terms of their influence on property value. The validity of the sales comparison approach depends on an adequate data sample representative of the market and comparable to the subject property and on the ability of the appraiser to measure the differences between the sales and the subject.

Various analysis techniques can be used to identify and measure adjustments. The techniques can be quantitative or qualitative. Several of these techniques are listed below:

*Quantitative*

- Paired sales analysis
- Statistical analysis
- Graphic analysis
- Trend analysis
- Cost approach techniques

*Qualitative*

- Bracketing

*Comparative analysis* is the process in which one technique or a combination of techniques is used to process each comparable sale price into a separate indication of value for the subject property. The first step in comparative analysis is to identify which elements of comparison affect values in the subject property's market. Each relevant element of comparison is then considered in relation to each sale to decide if an adjustment is necessary.

### Quantitative Techniques

When quantitative techniques are applied, the appraiser uses mathematical processes to identify if an adjustment is necessary and to estimate the amount of the adjustment.

*Paired sales analysis* is a process in which two sales are compared to measure the amount of difference between them due to a particular element of comparison. For this technique to be accurate, the sales must be similar in every respect except the element being measured. When such sales are not available, dissimilar sales may be used in paired sales analysis if sufficient market evidence is available to adjust the sales beforehand for all dissimilar elements except the one for which an adjustment is sought. However, the more adjustments that have to be made prior to the paired sales analysis, the more suspect the results of the analysis become.

The best example of paired sales analysis is the sale and resale of the same property. Assuming there has been no material change in the character of the property, the only variation is the time that elapsed between the sales. The difference in the prices of the two sales should then indicate the market trend in prices over the time interval. When a sale and resale are not available, the appraiser can look for otherwise similar properties that were sold at different times. With rural properties, the difference in the proportions of various land classes (land mix) generally constitutes the most significant difference in properties. It is relatively easy to adjust sales for land mix. Once this adjustment has been accomplished, paired sales analysis becomes much more feasible.

Table 14.1 shows a paired sales analysis to estimate a market conditions (time) adjustment using two sales similar in all respects except land mix and time. In the discussion of sales and data analysis in Chapter 12, several sales were presented to illustrate the sales analysis process. Table 14.1 shows how paired sales analysis can be applied to two of those sales, Unimproved Sales 6 and 12, to estimate a time adjustment. Sales 6 and 12, which are described in Chapter 12, are similar in every respect except time and land mix. Ratio analysis, which was also discussed in Chapter 12, has been used to allocate the sale price of each property to the various land classes. If the resulting value per acre for each class of land in Sale 6, the current sale, is applied to the acreages of the land classes in Sale 12, the older sale, then the two sales can be adjusted to reflect the same land mix. This is done by adjusting Sale 6 for land mix and then adding to the sale price of Sale 6 the difference in value per acre between the sale prices of Sale 12 and Sale 6. The adjustment is $220/acre and Sale 6 is adjusted upward from $750 per acre to $970 per acre. Now, by comparing the sale price of Sale 12 to the adjusted price of Sale 6, paired sales analysis can be applied to estimate the adjustment for time. This adjustment is estimated on a compound monthly basis by using a financial calculator and entering the Sale 12 price of $1,094/acre as the present value *(PV)*, the adjusted Sale 6 price of $970/acre as the future value *(FV)*, and 12 months as the time period *(n)* and then solving for the rate *(I)*. (Note: One of the values must be entered as a negative number.) This calculation results in a compound monthly rate of change of minus (–) 1.00% for the last 12 months.

**Table 14.1** Paired Sales Analysis for Time

| | Land Adjustment— Unimproved Sale 6 | | | | Unimproved Sale 12 | | |
|---|---|---|---|---|---|---|---|
| Cropland | 300 ac.@ | $1,045 | $313,500 | Cropland | 509 ac.@ | $1,045 | $531,905 |
| Pasture | 390 ac.@ | $523 | $203,970 | Pasture | 85 ac.@ | $523 | $44,455 |
| Total | 690 ac.@ | $750 | $517,470 | Total | 594 ac.@ | $970 | $576,360 |
| **Price Adjustment** | | **Sale 6** | | | | **Sale 12** | |
| Acres | | 690 | | | | 594 | |
| Date of sale | | Current | | | | 12 mos. | |
| Sale price | | $750 | | | | $1,094 | |
| Terms | | Cash | | | | Cash | |
| Time | | 0 | | | | ? | |
| Location | | Good | | | | Good | |
| Land | | $220 | | | | 0 | |
| Indicated value | | $970 | | | | $1,094 | |

Difference due to time: $124/acre or minus 1%/month (compound rate)

Paired sales can be used in the same manner to measure changes due to any other factors. The most effective way to use the technique with unimproved sales is to first adjust the total sales data set to the subject's land mix. With all sales on an equal footing as to land mix, the entire unimproved sales data set can then be examined to isolate other pairing opportunities. Similarly, if all improved comparable sales can be adjusted to the subject's land and building mix, paired sales analysis can be performed on this entire data set. A detailed example of such an analysis will be presented later in this chapter.

Whenever possible, pairing sales is the recommended method for estimating the amount to adjust for various elements of comparison. Occasionally the technique is impractical because only a few, sufficiently similar sales are available. An adjustment derived from a single pairing is not necessarily conclusive. Nevertheless, even when data is limited, if a number of single pairs of sales are analyzed for individual elements of comparison and the results fit together in a pattern and make logical sense, it is stronger evidence of market behavior than purely subjective judgment.

*Statistical analysis* involves applying methods such as statistical inference and regression analysis to the sales data set. Linear and multiple regression are good tools for analyzing certain market situations. The effectiveness of regression analysis is limited when a large number of factors influence sale prices. A discussion of statistical analysis is beyond the scope of this text. Many texts on statistics are available to appraisers who want to review statistical techniques. *Graphic analysis* is a type of statistical analysis in which sales data is plotted on a graph and the appraiser arrives at a conclusion by visually interpreting the graphic display.

*Trend analysis* involves the use of large amounts of data. It is useful when there is a shortage of good comparable sales, but a large amount of general data is available. Trend analysis is especially useful for analyzing trends in value levels over time. Total sales activity for regions or states can be entered into a computer sales database and cataloged according to various market characteristics, including highest and best use.

*Cost approach techniques* are sometimes used to make adjustments relating to improvements when improved sales are lacking. This is an acceptable technique, assuming the improvements do not represent the major portion of value and an adequate market search was attempted to uncover similar improved sales. However, the sales comparison approach should be an independent approach. If major adjustments for the sales comparison approach come from the cost approach, sales comparison ceases to be an independent indicator of value. If improvements are the major part of property value and the only way the sales comparison approach can be applied is to use cost approach techniques, the appraiser is probably justified in not applying the sales comparison approach at all.

### Qualitative Techniques

*Bracketing* is a way to study the relationships indicated by the market data without making mathematical adjustments. This technique is sometimes used in complex real estate markets. To apply this technique, elements of comparison are identified and the comparable sales' characteristics are compared with the subject and ranked as superior, inferior, or equal. This analysis is very similar to paired sales analysis except the adjustments are not expressed as specific dollar or percentage amounts. Bracketing is especially useful when productivity factors (e.g., price per animal unit, milking cow, sow, productivity index point) are used as units of comparison.

A simple example of bracketing taken from a hog facility appraisal is shown as Table 14.2. The unit of comparison is per sow unit. All sales are compared to the subject for these elements of comparison: time (date), size (number of sows), effective age of facilities, and location. To estimate the net adjustment when some factors are inferior and others are superior, subjective weights must be placed on the factors that are most important in the specific market. Once net adjustments are estimated, the comparable sales are arrayed with the sale prices requiring negative adjustment listed above the subject and those requiring positive adjustment shown below the subject. In this example, the subject is nicely bracketed between Sale C ($1,210/sow) and Sale D ($990/sow).

Bracketing is used when the market data is not sufficient to derive dollar or percentage adjustments. If this method is used, it is essential that the appraiser fully explain the logic employed in selecting the elements of comparison and in determining the relative rankings of superior, inferior, and equal.

**Table 14.2** Bracketing

| | Subject | Sale A | Sale B | Sale C | Sale D |
|---|---|---|---|---|---|
| Date | | Current | Current | Current | Current |
| Price | | $1,584,000 | $792,000 | $1,452,000 | $1,386,000 |
| No. of sows | 1,200 | 1,200 | 900 | 1,200 | 1,400 |
| Price/sow | | $1,320 | $880 | $1,210 | $990 |
| Effect. age | 10 years | 8 years | 12 years | 9 years | 11 years |
| Location | Good | Good | Average | Good | Average |
| **Comparison Grid** | | | | | |
| Time | | Equal | Equal | Equal | Equal |
| Size | | Equal | Inferior | Equal | Equal |
| Effect. age | | Superior | Inferior | Superior | Inferior |
| Location | | Equal | Inferior | Equal | Inferior |
| Net Adjustment. | | Negative | Positive | Negative | Positive |
| **Summary of Analysis** | | | | | |
| | | | **Price/Sow** | | **Net Adjustment** |
| Sale A | | | $1320 | | Negative |
| Sale C | | | $1210 | | Negative |
| Subject | | | | | |
| Sale D | | | $990 | | Positive |
| Sale B | | | $880 | | Positive |
| Conclude estimated value of subject: $1100/sow | | | | | |

## Steps in Comparative Analysis

In the comparative analysis the appraiser employs either paired sales analysis or bracketing (or a combination of the two) to derive a value indication. When quantitative adjustments are applied, they are expressed either as dollar or percentage adjustments. In the adjustment grid, dollar adjustments are preferred, although the analysis to estimate the adjustment may be based on percentages. When qualitative analysis (bracketing) is applied, terms such as inferior, superior, greatly inferior, greatly superior, and equal are used. Regardless of which technique (quantitative or qualitative) is used, there are four general steps in the comparison process.

1. Through analysis of the subject's market, identify the elements of comparison—i.e., factors that are considered by buyers as they negotiate to purchase properties similar to the subject.
2. Compare each comparable sale with the subject property and measure the difference in each element of comparison to make an adjustment. The measurement can be either quantitative or qualitative, but the adjustment must be adequately explained so that a third party who is not an appraiser can understand the reasoning behind it.
3. Derive a net adjustment for each comparable and apply it to the sale price to indicate a value for the subject. The net adjustment is the sum of positive and negative adjustments for each element of comparison. The net adjustment will be a number (quantitative) or a comparative term (qualitative).
4. Reconcile the results of the individual indications of value for each comparable sale and develop a final value opinion for the subject property.

The comparable sales most similar to the subject are given the most weight in the final value opinion. In the quantitative analysis, this may be the value indication derived from the best comparable or the most probable value within a range of values for the best comparable sales. In qualitative (bracketing) analysis, the sales are divided into two groups—those superior and those inferior to the subject. These two groups will bracket the value of the subject.

When the two techniques are used in combination, adjustments that can be estimated from paired sales are made to narrow the variation between the comparable sales and the subject. Then the adjusted sale price of each sale is rated for other elements of comparison qualitatively, as equal, inferior, or superior, and bracketed around the subject.

## Application of Quantitative Comparative Analysis

To demonstrate the application of the sales comparison approach, the sample analysis in Chapter 12 is continued. The following activities were completed in the sales analysis presented in that chapter.

1. A market search was conducted resulting in 13 verified sales.
2. The 13 sales were divided into unimproved sales (the eight sales shown in Table 12.1 and Table 12.2 of Chapter 12) and improved sales (the five sales shown in Table 12.3 ).
3. Ratio analysis was applied to the eight unimproved sales to allocate their sale prices to the land classes present (cropland and pasture). A cropland-to-pasture value ratio of 100% to 50% was estimated and used in this allocation.

4. The land under the five improved sales was "appraised" using the most comparable unimproved sales, thus letting the value of the improvements "fall out" of the analysis. The value remaining after the underlying land value is deducted from the sale price is the contributory of the improvements. Tables 12.3 and 12.4 in Chapter 12 show these values. These two tables are incorporated into this chapter as Tables 14.3 and 14.4.
5. The contributory value of the improvements that resulted from the analysis described above was allocated to the individual buildings on each sale using a procedure very similar to land ratio analysis. Sales with only one building were used as a key to estimate the value in mixed sales of similar buildings, again letting the remaining value "fall out" to the other buildings. By working through the sales and using known building values to derive unknown building values, the total improvement contributory values were allocated.
6. The five fully analyzed improved sales shown in Tables 14.3 and 14.4 represent the data to be processed in this example of quantitative comparative analysis.

When the appraiser has arrived at the adjusted land and building values for each improved sale, several paired sales analyses can be conducted to isolate elements of comparison such as terms, time, size (land area), location, and other characteristics affecting property value in the subject's market.

For the purposes of this presentation, assume that the subject has the following basic characteristics, which were described in Chapter 12.

| The Subject Property | | |
|---|---|---|
| Date of appraisal | | Current |
| Location | | Good |
| Acres | | 510 |
| Cropland | | 400 |
| Pasture | | 110 |
| Buildings | | |
| Dwellings: | average condition, good utility, 2,250 sq. ft. | |
| Equipment shed: | average condition, average utility, 4,200 sq. ft. | |
| Grain bins: | fair condition, fair utility, 10,000 bushel capacity | |

**Table 14.3** Improved Sale Summary

| | | | | | | Land Class Acres | | | | | |
|---|---|---|---|---|---|---|---|---|---|---|---|
| Sale No. | Sale Date | Sale Price | Acres | Location | $/Acre | Crop Acres | Crop $/Acre | Pasture Acres | Pasture $/Acre | Bldg. Acres* | Bldg. $/Acre |
| 7 | 6 mos. | $300,000 | 300 | Avg. | $1,000 | 200 | $1,000 | 100 | $500 | House | $167 |
| 8 | Current | $391,200 | 446 | Avg. | $877 | 346 | $950 | 100 | $475 | Shed | $34 |
| 9 | 12 mos. | $362,150 | 330 | Good | $1,097 | 255 | $1,180 | 75 | $590 | s/bns | $52 |
| 10 | 6 mos. | $457,500 | 450 | Good | $1,017 | 300 | $1,100 | 150 | $550 | h/s/b | $100 |
| 13 | 12 mos. | $538,000 | 500 | Avg. | $1,076 | 392 | $1,070 | 108 | $535 | h/s/b | $122 |

* h = house, s = shed, b = barn

**Table 14.4** Allocation of Improved Sales' Buildings

| **Sale no.** | **7** | | **8** | | **9** | | **10** | | **13** | |
|---|---|---|---|---|---|---|---|---|---|---|
| Sale Date | 6 months | | Current | | 12 months | | 6 months | | 12 months | |
| Total improvement value | $50,000 | | $15,000 | | $17,000 | | $45,000 | | $60,780 | |
| **Improvements** | **Cond./Utility*** | **$** | **Cond./Utility** | **$** | **Cond./Utility** | **$** | **Cond./Utility** | **$** | **Cond./Utility** | **$** |
| | **Size** | **$/Unit** | **Size** | **$/Unit** | **Size** | **$/Unit** | **Size** | **$/Unit** | **Size** | **$/Unit** |
| Dwelling (in sq. ft.) | G-G | $50,000 | | None | | None | A-A | $36,000 | G-G | $52,500 |
| | 2,500 | $20 | | | | | 2,400 | $15 | 2,625 | $20 |
| Equip. shed (in sq. ft.) | | None | G-G | $15,000 | G-G | $12,000 | F-F | $4,500 | A-A | $8,280 |
| | | | 4,000 | $3.75 | 3,200 | $3.75 | 4,000 | $1.13 | 4,140 | $2.00 |
| Grain bins (in bus. cap.) | | None | | None | G-G | $5000 | G-A | $4,500 | P-P | $0 |
| | | | | | 5,000 | $1.00 | 6,000 | $.75 | 5,000 | $0 |
| Total imp. contrib. | | $50,000 | | $15,000 | | $17,000 | | $45,000 | | $60,780 |

G = good, A = average, F = fair, P = poor

### Step 1—Adjust All Comparables to Subject's Land and Building Mix

Once the improved sales are adjusted so that the land mix and buildings of each are similar to the subject, then the sales are arrayed in chart form and examined for other elements of comparison. This first step was accomplished in Chapter 12 (pp. 178–183). Sales 7, 8, 9, 10, and 13 were adjusted to the subject's land mix and buildings. When the improved sales were compared to the subject for land and buildings, all other elements of comparison were ignored or assumed to be equal between the sales and subject. In Step 2 below, adjustments are made for these other elements of comparison.

### Step 2—Compare Improved Sales Adjusted for Land and Building Mix to Each Other and to the Subject (Paired Sales Analysis)

Now the improved sales can be compared directly to one another and to the subject for other elements of comparison such as time, location, and size. Table 14.5 shows a final sales comparison grid. Using this grid, the appraiser gradually makes each improved sale more like the subject, adjusting for each successive element of comparison as one proceeds down each sale column.

The first adjustments shown in Table 14.5 are the land and buildings adjustments derived in Chapter 12. There is considerable debate about *when* it is technically correct to make a time adjustment. Traditionally, property rights conveyed and financing terms come first, then time. All other adjustments generally follow the time adjustment.

However, in some markets land values may be changing at a different rate, and possibly in a different direction than building values. If this is occurring, then deriving a time adjustment from unimproved land sales first will definitely not produce an appropriate rate to apply to improved sales. Similarly, if all of the improved sales are not almost equal to the subject in land and building ratios, then deriving a time adjustment directly from the pairing of two or three improved sales may not give an accurate reading of the time influence.

Thus far, all land allocations and adjustments and all building allocations and adjustments have been made "as of the date" of each respective sale. By taking the land mix and buildings adjustments derived and applying them in Table 14.5, all improved sales are made equal to the subject in land and buildings "as of the date" each sale occurred. Once all sales have been made equal in land mix ratios and buildings, a more accurate assessment of the time influence can be made.

The financing terms adjustment, if any, should be applied to the gross sale price of a particular sale prior to any land or building allocations, thereby reflecting the cash equivalent value of the entire property. The sales shown in this example all reflect cash equivalent sale prices, so no terms or financing adjustments are required.

The detailed discussion that follows the adjustment analysis in Table 14.5 describes how the time, size, and location adjustments were derived. This direct comparison process serves as a multiple paired sales presentation. Note that as one proceeds down the columns for each sale, the adjusted values become more tightly clustered. This reflects the fact that more of the variations or differences between the subject and the sales are being accounted for with each successive adjustment as one moves toward the final indicated value for each sale. In other words, the data are analyzed on a "running total" basis.

**Table 14.5** Direct Sales Comparison Grid

| Sale No. | 7 | 8 | 9 | 10 | 13 |
|---|---|---|---|---|---|
| Date | 6 months | Current | 1 year | 6 months | 1 year |
| Size | 300 ac. | 446 ac. | 330 ac. | 450 ac. | 500 ac. |
| Financing | Cash equiv. | Cash equiv. | Cash equiv. | Cash equiv. | Cash equiv. |
| $/Acre | $1,000 | $877 | $1,097 | $1,017 | $1,076 |
| Adjustments | | | | | |
| Land | + 59 | + 5 | + 7 | + 64 | 0 |
| Buildings | – 61 | + 76 | + 60 | 0 | – 18 |
| Adj.$/acre | $998 | $958 | $1,164 | $1,081 | $1,058 |
| Market conditions | 6 months | Current | 1 year | 6 months | 1 year |
| Time adj. | – 47 | 0 | – 107 | – 51 | – 97 |
| Adj.$/acre | $951 | $958 | $1,057 | $1,030 | $961 |
| Location | Avg. | Avg. | Good | Good | Avg. |
| Loc. adj. | + 85 | + 85 | 0 | 0 | + 85 |
| Adj.$/acre | $1,036 | $1,043 | $1,057 | $1,030 | $1,046 |
| Size (in acres) | 300 ac. | 446 ac. | 330 ac. | 450 ac. | 500 ac. |
| Size adj. | 0 | 0 | 0 | 0 | 0 |
| Adj.$/acre | + $1,036 | + $1,043 | + $1,057 | + $1,030 | + $1,046 |
| Final indicated value | $1,036 | $1,043 | $1,057 | $1,030 | $1,046 |

### Market Conditions Adjustment

A comparison of the values of Sales 7 and 8 adjusted for land mix and buildings gives an indication of the possible value influence of time. These two properties are relatively similar in size and location, but they were sold six months apart. The $40 difference in their land-and-building-adjusted values can therefore be attributed to the six months' difference in time. Using a financial calculator, a rate of loss in value of approximately 0.7% compounded monthly is estimated.

Sales 8 and 13 are similar to one another in size and location, but their sale dates differ by 12 months. A comparison of their land-and-building-adjusted sale prices suggests a $100 per acre loss in value, or about 0.8% compounded monthly.

Sales 9 and 10, once they are made equal to one another and the subject in land mix and buildings, are relatively equal in all other elements of comparison except time. Sale 9 sold 12 months ago and Sale 10 sold six months ago. The $83 difference in their respective land-and-building-adjusted values can be attributed to the time difference. A 1.2% compound monthly loss in value is calculated.

These three paired sales analyses suggest that a downward time adjustment of 0.7% to 1.2% compounded monthly is warranted. The analysis of Sales 8 and 13 spans the entire 12-month period represented by all five sales, and it suggests a negative time adjustment of 0.8% compounded monthly. Based on these findings, a negative time adjustment of 0.8% compounded monthly is applicable to the land-and-building-adjusted values of all non-current improved sales shown in Table 14.5.

### Location Adjustment

Sales 7 and 9 are within 30 acres (10%) of being the same size, and once all sales have been adjusted toward the subject for land, buildings, and time, these two sales are equal in all other elements of comparison except location. Sale 7 has an average location, while Sale 9 and the subject have good locations. The $106 difference in the time-adjusted values of Sales 7 and 9 suggests that properties with average locations require a $106 per acre adjustment to make them more like properties with good locations.

Sales 10 and 13 are equal to one another in land, buildings, and time once all sales have been adjusted toward the subject for these elements of comparison. However, Sale 10 has a good location and Sale 13 has an average location. The $69 difference in the time-adjusted values of Sales 10 and 13 suggests that properties with average locations require a $69 per acre adjustment to make them more like properties with good locations.

Based on these two pairings for location, all properties with average locations require a positive $85 per acre adjustment to their time-adjusted values when compared to the subject property, which has a good location.

### Size Adjustment

A review of the location-adjusted values in Table 14.5 suggests that size does not appear to significantly influence per-acre sale prices of properties in the 300-acre (Sale 7) to 500-acre (Sale 13) range. Sales 8, 10, and 13 are the largest of the five sales, with a mean size of 465.33 acres. Their mean location-adjusted value is $1,040 per acre. Sales 7 and 9 are the smallest sales with a mean size of 315 acres and a mean location-adjusted value of $1,046 per acre. This analysis shows an insignificant difference in value of about $6 per acre in favor of the smaller properties. Based on these findings, no size adjustment is warranted in the comparison process.

### Step 3—Derive a Final Value Conclusion

The five indicated values suggested by the improved sales shown in Table 14.5 can now be summarized. The two sales requiring no location adjustment suggests a value of $1,030 to $1,057 per acre. The sale requiring no time adjustment suggests a value of $1,043 per acre. The three sales requiring the lowest gross dollar adjustment suggest a value of $1,030 to $1,047 per acre. The mid-range of the two extreme value indications is $1,044 per acre. The mean of the five indicated values is $1,042 per acre.

The mean exclusive of the two extremes is $1,043/acre
The median indicated value is $1,043/acre
Indicated value via the sales comparison approach
510 acres @ $1044 per acre = $532,440.00
Rounded to $530,000.00

This complete sales comparison approach analysis has been presented to give the appraiser another tool to use in presenting a methodical analysis of the available comparable sales and arriving at a well-supported final opinion of value for the subject property. An appraiser who chooses this method of analysis, or something similar, can feel comfortable even with data that are erratic. The key to an accurate sales comparison approach lies in allowing the market to do the talking. While judgment is always necessary, "rules of thumb" and preconceived notions should be eliminated from the appraisal process. Proper data analysis places clear parameters on the appraiser's judgment.

### Quantitative Comparative Analysis Using Productivity-Based Units of Comparison

In the previous example, the unit of comparison applied, acres, was based on physical size. The same technique can be applied using productivity-based units of comparison (e.g., per animal unit, per sow, per dairy cow, per bushel or barrel). The land mix and buildings of all the sales in the data set are first adjusted to be similar to the subject. Table 14.6 shows the land mix adjustment on an animal unit (AU) basis for the comparison of a ranch sale property to a ranch subject property. The carrying capacity of irrigated pasture is 2 acres per animal unit. Rangeland has a carrying capacity of 25 acres per animal unit. The adjusted value of the subject's land is derived by applying per acre values, but the actual land mix adjustment is based on dollars per animal unit.

The final building totals in the building adjustment grid would also be divided by the animal unit capacity of the sale and subject and these adjustments would be made on an animal unit basis. Working down a sales comparison grid for this ranch property (similar to Table 14.5), the paired sales and adjustments for each successive element of comparison would reflect the animal unit (AU) unit of comparison. This same procedure can be used to apply any productivity-based unit of comparison.

**Table 14.6** Animal Unit Land Adjustment Analysis (Productivity-Based Unit of Comparison)

| Sale X (650 Animal Units) | | | | Subject (700 Animal Units) | | | |
|---|---|---|---|---|---|---|---|
| Irr. pasture | 500 ac.@ | $600 = | $300,000 | Irr. Pasture | 1,000 ac.@ | $600 | $600,000 |
| Rangeland | 10,000 ac.@ | $100 = | $1,000,000 | Rangeland | 5,000 ac.@ | $100 | $500,000 |
| Total | 650 AUs@ | $2,000* = | $1,300,000 | Total | 700 AUs@ | $1,571† | $1,100,000 |
| Indicated land adjustment to Sale X = minus $429/animal unit (AU) | | | | | | | |

* $1,300,000 total divided by 650 AUs

† $1,100,000 total divided by 700 AUs

## Alternative Market Measures

Other market measures that truly reflect the actions of buyers and sellers in the subject's market can also be used. Income-based measures such as gross income multipliers *(GIMs)* are frequently used to support other sales comparison approach conclusions. Moreover, productivity-based units of comparison are not always adjusted as shown in the previous example. Gross income multipliers are never adjusted. Generally, to perform an analysis using *GIMs* and a per-unit productivity measure, the appraiser reviews the dollars per unit or gross income multipliers for a number of sales and, based on judgment and reasoning, selects a unit value for application to the subject's gross income or number of productivity-based units. The key to this type of analysis is the number and quality of sales and the reasoning behind the selection of the *GIM* or final dollar-per-unit value.

Another technique used in the sales comparison approach is the total property method. Here the total sale price for each comparable sale is adjusted for each element of comparison on a total dollar basis.

## Summary

The sales comparison approach is generally recognized as the most persuasive indication of market value because it is based on actual sales of similar properties. The sale price of a comparable property is a fact. The physical characteristics of the sale property can be identified by careful inspection and accurately compared in substantial detail to the property being appraised. The circumstances surrounding each sale, including the sale price and terms of sale, can be learned by carefully questioning the buyer or the seller or both. The appraiser is therefore able to develop a sound opinion as to whether the sale represents a true, open-market transaction and is prepared to draw a dependable market value conclusion from the facts of the sale.

The steps in the sales comparison approach are:

1. The appraiser researches the market to obtain information about transactions, listing, and other offerings of properties similar to the subject property. This step was accomplished as part of the sales analysis discussed in Chapter 12.
2. With care, the appraiser verifies the information obtained in Step 1, ascertaining that the data obtained are factually accurate and the transaction reflects arm's-length, market considerations. Verification is accomplished by obtaining the same information from a second knowledgeable source, usually one of the participants in the transaction.

3. The appraiser then identifies relevant units of comparison (e.g., acre, square foot, animal unit, bushel per acre) and develops a comparative analysis. In selecting a units of comparison, the appraiser is guided by the type and characteristics of the property and the unit of comparison commonly applied in the market.
4. In quantitative comparative analysis, the appraiser adjusts the sale price of each comparable property, as appropriate, or eliminates the property as a comparable. The subject and the comparable sales are compared on the basis of the elements of comparison: financing terms, conditions of sale, market conditions (time), location, and physical/economic characteristics. Using an appropriate unit of comparison, the appraiser adjusts each sale price, increasing or decreasing the sale price per unit for each significant variation between the comparable and the subject property. Paired sales analysis is the preferred method for estimating the amount of adjustments. By adjusting all the sales in the data set to the subject's land mix and building values, the entire data set can become part of the matched pair analysis.

   In qualitative comparative analysis the same procedure is applied, except instead of dollar or percentage adjustments for each element of comparison, the element is evaluated as being equal, superior, or inferior to the subject. Then overall comparisons for the sales are derived and bracketed around the subject.
5. From the pattern developed, the appraiser reconciles the multiple valuation indications resulting from the comparable sales into a single value indication. If the analysis is done on a per-unit basis, the appraiser multiplies the subject's indicated value per unit by the number of units in the subject to develop a final opinion of value.

There are many techniques that can be used in the sales comparison approach. The key to the validity of a technique is its ability to represents the way buyers and sellers interact in the subject's market.

CHAPTER • 15 •

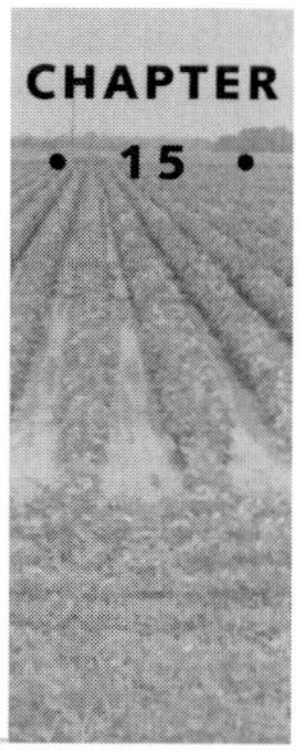

# The Income Capitalization Approach

The income capitalization approach is one of the three primary approaches used to arrive at an indication of market value. This approach is predicated on the principle of anticipation, which holds that present value is indicated by the expectation of future benefits. In this chapter two methodologies are explained: direct capitalization and yield capitalization. In both methods, the income studied is rental income, not business or entrepreneurial income.

The distinction as to the type of income is important and must be briefly explained. Rural properties are either owner-operated or rented for cash or a share of the crop. Owner-operators look to the farming enterprise as the income source; these properties are not developed to collect income in the form of rent. Consequently, owner-operator income and expense analysis includes other components which contribute to income, such as equipment and entrepreneurial ventures. Analysis of owner-operator income and expenses is an excellent tool for testing the feasibility of the enterprise, but due to its complexities and subjectivity, it is a difficult method to employ in the valuation of real estate. Cash rental or cropshare rental arrangements are used by individuals who choose not to farm their land. Instead they lease the land to another individual who operates the farming enterprise. In preparing an appraisal, it is important to ascertain the applicability of the income capitalization approach to the type of property being appraised. The approach is certainly appropriate in appraising rural properties that are typically leased. When valuing properties that are predominantly owner-operated such as orchards and vineyards, an owner-operator analysis may be appropriate. An analysis of owner-operator income and expenses customarily employs capitalization rates and yield rates abstracted from owner-operator income and expense statements.

## Income Estimates

The primary purpose of estimating the gross income of a property is to project the probable net income benefits from its ownership. To apply direct or yield capitalization, a reliable estimate of income expectancy must be developed. Rental income may be in the form of cash or a share of the crop, depending on local customs and the crops grown. Cash rent is generally a specified amount per acre. In a cropshare arrangement the owner receives a percentage of the commodity produced on the property. Before the rural appraiser can estimate the potential gross income from which to develop a reasonable net income projection, two unique aspects of farm income must be considered

1. The income stream of an agricultural property may be cyclical due to the variation in cropping patterns, including rotation.
2. Lease periods are generally shorter than they are for urban properties. If the property being appraised is leased, the appraiser must be aware of all the lease terms. Leases will vary in type, terms, and conditions from community to community. Our society is increasingly complex, and the appraiser should not make general assumptions as to the terms of a lease.

Income and expense forecasts begin with lease analysis. *Gross rental leases*[1] are leases in which the lessor pays all operating expenses of the real estate. In *net rental leases,* the tenant pays all such expenses. Leases can be classified as *flat rental leases,* with a specific level of rent that continues for the duration of the lease; *graduated rental leases,* in which the rent undergoes specified changes at one or more points during the lease term; *revaluation leases,* in which rent is adjusted at periodic intervals based on a revaluation of the real estate; *index leases,* in which periodic rent adjustments are tied to a specific index such as commodity indexes or the general inflation rate; and, of course, *percentage* or *cropshare leases.*

Typical lease data include the lease date, a legal description of the leased property, the names of the lessor and lessee, the lease term, the dates of occupancy and the commencement of payments, the amount and source of the rent, escalation clauses, rent concessions, land and tenant covenants regarding the division of expenses, the right of assignment or right to sublet, renewal options, purchase options, escape clauses, stipulated advance rent or bonding, provisions for casualty or condemnation loss, continued occupancy clauses, specifications concerning tenant improvements, and revaluation clauses.

Five different types of rent can contribute to income: *contract rent,* the actual rental income specified in a lease; *market rent or economic rent,* the rental income a property would most probably command in the open

---

1. Most of the definitions in this chapter are taken from *The Appraisal of Real Estate,* 11th ed. (Chicago: Appraisal Institute, 1996).

market; *excess rent,* the amount by which contract rent exceeds market rent at the time of the appraisal; *percentage rent,* rental income received in accordance with the terms of a percentage clause in a lease: and *overage rent,* rent paid over and above the guaranteed minimum rent.

The essential steps in estimating the gross income of a farm property are:

1. Determine the typical income arrangement for the area (cash rent, cropshare, or owner operation).
2. Ascertain the crops most likely to be produced.
3. Estimate yields based on property-specific data or local area data.
4 Estimate commodity prices.
5. Determine the crop share for percentage rents.
6. Determine property-specific terms and market rent terms for cash rents.

## Cash Rent

Cash rent is generally fixed for a specified period of time. Many cash rents are established on a year-to-year basis for agricultural properties, but rents can be specified for extended periods. Cash rents can be based on total acreage, crop acreage, base acreage, or any other basis typical for the area. They can be paid in advance or in arrears or payment can be split between the beginning and end of the crop year. Cash rentals for specialized agricultural buildings often have the same characteristics as rentals of typical commercial building. The appraiser must take care to analyze the local market to ascertain the cash lease arrangement common to the area. Long-term leases must be analyzed because there will often be a substantial difference between contract rent and market rent.

Often buildings on farms are rented to tenants other than those who rent the land. Rental arrangements for these buildings can take the form of simple residential rentals or more complicated arrangements in which the home, barns, and grain bins are rented separately. Another source of farm income that is becoming common is rent from off-season use of the agricultural property for recreation such as hunting, fishing, or resort-related activities. These arrangements are typically specified in cash leases.

Table 15.1 represents an income analysis for a hypothetical property. In this case differing cash rentals are charged for the different land types on the farm and the buildings are rented separately to an individual who is not involved in the farming operation. Other land not associated with the buildings is leased to a recreational tenant, who uses the property for

**Table 15.1** Cash Lease Income Analysis

| Land Type | Acreage | Rent Per Unit | Total Cash Rent |
|---|---|---|---|
| Cropland A | 120 | $30/acre | $3,600 |
| Cropland B | 300 | $20/acre | $6,000 |
| Pasture | 100 | $10/acre | $1,000 |
| Buildings | 10 | $500/month | $6,000 |
| Recreational lease | 520 | $10/acre | $5,200 |
| Total | 530 | | $21,800 |

hunting. Hunting is typically confined to the fall and winter and does not interfere with the agricultural enterprise.

## Percentage or Cropshare Rental

In many areas percentage or cropshare rentals are common. Consequently, annual income per acre fluctuates. It is not a stable, guaranteed annual sum as is typical in urban leases or in the cash rent arrangements previously described.

When dealing with cropshare rentals, the rural appraiser must have a working knowledge of production practices. The rural appraiser must prepare an estimate of gross income on an average annual basis over a period of time. A five-year period is commonly used. By using an extended time period and making necessary adjustments for economic realities, the appraiser can level out the cyclical effect of crop rotations and commodity prices and estimate an average income stream.

There are accepted procedures for developing the income and expense projections needed to derive a gross income estimate. The gross income estimate is based on the amount of acreage and the type of crop grown on the property in a rotation cycle on an annual basis. An entire agricultural property may be devoted to a single crop (e.g., dryland wheat production) in any single year. However, over the rotation cycle, a variety of crops can be grown on the property. For example, irrigated land may have a typical rotation of potatoes, wheat, corn, and alfalfa. To estimate the gross income over the rotation cycle, the appraiser ascertains the amount of acreage that will be devoted to a particular crop on an average annual basis and estimates the probable production (yield) of the crop, the probable price that will be received for the crop, and the percentage of the crop that will be paid to the landowner. The appraiser should employ assumptions that are typical for the area and property type and analyze the historical income and expenses of the subject property.

To complete the gross income estimate, the appraiser includes other actual or potential income from the property. Such income may be derived from the rental of improvements separate from the farm operation and from recreational leases. The income will typically be specified in cash leases. When there are different tenants on a farm, the cropshare lease will actually be a combination of percentage and cash leases. If this is the typical rental method in the area, the appraiser should have no problem with the mixture of percentage and cash leases.

Table 15.2 describes a cropshare rental of the same hypothetical property presented in Table 15.1. On this farm 200 acres are typically planted to crop A, and the typical share is one-fourth of the total production to the owner as rental; 300 acres are typically planted to crop B, and the typical share is one-third of the total production to the owner as rental. The pasture is typically rented for cash, possibly to a different tenant, and the buildings are rented to yet another tenant for a

**Table 15.2** Cropshare Income Analysis

| | Acres | Yield Per Acre | Total Yield | Price Per Unit | Share | Owner Income |
|---|---|---|---|---|---|---|
| Crop A | 200 | 100 bu. | 20,000 | $ 4.00 | ¼ | $20,000 |
| Crop B | 300 | 40 bu. | 12,000 | $ 3.00 | ⅓ | $12,000 |
| Pasture | 50 | N/A | N/A | $10.00 | All | $500 |
| Buildings | 5 | N/A | N/A | $400/mo. | All | $4,800 |
| Recreation | 550 | N/A | N/A | $ 5.00 | All | $2,750 |
| Total | 555 | | | | | $40,050 |

monthly rental. The recreational lease is also assigned to another tenant for cash.

## Owner-Operator Income Statement

To analyze owner-operator income, the entire income attributable to the operation of the farm must be estimated (see Table 15.3). It is assumed that the owner operates the farm and no rental income is received, except possibly income from a recreational lease. The home and buildings are occupied by the owner and the rental attributed to these buildings is actually the rental savings the owner is receiving because no financial outlay is required for a separate residence or base of operations.

Owner-operator income estimation may be required in appraising a standard crop operation if leases are not available for analysis. Because this form of income estimation is the most difficult and most susceptible to error, it is usually not attempted when lease information is available. For specialized agricultural endeavors such as permanent plantings, poultry operations, grain elevators, and feed lots, lease information is not readily available, so the owner-operator method of income estimation must be used.

In the following example, 200 acres of the farm are typically planted to crop A and 300 acres are planted to crop B. The pasture is operated by the owner for hay production, and the buildings are occupied by the owner. There is a recreational lease to a tenant who pays in cash.

## Effective Gross Income

*Effective gross income* is the portion of income that remains after an allowance has been made for vacancy and collection loss. Although vacancy is not typically considered in the income statement of a farm,

**Table 15.3** Owner-Operator Income Analysis

| | Acres | Yield Per Acre | Total Yield | Price Per Unit | Owner Income |
|---|---|---|---|---|---|
| Crop A | 200 | 100 bu. | 20,000 | $4.00 | $80,000 |
| Crop B | 300 | 40 bu. | 12,000 | $3.00 | $36,000 |
| Pasture | 50 | 1 ton | 50 | $20.00 | $1,000 |
| Buildings | 5 | N/A | N/A | $400/mo. | $4,800 |
| Recreation | 550 | N/A | N/A | $5.00 | $2,750 |
| Total | 555 | | | | $124,550 |

the appraiser should be aware that vacancy can be a problem for some types of agricultural properties. If vacancy or collection loss is typical for a particular type of property, then a charge should be shown in the expense statement.

Cash leases are typically paid in advance or a portion of the annual rental is paid in advance, so vacancy and collection loss does not occur. In the case of a crop failure in a cropshare arrangement, vacancy is not the appropriate term.

Although vacancy is not a concern in the typical crop production agricultural unit, vacancy can occur in other forms of agricultural endeavors and must be considered. Vacancy may occur for a variety of reasons, such as disease in the animals housed in specialized buildings, the inability to rent rural buildings, a lack of quality irrigation water, poor drainage, or an irrigation system in disrepair. Credit loss stems from the inability of the tenant to pay the agreed-upon rent.

An appropriate percentage factor may be applied to estimate vacancy. The appraiser should ensure that the percentage factor reflects current market conditions. If the property being appraised has a history of rental income, this is excellent property-specific data from which to estimate the gross and actual income received. The appraiser should also check with data sources such as area farm managers, other landlords, tenants, and lenders. Table 15.4 illustrates the procedure for deducting an anticipated 5% vacancy and collection loss from potential gross income to arrive at an estimate of effective gross income.

**Table 15.4** Estimating Effective Gross Income

| | |
|---|---|
| Potential gross income | $100,000 |
| Less vacancy and collection loss @ 5% | − 5,000 |
| Effective gross income | $95,000 |

## Analyzing Expenses

To derive the net income from the operation of a rural property, the appraiser begins with the gross income estimate and deducts all expenses required to secure and produce the income stream and maintain and manage the property. In analyzing the expense statement, the appraiser should ensure that the charges included by the landowner are directly related to the operation and management of the property. Deductions for depreciation and mortgage interest expense may be proper on the owner's income tax statement, but they are not included in the net income from operations.

The expenses deducted from income should be consistent with the nature of the income—i.e., cash rental, cropshare rental, or owner-operated expenses. Under some rental agreements, most notably cash leases and cropshare arrangements, expenses may vary from crop to crop, area to area, or year to year. The expenses deducted should be typical for the area at the time of the appraisal.

A landlord-tenant agreement (income in the form of rent) is the preferred basis for the income and expense estimate. The landlord's income and expense statement isolates that part of the net income that is directly attributable to the real property.

Whether the rental terms specify cash or cropshare rental, the landlord expenses in a gross rental lease, also called a *full-service lease,* usually include basic items such as management, real estate taxes, insurance, and building maintenance. Other expenses for irrigation assessments, electrical power, and irrigation system maintenance may or may not be included depending on the typical lease arrangement in the area. In certain areas and for certain improvements, a reserve for replacement of these items must be included in addition to standard maintenance expenses.

The appraiser should be aware of all the lease terms applicable to the property being appraised as well as the lease terms for the comparable sales and rentals. All data must be handled in a consistent manner. Obviously, not all expense items will be included in every situation. However, consistent data on the subject and comparable properties must be collected. Sources of expense data vary, and the records of the property being appraised are a good place to start.

Table 15.5 presents an income and expense statement for the cash-rented farm described previously.

Management fees are incurred whether the property is professionally managed or managed by the owner-operator. In any case, a property that is leased generates the need for management services. Management fees can be estimated with information from area farm managers and trust departments.

Real estate taxes are recorded at the county courthouse. Often the property owner has a current tax statement. Obtaining the assessor's parcel numbers helps to expedite the process. Typically real estate taxes are estimated annually. Some counties tax personal property such as machinery, tools, and livestock at the same time, and the appraiser should separate the two types of taxes. Irrigation, drainage, and weed assessments should also be segregated when applicable. The real estate tax history of the property should be analyzed to help identify trends since the appraiser must project this expense, especially when yield capitalization is applied. Many states allow for real estate tax deferral or abatement on qualified agricultural properties

**Table 15.5** Income and Expense Summary

| Land Type | Acreage | Rent Per Unit | Total Cash Rent |
|---|---|---|---|
| Cropland A | 120 | $30/acre | $3,600 |
| Cropland B | 300 | $20/acre | $6,000 |
| Pasture | 100 | $10/acre | $1,000 |
| Buildings | 10 | $500/month | $6,000 |
| Recreational lease | 520 | $10/acre | $5,200 |
| Total gross income | | | 21,800 |
| Expenses | | | |
| Taxes | | | $2,020 |
| Insurance | | | $1,760 |
| Maintenance | | | $2,400 |
| Management | | | $1,744 |
| Total expenses | | | 7,924 |
| Net operating income | | | 13,876 |

(e.g., open space assessed). A change in use or failure to comply with state law may cause a farm to lose its favorable tax status. In some instances, taxes that have been waived or deferred for a specific number of years can become a potential lien on the property. These accumulated deferred taxes may be an adjustment item in the sales comparison approach, particularly if the appraiser is dealing with transitional properties.

Insurance rates for various classes of buildings are available from local insurance agents. Insurance documents show a value estimate for the buildings, but this may reflect replacement cost, which could exceed the contributory value of the improvements. The appraiser should determine if the buildings are or can be insured. It is frequently difficult to insure unused or unoccupied farm buildings. The expenses attributable to property improvements, including insurance and building maintenance, depend on local practice. In some areas, typical rental agreements call for the tenant to provide a specified level of insurance on improvements.

Depending on the extent of improvements and community practice, building and other maintenance expenses can range from almost zero to a significant portion of the landlord income. *These expense estimates must reflect rental terms.* The tenant may assume all maintenance duties or be responsible for repair only, with the general maintenance burden falling on the landlord. For example, the rental contract may require the tenant to replace a specified number of fences and irrigation system components annually.

A reserve for replacement expense is often applicable to short-lived items or components of long-lived items, but it is not a typical expense in agricultural appraisal. A center-pivot irrigation system, for example, may be owned by the tenant or the landowner. Establishing a reserve for replacement of this type of asset may be appropriate, but often in agricultural appraisal the expense item remains in the income statement and is therefore left in the capitalization rate. The key is local custom and consistency in the derivation and application of the expense estimate.

When estimating maintenance expenses, the appraiser must consider all property improvements: buildings; irrigation and drainage facilities, including pumps, motors, and sprinkler equipment; specialty improvements such as trellises or permanent plantings; and other non-building improvements such as fences, corrals, and livestock watering facilities. The appraiser must ascertain whether the maintenance required is sufficient and, if not, whether additional cost must be included to meet market expectations. In the preceding example a percentage of the building value is allocated to annual maintenance.

Irrigation expenses are generally available from the irrigation district, county offices, or individual irrigation company secretaries. These charges will usually cover annual operation and maintenance as well as any debt repayments. The charge can be expressed per acre, per acre foot of water delivered, or per farm unit, or a combination of these bases can be used. For some irrigation systems, the power charges for pumping must be added to a district charge or considered in the irrigation expense, as is the case with irrigation wells or pumping stations on lakes or rivers. The division of the irrigation charge between the owner and the tenant may vary with the specific rental terms and should be identified.

In a cropshare situation, the landlord may participate in growing cost. Growing costs vary by crop and community and specific farm records are useful. The suppliers and dealers of inputs such as seed and fertilizers should have current product costs as well as delivery and application charges. Under a cropshare rental agreement, the landlord and the tenant may share the cost of fertilizer, herbicide, insecticide, harvesting, and hauling. The landlord may participate in labor costs as well.

Some expense items frequently encountered in an owner's expense statement should not be included in the estimate of expenses related to the income-producing capability of the property. These expenses include depreciation, mortgage interest, amortization, income tax, additions to capital, and unusual owner expenses not directly related to income production and operations.

### Net Operating Income

The next step in the income capitalization approach is to derive net operating income. Table 15.5 itemizes the income and expenses for the subject farm property. The effective gross income minus expenses equals the net operating income. This is a good point for the appraiser to test the reasonableness of the estimated expenses. The expense ratio for the sample property is 36%—i.e., expenses amount to 36% of effective gross income. Is this information consistent with the income and expense and rental data gathered from comparable sales? Are the expenses treated consistently? Consistency in the treatment of income and expenses for the subject and comparable sales is imperative in the income capitalization approach. It is critical in the derivation of an overall capitalization rate.

## Direct Capitalization

Direct capitalization is a method used in the income capitalization approach to convert a single year's income estimate into a value indication. This conversion is accomplished in one step, either by dividing the income estimate by an appropriate income rate or by multiplying it by an appropriate income factor. In direct capitalization no precise allocation is made between the return *on* and the return *of* capital because the method does not simulate investor assumptions or expectations concerning the holding period, the pattern of income, or changes in the value of the original investment. However, a satisfactory rate of return for the investor and recapture of the capital invested are implicit in the rates or factors applied in direct capitalization because they are derived from similar investment properties.

Direct capitalization may be based on potential gross income, effective gross income, net operating income, equity income, mortgage income, land income, or building income. The income rates used in direct capitalization include the overall (property) capitalization rate *($R_O$)*; the mortgage capitalization rate *($R_M$)*; the equity capitalization rate or equity dividend rate *($R_E$)*; the land capitalization rate *($R_L$)*, and the building capitalization rate *($R_B$)*. Income factors include the potential gross income multiplier *(PGIM)*, the gross rent multiplier *(GRM)*, and the effective gross income multiplier *(EGIM)*.

Income rates and factors reflect the relationship between income and value and are derived from market data. It is essential that the properties used as comparable sales reflect risk, income, expense, and physical and location characteristics that are similar to the property being appraised. Consequently, income multipliers and rates must be extracted from properties that reflect similar income-to-expense ratios, land-to building ratios, risk characteristics, and expected changes in income and value over a typical investment holding period.

## Derivation of Overall Capitalization Rates

Any interest in real estate that has an income stream can be valued by direct capitalization, but the interest most commonly appraised is the fee simple estate, which includes all property rights in the real estate. The direct capitalization formula that applies to this type of valuation is:

$$\text{Value} = \frac{\text{Net Operating Income}}{\text{Overall Capitalization Rate}}$$

An appraiser can estimate an overall capitalization rate with various techniques. The techniques applied depend on the quantity and quality of data available. Accepted techniques include 1) derivation from comparable sales, 2) derivation from effective gross income multipliers, 3) band of investment—mortgage and equity components, 4) band of investment—land and building components, and 5) application of the debt coverage formula.[2] In rural property appraisal, the most common technique for derivation of capitalization rates is direct extraction from comparable sales.

Deriving capitalization rates from comparable sales is preferred when sufficient data on sales of similar, competitive properties is available. Data on each property's sale price, income, expenses, financing terms,

2. All five techniques are described in *The Appraisal of Real Estate,* 11th ed. (Chicago: Appraisal Institute, 1996), 514–520.

and market conditions at the time of sale is needed. In addition, the appraiser must be certain that the net operating income of each comparable property is calculated and estimated in the same way that the income of the subject property is estimated and that non-market financing terms or different market conditions have not affected the prices of the comparable sales. When these requirements are met, the appraiser can estimate an overall rate by dividing each property's net operating income by its sale price. The formula is:

$$NOI = \frac{\text{Overall Rate}}{\text{Sale Price}}$$

Data requirements for direct capitalization are stringent, and these specifications are difficult to meet in all appraisal situations. However, because rural property uses (e.g., crop farms, dairy farms, cattle ranches, orchards) tend to be consistent in a given area, buyers and sellers have similar expectations about the direction and character of income changes, resale prices (or property appreciation rates), tax benefits, and holding periods. Therefore, direct extraction of rates from sales is a viable method of capitalization for rural properties.

Table 15.6 illustrates the derivation of overall capitalization rates from comparable sales. If all three transactions are equally reliable and comparable, the appraiser might conclude that an overall rate of 4.70% to 5.10% is reasonable. However, considering the larger size of Sale 3, a narrower range of 4.90% to 5.10% is more appropriate. The final rate selected depends on the appraiser's judgment as to how comparable each sale is to the subject property.

If there are differences between a comparable property and the subject property that could affect the overall capitalization rate, the appraiser must account for those differences in the selection process. When rates derived from comparable sales are used, the overall capitalization rate is applied to the subject property in a manner consistent with the manner of rate derivation. In other words, if the subject property is appraised using its cash rental income, the capitalization rates must be derived on the same basis. Rates derived from cropshare income cannot be used to value a cash rental property and a combination of rates is not appropriate.

**Table 15.6** Capitalization Rate Derivation

| | Sale 1 | Sale 2 | Sale 3 |
|---|---|---|---|
| Sale price | $300,000 | $270,000 | $500,000 |
| Net operating income | $15,300 | $13,230 | $23,500 |
| Indicated overall rate | 0.0510 | 0.0490 | 0.0470 |

Table 15.7 illustrates the final step in arriving at a value opinion by direct capitalization. The net operating income is divided by the overall capitalization rate. A 5% rate was selected based on the rates extracted from the comparable sales. The net operating income calculation is $13,876/.0500 = $277,520 and the estimate is rounded to $277,500.00.

**Table 15.7** Value Indication by the Income Approach

| Land Type | Acreage | Rent Per Unit | Total Cash Rent |
|---|---|---|---|
| Cropland A | 120 | $30/acre | $3,600 |
| Cropland B | 300 | $20/acre | $6,000 |
| Pasture | 100 | $10/acre | $1,000 |
| Buildings | 10 | $600/month | $6,000 |
| Recreational lease | 520 | $10/acre | $5,200 |
| Total gross income | | | $21,800 |
| Expenses | | | |
| Taxes | | $2,020 | |
| Insurance | | $1,760 | |
| Maintenance | | $2,400 | |
| Management | | $1,744 | |
| Total expenses | | | $7,924 |
| Net operating income | | | $13,876 |
| Overall capitalization rate—5% | | | |
| Indicated value by the income approach | | | $277,520 |
| Rounded to: | | | $277,500 |

This simple technique for estimating overall capitalization rates is preferred and will produce a reliable indication of value by the income capitalization approach if three conditions are met:

1. Income and expenses must be estimated on the same basis for the subject property and all comparable properties.
2. Market expectations concerning resale prices, tax benefits, and holding periods must be similar for all the properties.
3. The financing terms and market conditions that affect the comparables must be similar to those affecting the subject property or an adjustment must be made for any dissimilarities

## Yield Capitalization

When yield capitalization is used, future benefits are converted into present value by applying an appropriate yield rate. To select an appropriate yield rate for a market value appraisal, the appraiser analyzes market evidence of the yields anticipated by typical investors. When investment value is sought, the yield rate used should reflect the individual investor's requirements, which may differ from the requirements typical in the market.

To perform yield capitalization, the appraiser 1) selects an appropriate holding period; 2) forecasts all future cash flows or cash flow patterns; 3) chooses an appropriate yield, or discount, rate, and 4) converts the future benefits of property ownership into present value by discounting each annual future benefit or by developing an overall rate that reflects the property's income pattern, value change, and yield rate. The application of capitalization rates that reflect an appropriate yield rate, the use of present value factors, and discounted cash flow analysis are all features of yield capitalization. Mortgage-equity formulas and yield rate or value change formulas may be used to derive overall capitalization rates. Like direct capitalization, yield capitalization may employ residual techniques. To apply the discounting procedure, the appraiser must be familiar with income patterns, capital return concepts, the mathematics of discounting, and investor requirements or assumptions (e.g., holding period, anticipated market growth, inflation) and select appropriate discount rates.[3]

## The Nature of Annuities

Although the word *annuity* means an annual income, it is used to refer to a program or contract specifying regular payments of stipulated amounts. Payments need not be annual, but the interval between payments must be regular. An annuity can be level, increasing, or decreasing, but the amounts must be scheduled and predictable. Income that has the characteristics of an annuity is expected at regular intervals and in predictable amounts. Obviously, real estate income or rental income can have the characteristics of an annuity.

The income expected from a real estate investment, in this case a rural property, may take the form of a regular or irregular pattern. Various capitalization techniques can be applied to different patterns of income streams.

## Discounting

Periodic income and reversions are converted into present value through discounting, a procedure based on the assumption that benefits received in the future are worth less than the same benefits received today. The return on an investment compensates the investor for foregoing present benefits (i.e., the immediate use of capital) and accepting future benefits and risks. The discounting procedure includes the assumption that the return of capital will be accomplished through periodic income, the reversion, or a combination of both. In some instances, tax benefits may contribute to the investment yield.

An investor seeks a total return that exceeds the amount invested. The present value of a prospective benefit must be less than its expected future value. A future payment is discounted to present value by calculating the amount that, if invested today, would grow with compound interest at a satisfactory rate to equal the future payment. The standard formula for discounting future value to present value is

$$\text{Present Value} = \frac{\text{Future Value}}{(1 + i)^n}$$

where $i$ is the rate of return on capital per period that will satisfy the investor and $n$ is the number of periods that the payment will be deferred. The two examples that follow illustrate applications of the discounting formula.

3. A more detailed discussion of yield capitalization can be found in *The Appraisal of Real Estate,* 11th ed. (Chicago: Appraisal Institute, 1996), 529–559.

***Problem 1. What is the present value of $115 due one year hence discounted at 15% per year?***
To solve this problem, calculate the amount that should be invested now to grow to $115 with compound interest at a rate of 15% per year.

$$\text{Present value} = \$115/1.15^1 = \$100$$

***Problem 2. What is the present value of $1,000 due in three years discounted at 10% per year?***
This problem can be solved by calculating the amount that should be invested now to grow to $1,000 in three years with compound interest at 10% per year.

$$\text{Present value} = \$1{,}000/1.10^3 = \$751.31$$

If a series of future payments is expected, each payment is discounted with the standard formula and the present value of the payments is the sum of all the present values. This standard discounting procedure is the foundation for all yield capitalization techniques, which include discounted cash flow analysis.

The amount deposited or received can be in the form of a single lump sum, a series of periodic installments such as rental income, or a combination of both. When amounts are compounded or discounted, the rate used is the effective interest rate; on an annual basis, this rate is identical to the nominal interest rate. If amounts are compounded or discounted more often than annually (e.g., semiannually or monthly), the nominal interest rate is divided by the number of compounding or discounting periods. For example, a nominal rate of 12% is an effective rate of 6% for a semiannual conversion period, or an effective rate of 1% for monthly conversions. Standard tables of factors or preprogrammed financial calculators can be used to facilitate the application of factors, but the user must select the appropriate conversion frequency (i.e., monthly, quarterly, or annually).

Each precomputed factor and yield capitalization formula has specific built-in investment assumptions that are implied when the table is used or the factor is calculated. Therefore, the appraiser must identify the assumptions applicable to the property being appraised and use the factor table and capitalization formula that corresponds to these assumptions. Thus, to apply compounding or discounting procedures, the appraiser must know the basic formulas, how the various factors relate to one another, and how they may be used or combined to simplify income capitalization.

## Discounted Cash Flow Analysis

Discounted cash flow (DCF) analysis can be used both to estimate present value and to extract a discount rate from a comparable sale. DCF analysis also serves as an excellent tool in analyzing the feasibility of an enterprise. In DCF analysis the quantity, variability, timing, and duration of cash flows are specified. *Cash flow* refers to the periodic income attributable to the interests in real property. Each cash flow, including any reversion, is discounted to present value and then all the present values are totaled to obtain the value of the real interest being appraised. The future value of that interest, the reversion, is forecast at the end of the projection period (i.e., the holding period or remaining economic life) and is also discounted. The cash flows discounted with the DCF formula may be the net operating income to the entire property or the cash flows to specific interests—e.g., the pre-tax cash flows to the equity interest (equity dividends), the after-tax cash flows to the equity interest, or debt service for the mortgage interest.

The DCF formula is a yield formula expressed as

$$PV = \frac{CF_1}{1 + Y} + \frac{CF_2}{(1 + Y)^2} + \frac{CF_3}{(1 + Y)^3} + \ldots + \frac{CF_n}{(1 + Y)^n}$$

where $PV$ = present value; $CF$ = cash flow for the period specified; $Y$ = the appropriate periodic yield, or discount, rate; and $n$ = the number of periods in the projection.[4]

With the DCF formula, an appraiser can discount each payment of income separately and add all the present values together to obtain the present value of the property interest being appraised. The formula treats the reversion as a cash flow that can be valued separately from the income stream. The formula can be used to estimate total property value ($V_O$), loan value ($V_M$), equity value ($V_E$), leased fee value ($V_{LF}$), leasehold value ($V_{LH}$), or any other interest in real property.

When a series of periodic incomes vary in an irregular pattern, the basic DCF formula is used for analysis and valuation. Any series of periodic incomes, with or without a reversion, can be valued with the basic formula.

## Derivation of Yield Rates

One of the critical factors in discounted cash flow analysis is the yield rate. The best source of the rate is extraction from the market. A DCF analysis is used to extract an internal rate of return (discount rate) from comparable sales. To accomplish this, the appraiser must estimate the income and expenses for each sale, the timing of the expenses, and the holding period.

*Internal rate of return (IRR)* refers to the yield rate that is earned or expected from a given investment over the ownership period. In other words, the internal rate of return for an investment is the yield rate that equates the present value of the future benefits of the investment to the

---

4. DCF analysis is covered in *The Appraisal of Real Estate,* 11th ed. (Chicago: Appraisal Institute, 1996), 561–578.

amount of capital invested. The internal rate of return applies to all expected benefits, including the proceeds from resale at the termination of the investment. It can be used to measure the return on any capital investment before or after income taxes.

It is important that the appraiser's DCF analysis of the sale be conducted in the same manner that the DCF will be applied in the appraisal of the subject property. For example, if management and real estate taxes are line item expenses in the analysis of the sale, they should also be line item expenses in the analysis of the subject. If expenses for management and real estate taxes are not estimated separately, but are left in the rate, no expenses for these items should be deducted from the subject's gross income.

## Sample DCF Analysis

The property being appraised is a permanent planting. The appraiser's first task is extract a market-derived internal rate of return. Table 15.8 illustrates the extraction of an internal rate of return from a comparable sale property that was purchased and developed as a permanent planting. The analysis is performed on a before-tax basis and with no consideration of debt service. The comparable sale property was purchased for $544,000. The costs to establish the permanent planting were $844,800 in Year 1 and $627,200 in Year 2. Year 3 is the first year in which a crop is produced. By Year 5 full production is reached. Expenses are incurred during Years 3 through 5 for operation, maintenance, and harvesting. The property is resold at the end of Year 5 for $3,200,000.

**Table 15.8** Discount Rate Derivation from Comparable Sale

| | Year 0 | Year 1 | Year 2 | Year 3 | Year 4 | Year 5 |
|---|---|---|---|---|---|---|
| Gross income | | $0 | $0 | $192,000 | $384,000 | $576,000 |
| Expenses for development | | $844,800 | $627,200 | $0 | $0 | $0 |
| Production exp. | | $0 | $0 | $206,720 | $237,920 | $269,120 |
| Land acquisition | -$544,000 | | | | | |
| Net operating income | -$544,000 | -$844,800 | -$627,200 | -$14,720 | $146,080 | $306,880 |
| Reversionary value | | | | | | $3,200,000 |
| Cash flow | -$544,000 | -$844,800 | -$627,200 | -$14,720 | $146,080 | $3,506,880 |
| *IRR* | 16.04% | | | | | |

Table 15.9 illustrates application of the market-derived internal rate of return to the cash flows of the subject property. The subject property is a permanent planting with establishment costs of $680,000 in Year 1 and $105,600 in Year 2. The first harvest is in Year 3 and mature production is reached in Year 5. Expenses include operating, maintenance, and harvesting costs. The reversionary value at the end of Year 5 is $4,000,000.

**Table 15.9** Discounted Cash Flow—Subject Property

| | Year 1 | Year 2 | Year 3 | Year 4 | Year 5 | Total |
|---|---|---|---|---|---|---|
| Gross income | | | $240,000 | $480,000 | $720,000 | |
| Expenses for development | -$680,000 | -$1,056,000 | $0 | $0 | $0 | $0 |
| Production exp. | $0 | $0 | -$258,400 | -$298,000 | -$336,400 | |
| Net operating income | -$680,000 | -$1,056,000 | -$18,400 | $182,000 | $383,600 | |
| Reversionary value | | | | | $4,000,000 | |
| Cash flow | -$680,000 | -$1,056,000 | -$18,400 | $182,000 | $4,383,600 | |
| PV factors @ 16% | 0.862069 | 0.743163 | 0.640658 | 0.552291 | 0.476113 | |
| Present value | -$586,207 | -$784,780 | -$11,788 | $100,517 | $2,087,089 | $804,831 |

## Summary

The income capitalization approach is primarily used in appraising income-producing property. In income analysis a distinction is made between rental income and owner-operator income. To apply the income capitalization approach, the appraiser must research the expected gross incomes, vacancy and credit losses, operating expenses, patterns and/or durations of income streams, and anticipated resale values or appreciation rates for properties similar to the property being appraised. Whenever an income estimate is used in deriving a rate to value property, the rate must be appropriate and consistently applied.

In the income capitalization approach, the appraiser converts the anticipated monetary benefits of property ownership into a value estimate by discounting anticipated future income into a present value using direct capitalization or yield capitalization. The essential steps in estimating the gross income of a rural property are

1. Determine the typical income arrangement for the area (cash rent, cropshare, or owner operation).
2. Ascertain the crops most likely to be produced.
3. Estimate yields based on property-specific data or local area data.
4 Estimate commodity prices.
5. Determine the crop share for percentage rents.
6. Determine property-specific terms and market rent terms for cash rents.

Direct capitalization is the method used to convert an estimate of a single year's income expectancy into an indication of value in one direct step, either by dividing the income estimate by an appropriate rate or by multiplying the income estimate by an appropriate factor. The capitalization rates and multipliers applied are extracted from sales. Only the first year's income is considered. Yield and value change are implied, but not specifically identified. Direct capitalization may be based on potential gross income, effective gross income, net operating income, equity income, mortgage income, land income, or building income.

Yield capitalization is the method used to convert future benefits into a present value by discounting each future benefit at an appropriate yield rate or by developing and applying an overall rate that explicitly reflects the property's income pattern, value change, and yield.

This chapter has provided an overview of the income capitalization approach as it is applied to rural properties. For a more detailed presentation of the approach, see *The Appraisal of Real Estate,* eleventh edition, Chapters 20 through 24.

CHAPTER • 16 •

# *Reconciliation and the Final Opinion of Value*

In reconciliation the appraiser reviews the appraisal for errors, inconsistencies, and omissions and reaches a final value opinion. The process of reconciliation includes reviewing each step in the appraisal process, assessing the quality and quantity of the data available, and presenting the final value opinion. In the course of the assignment, the appraiser identifies, collects, and analyzes data about the property; determines the property's highest and best use; and applies the cost, sales comparison, and income approaches as warranted by the assignment and suggested by the facts.

Typically each valuation approach results in a different indication of value. Often more than one value indication is derived in a single approach. In the sales comparison approach, for example, the application of different units of comparison may produce different value indications. For example, a Midwest grain farm will be analyzed on a per-acre basis, but it may also be analyzed in terms of price per productivity index point. Livestock facilities may be analyzed based on price per-square-foot or per-head capacity. In the analysis of income, different indications of value may result from applying a gross income multiplier, direct capitalization, and discounted cash flow analysis.

When value indications differ, the appraiser must reconcile them. Multiple value indications are often resolved within an approach. In other cases, the appraiser may choose to resolve these differences after reviewing the entire appraisal. Either method is acceptable as long as the integrity of each approach is maintained.

## Review

Before reconciling the value indications, the appraiser reviews the appraisal and makes sure that the data collected, the analytical techniques applied, and the logic employed are consistent with the assignment and reflect consistent judgments. Data are reviewed to ensure that they are authentic, pertinent, and sufficient. The value defined, the property rights appraised, and the qualifying conditions imposed are carefully reviewed to confirm that they have been properly addressed. The appraiser also reviews the work for inconsistencies. For example, are the effective ages of the buildings used in the cost approach consistent with the age and physical condition reported in the description of the improvements? Are productivity differences between the subject and sale properties reflected in the sales comparison and income approaches? Is each approach to value consistent with the stated highest and best use?

All calculations are checked, preferably by someone other than the appraiser who prepared them. Significant errors can lead to incorrect value indications, and even minor errors can destroy the client's confidence in the appraisal. Finally, the logic used throughout the appraisal is scrutinized. Do the approaches and techniques employed make use of all the available data and systematically lead to a meaningful conclusion consistent with the purpose of the appraisal? Does the appraisal provide the information needed to solve the client's problem?

## The Reconciliation Process

Although review of the appraisal substantiates the accuracy, consistency, and logic leading to value indications, it is in reconciliation, more than in any other part of the valuation process, that the appraiser's professional experience, expertise, and judgment are exercised. In reconciliation the appraiser evaluates alternative conclusions and selects from among them to arrive at a final opinion of value. The appraiser weighs the relative significance, applicability, and defensibility of each indication of value and relies most heavily on those that are most appropriate to the purpose of the appraisal. All aspects of the assignment are considered in relation to the appraisal problem, which guides the appraiser's deliberations in reconciliation.

## Criteria

As evidence is evaluated and judgments are made in reconciliation, a meaningful, defensible conclusion is formed based on the appropriateness, the accuracy, and the quantity of evidence.

### Appropriateness

How appropriate is each approach to the purpose and function of the appraisal? Appropriateness relates to the type of property, its market, the valuation technique, and the function of the appraisal. For example, in an estate appraisal of an unimproved irrigated grain farm in a market with ample comparable sales, the sales comparison approach would receive the greatest weight in reconciliation. Because there are no buildings, the cost approach would be much less appropriate and might be omitted. The income capitalization approach would be appropriate, but would be given less weight due to the preponderance of sales data and the manner in which the appraisal is to be used. If the appraisal was for an investment group considering a purchase, the income capitalization approach would be the most appropriate and therefore given the greatest weight in reconciliation.

How appropriate is the data analyzed in each approach? Appropriateness relates to the available data and the property. Highly comparable sales requiring few adjustments are usually very appropriate and deserve strong consideration in reconciliation. How appropriate is each step in each approach? Are the sales selected from the available data the most appropriate indicators of value? Are the adjustments applied in the sales comparison approach appropriate for the market where the property is located? Are the estimates of accrued depreciation based on an appropriate comparison of comparable costs and sales?

For example, if a reliable local builder has prepared a detailed replacement cost estimate of the major buildings on a well-maintained hobby farm located in a neighborhood of similar properties, this should be very appropriate data. A cost estimate of the same type and quality is not as appropriate in the appraisal of a remote farm with many dilapidated buildings. Income and expense figures based on averages from a local professional farm manager are more appropriate than figures based on averages for the entire state prepared by a university. The data used must be appropriate for the stated highest and best use of the property. The income approach may be more appropriate in the appraisal of a grain farm located fifty miles from a town of 5,000 than a similar farm three miles outside a growing town of 100,000.

An analytical technique that is appropriate for measuring a specific characteristic of a property may not be appropriate to derive the total value of the property. One example would be the cost to replace a leaking roof on a dwelling that is 30 years old but otherwise in average condition. Estimating the cost to replace the roof is a very appropriate technique for measuring that item of depreciation. In valuing the 30-year-old dwelling, however, the cost approach is less appropriate than a sales comparison approach analysis based on comparable sales. Similarly, the capitalized value of a rent loss may be a very appropriate technique for measuring incurable functional obsolescence in an appraisal although the income approach is the least appropriate approach for valuing the entire property.

## Accuracy

The accuracy of an appraisal is measured by the confidence the appraiser has in the correctness of the data, the valuation techniques, and the calculations presented. The appraiser typically has more confidence in the accuracy of the data used and the techniques applied in one approach than another. For example, the appraiser may have more confidence in the accuracy of a sales comparison approach analysis that includes a few small adjustments than an income approach analysis that includes a capitalization rate based on a broad range of market-derived rates.

In an appraisal of a nearly new confinement dairy located near other, similar properties, a cost approach analysis that includes detailed cost estimates from two local contractors, market-derived depreciation estimates, and very comparable vacant land sales would be more accurate than a cost approach with cost figures from a cost manual and land values from grain farm sales. An income approach analysis using crop yields based on a five-year history of the subject property is more accurate than an analysis with yields based on county averages. Finally, a sales comparison approach analysis with eight sales and adjustments based on paired sales is more accurate than one with four sales and adjustments made with little support.

In the sales comparison approach, the accuracy of the approach is indicated by the number of comparable properties and their similarity to the property being appraised. In the income approach, accuracy is indicated by the strength of the data supporting each income and expense item, the comparability of the sales used to derive the capitalization rate, and the pattern of the indicated rates. The accuracy of the cost approach is indicated by the number and comparability of land sales, the strength of the cost estimates, the comparability of the sales used to derive depreciation estimates, and the consistency of the indicated depreciation rates. The availability of a large number of comparable properties for one approach may suggest greater accuracy and, therefore, greater reliability than the other approaches. For example, if the appraiser valuing a large feedlot can find few comparable sales, but a number of properties are available from which to extract income, expenses, and capitalization rates, greater accuracy might be attributed to the income capitalization approach than to the sales comparison and cost approaches.

If one or two comparable sales within an approach require fewer total adjustments than others, greater accuracy might be attributed to the value indications obtained from the sales with fewer adjustments, if the adjustments are of relatively equal magnitude. Even if the number of adjustments is similar, the gross or net dollar amount of total adjustment might vary greatly. For example, five comparable properties in the sales comparison approach might each have six adjustments. However, for one sale the gross dollar amount of adjustment might total 15% of the sale price, while the gross dollar amount of the others might be less than 5% of their sale prices. All else being equal, the sale with the largest percentage of adjustment would be considered the least accurate.

The net dollar amount of adjustment applied to each sale in a sales comparison is usually not a reliable indicator of accuracy. The figure is obtained by totaling the positive and negative adjustments. It cannot be assumed that inaccuracies in adjustments will cancel each other out when the positive and negative differences are added. Several adjustments that are all positive (or negative) may be more accurate and produce a smaller total gross adjustment than a combination of larger positive and negative adjustments. Also, inaccuracies may be compounded if the adjustments are applied as percentages.

### Quantity of Evidence

Appropriateness and accuracy reflect the quality of the data and analysis; they indicate how relevant and correct the value indication derived from a comparable or an approach may be. Appropriateness and accuracy are important in reconciliation, but each must be considered in relation to the quantity of data available in each approach. An inadequate quantity of data weakens any approach regardless of how appropriate and accurate the information and techniques are.

For example, consider the extraction of capitalization rates from the sale of three comparable ranches. The three are equally comparable in regard to terms of sale, physical characteristics, and location. Information about each is verified and considered reliable and each sale produces an accurate indicated rate. Now assume that income and expense information for two of the comparable sales is available for only the last two years. The income and expense information on the third sale extends back six years. The accuracy and appropriateness of the data for all three comparable sales are equal, but the greater quantity of data on the third sale gives the appraiser greater confidence in its indication than the indication derived from the other two.

## Final Value Opinion

An appraisal is an opinion of value. Although it is an impartial, expert, and reasoned conclusion formed by a trained professional based on an analysis of all relevant evidence, it is still an opinion. It represents the appraiser's perception of the most probable price reached in an arm's-length transaction for the appraised interest, subject to the qualifying conditions. The final value opinion is not derived simply by applying technical and quantitative procedures; rather, it involves the exercise of judgment. The value conclusion reached must be consistent with market thinking, and the quantity of data used should correspond to the amount of data the market considers relevant to the problem.

Although there may be a large quantity of comparable sales information that is accurate and appropriate, this information reflects events that took place before the appraisal. Improved and vacant sales, rentals, expenses, vacancy rates, and absorption rates all reflect the *past*. These are relevant only insofar as they help the appraiser estimate the *future* benefits of the property and the *present* value of these benefits.

The final opinion of value may be stated as a range or as a single figure. Or it can be shown as a single figure within a range. Rounding to a single figure may suggest that the value conclusion is an estimate. A value opinion reported as a single dollar amount is called a *point estimate* and represents the appraiser's best opinion of the value of the property. A point estimate is required for many purposes, e.g., estimating compensation in condemnation cases, determining value-based rent, and making property transfer decisions. Furthermore, because of legal requirements, many clients expect a point estimate of value.

A point estimate should be rounded to reflect the degree of precision associated with the value opinion, but even a rounded figure may imply greater precision than is warranted. Because an appraised value is an opinion, it implies a range in which the value may fall. That value range usually reflects the range of indications from two or more approaches.

## Summary

Reconciliation is the analysis of alternative conclusions to arrive at a final value estimate. It is required because each approach to value typically results in a different value indication. Various indications often result from different methods applied within a single approach.

In reconciliation the appraiser reviews the entire appraisal to determine that the data, techniques, and logic used are valid and consistent. Inconsistencies are reconciled. The appraiser confirms that the approaches and methods employed relate to the property interests being appraised, the definition of value, and the purpose and intended use of the appraisal. All mathematical calculations are checked, by another individual if possible.

The appraiser produces a meaningful, defensible value opinion by considering three criteria: appropriateness, accuracy, and quantity of evidence. Appropriateness relates to the data and the manner in which it is processed to conform to the market for the property or the function of the report. All data, calculations, estimates, and adjustments are examined for accuracy. The quantity of evidence is examined to determine if the conclusion is adequately supported. Market data usually reflects the past, but the appraiser must rely on such data when estimating value, which is the present worth of future benefits.

The final value opinion is based not only on technical and quantitative analysis, but also on the experienced judgment of the appraiser. The conclusion must be consistent with and representative of market thinking concerning the property and the assignment. The final value opinion may be stated as a single figure, a range, or a figure within a range.

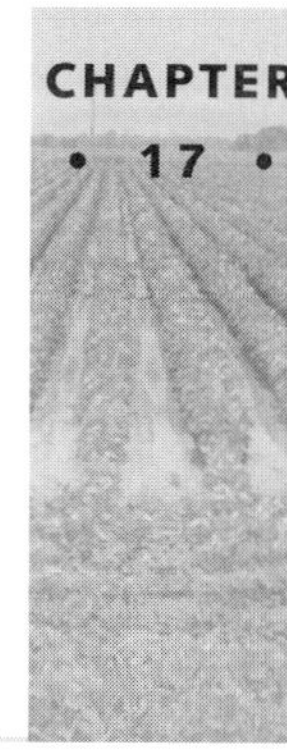

CHAPTER • 17 •

# *Writing Appraisal Reports*

For the typical appraiser, writing appraisal reports is the least pleasant part of appraisal work, but the report is one of the most crucial components of the appraisal process. When readers cannot follow the reasoning in the appraisal report, do not understand what they read, are alienated by poor grammar and writing style, or are not convinced of the appraiser's analysis of the information presented, the appraisal is not effective. The collection and proper analysis of extensive data accomplishes little if readers cannot understand the appraisal report or react negatively to its poor construction.

The Uniform Standards of Professional Appraisal Practice (USPAP)[1] serves as a guide for determining what must be included in appraisal reports. Often tradition and local custom determine how reports are organized and presented to clients. The needs of readers are considered after USPAP requirements have been satisfied. Good appraisal reports satisfy the report requirements specified by USPAP, provide readers with the information they need and desire, and communicate the appraiser's opinion of value in a form that is easily understood and engenders confidence.

Because appraisal reports are technical and persuasive documents, the appraiser's goal is to convince readers that the data collected, the analyses performed, and the conclusions reached are appropriate. In addition to meeting USPAP requirements, the report must satisfy the procedural and computational norms expected of professional appraisers. An appraiser who cannot persuade readers of the correctness of the procedures followed and the results presented may have little future in the appraisal profession.

---

1. Appendix A outlines the requirements of the Uniform Standards of Professional Appraisal Practice. Each appraiser is responsible for complying with the specific requirements stated in USPAP.

This chapter summarizes the requirements for successful appraisal reports. To prepare good reports, the appraiser must possess good writing skills and know what information to include and how to present it to readers. This chapter is divided into three parts. The first part summarizes communication concepts. The second discusses report organization and the topics to be included in each part of the appraisal. The final portion contains guidelines for preparing tables and figures.

## Successful Communication

Appraisal documents are typically the only means for the appraiser to communicate with users of the appraisal report. While the appraiser may meet with clients, he or she will seldom have a chance to meet with each reader of the appraisal report and explain in detail the data collected, the analyses performed, and the factors leading to the value opinion. Thus, the appraisal must provide all the information that readers might need. Part of preparing appraisals is deciding what information must be included to convince readers of the appraisal's validity.

Five steps for successful report writing are

1. Collect all needed data.
2. Do the research and analysis required.
3. Organize the data and results.
4. Write a draft.
5. Revise the draft until it is acceptable.[2]

Previous chapters in this book have covered the collection and analysis of data. This chapter will focus on organizing the data, writing a draft, and revising the draft until it is acceptable.

Appraisals are persuasive documents and one key to writing persuasively is understanding the reader. Understanding the readers of appraisal reports is difficult when you don't know who the reader is and how many readers there may be. Some readers will be persuaded by the appraiser's qualifications. They will accept the values presented regardless of the form in which they are provided. Other readers will be persuaded only after they have read and critically evaluated the analyses presented. Appraisal reports must be written to hold the attention of readers and convince them to accept the arguments presented. Critical readers will not be motivated to accept the appraiser's conclusions unless the report is readable, clear, concise, and persuasive.

To write appraisals that communicate effectively, the appraiser should know what readers want. Some readers will find and read value conclu-

---

2. For a detailed discussion of successful report writing, see Charles T. Brusaw, Gerald J. Alred, and Walter E. Oliu, *Handbook of Technical Writing*, 4th ed. (New York: St. Martins Press, 1993). The writing of appraisal reports is specifically addressed in *The Appraisal Writing Handbook*, by Alan Blankenship, PhD (Chicago: Appraisal Institute, 1998).

sions only; others will read every word to make sure they can accept the assumptions and analyses on which the conclusions are based. Good organization will help readers find the information they need. One technique is to divide the appraisal into sections and subsections and provide a table of contents to direct readers to the information they want to read. The letter of transmittal may state the value opinion. In this case, the letter of transmittal is considered an integral part of the appraisal and is bound with the appraisal report. Effective letters of transmittal communicate value conclusions and any other information the appraiser is obligated to provide. (See Appendix F for a list of the items that should be included in the letter of transmittal.) Besides dividing the appraisal into sections, the report writer should organize and develop each paragraph so that readers can logically follow the points being made. The report should clearly indicate which items are most important, how the data was analyzed, and what factors were considered in reaching conclusions. An effective report tells readers what is coming and how they should interpret the information provided.

The process of writing varies from one appraiser to another. Some appraisers dictate major parts of their appraisals; use a considerable amount of standardized material, or "boilerplate," from past appraisals; and rely on staff to assemble the first draft. Another appraiser may complete the entire report himself, dictating it, writing it out in longhand, or keying it into a word processor. Use the procedure that works best for you and remember that your goal is effective communication. Some guidelines for writing appraisals follow.

- Write documents for expected readers, not other appraisers. Avoid using appraisal jargon that only other appraisers understand. Assist readers by defining all unusual and technical terms.
- Whenever possible, use the active voice. Writing in the active voice usually requires fewer words and is less confusing.[3] The passive voice sentence, "It was reported by the engineer that the water well needed extensive repairs." can be recast in the active voice as "The engineer reported that the water well needed extensive repairs." Writing in the active voice makes appraisal reports more lively and less wordy.
- Use correct grammar and spelling. Spell checkers make sure words are spelled correctly, but they do not ensure that the context is correct. Incorrect grammar and spelling offend many readers and can undermine the writer's credibility. If grammar is a weak point, another individual should review the draft for problems.
- Avoid using several words when one will do. Wordiness obscures meaning and causes misunderstandings.[4]

3. Brusaw et. al., *Handbook of Technical Writing,* 719–722.
4. Ibid., 129–134.

- Be accurate in calculations. If readers get different answers when they perform calculations, they will question whether other mistakes were made that could invalidate the appraiser's results. Double check all calculations for accuracy. Judgment will dictate how many significant digits are included when presenting numbers. Rounded figures should be identified as such to avoid the appearance of error.
- Revise each appraisal. It is virtually impossible to produce an acceptable appraisal report on the first draft. It may be helpful to have an outside writer or editor proofread the appraisal.
- Use an attractive, business-like cover. A logo or unique lettering will help distinguish an appraisals from others. Use professional binding and covers made of vinyl or quality cover stock. Remember that readers form an opinion about a report based on its appearance.

The main requirement of an appraisal is that it communicates the appraiser's opinion of value to clients and other readers.

## Content of Narrative Rural Appraisal Reports

Appraisal reports are traditionally divided into four parts: introduction, factual data, analysis and conclusions, and addenda. Appendix F provides an outline of the items typically included in each part. Although all appraisers do not organize their appraisals in the same way, most appraisals provide the same information. Note that appraisers do not prepare reports by first collecting all the required information, then completing the analysis, and finally presenting the conclusions. In many cases appraisers cannot collect relevant information until they have done some analysis to determine exactly what information is relevant. For example, the appraiser must determine highest and best use, which is part of data analysis, before appropriate comparable sales can be collected. Organize appraisal reports to inform readers of value opinions and how these opinions were reached; the report does not record the sequential steps involved in preparing the appraisal.

### Introduction

The introduction to an appraisal report does more than identify the client, the appraiser, the property being appraised, and the date of appraisal; it also informs clients and other readers of the final value opinion. Depending on their needs, some readers never read beyond the introduction. Therefore, the introduction must include all the points relevant to the value conclusion reached. Of particular importance are the assumptions and limiting conditions used to prepare the appraisal. In effect, the introduction provides a discussion of salient facts about the appraisal problem and a summary of the value conclusion reached.

The typical introduction has these components: title page, letter of transmittal, summary of salient facts and conclusions, and assumptions and limiting conditions. The letter of transmittal is traditionally considered part of the appraisal. It should provide complete and accurate information so those who read only the letter of transmittal are not misled. Much of the information in the letter of transmittal duplicates information provided in the summary of salient facts and conclusions. Another type of introduction may include only a title page, executive summary, and assumptions and limiting conditions. Under this organization there is no letter of transmittal; a cover letter is used to transmit the appraisal to the clients. Cover letters are not part of the appraisal. They merely convey documents to clients. When the alternative organization is used, the "summary of salient facts and conclusions" becomes an "executive summary."

The purpose of an executive summary is to consolidate the principal points of the report in one place.[5] This is essential because clients often read executive summaries in place of full reports. Readers of executive summaries must be able to understand the conclusions reached, what analyses were performed, and what data was used in the analyses. The major advantage of using an executive summary is that the format is familiar to most readers of technical documents.

## Factual Data

Appraisals are based on facts and relevant information. The factual data portion of the appraisal report informs the reader of all the facts assembled as relevant to the appraisal problem. Often appraisals are used by individuals other than the client and it is important for these readers to know the exact property rights being appraised, the appraisal objectives, the definitions of value used, and the scope or extent of investigation by the appraiser.

The factual data section should include a complete description of the area and neighborhood where the property being appraised is located. Good community descriptions provide readers with a basis for understanding the environment surrounding the subject property. Descriptions of the community environment and the property characteristics are essential to the valuation process. Trends affecting property values, the transportation available, and other neighborhood characteristics should be discussed in detail. (A more complete list of area and neighborhood topics is presented in Appendix F.) It is often efficient and convenient to use boilerplate for much of the factual data. However, each item of factual data should be carefully examined to ensure that it is relevant to the appraisal problem. The area and neighborhood description section of an appraisal of a vegetable farm with a roadside market requires different information and emphasis than the appraisal of a dairy farm in the same neighborhood.

5. Ibid., 232.

The final component of factual data to be included is a description of the property being appraised. The property description should give readers complete knowledge of the physical, environmental, and economic characteristics of the property. Any past or expected governmental or legal actions affecting the property should also be identified. Maps, sketches, and photographs should be used to aid understanding. After reading the factual data portion of a report, readers should understand the uses and characteristics of the property being appraised and how they relate to the surrounding communities.

### Analysis and Conclusions

The appraiser's analysis and conclusions are presented after the factual data. Traditionally comparable sales data are part of this section of the report. Provide readers with sufficient information about comparable sales so that they know the differences and similarities between each sale and the property being appraised. One way to present comparable sales information is to provide verbal descriptions of each comparable sale along with an analysis of each property's rate of return and a summary of the property's component values (cropland values, pasture values, improvement values, etc.). If including the description and analysis of each sale in the body of the report disrupts the continuity of the presentation, include all or part of the sales analysis in the addenda. The advantage of presenting all comparable sales analysis in one location is that readers can easily follow each step in the process.

A second alternative is to present the comparable sales information used in each approach to value with the discussion of that approach. This organization ties the information on comparable sales to the approaches to value. Use the alternative that best fits the appraisal problem.

Do not introduce new information about the subject property or the community in the analysis and conclusion section of the report. Except for comparable sales data, all information used in the approaches to value should be presented in the factual data part of the appraisal. When necessary, refer readers to the factual data portion of the appraisal to prevent unnecessary duplication.

The discussion of the approaches to value included in the report should persuade readers that appropriate procedures were applied to obtain correct results. Clearly present the rationale behind each approach, the mechanics of the analysis, and the conclusion reached. The goal is to convince readers that the best possible value estimate has been reached in each approach. The Uniform Standards of Professional Appraisal Practice (USPAP) require that any departures from normal appraisal practice be justified. Discuss such variations here and, if necessary, explain why one or more of the three approaches to value was not used.

After the cost, income, and sales comparison approaches to value are presented, the value indications obtained should be reconciled into a single value or a range of values. The appraiser should include logical and concise statements designed to persuade readers that a reasonable value was selected. Developing final value opinions requires more experience, expertise, and judgment than any other part of the appraisal. Because there is no formula for choosing a final value, the appraiser must effectively communicate his or her thought processes to support the value conclusion. Write persuasively when presenting final value determinations because readers must be convinced that the conclusions are correct.

The final component of the analysis and conclusion section is the certificate of value. Prepare the certificate of value as described in USPAP Standards Rule 2-3. State without equivocation that the appraisal was prepared professionally, is in compliance with the requirements of USPAP, and does not violate any of the ethical standards required of professional appraisers.

### Addenda

The addenda consists of one or more sections of data and analysis supporting the analysis in the text. Information is placed in the addenda because including it in the text would disrupt the flow of the narrative. Information essential to determining value conclusions should be presented in the body of the appraisal rather than the addenda.

## Illustrating the Appraisal

Include illustrations such as tables, graphs, photographs, maps, and drawings in an appraisal report helps readers understand. When used appropriately, illustrations can communicate points and relationships that words cannot make clear. Use illustrations to enhance communication, not to make the appraisal appear more impressive. Some of the attributes of good writing—simplicity, clarity, and conciseness—are equally important in illustrations.

General guidelines for the use of illustrations follow.[6]

- Keep the information as brief and simple as possible.
- Try to present only one type of information in each illustration.
- Label each illustration carefully.
- If possible, make lettering horizontal for easy reading.
- Keep terminology consistent. For example, do not use the word "proportion" in the text and "percentage" in an illustration.
- Provide adequate white space.

6. Ibid., 310

- Position the illustration as close as possible to the text where it is referenced. Normally the illustration appears immediately after the first text reference.

Refer to all illustrations as either tables or figures. Tables present specific, related facts in a small space using a row and column arrangement; other illustrations—photographs, graphs, drawings, sketches, and maps—are identified as figures.

## Guidelines for Tables

Tables facilitate comparisons among statistics through the arrangement of rows and columns. Tables are precise, but they do not show trends or comparisons as effectively as graphs. Table 17.1 illustrates the proper format for a table.

**Table 17.1** Gross Crop Income for the Blackstock Farm*

| Crop | Units | Acres | Yield Per Acre | Total Production | Price Per Unit | Value of Production |
|---|---|---|---|---|---|---|
| Wheat | Bushels | 365 | 38 | 13,870 | $3.00 | $41,610.00 |
| Grain sorghum | cwt. | 124 | 33 | 4,092 | $3.60 | $14,731.20 |
| Alfalfa | Tons | 106 | 4 | 424 | $85.00 | $36,040.00 |
| Total | | 595 | | | | $92,381.20 |

* Average for the years 1989 through 1997

Source: Blackstock farm records obtained from the current operator.

Specific presentation guidelines follow.

- If there are more than five tables in a report, assign each table a sequential Arabic number. Refer to each table by number, not as "the following table," and include a "List of Tables" on a separate page following the table of contents. Place tables as close to their first reference as possible. If there are five or fewer tables, they need not be numbered. Integrate them into the text and refer to them as "the following table" or use another descriptive term.
- Each table title should describe the scope and objective of the table so the reader can determine whether or not to read it. Table titles should not include abbreviations or acronyms.
- The column headings describe what each column contains. Make sure units of measurement are clear by identifying them in a separate column (e.g., the "Units" column in Table 17.1) and use a dollar sign when dollars are indicated. Alternatively, units can be shown in parentheses beneath a column head.
- The *stub* is the left-hand vertical column that identifies a horizontal row. Stubs must be concise, but clear, so that readers understand the contents of each row.

- The body of a table is the data below the column heads and to the right of the stub.
- The horizontal lines in Table 17.1 are *rules*. Place horizontal rules under the title, the column heads, and the body of the table. Do not use vertical lines on the sides of tables, but vertical lines can be used to separate columns for clarity.
- Table notes explain items in the title, column headings, or body of the table. Use symbols or lowercase letters as note references because readers might mistake numbers for data.
- Include source lines when tables are taken from other publications. Source lines can also be used to tell readers the source of the data used to build the table.
- When tables are continued on a following page, this should be noted in the table title (e.g., Table 1—continued) and the column heads should be repeated.

## Guidelines for Other Illustrations

Photographs, graphs, drawings, flowcharts, schematic diagrams, and maps in a report must also follow certain guidelines. Guidelines for building illustrations and incorporating them into appraisal documents follow.

- If there are more than five illustrations (not including tables) in a report, they are all identified as figures. Assign each figure (e.g., graph, map, sketch, photograph) a sequential Arabic number and include a "List of Figures" on a separate page following the table of contents. When both a "List of Tables" and a "List of Figures" are required, they may be on the same page with the "List of Tables" at the top. Place figures as close as possible to the first reference in the text.
- Assign concise titles that clearly describe the content of each illustration.
- When there are more than five illustrations other than tables, refer to them in the text by number and title—e.g., Figure 1. Plot Plan . When there are five or fewer illustrations, you can integrate them into the text and refer to them as "the following" graph, chart, or sketch.
- If figures are central to the discussion, place them in the text. If the figures are supplemental, place them in an appendix or addenda. Do not include figures as filler. If there is no reference to the figure in the text, the figure is probably unnecessary.
- The amount of explanation required for each illustration depends on its complexity and the level of understanding expected of readers. Readers who are not experts generally require more lengthy explanations.

- Photographs are excellent for showing the improvements and unique features of real estate. The photographer should stand close enough for the object to fill the frame. If it is important to show relative size, place a familiar object near the item being photographed. Appropriate captions should accompany each photo.
- Graphs present numerical data in visual form. They show trends, distributions, and cycles better than tables, but with less precision. There are many different types of graphs, but most graphs divide the vertical axis into equal increments, showing the level of the item being measured, while the horizontal axis is divided into equal increments showing time. Clearly label the vertical and horizontal axes. If the vertical axis does not originate at zero, include a break in the scale to reduce possible misunderstanding. Place the information source and any explanatory notes below the figure along the left margin. If possible, labels should appear in horizontal type.
- Drawings and sketches illustrate how objects are related. Sketches of homes and farmsteads are commonly included in appraisals. The appraiser should indicate whether or not sketches are drawn to scale. Include clear, concise titles with drawings and place any needed source references beneath the figure on the left margin. Label drawings to inform readers and show words in horizontal type.
- Maps show specific geographic features and the spatial relationships between the properties being appraised and the properties used as comparable sales. Label maps clearly and be sure to show the scale and which direction is north. Clearly identify boundaries on maps and eliminate any unnecessary boundaries and extra information. Emphasize features using shading, dots, cross-hatching, or another appropriate technique and be sure to provide needed references. If necessary, provide source information below the map.

## Conclusions

Collecting the proper data and deriving a final value opinion are critical to the appraisal process. Presenting the appraiser's final conclusions to readers in a clear, persuasive manner is also essential. If readers cannot understand the discussion and comments presented in the appraisal report, they are not likely to accept the appraiser's results. Effective writing is difficult and depends on proper grammar, style, and punctuation as well as clear presentation of facts. Some appraisers are proficient at developing final value opinions, but pay so little attention to organization and presentation that readers find their appraisals impossible to follow. Other appraisers, perhaps relatively few, are excellent writers who can convince readers that their conclusions are correct even though their data collection and analysis are inadequate. Unfortunately, these latter individuals are likely to get more than their share of appraisal assignments just because readers can understand their appraisal reports. Truly successful appraisers do not fall into either of these categories. They place equivalent emphasis on collecting data, performing analyses, and presenting results.

There are many ways to improve your writing. Basic writing skills—grammar, punctuation, spelling—are taught at high schools, vocational-technical schools, and two- or four-year colleges. To improve your ability to organize your thoughts and communicate with clients, take a course in technical writing. Numerous books on grammar, style, and other aspects of report preparation are available. The selected bibliography that concludes this chapter lists a number of books that may be helpful. Most appraisal organizations offer courses in appraisal report writing that can help you determine what and how to communicate to readers.

It is possible to improve the way you write appraisal reports, but it is up to you. The only way to really alter your writing is to spend the time necessary to improve your skills. Ask someone knowledgeable about writing to edit your reports. If the editor you choose is not familiar with appraising, he or she will quickly find the passages in which you do not define terms clearly and communicate what you intend. In addition, the editor can identify where you use the passive voice, incorrect grammar, and inappropriate style. Improving your technical writing can be a frustrating and humiliating experience, but it can improve your work products and help your clients understand your documents. With regular editing and review, your writing will improve and the editor will have fewer changes to recommend. Even after you are comfortable with the quality of your reports, it is beneficial to have an editor review your work occasionally. A periodic review will keep you conscientious about your work and prevent you from acquiring bad writing habits.

There are two major components of appraisal report writing—determining what to write and deciding how best to write it. The main objective of an appraisal report is to communicate the appraiser's opinion of value to readers. USPAP requirements dictate a number of items that must be included in appraisals. As the report writer, your task is to convince your readers that you have assembled the correct information, have analyzed the information correctly, and have developed a reasonable opinion of value. After you accomplish these tasks, your appraisal is complete.

## References

The following sources are recommended for further information on report writing.

Blankenship, Alan. *The Appraisal Writing Handbook.* Chicago: Appraisal Institute, 1998.

Bore, Tony, et al. *The Art of Desktop Publishing: Using Personal Computers to Publish it Yourself.* 2d. ed. New York: Bantam Books, 1987.

Brusaw, Charles T., Gerald J. Alred, and Walter E. Oliu. *Handbook of Technical Writing.* 4th ed. New York: St. Martins Press, 1993.

Lefferts, Robert. *How to Prepare Charts and Graphs for Effective Reports.* New York: Barnes and Noble, 1981.

MacGregor, A. J. *Graphics Simplified: How to Plan and Prepare Effective Charts, Graphs, Illustrations, and Other Visual Aids.* Toronto: University Press of Toronto, 1979.

Quirk, Randolph and Sidney Greenbaum. *A Concise Grammar of Contemporary English.* New York: Harcourt, Brace and Jovanovich, 1973.

Robertson, Bruce. *How to Draw Charts and Diagrams.* Cincinnati, OH: North Light Books, 1988.

Strunk, William, Jr. and E. B. White. *The Elements of Style.* 3d ed. New York: Macmillan, 1979.

Uniform Standards of Professional Appraisal Practice (USPAP). 2000 ed. Washington, DC: The Appraisal Foundation, 1992.

Watkins, Floyd C., et al. *Practical English Handbook.* 6th ed. Boston: Houghton Mifflin, 1982.

Williams, Joseph M. *Style: Ten Lessons in Clarity and Grace.* 3d ed. Glenview, IL: Scott, Foresman and Company, 1989.

PART
• IV •

# The Valuation of Selected Rural Properties

# The Valuation of Dairy Farms and Facilities

## Introduction

American consumers spend between 10% and 12% of their food dollars on milk and other dairy products. Total sales of dairy products amounted to $76.3 billion in 1998. The per capita consumption of milk products for that year was 583 pounds, valued at $305 per person.

The dairy industry in the United States has demonstrated these trends: decreasing numbers of individual dairy farms and total milk cows and increasing herd size, milk production per cow, and total milk production. Beginning with the introduction of the free-stall barn in the 1950s and 1960s, the industry has placed emphasis on increasing milk production and efficiency. Total milk production for 1998 was 157 billion pounds, from 9.2 million cows, with an average of more than 17,000 pounds of milk per cow.[1]

Although dairy cattle are raised in all 50 states, the major concentrations of dairy farms are in the states bordering the Great Lakes, the Northeast, the Southwest, and Florida. The top six states in dairy production are: Wisconsin, California, New York, Pennsylvania, Minnesota, and Texas. This chapter includes several examples provided for illustrative purposes only. They are not meant to cover all cases or all situations. The appraiser is encouraged to investigate the market for specific trends and value levels relevant to each appraisal assignment.

## Historical Background

The dairy cow can efficiently convert vegetable food sources into high-quality food that is palatable to humans and highly nutritious. Throughout history, people have kept domestic cows as a convenient

1. USDA, *Dairy Yearbook,* stock no. 89032.

source of milk and have turned milk into other, longer-lasting dairy products such as butter, cheese, ice cream, and yogurt. Scientific developments have made storing and transporting milk easier and more hygienic. These innovations include the condensing and evaporating of milk, pasteurizing milk to improve preservation and destroy harmful bacteria, and homogenizing milk to render it more appealing and easily digestible. In addition, improvements in refrigeration and the speed of transportation have allowed milk producers to distribute more milk over greater distances and preserve it longer on grocers' shelves. New techniques are continuously being explored, including the development of longer shelf lives and nonrefrigerated storage.

Veterinary science has contributed to the improved health and breeding of dairy cattle. Vaccines are now used to prevent such diseases as brucellosis, leptospirosis, and blackleg. Effective medical treatments decrease losses from milk fever, mastitis, and other diseases. Careful selective breeding using production monitoring and component testing has greatly increased the productivity of dairy cows.

Genetic improvements have been facilitated by the use of artificial insemination. Semen from proven bulls can be frozen and shipped anywhere in the world. This has helped both small and large dairy enterprises to upgrade desirable genetic traits. New techniques, such as superovulation and embryo transplants, also help produce more profitable offspring for future dairy herds.

Advances in the science of animal nutrition have provided better feeding and increased milk production at the most economical level. Increased automation facilitates feeding large herds of cattle and milking through the use of milking systems, pipelines, and bulk storage. In addition, the use of bovine somatotropin (BST) or bovine growth hormone (BGH), which was recently approved by the USDA, could result in an annual increase of 1,500 to more than 3,000 pounds of milk produced per cow.

## Dairy Management

This section could be subtitled, "Managing a Dairy Farm's Resources Profitability" since making a profit is, or at least should be, the primary goal of the dairy farm operator. A prospective dairy farm owner will assess the resources available (farm real estate and facilities) and adapt management practices that will result in maximum profits. Any appraisal of a dairy farm should assess whether its available resources will make it difficult for the manager to adopt the technology and practices required by current industry economics.

For example, a manager may realize that the dairy farm to be operated has shallow and poorly drained soils. To adapt the soils to a roughage program that will support milk production, the operator may need to establish a pasture feeding system rather than an intensive crop production program. In some instances the manager may have to purchase concentrates and roughage because they cannot be grown on the property. Occasionally marginal soils can be too much of a handicap. If so, the highest and best use of the farm may change from dairy to some other farming enterprise. Often the property use changes from dairy use to rural residential and/or recreational uses.

When appraising a property for transfer purposes, the appraiser does not know who the manager or purchaser of the dairy farm will be. The appraiser must assume that the property will likely be purchased by someone with the ability to adapt the resources to realize maximum profitability. In describing the resources, therefore, their adaptability must be described. An appraiser must be prepared to consider and evaluate the adaptability of both the land and building resources.

Today's dairy industry is highly competitive and minimum returns are produced by average management. The industry provides the highest level of reward to the highest levels of management, much like other agricultural enterprises. In many parts of the United States, dairy farms have ceased operations because the available resources are not adaptable. In these areas the highest and best use of the farm has become something other than dairy use, regardless of the level and ability of management. It is incumbent on the appraiser to evaluate the trends in the market area thoroughly. Key questions are

- What is the trend for dairy units in terms of size and capacity?
- What is happening to the infrastructure supporting the industry, including processors, suppliers, and equipment dealers?
- What factors are impacting these changes—e.g., milk pricing, transportation and marketing costs, changing consumption habits?

The size and capacity of a dairy is more important in today's market than it was as recently as 20 years ago. Dairy farms can be generally described as falling into one of four different size groups: under 100 cows, 100 to 500 cows, more than 500 cows, and feedlot dairies.

Dairies with fewer than 100 cows are typically operated with a family labor force supplemented with part-time help. In this segment of the industry, properties of 200 to 300 acres with facilities will adequately support a herd of 50 to100 cows. Farms of this size generally have adequate land resources for pasturing the animals and raising a significant amount of the herd's roughage requirements. These properties do not usually have excess houses since they rely on a family labor force. Smaller farms like this attract buyers looking for a "rural experience." The income stream produced on these farms is often subsidized by the off-farm income of one or more family members. Properties of this size may be purchased for use as a rural residence with rural subsistence as a secondary objective. The farms are generally located in transitional areas and have alternative uses for other livestock farming. The buildings on these dairy operations will generally exhibit some functional obsolescence.

Dairy farms with 100 to 500 cows have utility if they are profitable at their current capacity or can be expanded through the addition of smaller farms located in the immediate area. They can be profitable if there are custom crop growers in the area from whom operators can purchase feed for their facilities at a reasonable price. It is difficult to make small changes to these farms to increase their efficiency of operation. When appraising a dairy of this size, the appraiser must assess whether the facilities have the capacity to handle crops and cows efficiently. If they do not, the unit becomes less valuable. When the land resources and production facility are out of balance, major facility changes may be required to maximize profits. Or specialized management may be required to correct the imbalance and maintain profitability. For example, the manager may have to emphasize herd management rather than crop management if the land base is deficient and feed must be purchased to support the dairy operation.

A third segment of the industry consists of dairy farms that have more than 500 cows and a sufficient land base to support all or most of the herd. These farms range in size up to 1,500 to 2,000 milking head. A significant land base is required for waste and nutrient management to balance profitable milk production and environmental concerns.

Management of animal wastes is of particular concern in cold climates, where it is difficult to move manure off the farm during winter months. With large concentrations of livestock, the storage, processing, and distribution of wastes can be a significant concern. Generally a land base of 1.5 to 2 acres per animal unit is required. By this measure, a farm with 600 cows on 600 acres is out of balance and it would have a lower value than a farm with similar capacity but 900 to 1,000 acres of cropland. Cropland requirements can be met either by ownership or by leases. In some areas, long-term leases are executed so that a sufficient land base is included with the dairy.

If the capacity of the farm buildings exceeds the land base by a significant amount, the appraiser must identify and measure any functional obsolescence resulting from this superadequacy. It is not uncommon to find a dairy in this size range that has limited land resources, inadequate waste disposal and/or storage facilities, and a great amount of physical deterioration.

The fourth general group within the dairy industry consists of "feedlot" dairies. Typically found in the Southwest, California, and parts of the Southeast, these dairies are characterized by high-density confinement of dairy animals, limited capital investment in dairy housing, limited land resources, and almost exclusive dependence on purchased feed. These dairies may have 5,000 or more head at one location with no land base on which to grow feed. Herd production is typically high, as is the cull rate in the herd. Management of animal wastes is of great concern on these farms. Many farms will have either contracts or leases which allow for the disposal of animal wastes. Management must pay careful attention to nutrient management on the lands where wastes are disposed. Sometimes plant varieties that consume high levels of nitrates, nitrites, phosphorous, and trace elements must be grown to avoid oversaturation of the soil with nutrients.

As previously indicated, the primary motivation for owning a dairy is the production of income. Therefore, factors related to profitability must be considered by the appraiser. In general, these factors relate to climate and farm size. Attributes of dairies in warm versus cold climates must be considered as well as the attributes of small versus large dairies.

## Climate Issues

Dairy facilities, including feedlot and milking parlor operations, need to be labor-efficient in their design and operation. In general, the more animals that can be managed by one individual, the more efficient the unit. Under average management, one worker can be expected to manage at least 75 to 100 cows. Milking parlors, for example, should have a capacity of about 100 cows per worker per hour.

In warm climates cow comfort is critical during the stressful periods of high humidity and high temperatures. Facilities that have been modernized to cope with heat stress have significant advantages. Shades for cattle in feed lots, misting systems in holding areas, and automatic cow preparation washes in the holding area are representative of features that add value to the improvements.

Infrastructure, i.e., the overall support system in the area, is critical. An ample supply of labor is needed for milking, cleaning the facility, and performing other tasks around the farm. Skilled and unskilled workers for the milking crew must be available near the farming unit.

The farm infrastructure also includes veterinarian and feed service personnel to assist with the nutrition and health needs of the herd. Active dairy extension programs administered through state land grant colleges are generally found in states where dairies are an important part of the agricultural economy. Other services that must be available are: feed and nutritional supplement suppliers, equipment dealers and repair facilities, processors, and a transportation system to move raw products such as feed and fertilizer to the farm and milk from the farm to the processor.

Feed must be readily available at reasonable cost. Since feed comprises nearly 50% of the total cost of production in a warm climate area, an adequate supply of feed must be available on an ongoing basis.

The environmental regulations in the area should be researched by the appraiser. Local zoning, health, and animal density requirements must be reviewed. Are test wells required and do they need to be monitored? Can manure be disposed of off site or must it be stored and treated on site?

Cold climate areas tend to have a large number of dairies that are sometimes referred to as "picture book" or "traditional" farms. These are well-kept farming operations with upright silos. The cows are usually housed in tie-stall barns, milked with a pipeline milker, and fed inside the barn. Some smaller farms may have free-stall barns and milking parlors. The common trait for all of these farms is herd size, generally 150 animals or less. These farms may be competitive in an area as entry-level, or starter, farms.

### Small Family Operations

Small farms are not equipped to handle a large number of cows per worker, nor do they typically have high levels of production per worker. Buyers of these properties are not interested in milking large herds, do not want a large work force, and will generally supply a majority of the farm labor themselves. To be financially viable, these farms require specific management skills. The farming unit is often situated in a transitional market. The farm may change in use from full-time dairy to part-time dairy/rural residential to another livestock use. Some of the factors to be evaluated on these farms include:

1. Animal housing. Some of these units have tie stalls while some may have free stalls. The size of the stall is critical. Most milk cows on modern dairies are large, often weighing up to 2,000 lbs. The stalls should be large enough to accommodate these animals. Occasionally a barn will be designed for "colored" cows, that is for Jersey, Guernsey, or other breeds that are considerably smaller than the large Holstein, Ayrshire, and Brown Swiss breeds. Stalls and facilities designed for the smaller breeds may not be adaptable to larger animals.
2. Ventilation. Barns should have adequate ventilation systems which force outside dry air in and moist air out. A significant amount of research has been done on the need for ventilation in dairies. The moist air in the building must be removed and replaced with dry air from the outdoors. Failure to ventilate a barn adequately may result in herd health problems, deterioration of the buildings due to moisture concentration, or both.
3. Feeding. The profitability of a dairy farm is partly a result of feeding. The system on a small farm must be able to feed as quickly as possible, as efficiently as possible, and as often as demanded. The system can be either an automatic feeding system located in the barn or a feedbunk located outside the barn or in the common areas of a free-stall barn. An automated feeding system that relies on belts, augers, and motors can necessitate additional operating and maintenance expense. The appraiser must be aware of the efficiency, or lack of efficiency, provided by the feeding system on the property.

4. Feed storage. On small dairy farms feed storage is a critical issue. When evaluating upright silos, the appraiser must assess the age of the silo, the condition of the unloaders, if present, and whether maintenance and repair will be costly. Many upright silos offer sealed storage capabilities. The appraiser must assess the utility of these units and determine the market's acceptance of the structures.
5. Young stock. Most small farms have housing for young stock that is separate from the housing for the milking herd. These buildings must be well-ventilated, allow for easy, efficient feeding, and have adequate capacity. On many smaller farms, up to 100% of the milking cow population is maintained as young stock. If young stock facilities are too small or overcrowded, they may become obsolete. The appraiser needs to consider whether there are other barn complexes in the area available for use as housing for young stock. These facilities may be readily available in a transition area where some dairy farms have ceased operation. One alternative to on-farm housing of young stock is use of a custom heifer grower. In some areas this practice is commonplace.

## Large Dairies

The objective of a large dairy is to operate on a lowest-cost-per-hundredweight basis. This requires maximum efficiency. Efficiency is required in all facets of the operation: moving feed from the field to the cows; grouping cows according to production, stage of lactation, etc.; and operating milking systems. The standard criteria in a cold climate, large dairy is to handle at least 50 cows per worker and produce a million pounds of milk per worker. Herd averages are typically above 18,000 pounds per cow and can be as high as 25,000 pounds per cow or more. With such high levels of production, facilities must be designed for maximum efficiency without sacrificing herd health and comfort. The criteria to be evaluated in large diary units include:

1. Milking parlors. Free-stall barns with milking parlors are the standard for large dairy units. Milking parlors are generally operated by one person and feature automatic crowd gates. The cows can be changed with a minimum of movement on the part of the operator. For example, a double 32 parlor is generally constructed so that all 16 cows on one side can be moved in and out at one time. A modern dairy barn with an older parlor can have a significant amount of physical and functional obsolescence. It is expensive to replace and/or update a parlor. The value of a dairy facility with an old, inefficient parlor is generally lower than one with a new, modern parlor.

2. Ventilation. Free-stall barns must be well-ventilated. Warm free stalls with solid sidewalls should have a fan system for use during summer months. In some moderate climates, the ventilation systems are used year round. Cold free stalls, which have open or partially open sidewalls, should have a system that moves the air rapidly. The eave height on most free-stall barns constructed in the last several years is 12 feet. However, barns constructed since 1995 have an eave height of 14 feet or higher to accommodate modern equipment.
3. Animal grouping. Large free-stall units should be designed so that cows can be segregated into groups. Large commercial dairy herds typically are divided into as many as five groups based on factors such as stage of lactation, production, and size. If a barn does not allow for the grouping of milking animals, the appraiser should consider the potential for functional obsolescence.
4. Waste. Waste disposal and management are becoming more critical. Daily spreading of manure is still the most common method of waste management and disposal. However, new environmental regulations may make it difficult to apply manure to nearby fields on a daily basis year round. Complementary waste management facilities and equipment are becoming more common on large dairies. Many farms have a strategically placed lagoon; a lagoon that is centrally located and has relatively easy access to fields is preferable. Alley scrapers and storage pits in the middle of the barn can be efficient methods of waste management.

   Many local and regional authorities are regulating waste management through land use and/or zoning regulations. The density of livestock allowed on the acreage may be regulated through zoning. Wetland regulations may dictate what wastes can be disposed of and where. When large concentrations of livestock are housed, there may be requirements to monitor aquifer water quality. The appraiser should thoroughly review all land use, environmental, and zoning regulations as a part of the appraisal process.
5. Feed storage. Cattle on large dairy farms consume large amounts of feed and commodity storage is required on all farms. Most large dairy farms have a centrally located storage complex where silos (generally bunker silos), grain bins, and commodity storage form a feeding center. With this type of system, mixer wagons can be used efficiently. The wagons are used to collect the feed ingredients, mix the ration, and then distribute it to the animals at feedbunks located either in the free-stall barn, along the side of the barn, or in a feeding/exercise lot.

## Characteristics Common to All Dairy Farms

Milk marketing is the most critical factor to be considered in evaluating a dairy. In assessing the market area, the appraiser must determine the long-term viability of the milk market. A viable market will have several handlers or dairy cooperatives, which provide a degree of competition and offer alternatives to the producer. In addition to the number of outlets, the access to these outlets must be considered. Modern milk haulers use large trucks, often tractor-trailers, to move the milk from the farm to the plant. A dairy farm located on roadways that cannot accommodate these large vehicles is at a disadvantage in the marketplace.

Infrastructure, the availability of support services and industry, and an adequate concentration of similar dairy farms are required for a successful dairy farm and dairy industry. As herd size increases, farms become more equipment-intensive, and farming practices become more specialized. Dairy farm operators are increasingly dependent on the services of nutritionists, feed suppliers that can provide specialized ingredients, dealers who can supply and service the specialized equipment needed to operate a modern dairy, and sufficient competing farms to attract and maintain the infrastructure. In the areas where feedlot dairies dominate, an adequate supply of quality roughage must also be readily available.

Dairy facility and farmstead layout have a significant impact on profitability. The more efficiently the site can be configured, the more cost-effective the operation of the unit becomes. Herd housing, feed storage, milking facilities, and milk storage must be designed with efficiency of movement in mind.

## Land Resources

The physical requirements for operating a dairy farm fall into two basic categories: buildings and land. The characteristics of the building improvements vary, depending on environmental limitations and local requirements and trends. (Dairy farm buildings will be discussed later in this chapter.) Land requirements remain constant regardless of where the farming activity is located. The basic land needs are for crop production, animal exercise, waste disposal, and water resources. Land use is generally classified as tillage or cropland, pasture or grazing, and building sites or farmstead. Some farms also have woodland, brushland, wetlands, and other miscellaneous land categories.

The building site(s) on dairy farms can represent a significant contribution to the property, both in terms of value and utility. The following factors must be considered in the evaluation of building sites: location (relative to population centers, community needs, schools), access (roads, transportation arteries, driveways), frontage (exposure), topography, and shape (regular, divided by road, multiple sites). Each of these characteristics can influence the utility of the site by either positively or negatively affecting the maneuverability of vehicles and equipment, the flow of operations between buildings, and the potential for expansion.

Tillage or cropland provides the primary land base for production of the roughage needed to feed the livestock. The quality of the soils that comprise the tillable land is critical. On a dairy with good-quality soils, as little as one acre of cropland may be needed per cow. By contrast, if the soils are of poor quality, the requirement may be as high as five acres or more per cow. Factors that affect tillable land and its desirability for a dairy farm include location; soil types (especially the availability of Class I and II soils); topography; field shape, configuration, and size; surface and subsurface drainage; natural fertility and tillage; the ability of the soil to utilize nutrients; conservation practices; and the proximity of the cropland to the building site. Crop rotation, strip cropping, alternative cropping, and no-till or minimum tillage practices are relatively modern techniques which have enhanced crop yields and reduced soil erosion.

Pasture and grazing land is becoming more important on many dairy operations. The confinement housing trends of the 1970s and 1980s are being reversed in some areas. Many dairy farmers are recognizing that they may need to provide more exercise for the herd and, at the same time, provide a supplemental feed source. Traditionally, pasture for mature animals has been located close to the barn and included minimal land area—just enough to permit some exercise. Pasture areas for herd replacements and dry cows were established farther from the building complex, mainly because these animals required less observation and management than mature, milking animals.

Many dairy farmers have determined that profitability can be enhanced by converting tillable acreage near the farmstead into rotational grazing land. This grazing plan rotates the herd through a series of pastures so that the growth potential of the vegetation is not exceeded by grazing. Originally developed in New Zealand, rotational pasturing is gaining acceptance in many dairy areas of the U.S. Some results of this practice have been a reduction in the production, storage, and purchase of other supplemental feed; the ability to sell excess forage; and increased profitability. Some farmers have experienced slight reductions in production, but this appears to be offset by an increase in overall profitability.

Grazing land is sometimes measured in two productivity units: animal unit months or cow years long. These measures indicate the capacity of a particular plot to provide adequate nourishment for mature animals for one month or year. These terms are most commonly used in the western portions of the U.S.

Most dairy farms have some woodland. Historically, the trees were used to provide lumber for farm use, timber sales, firewood, fence posts, sugarbush (in maple-producing areas), and fruit or nut production. Many of these uses are no longer relevant to modern farming needs.

Therefore, the contribution of woodland to the value and utility of a dairy has been reduced.

Occasionally wetlands may be included in the land base of a dairy farm. Because large acreages are associated with dairies, wetlands are often situated between and within the more productive tracts. While some wetlands areas may once have been drained and converted into cropland, this practice is now restricted. Wetlands do provide areas for water retention and collection, flood control, filtration, and recreational uses.

One of the most important uses of the land base of a dairy farm is to serve as a site for waste disposal. While manure can be used to supplement the fertility of the soil, there are limitations to the quantities that can be applied. Application of animal wastes on an ongoing basis may require intensive crop management. Crops that can better utilize the nutrients found in animal wastes may be planted; crop rotation becomes more important as a management tool; and the timing of waste applications may have to be adjusted. Recent legislation and public perceptions may further restrict where, how, and how much animal waste can be applied to the land.

Historically, dairy farms were balanced with all, or most of, the required land base included as part of the farm. While some acreage may have been rented, the farm unit was typically self-contained. Rented land represented a very small portion of the total land base. Now, however, many rural areas have experienced significant land use changes resulting from pressure to develop residential, commercial, and industrial uses. As a result, land values have been increasing beyond the levels that can be supported by continued operation as dairy support land. New residential areas are dotted with former farms or rural residences built on large parcels that have been split off from former farms. Owners of these residential units often want to maintain the agricultural atmosphere that first attracted them to the community. Therefore, they are often willing to lease their land to area farmers at rental levels that reflect the land's agricultural value. This provides existing dairies with the use of supplemental land without the large capital expenditures required for its purchase. Both parties gain from such an arrangement. These rental agreements may include many different types of consideration, including payments in services rather than cash (e.g., garden plowing, driveway snow plowing) which must be considered when the appraiser evaluates the subject's land base. It is not uncommon to find dairy farms that rent more than half of the land used in their operation.

Some dairy farms may purchase all or most of their feed rather than raising it on the farm. Some of these farms are small units, but they can include some very large farms and most feedlot dairies. These farming operations balance the costs of purchasing feed from other sources with the cost of owning, maintaining, and operating equipment lines and the cost of owning large tracts of land. These farms may be able to purchase the feed for less than it would cost them to raise it.

When land is rented rather than owned, waste disposal is again a major consideration. Without a land base on which manure can be spread, alternative methods of waste handling and disposal must be available. The appraiser must assess market acceptance of these operations, their economic viability, and what, if any, impact they have on property value.

Improved transportation networks, home businesses, migration from urban/suburban neighborhoods, and a desire for rural living have resulted in a migration of population from urban and suburban areas into rural communities. Many of these new residents do not realize that farming activities may have some "unpleasant" side effects: noises, odors, visual nuisances, long work hours on neighboring lands, dirty road surfaces, "escaped" livestock, etc. Occasionally, new rural residents are successful in enacting local regulations aimed at curtailing or inhibiting traditional agricultural activities that result in these "nuisances."

"Right to Farm" laws of various sorts have been adopted throughout the country in response to changing demographics. These laws may fall under the categories of transferable development rights (TDR), purchase of development rights (PDR), agricultural districts, agricultural assessments, and others. Some governmental entities and not-for-profit land trusts such as the American Farmland Trust have been acquiring conservation easements on farmland in an attempt to preserve its agricultural utility and integrity. Many of these programs call for special or reduced assessments if the land is kept in agricultural production. The appraiser should be aware that they can also impose penalties if the land is converted to nonagricultural use.

The appraiser should investigate the existence of easements, deed covenants, and other legal use restrictions affecting the property being appraised. Easements are acquired for a variety of purposes, including conservation, fishing, hunting, trapping, recreational use, timber, development, mining, and scenic preservation. They can be short-term or remain in effect into perpetuity. Easements typically affect the value of the land, but the land's utility as the primary land base or support land for a dairy may not be affected. The appraiser must determine what, if any, impact on value legal restrictions have on the farm being appraised.

## Improvements

The value of improvements relative to the total value of a dairy is significant. The total value of the improvements may exceed the value of the land in many instances. To determine the value of dairy farm buildings and improvements, the appraiser must understand the purpose of each structure as well as its overall utility and contribution to the whole enterprise.

Building improvements on a dairy will typically fall into six general categories: herd housing; milking facilities; feed storage; equipment, fixtures, and personal property; machinery storage; and waste storage and handling.

## Herd Housing

The primary purpose of herd housing facilities is to protect the herd from the natural elements. Because climates vary there is a wide variety of improvement styles. In the most extreme northern climates, shelter from wind, cold, snow, rain, and heat must be provided. Farm buildings are typically fully enclosed and of wood frame or pole construction. The design of buildings must provide for adequate ventilation and air movement in all seasons. The building must not be a source of stress for the animal. Styles commonly found in northern climates include stanchion or tie-stall barns, which are generally found on registered and purebred farms and on older, small farms; stanchion or tie-stall barns converted to loose housing or free-stall operation; and free-stall barns, which are designed for, and used by, most commercial farms.

In the moderate climates of the Southeast and Southwest, animals may only need shade and shelter from extreme sun and heat. In these areas the housing facilities may be open on one or more sides and have large fans to facilitate air movement during very hot periods. Some facilities may even have cooling equipment or sprinkler systems that provide constant temperature control. Again, stress on the animals must be minimized and adequate ventilation must be provided.

Regardless of the style of housing provided, certain common attributes are required. Efficient movement of the herd is essential. The location of feeding areas; the proximity of housing to milking facilities; the presence of hospital and calving facilities and areas for breeding, clipping, and hoof trimming; and waste removal should be considered in evaluating a building's efficiency and utility. Stall size must be sufficient to accommodate modern dairy animals. As mentioned previously, breeds such as Guernsey and Jersey cows have a smaller stall space requirement. Larger animals such as Holsteins can weigh up to 2,000 pounds and need significantly more stall space to ensure their comfort and to protect them from udder injury.

Concrete and paved walkways must be designed without steep grades or slippery surfaces. Many exercise yards and barn alleys are grooved. Older facilities are typically regrooved on a regular basis.

Unless the farm operator is purchasing all replacements (first calf heifers), housing for young stock (calves and yearlings) and replacement animals is required. Nursery facilities for newborn calves, housing for young stock, and buildings to house bred heifers and dry cows are needed on most dairy units. Young stock may be housed in individual calf hutches, which are mobile units that can be moved around the farmstead, or their housing may be included in the design of a free-stall barn. Sometimes the housing of young stock is provided through the use of low-cost, aluminum hoop structures similar to greenhouses. Efficient feeding, cleaning, and animal care should be considered in the evaluation of the young stock barns. Ventilation and temperature control are requirements in all young stock and replacement housing.

### Milking Facilities

Most modern milking facilities consist of two components: the milking area and the milk room or milkhouse, where milk storage, the cleaning of milking equipment, office operations, and other support functions take place. In some barns milking is conducted where the animals are housed. Typically these are stall barns with milk pipelines and/or milk transfer systems. Most free-stall barns have a parlor/milkhouse attached to the housing unit. The animals are transferred from the housing area to the milking area without leaving the main building. Some dairies have separate milking centers. Animals leave the housing area via covered and/or enclosed walkways or holding areas and are milked in a parlor that is detached from the herd housing unit.

Milking parlors are designed to accommodate the milking equipment they contain. Parlor styles range from in-line systems and herringbone styles, to rotary parlors, trigon parlors, and/or polygon parlors. The type of parlor generally influences the design and construction of the milking center. Size, layout, location (relative to housing), cow traffic patterns, milking routes, and the amount of mechanization required are all related directly to the type and style of the milking system. Other factors influencing milking facilities include herd expansion plans, available labor, available capital, available milking time, herd production levels, and personal preference on the part of the farmer.

The support portion of a milking center should include the milkhouse and other supplemental areas. The design of the milking center should allow adequate space for bulk storage tanks, compressor units, washing systems, a farm office, general utility rooms, lavatories, and general storage. Because this area houses the milk storage facilities, it is imperative that the building be conveniently located within the farmstead. Bulk hauling trucks, generally semi-trailers, must be able to get to the building with a minimum of effort. If trucks must back into the facility through a maze of other buildings, the placement of the structure is not efficient. Ideally, the building should be situated so that a large truck can drive into the area of the milkhouse, turn around conveniently to load the milk, and exit smoothly from the farmstead.

A lavatory, and preferably a shower and locker facility on large farms, should be conveniently located in the milking center. Because farm records are increasingly being kept electronically, the farm office should be located in or near this building. Herd production records, feeding records, breeding records, and financial records can all be consolidated in this one location.

## Feed Storage

Feed for the dairy should be stored as close to the herd as possible. Feed may include roughage in the form of hay, silage, concentrate, feed ingredients, and/or feed supplements. As the size of dairy farms increases, the traditional upright, or tower, silo seems to be losing its acceptance. While sealed storage (usually steel silos with glass linings and bottom unloading capabilities) may be found on many dairies, thorough analysis of sales is required to determine their level of market acceptance. Silage and ingredients may be stored in bunk silos, which are also called *pit, trench,* or *bunker silos.* These structures typically have concrete sides and floors, although occasionally they may have asphalt floors, treated lumber sides, or even dirt sides dug into a hillside. Bunk silos have sides approximately eight to ten feet high and no roofs.

Buildings used for ingredient storage typically have roofs and are located near the silage and roughage storage facilities. Supplements, both vitamin and mineral, may be stored in elevated metal bins. This configuration allows for efficient mixing of rations using mixer wagons. Feed storage should be located on the farmstead in an area that allows adequate space for ingress and egress for filling the storage units and removing the feed. Access for large trucks, bucket loaders, tractors, and other equipment must be considered. The capacity of the feed storage must be sufficient to store feed for an extended period of time. In areas such as the Northeast, dairies require sufficient silage storage for a 12-month period. Many farmers in that area prefer to have excess storage in case crop yields are below normal.

## Machinery Storage

Most dairy farms require a considerable amount of machinery and equipment for day-to-day operations. Storage and repair facilities for the equipment should be located at the farmstead or near the farm unit. In evaluating machinery storage, eave height is a primary consideration. A 12-ft. to 16-ft. eave height is adequate for most farm equipment. The width of access doors is critical, especially if large, articulated tractors and harvesting equipment capable of multirow harvesting are used.

Machinery storage should include provisions for a farm shop. It is preferable to perform tractor and equipment repair inside, out of the elements. The area's climate will dictate the degree of insulation required in the building, and whether heating or cooling facilities must be installed. The shop should be located in an area that is accessible to all equipment and close enough to the milking and housing facilities so that the equipment in those buildings can be repaired efficiently.

Utilities must be available at the shop site. Three-phase electrical service may be required for some of the large welding equipment found in farm shops. Provisions for discarding and/or recycling waste materials such as oils and paints must be considered. Fuels and lubricants will generally be stored near the shop. The method of storage must comply with local, state, and federal requirements. Permits and registration certificates for the storage facilities should be obtained and reviewed by the appraiser.

Provisions must be made to store agricultural chemicals used in crop production in and around the machinery storage and shop facilities. These storage units must meet local and regional standards as well as state and federal regulations. The appraiser is responsible for reviewing the suitability of the storage facilities.

### Waste Storage and Handling

Waste storage and handling improvements are becoming increasingly important on a dairy farm. Increased scrutiny of agriculture and the environmental impact of large concentrations of livestock have made the storage and disposal of manure and other wastes an important consideration in dairy farming. The typical dairy will generate wastes from the milking parlor, the herd housing units, the shop facilities, and the treatment of ill animals. All of these wastes must be carefully stored and disposed of.

Many jurisdictions strictly regulate the waste removal and storage functions. Local, regional, and/or state law may regulate the maximum number of animals on tracts. Disposal of animal waste may be regulated with requirements that the waste be "turned under"—i.e., plowed into the soil within a certain period of time after application.

Mechanization of waste disposal is a necessity. Waste may be pushed by tractor or directed through large, sewer-like pipelines from the dairy barns and parlors to a lagoon built to hold waste. Many parlors and milking centers have separate septic facilities that must be pumped out at regular intervals. It is not uncommon to find injection and/or irrigation systems that use the liquid effluent or excess water from lagoons and storage units. Some farms may even incorporate advanced technology for methane generation of electricity from byproducts and wastes; composting of solid wastes; sophisticated water removal systems to separate liquid and solid wastes for ease of handling; or irrigation systems utilizing waste effluent.

The disposal of used medical syringes, medications, lubricants, paints, and other specialized waste material used in the normal course of business is closely regulated. Many areas require that these items be collected separately and taken to specialized collection and recycling facilities. On-farm storage for short periods of time may still be required, however.

To review the adequacy and utility of a dairy's waste handling and storage capacity properly, the appraiser should be familiar with local, regional, state, and federal environmental requirements. While the appraiser need not be an expert in this field, a general knowledge of the requirements and constraints imposed on the industry is required. A lack of familiarity with regulations could result in a misleading value conclusion. If strict control of wastes is required and facilities or provisions for this function are not on the farm, this fact could be inadvertently ignored or overlooked by the appraiser in the valuation process.

### Equipment, Fixtures, and Personal Property

Equipment for milking, milk storage, feeding, and manure movement is critical. It is the appraiser's responsibility to differentiate between the equipment considered personalty and the items considered fixtures required for the ongoing farm operation. The market will indicate which items are required to maximize the value of the dairy improvements. To reflect market preferences, the appraiser must account for any and all equipment included in the property value and also identify any non-real estate items included in the final valuation. The identification should describe the item and assess its value contribution.

## Highest and Best Use

The determination of highest and best use is the key to successful valuation of a dairy farm. Chapter 11 includes a complete discussion of the steps followed in determining highest and best use. Issues that relate specifically to dairy farm appraisal will be presented here.

To meet the test of *legally permissible* uses, a dairy must conform to local zoning and land planning requirements. The location of improvements, especially those that are "to be built," must conform to local building and zoning laws.[2] The use of the land and building improvements must be allowed under the same regulations and comply with any regional or statewide requirements for waste management and animal density/housing.

To be a *physically possible* use, the infrastructure required to support dairying must be considered. The distances to, and availability of, equipment dealers, dairy suppliers (of specialized ingredients, tools, and equipment), feed suppliers and feed mills, large animal veterinarians, and inseminators should be evaluated. The questions to be answered include: What is the state of the infrastructure in the area? Is it expanding, contracting, or static? Does the local dairy farmer have to travel farther and farther to find an implement dealer, source of concentrates, or a commercial hoof trimmer or are these services available nearby?

The land base must be evaluated for its ability to support a dairy unit. Can organic wastes be disposed of on the soils without the threat of contaminating water supplies? Similar consideration must be given to the location of the herd housing and waste storage facilities. Soil resources must be available in sufficient quantities to provide adequate feed for the

2. See Uniform Standards of Professional Appraisal Practice, Statement of Appraisal Standards No. 4, July 8, 1991.

size of the facility that exists or is planned. The land required to sustain an animal varies from region to region, but sufficient land must be available through ownership or rental to sustain a profitable operation. Some dairy farms purchase 100% of their feed rather than grow it. In this case, the physical considerations center on waste storage and disposal and the availability of sufficient facilities to store feed.

The *financial feasibility* of a dairy and the *maximally productive* use of the land and building complex require careful analysis and consideration. Appraisers are charged with determining property value under "typical" or "average" management. Neither good management nor poor management are to be considered in the valuation process. However, the success and profitability of a dairy depend on the management skills employed in its operation.

An appraiser needs to collect as much information as possible about the dairy's income and expense characteristics. The type and size of the dairy farm being appraised will determine the information required. If the dairy being valued is in an intensive dairy area and is similar to most other area farms, the economic data required to evaluate financial feasibility should be relatively easy to gather. An appraiser who regularly works in the area should have a significant amount of data on hand. If, however, the farm being appraised is much larger than is typical in the area, the scope of information-gathering may have to be expanded. Rather than a small area of consideration, investigation of the market may cross regional and/or state lines.

In developing this part of the highest and best use analysis, it is wise to consider the trends in dairy economics. The industry has a cyclical nature, so placing the date of the appraisal within the observable business cycle may assist the appraiser in evaluating feasibility. Averaging financial data, either as the mean, median, or mode, and using weighing techniques can help smooth out the normal peaks and valleys in the dairy cycle.

If financial data is insufficient or unreliable, an alternative method of analysis may be considered. The four points discussed below, which relate to financial feasibility and the maximization of income potential, may be investigated.

- Recent developments in the area. Are dairy farms being used as dairies after their sale or are they being put to an alternative use? In some areas the number of dairy farms may be falling, but the number of cows being milked may remain constant or be increasing. This could suggest that small units are being combined into larger, more efficient units. Typically, the changing concentration of a particular use, in this case dairy farming, is a direct reflection of the financial feasibility of the business.
- The supply of and demand for dairy farms as evidenced by the number of property transfers (sales of dairies for use as dairies),

the number of dairy farms offered for sale (listings), and/or the occupancy of existing units. If it is feasible to operate a dairy farm, the available units will be in operation. If profit margins are declining, the supply of farms for sale and the number of vacant units may rise.

- The purpose to which dairy farms are put after their most recent transfer. This observation provides an indication of whether or not use as a dairy is maximally productive. A dairy may generate positive cash flow (and thus be financially feasible) but not generate enough cash flow to maximize net income. In such a case, the farm may become absorbed by a larger dairy farm unit or may be put to another, unrelated agricultural use.
- Comparison of the (vacant) site value to the value of the improved property less building (improvement) demolition costs. If the site would be more valuable after the improvements are demolished, then the improvements make a negligible contribution to the value of the whole property. In other words, the improved property is legally permissible, physically possible, and financially feasible, but fails the test of maximum productivity.

## Valuation Procedures

After a thorough investigation of the market area and the trends impacting value in the industry, the appraiser must define the appraisal problem: What is being appraised? Usually it is the real property, which consists of the land, site improvements, and buildings plus dairy fixtures that are attached or built into the buildings—e.g. bulk tanks, compressors, automatic washing equipment, feeding equipment, and waterers. The ownership of the specific fixtures must be clearly determined.

Because improvements contribute significantly to the value of a dairy farm and they tend to be highly specialized, the appraisal report usually presents a detailed description of all improvements. The typical report also includes a discussion of soils, property balance, and the unit's overall utility. The appraisal report should also include soil maps; plats showing boundaries, drainage, and different uses of acreage; and photographs and sketches with measurements of all the significant improvements.

In an appraisal the choice and application of the approaches to value depend on the purpose and use of the appraisal and the data available. The sales comparison approach has traditionally been the most reliable, and understandable, method for valuing dairy farms. More emphasis is now being placed on the income capitalization approach, especially for valuing larger operations where the number of comparable sales is limited. The cost approach is recognized as a viable valuation technique because of the extensive improvements on dairy farms. It is especially appropriate when the subject property has relatively new building improvements. Depreciation can be difficult to measure, but it can be verified by applying market extraction techniques to the sales of other dairy farms.

## Sales Comparison Approach

To apply the sales comparison approach, comparable sales of dairy farms must be gathered and analyzed. The sales must be similar to the property being valued in terms of size, capacity, use, adaptability, and utility. A small dairy with a stanchion or tie-stall barn that has a limited tillable land base should not be appraised using sales of large farms with free-stall barns and excess land.

The comparability of the sales should extend to size (both herd capacity and size in acres of total land); housing facilities; the method of feed supply (purchased feed or grown feed); physical location; the type of milk marketing (fluid use, manufacturing use, a combination of the two); soil quality; quality of improvements; and overall farm balance.

### Sales Verification

Appraisers must verify many different elements of a dairy farm sale. Information on the improvements, including their quality, number, utility, age, configuration, and capacity must be gathered. The balance of the unit must be identified, and the present and intended uses of the farm should be evaluated. When gathering data on a dairy farm, attention should be directed to its income characteristics as well as to the costs of the different building improvements on the property. Table 18.1 illustrates a format for data collection.

Improvements on a dairy farm generally make a major contribution to the overall sale price and, hence, to the overall value of the property being appraised. The appraiser must have sufficient information to analyze adequately the improved sales that have occurred in the market. This necessitates the collection of sales of vacant land suited for use as a dairy farm; sales of farms with few improvements; and sales of farms with many improvements. By estimating the contributing market value of the land on each improved sale and subtracting that value from the total sale price, the remainder of the purchase price will necessarily represent the depreciated value of the improvements.

Analyzing the sales in order of the degree of improvement allows the appraiser to determine the value relationships of the different components. In some instances, mathematical relationships may emerge. For example, herd housing, heifer housing, and young stock housing facilities may demonstrate a relationship that can be expressed in ratios. These ratios, which are determined through analysis of sales with only a few improvements, can then be applied to highly improved properties to create order out of data that initially seems to be chaos. When a large array of dairy sales is analyzed, the appraiser develops an understanding

**Table 18.1** Data Collection

| | | | | | |
|---|---|---|---|---|---|
| Sale no. ____ | | | | | |
| Grantor: ____ | | | | Volume (book): ____ | |
| Address: ____ | | | | | |
| Grantee: ____ | | | | Page: ____ | |
| Address: ____ | | | | | |
| Deed date: ____ | Transfer tax: ____ | | | Financing: ____ | |
| Recording date: ____ | | | | | |
| Sale verified with: ____ | Sale verified by: ____ | | | | |
| Purchase price: | **Total $** | | **$/Cow** | Property rights conveyed: ____ | |
| Total | ____ | | ____ | ____ | |
| Personal property | ____ | | ____ | ____ | |
| Net to real estate | ____ | | ____ | ____ | |
| **Land:** | **Acres** | **$/Acre** | **$/Cow** | **Total Contribution** | |
| Total | ____ | ____ | ____ | ____ | |
| Tillable | ____ | ____ | ____ | ____ | |
| Pasture | ____ | ____ | ____ | ____ | |
| Woods/waste/etc. | ____ | ____ | ____ | ____ | |
| Total | ____ | ____ | ____ | ____ | |
| Improvements: | ____ | ____ | ____ | ____ | |
| **Description** | **Est. Eff. Age** | **Est. Econ. Life** | **RCN** | **Contribution** | **Total Acc. Depr.** |
| Dwelling(s) | ____ | ____ | ____ | ____ | ____ |
| Dairy barns | ____ | ____ | ____ | ____ | ____ |
| Milking parlors | ____ | ____ | ____ | ____ | ____ |
| Heifer/young stock housing | ____ | ____ | ____ | ____ | ____ |
| Machinery/equipment sheds/shop | ____ | ____ | ____ | ____ | ____ |
| Feed storage | ____ | ____ | ____ | ____ | ____ |
| Total | ____ | ____ | ____ | ____ | ____ |
| Total per cow | ____ | ____ | ____ | ____ | ____ |
| Facility capacity: ____ | | | | Comments: ____ | |
| Est. gross income: $income/cow: $____ | | | | ____ | |
| Est. expenses: $____ | expenses/cow: $____ | | | ____ | |
| Est. *NOI:* $____ | *NOI*/cow: $____ | | | | |

of the contribution of specialized improvements to the total sale price of operating dairy farms that have been sold recently in the marketplace.

The comparison process requires identification of both elements of comparison and units of comparison. An element of comparison is a factor recognized by the market as either adding to, or subtracting from, total value. The elements of comparison for most dairy farm appraisals include land balance and mix; improvement balance and mix; property rights conveyed; financing terms and conditions; conditions of sale; market conditions; location to market and within a region; size of farm in total acres as well as capacity, expressed in total cows or total animal units; quality of the dairy unit; and any other physical and/or economic characteristics that the market may recognize.

Units of comparison refer to the units into which a property can be divided for comparative purposes. Many appraisers of rural property are conditioned to consider only acres as a unit of comparison. However, a particular market may or may not recognize this unit. Before determining the appropriate unit to apply, an appraiser should analyze the array of sales to determine what unit the market is using to evaluate property. For example, the sales can be compared to one another using not only acres, but also capacity of milking animals, total animal capacity, mature animal equivalent capacity, unit carrying capacity, and total building capacity.

Generally if a market is buying and selling dairy farms based on per cow units, recent sales will reflect that fact. The better properties will have higher per-unit sale prices expressed in dollars per cow; the poorer properties will have lower per-cow-unit sale prices. Comparing the sales to each other on a per-cow basis will help the appraiser extract appropriate elements of comparison as well as the specific amounts of adjustment that may be required.

Table 18.2 illustrates the presentation of unadjusted sales to be used in the valuation process.

**Table 18.2** Unadjusted Sales Chart

| Sale No. | Sale Date | Sale Price | Size (Ac.) Total/Crop | Capacity (Cows) | Location | Condition Utility | Effective Age of Facility | Price Per Acre | Price/ Cow |
|---|---|---|---|---|---|---|---|---|---|
| 1 | 3 yrs. ago | $2,050,000 | 675/520 | 1,025 | Fair | Good | 3 yrs. | $3,037 | $2,000 |
| 2 | Current | $2,145,000 | 950/800 | 1,400 | Good | Fair | 10 yrs. | $2,258 | $1,532 |
| 3 | 2 yrs. ago | $2,566,000 | 850/800 | 2,000 | Fair | Good | 1 yr. | $3,019 | $1,283 |
| 4 | Current | $2,675,000 | 900/830 | 1,450 | Good | Good | 2 yrs. | $2,972 | $1,845 |
| Subject | Current | | 1000/850 | 1,200 | Good | Good | 2 yrs. | | |

After an appropriate unit of comparison is selected, improved sales that are similar to the property being valued are compared on an individual basis with the subject. Land is compared to land, in terms of balance and mix as well as quality and utility. Improvements are compared in much the same way, although the utility of the buildings relative to the total farm is addressed in a separate adjustment. Herd housing on the sale is compared directly to herd housing on the subject. Other improvements are compared in the same way. The comparisons are made systematically, as outlined in Chapter 12. Data Collection and Analysis.

Table 18.3 illustrates the process for deriving adjustments based on analysis of paired sales. Since no difference in financing terms exists in these sales, no adjustment is warranted. If financing terms did impact the value of any sale, however, this adjustment should be addressed.

The location adjustment reflects the transportation distance to markets for Sales 1 and 3, which are located 125 miles south of the subject area. The two comparisons indicated adjustments of $100 and $158 per cow, respectively, for this difference. In this instance $100 per cow is selected as the appropriate adjustment.

An adjustment must be considered for the overall utility of the farm unit and the unit's condition and quality, as reflected in the market's acceptance or lack of acceptance. In this example, Sale 2 has a milking parlor that is in good condition and functions well but is out of balance with the housing capacity of the farm. The parlor is much smaller than the size of the unit demands and an upward adjustment is required. Sale 1 is adjusted downward slightly because its overall condition and utility is excellent, a trait not typically found in the market.

**Table 18.3** Sales Comparison Approach Adjustment Grid

| | Sale 1 | Sale 2 | Sale 3 | Sale 4 |
|---|---|---|---|---|
| Sale date | 3 yrs. ago | Current | 2 yrs. ago | Current |
| Acres | 675 | 950 | 850 | 900 |
| Capacity | 1,025 | 1,400 | 2,000 | 1,450 |
| Financing terms | Market | Market | Market | Market |
| Location | Fair | Good | Fair | Good |
| | 125 mi. SW | 25 mi. NE | 125 mi. S | 5 mi. S |
| Condition/utility | Good | Fair | Good | Good |
| Price/acre | $3,037 | $2,258 | $3,019 | $2,973 |
| Price/cow | $2,000 | $1,532 | $1,283 | $1,845 |
| Units of comparison: $ per cow | | | | |
| Land adjustment | 251 | 83 | 159 | 111 |
| Improvements adjustment | -556 | -159 | 233 | -146 |
| Adjstd. $/cow | $1,695 | $1,456 | $1,675 | $1,810 |
| Location | | | | |
| | Compare # 1 | To # 4 | $115/Cow | Conclude |
| | Compare # 3 | To # 4 | $135/Cow | $125/Cow |
| Adjustment | $125 | 0 | $ 125 | 0 |
| Adjstd. $/cow | $1,820 | $1,456 | $1,800 | $1,810 |
| Condition/utility | | | | |
| | Compare # 1 | To # 2 | $364/Cow | Conclude |
| | Compare # 3 | To # 2 | $344/Cow | $350/Cow |
| | Compare # 4 | To # 2 | $354/Cow | |
| Adjustment | 0 | $ 350 | 0 | 0 |
| Adjstd. $/cow | $1,820 | $1,806 | $1,800 | $1,810 |
| Subject capacity (cows) | 1,200 | 1,200 | 1,200 | 1,200 |
| Indicated value of subject | $2,184,000 | $2,167,200 | $2,160,000 | $2,172,000 |

Appraisers can develop alternative market measurements using sales comparison techniques. Estimating gross and/or net income levels and dividing the results by the total sale price can provide the appraiser with a gross or net income multiplier. If this calculation is performed for all of the sales, a range will emerge. Multipliers can be compared to one another using the methods of comparison previously described. The appraiser should evaluate the merits of each multiplier in the range and select an appropriate "indicated" gross or net income multiplier (or range of multipliers) for the property being valued. This multiplier is then applied to the gross or net income of the subject to produce an indication of value.

Calculating a multiplier requires information developed in the income capitalization approach. Cash rentals are preferred in developing income multipliers, but since most dairy farms are owner-operated, it is necessary to develop an overall income and expense profile for each of the sales analyzed as well as for the property being valued. The derivation and application of alternative market measurements in the appraisal of a dairy farm can serve as support for the final value conclusion.

## Cost Approach

The cost approach is seldom used as the sole method for valuing dairy farms. It can, however, provide support for the sales comparison and income capitalization approaches. Use of the cost approach may be preferred when valuing new or relatively new buildings, to appraise a proposed development, or for allocating value between various improvements (e.g., eminent domain or insurance valuation). It might be relatively easy to calculate the reproduction or replacement cost of the different buildings, but accurately analyzing all forms of depreciation is difficult.

In the cost approach, a value opinion is reached by calculating reproduction cost, deducting depreciation as of the date of valuation, and then adding land value to produce a total property value. Refer to Chapter 13 for a thorough discussion of the cost approach.

### Land Value

The value of the land portion of a dairy property is estimated using undeveloped land sales that have a similar highest and best use. Good-quality farmland can be adapted to a variety of uses, including forage production for dairy facilities. The market recognizes that properties differ in their quality characteristics and suitability for a wide range of farming uses ranging from pasture and grazing land to the production of high-value crops. Typically, a dairy operation must compete against these alternative uses for land.

The ownership risk of land varies with its use. Typical unimproved farmland will generally carry relatively little risk, as reflected in low capitalization rates. The more intense and specialized the use, the higher the associated risk and the higher the capitalization rate.

The value of land, in its vacant state and available for development to its highest and best use, is measured from the market. There are six ways to value land: sales comparison, capitalization of land rent, allocation, extraction, land residual, and subdivision development. In rural appraisal, land values are typically developed using sales comparison techniques, capitalization of ground rent, or extraction. An appraiser should also be familiar with the other techniques, which may be applied in some cases.

The three commonly used methods all are based on the market. Sales comparison requires sales of similar properties to which the subject can be compared. Extraction is a variation of allocation in which land value is extracted from the sale price of an improved property by deducting the value contribution of the improvements. Ground rents can be capitalized at an appropriate, market-derived rate to indicate the value of the subject site. Sales comparison techniques are most commonly used to determine the value of land for dairy farms. Table 18.4 displays land sales that could be used to value a site.

**Table 18.4** Unadjusted Land Sales

| Sale | Sale Date | Sale Price | Total Acres | Total $/Acre | Crop Acres | Crop $/Acre | Other Acres | Other $/Acre |
|---|---|---|---|---|---|---|---|---|
| 1 | 1-Year | $670,000 | 800 | $838 | 725 | $900 | 75 | $233 |
| 2 | Current | $605,000 | 750 | $807 | 700 | $850 | 50 | $200 |
| 3 | 1-Year | $455,000 | 650 | $700 | 550 | $800 | 100 | $150 |
| 4 | Current | $660,000 | 800 | $825 | 775 | $850 | 25 | 0 |
| 5 | Current | $655,000 | 800 | $819 | 675 | $935 | 125 | $191 |

By analyzing verified sales data, the appraiser can develop an opinion of land value for use in the cost approach to value.

### Cost of Improvements

In the cost approach, each of the improvements on the property being appraised must be valued. The appraiser has the option of using either reproduction cost or replacement cost to determine the cost of the improvements. (These two terms are defined in Chapter 13.) Replacement cost is generally used in the valuation of dairy buildings.

An appraiser can use cost guides, local contractors, or market abstraction to obtain costs. If cost guides are used, the costs provided should be compared to local costs to ensure that the multipliers supplied with the guide are accurate for the area in which the property is located. In areas with an established dairy industry, cost figures can be obtained from contractors specializing in the construction of buildings for this type of farm. The appraiser is encouraged to establish relationships with these contractors so that the cost estimates used are as accurate and current as possible.

An example of the compilation of construction costs is shown in Table 18.5.

### Depreciation

The estimation of depreciation depends on the availability of sufficient, verified sales data. To estimate depreciation on dairies, most appraisers use market abstraction of depreciation rates. Use of this method requires proper analysis of sales data, including the estimation of the economic age of, and curable obsolescence in, the farmstead and/or

**Table 18.5** Replacement Cost of Improvements

| Description | Buildings | Equipment |
|---|---|---|
| Milking parlor/ milk house | $378,000 | |
| Milking equipment | | 750,000 |
| Free-stall barn | 775,680 | |
| Lockups | | 104,000 |
| Feed lanes | 202,680 | |
| Pens/corrals/gates | | 190,000 |
| Commodity/ feed barn | 121,500 | |
| Feeding machinery | | 50,000 |
| Site improvements | 260,000 | |
| Shop building | 42,750 | |
| Total | $1,780,610 + | $1,094,000 = $2,874,610 |

**Table 18.6** Market Abstraction of Depreciation Rates

| Comparable Dairy Facility | | Total | Cost/Cow | |
|---|---|---|---|---|
| Replacement cost— | | | | |
| Buildings | | $1,232,000 | 1,202 | |
| Equipment | | $1,200,000 | 1,171 | |
| Total | | $2,432,000 | 2,373 | |
| Indicated contributory value | – | 1,900,000 | 1,854 | |
| Total indicated depreciation | | $532,000 | $519 | |
| Estimated physical curable deterioration | – | 50,000 | $49 | |
| Estimated physical incurable— short-lived deterioration | – | 200,000 | $195 | |
| Remaining physical incurable long-lived deterioration | | $282,000 | | |
| Estimated % physical incurable—long-lived deterioration ($282,000/$1,182,000) | | | | 23.9% |
| Estimated effective age | | | | 4 years |
| Estimated annual rate of depreciation | | | | 6.0% |
| Estimated economic life (1 ÷ 0.06) | | | | 17 years (rounded) |

improvements. Table 18.6 illustrates the application of this method in analyzing the sale of an improved dairy farm.

A similar analysis can be conducted using the economic age-life method. Applying this method to dairy farms is complicated since they usually include a complex of several buildings. Either each building must be treated separately or the entire complex must be analyzed as a unit. Market abstraction of depreciation allows the appraiser to develop a range of depreciation rates for different types of buildings. These rates are then applied to the improvements on the property being valued. The market-derived rates provide empirical support for the rates that the appraiser uses in valuing the subject property components.

In the breakdown method, all types of depreciation are considered separately. Curable and incurable physical deterioration (short-lived and long-lived), curable and incurable functional obsolescence, and external obsolescence are all considered in sequence when the breakdown method is used.

While the physical deterioration and most forms of functional obsolescence found on a dairy farm can be estimated with relative ease, the estimation of external obsolescence can be somewhat difficult.

External forces including the condition of the general economy, the status of the dairy industry, the number of dairy farms available for purchase, highway networks, and distances to metropolitan centers and suppliers/markets all impact market value. In addition, changing environmental regulations exert external pressures on dairy property.

The easiest way to estimate external obsolescence is through analysis of sales with new improvements or improvements that are nearly new and exhibit minimal functional obsolescence. Table 18.7 illustrates the analysis of external obsolescence in a sale property improved with a free-stall barn and milking facility, feed storage, and shop/machinery storage.

**Table 18.7** External Obsolescence Estimate

| | | |
|---|---|---|
| Sale price | | $1,750,000 |
| Less: land value | | – 240,000 |
| Contribution of building improvements | | $1,510,000 |
| RC of building improvements | $2,187,381 | |
| Less: improvement contribution | – 1,510,000 | |
| Total depreciation | | $677,381 |
| Curable physical deterioration | $15,000 | |
| Incurable physical deterioration—short-lived | $120,000 | |
| RC of improvements less curable physical and cost of short-lived items ($2,187,382 – $15,000 – $1,079,300 = $1,098,381) | $1,093,081 | |
| Effective age of improvements | 3 years | |
| Annual depreciation rate from sales | 6% | |
| Physical deterioration—long-lived ($1,093,081 × 18%) | $196,755 | |
| Total physical deterioration | 331,755 | |
| Physically depreciated RC | | 1,855,626 |
| Less functional obsolescence | $0 | |
| Physically and functionally depreciated RC | | $1,855,626 |
| Total depreciation | $677,381 | |
| Identified physical and functional | $331,755 | |
| Excess identified as external obsolescence | $345,626 | |
| Percent of external obsolescence ($345,626/$1,855,626) | | 19% |

After depreciation rates have been derived from the market, these rates are applied to the cost of the subject property. Table 18.8 summarizes the results.

## Income Capitalization Approach

The income capitalization approach measures the present value of anticipated future benefits. The application of direct capitalization is appropriate for most dairy farms, especially those that are well-established and similar to other dairies in the area. The appraiser must use care in estimating the income and expenses for these farms since there may be income related to farming other than milk production. The breeding and sale of high-quality cattle, registered cattle, embryo transplants, and 4H animals and family living expenses all impact the income stream of a dairy farm. All dairy farm

**Table 18.8** Appraisal of Subject Property

| | | |
|---|---|---|
| Replacement cost | | $2,874,610 |
| Curable physical | $60,000 | |
| Incurable physical—short-lived | $100,000 | |
| Incurable physical—long-lived ([$2,874,610 – $60,000 – $1,094,000 = $1,720,610] × 12%) | $206,473 | |
| Total physical deterioration | | – 366,473 |
| Physically depreciated RC | | 2,508,137 |
| Functional obsolescence | $0 | |
| Physically and functionally depreciated RC | | $2,508,137 |
| External obsolescence ($2,508,137 × 19%) | | – $476,546 |
| Depreciated improvement cost | | 2,031,591 |
| Plus land value | | $ 250,000 |
| Total | | 2,281,591 |
| Indicated value by cost approach (rounded) | | $2,300,000 |

appraisals should include consideration of the income capitalization approach because a dairy farm is an income-producing entity. The appraisal report must discuss the reasons for excluding the approach if it is not used.

Dairy farms are typically owner-operated businesses. As such, the estimation of income and expenses is critical since cash rental data can rarely be obtained. The historical income and expenses of the subject property are important data for the income approach. Income and expense information for the dairy industry is available from many sources including accounting services, county extension agents supported by land grant agricultural colleges, lenders, and dairy cooperatives.

Typical revenue sources for a dairy farm may include

- Milk sales
- Calf sales
- Cull cow sales
- Government payments and fuel tax refunds
- Patronage dividends
- Fuel tax refunds
- Utility sales (i.e., the sale of power resulting from methane generation)
- Sales tax and insurance refunds
- Crop insurance proceeds
- Excess feed sales

On smaller farming operations, farm operators and other family members may work off the farm. This non-farm income is not included in the evaluation of the farm's net income potential.

Typical cash expenses may include, but are not limited to

- Purchased feed

- Chemicals
- Custom hire
- Fertilizer and lime
- Freight and trucking
- Fuel costs
- Insurance
- Repairs
- Lease payments
- Seeds and plants
- Supplies
- Farm portion of taxes and utilities
- Numerous miscellaneous expenses

These expenses may also be considered

- Purchase of replacement livestock
- Building, machinery, and equipment replacement
- Entrepreneurial risk and management charges

The development of overall capitalization rates requires the analysis of income and expenses for comparable sale properties. Rates can also be developed using income techniques such as the band of investment developed for appraising other rural properties.

Rental and lease information may be available for smaller farms. From this information, the appraiser can extract overall rates. However, the appraiser must be careful to select rates from homogeneous data—i.e., from cash rents or cropshare rents, but not from both. Typically, the overall rates abstracted from cash rentals are lower than those abstracted from cropshare rentals, which in turn are lower than those abstracted from owner-operator data. The overall rates reflect the level of risk inherent in each mode of operation.

Table 18.9 illustrates the development of capitalization rates using four sales. The bottom row of the last column shows the rate to be applied to the subject property. A more complete analysis of the deriva-

**Table 18.9** Sales Analysis—Calculation of Overall Capitalization Rates

| | Sale 1 | Sale 2 | Sale 3 | Sale 4 | Subject |
|---|---|---|---|---|---|
| Size | 1,025 Hd | 1,400 Hd | 2,000 Hd | 1,450 Hd | 2,000 Hd |
| Price | $2,025,000 | $1,800,000 | $2,155,000 | $2,115,000 | $2,150,000 (r) |
| Price/cow | $1,976 | $1,286 | $1,078 | $1,459 | $1,075 |
| Income | $2,615,685 | $3,360,000 | $4,900,000 | $3,625,000 | $4,900,000 |
| Expenses | $2,352,408 | $3,080,000 | $4,600,000 | $3,262,500 | $4,600,000 |
| Net operating income | $263,277 | $280,000 | $300,000 | $362,500 | $300,000 |
| Indicated overall rate | 13.0% | 15.6% | 13.9% | 17.1% | 14.0% |

tion of incomes, expenses, and capitalization rates can be found in Chapter 15.

## Reconciliation and Final Opinion of Value

The strengths and weaknesses of each valuation method must be discussed in the reconciliation section of the appraisal report. Appropriate weight should be given to each value indication derived based on the quality and quantity of market data available for analysis. For small- to medium-size dairy farms, the sales comparison approach may provide the best indication of value if a number of arm's-length sales are available in the area. For larger dairy farms, the income capitalization approach may be given more emphasis because buyers of and investors in larger operations base their decisions on income potential. The cost approach may be given the greatest weight for new or proposed construction, but a detailed depreciation analysis must be done since even new buildings can include functional and/or external obsolescence.

## Addenda to the Report

The addenda to a self-contained appraisal report prepared to communicate a complete appraisal should include:

- Subject property photographs
- Area and neighborhood data
- Comparable sale sheets with descriptions, analyses, photographs, and location maps
- Income and expense data
- Capitalization rate support data
- Reproduction or replacement cost estimates from contractors/engineers/architects or referenced pages from cost manuals
- Additional maps (e.g., location, tax, flood, historic district, soil, topography, vegetative cover, cropping sequence, wetlands)
- Building sketches, plots, and surveys
- Zoning map and regulations
- Copies of deeds and easements
- Assumptions and limiting conditions
- FIRREA checklist
- Copy of engagement letter
- Qualifications of appraiser(s)

# The Valuation of Livestock Ranches

A livestock ranch is a business enterprise for the raising of livestock on which forage grasses and forage crops are the main source of livestock feed. Although the term *ranch* is sometimes used synonymously with farm or homestead to describe a rural property, in this chapter the term refers to a livestock-producing property.

Livestock ranch appraisals can be simple or very complex, like the livestock industry. Regardless of the type of property, every appraisal is a research project. A vast amount of information is collected, analyzed, and reconciled into an opinion of market value. This chapter will identify the livestock ranch appraisal problem and provide an orderly procedure with which to examine the factors that influence ranch values. Both natural and man-made features and relationships affect the value of livestock ranches. It is most important for the ranch real estate appraiser to understand these elements and relationships to assess the land, its vegetation, the livestock that use the land, and to all the other factors that influence value.

## Types of Livestock Ranches

The types and styles of livestock ranches in North America, Australia, Africa, South America, and the European Continent vary greatly. Within the United States, a livestock ranch might be operated as a separate unit or in conjunction with various crop production or recreational endeavors. The appraisal of a livestock ranch may involve special considerations and knowledge if the ranch is located in the Pacific Northwest, the Sandhills of Nebraska, desert areas of the Southwest, fruit- and nut-producing areas on either coast, the Rocky Mountain region, the Corn Belt, or high rainfall areas of the Gulf Coast. The highest and best use of agricultural properties within these regions may be heavily influenced by the availability of feedstuffs such as wheat pasture and corn stalks, fertilization within high rainfall areas; cake and range versus meadow hay production; and the inclusion of land under government leases or permits in the Western states. These influences also affect the operation and highest and best use of individual properties within a subregion.

The operation of a livestock ranch for the production of cattle, sheep, or goats is usually influenced by the types of forage available and the desires of the current owner. Variations within or between regions may also have a significant impact on the ability of a property to be used for maximum productivity. A wide variety of factors typically impact the style and operation of a ranching unit and dictate the species or combination of species in that unit. Regional economics and lifestyles may require that the ranch have a primary function for recreation with livestock production as a sideline. The owners of many properties have options, choosing to use the property for year around cow/calf or farm flock production, for seasonal grazing of yearling or mature livestock, for either commercial or purebred production, or for a combination of uses.

In appraising a livestock ranch, the appraiser must become fully informed about the types and styles of ranching operations in the market area and determine the flexibility or limitations in the operation of the subject property and comparable sales. The appraiser must also recognize the recreational amenities that relate to the subject property and comparable sales.

## Land Base of Livestock Ranches

Throughout much of the United States and Canada, many livestock ranch operations are based entirely on deeded land held in fee simple title. Regardless of its geographical location, the operation of a deeded ranch may include both owned and leased land. Ranching operations in the 17 western states often include privately owned, deeded land as a base unit to provide forage production or base water combined with land leased from federal or state governments. Livestock operations throughout the United States include land leased from other entities such as municipalities, utilities, timber companies, and private owners. The terms of leases and permits will vary from annual to multi-year. To analyze the use and value of either deeded or leased lands properly, the appraiser must research and understand all the real estate used for the livestock ranching operation.

### State and Federal Leases

To stimulate development of the nation, the U.S. Government gave deeded land to many state governments. States usually received specific sections within each township, such as sections 16 and 36. Although some states have sold this land to private individuals, many still retain ownership. Most state-owned lands are leased to private enterprise, often for agricultural purposes.

State ownership usually involves a wide variety of land use types, including dry rangeland, subirrigated pasture or meadow, forest, and cropland. Some states allow multi-year leases in which state-owned land becomes an integral part of a livestock ranching operation. Over time, this leased land may become extremely important to the successful operation of the associated deeded land. Changing social and economic factors may heavily influence the utility and financial feasibility of state leases.

The U.S. Government owns and leases many millions of acres suitable for livestock ranching, particularly in the 17 western states. Since various divisions of the federal government may own the land, a variety of lease/permit restrictions or opportunities may be involved. For example, the professional appraiser will recognize that lands within national forests are administered by the U.S. Forest Service within the Department of Agriculture. Lands within the Department of the Interior are administered by the Bureau of Land Management (BLM). Both entities have a variety of lease terms, varying from annual to 7 to 10 years. Both entities apply strong controls on the use of leased land. Leases and permits will specify the species of livestock, number of livestock, and allowable grazing season.

Livestock grazing on either forest or BLM lands is considered to be a privilege. Although this privilege is held by a private individual and associated with specific deeded land, permits and leases are not automatically conveyed with the sale of a ranch. Permittees of forest or BLM lands must possess commensurate deeded land. That is, the permittee must have the ability to maintain the livestock during remaining portions of the year. Government land cannot make up the total enterprise.

Government leased land often becomes an integral part of the ranch enterprise. Upon sale of the deeded land, the permittee must request a transfer of the grazing privileges to the buyer of the land. Although permits typically remain with the base property to which they have been attached for many years, the permit technically reverts to the Forest Service or BLM and does not have to be granted to the new owner. A second alternative is for forest and BLM permits to be transferred with the owned cattle. A permittee may sell cattle to a new owner and request that the federal permit be transferred with the cattle. Again, the new owner/permittee must possess commensurate deeded land.

Grazing permits are closely supervised. U.S. Forest Service and BLM administrators have the right to set up strict pasture rotations, salting patterns, and periodic counting of the livestock. It is always sound appraisal practice to verify all details of a government grazing permit with the appropriate U.S. Forest Service or BLM personnel. Other government entities such as military reservations may grant grazing leases to private operators.

In addition to state and federal grazing leases, many ranches today incorporate additional lands through private leases. Private leased land may be owned by individuals, families, grazing associations, investment companies, railroads, or municipalities. Railroad ownership in some western states is checkerboarded with private or state ownership of adjoining sections. Like federal lands, land held through private leases may become an integral part of a ranching operation. Conversely, the market value of the real estate included in these private leases may be controlled by the lessee due to the physical configuration of the owned and leased land. Regardless of its ownership characteristics, the feasibility and value of leased land within livestock ranches will vary with the requirements of the specific administrator.

## Deeded Ranches

Real estate is not homogenous in nature, but the livestock industry divides livestock ranches into specific classes or types of ranches. Although a significant portion of the national production of grazing livestock is associated with farm or orchard operations, the characteristics of various types of ranches described below distinguish ranches from farm properties.

Ranch types within the 17 western states vary greatly. The term *mountain ranch* typically refers to properties at higher elevations and is often associated with government leased land. The base property may be included at the mountain location or be located miles away. The predominant grasses found on mountain ranches are cool season grasses. Forage production can vary from high feed production areas to irrigated and subirrigated lands to dry ranges with limited forage production.

On year-round mountain ranches, hay must be fed to the livestock during the winter period of five to six months. Hay is often produced on the ranch, but some hay may be purchased. In winter the cost of hay may be a high expense item for a mountain ranch. Although pregnancy rates are fairly high, most mountain ranches do not have high weaning weights because of late calving and limited forage for summer grazing. A mountain ranch can also experience significant losses from predators or poisonous plants.

Desert ranches can be found in several states. This type of ranch is characterized by very low forage production per acre and often a very small percentage of deeded land compared to leased land. In desert ranches water sites are often part of the private, deeded land and the extensive acreages of government or state lands have very limited water. The desert ranch often uses specific terrain or forages for the nongrowing season. Cattle are adapted to harvesting browse, including various woody plants. This off-season grazing of browse may be supplemented by feeding the livestock high-protein energy supplements, or cake. Most operators consider a desert ranch to have less expense per cow than a mountain (hay) ranch. However, the large acreages required for a desert ranch significantly reduce the ability to manage livestock.

Pregnancy rates and even weaning weights can be significantly lower than they are on other types of ranches.

Ranches in the Gulf Coastal regions enjoy very high carrying capacities per acre due to high rainfall and a long growing season. The high forage production in such areas usually depends on the production of Bermuda grasses and requires large amounts of costly fertilization. This type of ranch may use hay and/or concentrate cake for the short winter period. The Gulf Coast ranch allows for intensive management but has specialized requirements. High rainfall, heat, and other factors create excessive parasite problems. Routine treatment of livestock for both internal and external parasites including ticks and worms is required. This environment also is conducive to other types of disease and management problems. In the operation of a Gulf Coast ranch, it is important to select breeds of cattle resistant to parasites, disease, and heat problems.

The term *plains ranch* refers to properties in the central United States characterized by broad expanses of undulating to rolling terrain. From north to south, plains ranches vary in rainfall and forage production. They can be operated as either hay/cake or range properties. The style of operation can vary with the availability of hay or the desires of the operator. From east to west, plains ranches also differ in rainfall, soil type, and, therefore, type of grass. Individual areas in the western, drier plains produce limited forage with drought-tolerant grasses such as buffalo and grama grass and may include extensive amounts of cactus. Tallgrass prairie ranches, in the high rainfall areas of the eastern plains, produce excessive forage per acre from big bluestem, switchgrass, and Indian grass. Intermixed between are prairies of midsize grasses such as little bluestem, side oats grama, and wheat grass.

Livestock production in the plains can vary. Some ranches have intensive management with excellent pregnancy rates and weaning weights while ranches in drier areas are subject to less favorable factors. Northern portions of the plains can experience severe winter storms, while southern areas require heat- and parasite-resistant cattle.

Obviously, many areas in the West include ranching operations with combinations of the previously described ranch characteristics. Small geographic areas may also possess widely different types of ranches. The coastal areas of California enjoy a reverse of seasons in which winter is the productive period for livestock grazing. Government leased land is typically more important to mountain and desert ranches. Recently, significant public pressure has mounted concerning government leases. In each appraisal assignment it is incumbent on the appraiser to understand those factors that affect value of the ranch property being appraised.

## Function of a Ranch Property

The appraiser's analysis of highest and best use and the resulting estimate of value for a livestock ranch are typically based on the productivity and amenities provided by the ranch. For a typical livestock ranch, the basis of productivity and, therefore, market value is the owner or operator's ability to convert forage into income via livestock. In reality, a livestock operation is the "farming" of forage and "harvesting" by livestock. The value and highest and best use of a livestock ranch can be significantly influenced by amenities other than livestock production. Actual productivity may become a minor factor relative to the value of the property for scenic, recreational, or conservation purposes. A thorough study is needed to determine the amenities of a specific property, and if they contribute to value. The appraiser should realize that the highest and best use of a ranch property might not be the current use by the current owner.

## Converting Forage into Marketable Livestock

The economics of a ranch are directly related to forage production, usually measured as carrying capacity in animal units. Forage production, whether expressed directly through carrying capacity or indirectly on an per-acre basis, is the foundation that has to be established before a credible valuation can be completed. Most livestock ranches in the Western U.S. are marketed based on carrying capacity. The normal basis in other parts of the country is per acre. Analysis of ranches indicates that property price is directly related to forage production or carrying capacity.

Forage production for livestock use is typically expressed in animal units or animal unit months. The animal unit month measurement of productivity allows the appraiser to relate and compare properties using different kinds of feed and feed sources on a standard basis. Thus one property can be compared with another even if they do not have exactly the same kind of feed, timing of feed, or level of production per acre.

There are several different definitions of an animal unit (AU). An animal unit is defined in this chapter as one mature cow weighing 1,000 pounds either dry or with a nursing calf up to six months of age, or the equivalent. The cow is assumed to consume a standardize amount of forage, typically 26 pounds of dry material per day. When related to forage, the animal unit is the amount of feed necessary to sustain the 1,000-pound cow or its equivalent for one year, which is 9,490 pounds of dry matter. The AU measurement provides a yardstick which can be related to other cattle or sheep. If, for example, a 1,000-pound cow equals 100% of one AU, then a 500-pound calf is equal to one-half or 50% of an AU. Each 100 pounds of live weight is equal to 0.1 AU.

Regardless of the definition that is prevalent in the area, if the appraiser calculates the animal unit measurement consistently for the subject property and comparable ranch properties, reliable comparisons can be made. Table 19.1 provides animal unit equivalents for various classes of livestock.

Table 19.2 illustrates the calculation of animal units on a combined cattle-sheep operation when the number and class of livestock are known.

If the element of time is considered, a useful relationship can be described. The animal unit month, or AUM, is the quantity of feed needed for good growth or sustenance for one animal unit for one month. (Typically 26 pounds of dry matter feed per day)

AUM is actually a formula:

$$A \times U \times M = AUM$$

in which

$A$ = the number of head of livestock
$U$ = the size of the unit in decimals
$M$ = the number of months

Using this formula, four animals that are about 500 pounds each (0.5 AU or one-half of a mature animal unit of 1.0) grazing for six months equal 12 AUMs.

$$4 \times 0.5 \times 6 = 12 \text{ AUMs}$$

Table 19.3 uses the information included in Table 19.2 and introduces the element of time in determining animal unit months. Table 19.3 indicates that this mixture of 918 animal units will require 10,064 animal unit months annually. Appraisers must understand the distinction between the different methods of calculating animal unit months. The relationship between the number of head, number of cows, and number of animal units often varies in different regions as do local terminology and methods of calculation. These differences must be understood to estimate the forage production and value of livestock ranches accurately.

**Table 19.1** Animal Unit Equivalents

| Animal | Animal Units |
|---|---|
| Mature cow | 1.00 |
| Calf, four months to weaning | 0.25 |
| Weaner calf | 0.50 |
| Yearling | 0.60 |
| Coming two-year steer (long yearling) | 0.80 |
| Horse | 1.25 |
| Bull | 1.25 |
| Sheep | 0.20 |
| Goat | 0.20 |

**Table 19.2** Animal Units

| Animal Class | Number of Head | Animal Unit Equivalent | Animal Units |
|---|---|---|---|
| Mature cows | 500 | 1 | 500 |
| Calves | 475 | 0.25 | 119 |
| Yearlings | 80 | 0.6 | 48 |
| Bulls | 29 | 1.25 | 36 |
| Horses | 12 | 1.25 | 15 |
| Sheep | 1,000 | 0.2 | 200 |
| Total | 2,096 | | 918 |

**Table 19.3** Animal Unit Months

| Animal Class | Type or Number of Head | Animal Unit Equivalent | Animal Units | Time in Months | Animal Unit Months |
|---|---|---|---|---|---|
| Mature cows | 500 | 1 | 500 | 12 | 6,000 |
| Calves | 475 | 0.25 | 119 | 4 | 476 |
| Yearlings | 80 | 0.6 | 48 | 12 | 576 |
| Bulls | 29 | 1.25 | 36 | 12 | 432 |
| Horses | 12 | 1.25 | 15 | 12 | 180 |
| Sheep | 1,000 | 0.2 | 200 | 12 | 2,400 |
| Total | 2,096 | | 918 | | 10,064 |

## Feed Requirements

As previously mentioned, the value of a livestock ranch is based on the conversion of forage into marketable livestock. Both the operator and the appraiser must have an understanding of the nutrient requirements of livestock. The appraiser does not need a degree in nutrition, but he or she should understand how feed requirements vary with the class and age of livestock and with the digestibility and utility of the growing or harvested forage supplies. Individual nutrient requirements (protein, TDN, etc.) increase significantly for growing animals compared to mature livestock. Additionally, feed requirements are much higher during periods of inclement weather, during the later stages of gestation, and throughout lactation.

The availability of feedstuffs in both summer and winter has a significant impact on the weaning weights, daily gain, and reproductive capabilities of both female and male livestock. All of these items impact the productivity and profitability of individual ranches. Highly qualified ranch appraisers have a thorough understanding of the feed requirements for livestock and know how to identify the availability of feed on a specific ranch property. They also understand how the natural elements of feed availability can be altered by management. Feed availability on a specific ranch is heavily influenced by elevation, growing season, precipitation, soils, and management. The availability and utilization of browse, protein supplements, and harvested forage (hay or silage) will dictate the species of livestock, the season of use, and ranch balance.

Table 19.4 describes the feed requirements for livestock and is based on data from the National Research Council. Complete tables are available from most land grant universities.

Table 19.4 clearly illustrates that different livestock require different amounts of feed during different times of the year and different production periods. Notice, for example, the Cows Nursing Calves—First 3 Months class. The 1100-pound cow in the first row is considered to be an average milking animal while the second 1100-pound cow is a superior milking cow.

When the appraiser is estimating animal units, animal unit months, ranch capacity, and ranch balance, the actual amount of feed that is required by each livestock class must be correlated to the available forage production. Entire textbooks have been devoted to the daily nutritional requirements of livestock and this information is readily accessible to the appraiser. Books can also be found on feed ingredients and how to balance a ration.

The livestock ranch appraiser also needs reliable information as to the forage production on various grasses and range types. The best data is obtained from actual local livestock operations. Using this data the appraiser can calculate actual forage production. Another excellent source for estimates of the forage availability for the grazing of various grasses and range types is the USDA Natural Resource Conservation Service.

**Table 19.4** Daily Nutritional Requirements (NCR)

| Species, Class | Body Weight* | Daily Gain | Min. Dry Matter | Total Protein | Digest. Protein | TDN | Vitamin A 1,000 IU |
|---|---|---|---|---|---|---|---|
| Growing steer calves/ yearlings | 440 | 1.5 | 12.6 | 1.34 | 0.86 | 7.9 | 13 |
| | | 2.4 | 10.1 | 1.39 | 0.95 | 8.6 | 13 |
| Pregnant yearling heifers—last trimester | 880 | 0.9 | 15.4 | 0.97 | 0.46 | 8.7 | 21 |
| | | 1.8 | 25.6 | 2.22 | 1.25 | 14 | 33 |
| Dry pregnant mature cows—last trimester | 880 | 0.9 | 15.4 | 0.97 | 0.046 | 8.7 | 21 |
| | 1100 | 0.9 | 17.9 | 1.12 | 0.53 | 10 | 24 |
| Cows nursing calves—first 3 months | 1100 | 21.6 | 1.98 | 1.17 | 11.7 | 24 | |
| | 1100 | 26 | 2.84 | 1.67 | 14.8 | 38 | |
| Bulls, growth and maintenance—moderate activity | 1100 | 1.5 | 26.9 | 2.35 | 1.36 | 16.5 | 48 |
| | 1980 | | 25.1 | 2.18 | 1.21 | 13.9 | 44 |
| Ewes—last 6 weeks gestation | 120 | 0.37 | 4.2 | 0.34 | 0.19 | 2.2 | 2.7 |
| Ewes—first 8-10 weeks lactation | 120 | -0.07 | 5.1 | 0.42 | 0.23 | 2.9 | 2.7 |
| Rams | 120 | 0.2 | 4.2 | 0.32 | 0.18 | 2.1 | 1.5 |
| | 160 | 0.1 | 4.8 | 0.32 | 0.18 | 2.4 | 2 |

* All figures except last column are in pounds.

Individual grass species provide a range of available forage as do various range sites. Forage availability can vary from several animal unit months per acre in high rainfall areas to less than 0.1 animal unit month per acre in desert areas. U.S. Forest Service and BLM lands and western ranges typically vary from 0.20 to 0.65 animal unit months per acre. The forage that can reasonably be harvested per acre, measured in animal unit months, is influenced by dry matter content, palatability, and management (rotational grazing, seasonal grazing, or intensive grazing practices). The availability of forage is also influenced by conservation practices and by livestock waste (less than 100% uniform grazing). Proper long-term grazing practices are a requirement in all types of livestock ranching operations.

In addition to supporting domestic livestock, ranches may support a wide range of wildlife. In many areas of the Great Plains and western United States, fishing and hunting amenities may produce income equal to or exceeding the income derived from typical livestock operations. Therefore, the fish and wildlife population and habitat on a ranching unit are of utmost importance in estimating the ability of a ranch to produce income, which in turn affects property value and marketability.

Fishing leases are available on many ranches which include mountain streams, rivers, reservoirs, and lakes as well as earthen tanks. Fishing on mountain streams and rivers is typically on a per-day or per-week basis, while reservoir, lake, and earthen tank fishing is usually granted for a season.

Hunting leases are in effect in areas with upland game birds, waterfowl, and big game animals. In the plains the primary upland game bird species hunted are bobwhite quail, blue or scaled quail, and ringneck pheasant. Quail hunting is typically on a per-acre basis, whereas pheasant hunting is typically on a per-day basis. Waterfowl hunting includes wild ducks, geese, and sandhill cranes. They are hunted in grain fields, lakes, earthen tanks, and river bottoms. Hunting leases for deer, elk, bear, and mountain lions are found in all the intermountain regions as well as many areas of Texas, Oklahoma, New Mexico, and Arizona. Big game hunting leases come in various forms. One common hunting lease calls for a lump-sum amount to cover the property to be hunted and the time to hunt. Often the leases will call for the number and size of the animals to be taken to protect the supply of wildlife. Another form of big game lease is a per-gun shared lease. The outfitter pays the landlord a set portion of the fee paid to him. The risk to the outfitter is less because if some hunters cancel, the outfitter does not have to pay the full amount.

The amount of money an outfitter is willing to pay for a hunting lease is based on the known big game population, habitat, accessibility, and time of year the lease is negotiated. If a lease can be negotiated early in the year, the outfitter will pay more because there will be up to nine months to book hunters to pay for the lease.

## Regional Data

The valuation process for livestock ranches follows the pattern described in previous chapters of this book. Generally the information to be collected first is the regional data or the local and area data. This information is important to the ranching industry and the ranch appraiser. The ranch appraiser should consider livestock markets because the distance to and accessibility of livestock markets are of major importance. A ranch that runs stocker cattle that will later be sold as feeders should be located in reasonably close proximity to major suppliers to the high plains cattle feeding industry. However, not every producer or every ranch wishes to market through the larger auctions. Therefore, it could be advantageous to be near a smaller local auction or be in an area where many order buyers and country traders are available for direct marketing.

Transportation is important to a ranching operation. The availability of trucks and, in some cases, access to the railroad should be considered on each ranch appraisal. Climate is also important. Relevant factors include the growing season, the normal high and low temperatures, average annual precipitation, and major deviations from the normal precipitation pattern.

Local customs should be considered. Some areas specialize in cow-calf operations or yearling operations. Some areas specialize in high-quality bred cattle, whereas other areas may specialize in quality crossbred cattle. Still other areas are known for handling plainer cattle. Ranches in some areas must use specialized breeds of cattle and sheep that have the necessary natural resistance to parasites, disease, and heat problems.

To obtain information for a particular community or area, appraisers can contact the Natural Resources and Conservation Service, The Farm Service Agency, The U. S. Weather Bureau, livestock auctioneers, livestock inspectors, and cattle and sheep buyers.

## Property Data

Property data generally references the subject property. Comparable sales should be analyzed with the same intensity as the subject property.

Land inventory, which is the foundation of an agricultural appraisal, is the measurement of a ranch property. A land inventory will account for all of the different types of land on a property. The land inventory segregates the different qualities of land within the property being appraised from the highest to lowest productivity, based on soils, topography, water, irrigation systems, and use. Proper use and skilled interpretation of aerial photographs are needed to formulate the land inventory.

Obtaining an accurate map of the subject property is a high priority. An accurate map can indicate forage types, topographical features and elevations, water sources, fences and roads, livestock protection, and land use. An appraiser who can read maps, aerial photographs, and topographic maps will have a good idea of what the property looks like prior to the inspection. Forage types can be determined by looking at an aerial map. For example, on a high plains ranch, hard grass areas can be differentiated from sand grass areas by the topography. From an aerial map of a mountain ranch, it is possible to differentiate between north and south slopes. On both types of ranches the tree or brush canopy can be fairly accurately estimated.

Topographical features such as elevation, high points, deep canyons, and bottom land areas can be determined. Water sources such as streams, rivers, earthen tanks, wells equipped with windmills, and irrigation ditches can be pinpointed from a map. Property boundary fences and cross fencing as well as exterior and interior roads can be plotted. Livestock protection such as canyons and timbered areas can be located. The land use may also be identified on the map as rangeland, pasture, or cropland. All these physical characteristics are critical to the ranch operation because they affect ease of operation and the available forage production.

Studying the maps prior to inspection of the property will give the appraiser an idea of what is to be appraised. Landmarks, fences, roads, and watering facilities can be noted in advance and used as references during inspection of the property from the ground or air.

Aerial photographs can be obtained from the Farm Service Agency office that serves the area where the ranch is located. These maps can be ordered through the local FSA office or they can be ordered directly from the Cartographic Unit in Salt Lake City, Utah. The address is: U.S. Department of Agriculture, Farm Service Agency, Aerial Photography Field Office, 2222 West 2300 South, P.O. Box 30010, Salt Lake City, Utah 84130-0010. The cost of these maps is nominal.

The maps obtained from the Natural Resource Conservation Service office are essentially the same as those obtained from the FSA office; however, with written permission from the landowner, detailed maps of a particular ranch can be obtained from the Natural Resource Conservation Service. Often the maps show the fence layout, the acreage in each pasture or field, roads, livestock and irrigation water supplies, and other physical characteristics of the ranching operation. County assessor's maps are valuable in that they show the perimeter of the subject property as well as the boundaries of any comparable sales.

Topography maps prepared by the U.S. Geological Survey are fairly reliable. The maps show the elevation contour of an area, wooded areas, streams, fencing, roads, and bodies of water. These maps are available from various map companies in larger cities or they can be ordered directly from the U.S. Geological Survey.

After obtaining some good maps, the ranch appraiser may have to collect a significant amount of property data on ranch improvements. Ranch improvements may influence the productivity and value of individual comparable sales and the appraised property. As in any appraisal, the size, construction, age, and condition of building improvements determine their utility. The presence or absence of residences, lambing or shearing barns, calving facilities, and goat protection can impact the value of a livestock ranch.

Improvements or developments to the real estate typically have a significant impact on the utility of livestock ranches, particularly in the western states. Perimeter and cross-fencing together with physical access impact the grazing patterns and feed utilization of ranges. The availability, reliability, and distribution of livestock water sources restrict or improve grazing patterns and therefore forage utility and ranch value. The establishment of improved grass species, irrigation development, or other management practices may increase the length of the grazing season or forage availability and improve ranch balance and value. In some geographic regions, management practices such as grass fertilization, range burning, sagebrush and willow control, and control of pests and predators may be very important and part of the standard management and ranch improvement scheme for livestock ranches.

In all ranch appraisal assignments, the appraiser should obtain as much information as possible on the property history of the comparable sales and the subject property. Investigation of the number, species, class, and age of the livestock produced on a ranch is key to the analysis and understanding of the property. However, operating history and claims by the owner may or may not be an accurate indication of future operations. Improvements to fencing and water or development of improved grasses may increase carrying capacity. The loss of water rights; an increase in weeds, predators, or pests; or reductions in BLM or forest AUMs can substantially alter productivity and value. In addition to past operators, the BLM, the U.S. Forest Service, and state lease authorities should be contacted for information. Obviously, the probable future use should be determined, including carrying capacity. Factors such as the tons of hay produced, the grazing patterns utilized, and the amounts of saleable product (beef, sheep, wool, and mohair) are typically investigated. Care should be taken to analyze the livestock ranch based on typical management, not the superior or deficient management skills of the current operator. A thorough investigation of the subject community and the specific ranch should be completed to determine the feeding season, supplements required, bulls per 100 cows, level of replacement heifers, and marketing availability.

The historic use of non-fee (leased) land can usually be determined. The historic use and probable future use of the AUMs available from BLM, forest, state leases/permits, and other leased land is important to the carrying capacity, balance, and future utility of the total ranching operation. Season and level of use are typically dictated by government leases. The right to renew or transfer the lease to a new owner is of special importance. The physical location and ease of operation of leased land relative to the base property should be analyzed for comparable sales and the subject property. In addition to reductions in the permitted AUMs, leases of U.S. Forest Service and BLM properties should be viewed in light of recent pressure for duel uses. That is, the public's demand for recreational uses such as hunting, fishing, and camping may alter the utility of a livestock ranch.

## Forage Species

Forage species, their condition, and the percentage of the plant community that the various forage species occupy are key to the productivity of a ranch. Some factors that should be considered are climax or key species, the season of growth, the season of use, the canopy or overgrowth, and invaders.

The *climax* or *key species* is the major grass species located on the range site. *Range site* is an important term and should not be considered synonymous with the overall ranch. A range site can be defined as a soil or group of soils that produce the same kind and amounts of forage if the range is in similar condition. Each major range site in all probability will have a different key species and the key species will be one of the most desirable plants found on the range site that is capable of producing a substantial portion of the forage. This is the species to be managed by the ranch operator. For example, if blue grama makes up 20% of the plant population by weight on a hard grass range site and this is the species that the land user thinks is the best grass on the range site, the blue grama will be the key species even though another grass species might constitute 50% of the present plant community.

Forage species can be categorized into two basic seasons of growth: warm season and cool season. Warm season grasses typically grow in the late spring, through the summer months, and into the fall months. Cool season grasses usually begin their growth in the fall and grow into the early part of the winter, at which time they are dormant. They begin to grow again in early spring and grow until hot weather returns and they become dormant again. A browse plant such as a live oak is neither a cool season nor a warm season forage plant. It might be green year-round, but its major growth periods are in the spring and early summer.

The season of use of a forage species depends on the ranching operation as well as the type of livestock or wildlife that are going to use the forage. Short grasses and midgrasses that are typically found on high plains hard grass ranches normally cures out very well and can be used as a standing hay crop or grazed during the growing season. The season of use would depend on the ranching operation. Conversely, tall grasses typically found on sand grass ranches do not cure out well and are not as desirable for a standing hay crop. The better season of use for tall grasses is during the growing season. Note that some grasses or forage plants may be desirable to livestock and wildlife only during certain seasons of the year. For example, a forage may be very desirable in the early spring and of little or no value during the dormant season. To analyze the seasons of use for forage species, a forage inventory should be developed by individual pastures. Then a grazing plan should be outlined based on the forage species available. This could indicate whether a ranch could be considered a year-round ranch for a cow-calf operation or whether it would best be used as a stocker operation because for two or three months of the year the desirable forage is not available. An inexperienced ranch appraiser faced with this situation should seek assistance from a range conservationist with the Natural Resources Conservation Service.

The canopy or overgrowth should also be considered by the appraiser. The canopy cover can be defined as the percentage of the soil surface that is shaded by invading brush species when the brush is in full foliage. *Overgrowth* and *canopy* are terms that can be used synonymously. The percentage of canopy cover affects the type and species of grasses and forbs that can be grown under the canopy and also the amounts of forage that can be produced.

The appraiser's land use maps and plant inventory list also include invaders.[1] Invaders are generally undesirable plants. They can compete with plants in the climax community for moisture, nutrients, and light.

1. For definitions of these terms, see Chapter 8.

Invaders come in and grow along with increasers after the climax vegetation has been reduced by close grazing or natural causes. Many are weeds and shrubs of little or no grazing value like snake weed.

With the forage inventory in hand, the rural appraiser needs to check for poisonous plants such as larkspur. This plant is easy to identify in early spring when it grows above the grasses and is easily taken by the grazing cattle. Pastures containing a high percentage of larkspur can only be used for grazing in the late fall, preferably after the first killing frost.

The property map should be correlated to the forage inventory list to help identify problems with range site deterioration from wind or water erosion, rodents, and brush encroachment. Correlating the forage inventory and property maps will also help the appraiser recognize the acceptability of the livestock distribution. Often the forage around the primary water sources will become overgrazed. The livestock must be forced to seek the readily available forage away from the water. This can be accomplished with cross fences, additional water distribution, baiting with salt or feed supplements, and herding.

Over the years, many ranches have been built up, split off, and built up again. This has improved the convenience of operation on some ranches and made other ranches very difficult to operate. A good example would be a ranch with an assured grazing privilege that is a great distance from the base property and headquarters unit. Government control can also limit the convenience of operation. In some instances, the managing government body will dictate the number of animals that can be grazed on the deeded base property as well as the public land. Similarly, some common grazing allotments and grazing associations dictate the kind and class of livestock that can be grazed.

In summary, the analysis of forage species is one of the more critical parts of a ranch appraisal. All of the factors mentioned—i.e., the key species, the seasons of growth, the season of use, the canopy cover, invaders, poisonous plants, rodents, livestock distribution, and convenience of operation—are primary factors to consider in estimating the carrying capacity for a ranch. These same factors are also important in determining the highest and best use of the ranch and of the comparable sales. It is recommended that the appraiser seek assistance from range management specialists qualified in the area of the ranch being appraised.

## Ranch Capacity

The ranch appraiser should use all the information obtained in the land inventory and forage species study to complete the ranch capacity analysis. Although carrying capacity is much less important in the valuation of a recreational ranch or one with a nonagricultural use, a study of ranch capacity should be completed for the subject property and each comparable sale.

In the analysis of carrying capacity, the appraiser estimates the amount of forage available on a ranch. Units of measure and comparison vary in different geographical areas. The following units of comparison may be used:

- Head
- Cow unit
- Animal unit
- Animal unit month
- Cow year long
- Yearling year long

These different units representing total forage or carrying capacity are not synonymous. The per head unit represents the total number of head run, which would include cows, bulls, replacements, and other livestock normally measured in a fall count. The cow unit represents the number of cows run in a normal year on a particular ranch. The cow unit measure also includes the necessary bulls, replacement heifers, and weaners for operation of that ranch. In contrast, the animal unit uses the equivalent number of mature cows as a basis of measurement related to carrying capacity. This has created a significant amount of confusion within the industry. All of the different units, if used and represented correctly, will result in the same final estimates of forage production.

For example, in certain localities 1,000 AUs, 1,200 head, and 800 mother cows could represent the same amount of forage. The per head and mother cow units can vary from area to area based on local climate conditions and grazing patterns and adjustments would be required for each. The animal unit is simply AUMs (animal unit months) annualized. That is, it measures the total forage available on a specific property divided by 12 months based on the feed requirements for a 1000-lb. mature cow. It is recommended that carrying capacity be based on the total amount of forage available, normally expressed in AUMs. The ranch balance, economy of feed, and other factors should be addressed separately.

The first step in determining carrying capacity is to estimate the AUMs available from individual feed sources and then to summarize all the sources. Table 19.5 relates to a ranch unit with both fee-owned land and assured lease land.

Managers with the U.S. Forest Service and the BLM dictate the season of use and the number of animals that can be placed on a particular grazing privilege. The only math that is required is season of use (time) multiplied by the number of head equals the number of AUMs. On BLM permits, the appraiser must consider the percentage of public land use. Sometimes portions of the deeded land are not fenced separately from the BLM land and the BLM controls the stocking rates of both

**Table 19.5** Ranch Capacity Analysis

| Type of Crop or Pasture | Acres | Season of Use | Crop Yield Per Acre | AUMs Per Acre | Total AUMs Capacity |
|---|---|---|---|---|---|
| Fee-owned land | | | | | |
| Alfalfa hay | 130.00 | Winter | 3.50 | 8.85 | 1,151 |
| Meadow hay | 415.00 | Winter | 2.00 | 5.06 | 2,099 |
| Hay aftermath* | 545.00 | Fall | 0.00 | 0.33 | 180 |
| Native pasture | 4,853.00 | 4/1 to 11/30 | 0.00 | 0.40 | 1,941 |
| Crested wheat pasture | 2,000.00 | 4/1 to 4/30 | 0.00 | 0.25 | 500 |
| Site and roads | 0.00 | | | | |
| Total | | | | | 5,871 |
| **Classification** | **Acres** | **Season of Use** | **% Public Use Land (PLU)** | **Number of Head** | **Total AUMs** |
| Assured grazing lands | | | | | |
| BLM # 73062 | 6,066.00 | 5/1 to 6/30 | 100 | 645 | 1,290 |
| BLM # 73063 | 3,034.00 | 10/1/ to 11/30 | 80 | 406 | 650 |
| National forest | 11,300.00 | 7/1 to 9/30 | 100 | 408 | 1,224 |
| Total | | | | | 3,164 |
| Summary | | | | | |
| Total animal unit months (AUMs) capacity | | | | 9,035 | |
| Percent capacity on fee-owned land | | | | 64.98% | |
| Percent capacity on assured leases and permits | | | | | 35.02% |
| Total animal units = AUMs divided by 12 months | | | | | 753 |

* Hay aftermath must be deducted from the acreage to prevent double counting for acres.

types of land. For example, consider the mathematical calculation for the BLM permits from Table 19.5.

| Percent of Public Use Land | Season of Use | Number of Head | Total AUMs |
|---|---|---|---|
| 100% | 2 months | 645 | 1,290 |
| 80% | 2 months | 406 | 650* |
| Total | | | 1,940 |

* 2 months × 406 head = 812 AUMs × 80%.

Table 19.5 indicates that the sample ranch property can produce 9,035 animal unit months of available forage. It is most important that the reader understands how the number of AUMs was determined from each type of forage. Consider the alfalfa hay as an example. The estimated AUMs per acre is listed as 8.85. The 8.85 figure recognizes that each animal unit will eat (by definition) 26 pounds of hay per day on average throughout the feeding season. That is, 26 pounds × 365 days = 9,490 pounds per year ÷ 12 = 790.83 pounds per month. Livestock feeding is every day and does not consider an average month of 30 days; however, the math in the following chart has been simplified using an average month.

26.36 lbs. × 30 days (one month) = 790.83 pounds per month
3.5 tons × 2000 pounds per ton = 7,000 pounds per acre
7,000 pounds/790.83 = 8.85 AUMs per acre

The ranch appraiser must be very familiar with the length of the feeding season and the typical amount of supplemental feed such as hay and protein supplements that is required in the area where the subject property and comparable sales are located. Mathematically, it is easy to support the AUMs per acre on measurable feed such as hay. It is slightly more difficult to support the AUMs per acre on native or seeded pasture tracts. This information can be obtained from Natural Resource Conservation Service range scientists and from county soil books. When using the soil books, the total amount of dry matter produced is recorded. Adjustments are required for nonpalatable forage, range conservation, and waste. Local range professionals generally use rules of thumb for forage adjustments that they can scientifically support.

A scientific, detailed, mathematical ranch capacity analysis is a logical and supportable method for measuring the forage production of ranch property. This system of forage measurement has checks and balances. The projected forage production or the ranch capacity analysis should correlate with the actual forage production that is obtained from local, knowledgeable, unbiased ranch operators who provide "typical" management. The historical capacity of the subject property should be reviewed, but the appraiser should not let past operations bias the ranch capacity analysis. If there is a sizable difference between the historical capacity and the more scientific capacity analysis, the difference should be addressed by the appraiser.

The information typically provided by local operators is not the actual AUMs per acre, but the actual number of cattle that were allowed to graze a certain amount of pasture for a specific amount of time. Often this is referred to *acres per cow* or *acres per animal unit.* This information can be easily converted into AUMs per acre. For example, in the ranch capacity chart the native pasture was estimated at 0.40 AUMs per acre. In other words, 30 acres of pasture is required per animal unit.

One month/0.40 AUMs per acre = 2.5 acres per animal unit month
2.5 acres per AUM × 12 months = 30 acres per animal unit

## Ranch Balance

The culmination of the appraiser's analysis of the physical characteristics of a livestock ranch is the determination of ranch balance. The studies of forage types, level of productivity, lease utilization, and wintering requirements and the mathematical calculation of carrying capacity are precursors to the determination of ranch balance. The appraiser studies ranch balance to determine the carrying capacity on a year-round basis. The ranch may be correctly balanced for seasonal feed requirements, or it may be deficient or possess excess amounts of individual feed items. Ranch balance is important to the income-producing ability of a ranch property and to the application of the sales comparison approach on a unit basis.

Since the timing of feedstuff availability is critical to the feeding of livestock, an analysis of carrying capacity should be presented in a spreadsheet or graph. This format shows the available feed supply on a month-by-month basis. Table 19.6 indicates the available feed supply by month, broken down by feed source. The spreadsheet was developed using the AUM information obtained from the previous ranch capacity analysis.

**Table 19.6** Ranch Balance Analysis

| Month | BLM Pasture | BLM Pasture | National Forest | Native Pasture | Crested Wheat | Hay Aftermath Pasture | Meadow Hay | Total AUMs | Diff. AUMs |
|---|---|---|---|---|---|---|---|---|---|
| January | – | – | – | – | – | – | 750.00 | 750.00 | – |
| February | – | – | – | – | – | – | 750.00 | 750.00 | – |
| March | – | – | – | – | – | – | 750.00 | 750.00 | – |
| April | – | – | – | 250.00 | 500.00 | – | – | 750.00 | – |
| May | 645.00 | – | – | 105.00 | – | – | – | 750.00 | – |
| June | 645.00 | – | – | 105.00 | – | – | – | 750.00 | – |
| July | – | – | 408.00 | 342.00 | – | – | – | 750.00 | – |
| August | – | – | 408.00 | 342.00 | – | – | – | 750.00 | – |
| September | – | – | 408.00 | 342.00 | – | – | – | 750.00 | – |
| October | – | 406.00 | – | 222.00 | – | 122.00 | – | 750.00 | – |
| November | – | 406.00 | – | 71.00 | – | 58.00 | 250.00 | 785.00 | 35.00 |
| December | – | – | – | – | – | – | 750.00 | 750.00 | – |
| Total | 1,290.00 | 812.00 | 1,224.00 | 1,779.00 | 500.00 | 180.00 | 3,250.00 | 9,035.00 | 35.00 |
| Summer pasture | | | | | | | | | – |
| Excess fall pasture | | | | | | | | | – |
| Excess hay AUMs | | | | | | | | | 42.00 |
| Excess hay in tons | | | | | | | | | 16.61 |

The ranch balance table indicates that the subject ranch is well-balanced at 750 animal units. To support this conclusion, the appraiser must be confident that the subject ranch possesses the necessary physical characteristics to be operated as projected. A good point for further explanation would be the lack of fencing separating the hay ground from the pasture. If the hay ground was not fenced separately, then the pasture could not be used until after the crop-growing season, which could influence annual ranch balance.

The total 1,779 AUMs of native pasture in this analysis is different from the 1,941 found in the ranch capacity analysis (Table 19.5). Inclusion of deeded land within lands controlled by the BLM creates this apparent discrepancy. The difference of 162 AUMs is actually listed in the total 812 BLM capacity. Because deeded and BLM lands are fenced into the same pasture, the appraiser must be careful to avoid a double counting of the capacity.

The ranch balance table illustrates how land use, type of forage, and season of use can be correlated to determine ranch balance. Appraisal assignments are not textbook cases, however, and the ranch appraiser will encounter different scenarios related to ranch balance. Although a ranch may be out of balance, the appraiser is still required to analyze and determine carrying capacity. For example, consider a slight change in the land use described in Table 19.5, the previous ranch capacity analysis. In Table 19.7 it is assumed that only 100 acres of native pasture is converted to alfalfa hay.

**Table 19.7** Ranch Capacity Analysis

| Type of Crop or Pasture | Acres | Season of Use | Crop Yield Per Acre | AUMs Per Acre | Total AUMs Capacity |
|---|---|---|---|---|---|
| Fee-owned land | | | | | |
| Alfalfa hay | 230.00 | Winter | 3.50 | 8.85 | 2,036 |
| Meadow hay | 415.00 | Winter | 2.00 | 5.06 | 2,099 |
| Hay aftermath* | 645.00 | Fall | 0.00 | 0.33 | 213 |
| Native pasture | 4,753.00 | 4/1 to 11/30 | 0.00 | 0.40 | 1,901 |
| Crested wheat pasture | 2,000.00 | 4/1 to 4/30 | 0.00 | 0.25 | 500 |
| Site and roads | 0.00 | | | | |
| Total | | | | | 6,749 |
| **Classification** | **Acres** | **Season of Use** | **% Public Use Land (PLU)** | **Number of Head** | **Total AUMs** |
| Assured grazing lands | | | | | |
| BLM # 73062 | 6,066.00 | 5/1 to 6/30 | 100 | 645 | 1,290 |
| BLM # 73063 | 3,034.00 | 10/1 to 11/30 | 80 | 406 | 650 |
| National forest | 11,300.00 | 7/1 to 9/30 | 100 | 408 | 1,224 |
| Total | | | | | 3,164 |
| Summary | | | | | |
| Total animal unit months (AUMs) capacity | | | | | 9,913 |
| Percent capacity on fee-owned land | | | | | 68.08% |
| Percent capacity on assured leases and permits | | | | | 31.92% |
| Total animal units = AUMs divided by 12 months | | | | | 826 |

* Hay aftermath must be deducted from the acreage to prevent double counting for acres.

Considering the slight acreage change, the total available forage production changes from 750 to 826 animal units. Because of the first example we know that the 826 animal units are not balanced. Thus the ranch balance analysis must also be changed.

Based on this ranch balance analysis, now the ranch has 913 animal unit months of excess hay in the month of April. It is most logical to consider the ranch a balanced ranch at 750 animal units with excess hay base. The appraiser must determine from the marketplace if the excess hay base (per animal unit or acre) contributes the same proportional value as the units required in the ranch balance. Generally, excess hay base will sell for the same amount because of the market demand for hay, and additional income may be generated. In the above example, the 913 animal unit months of excess hay is equivalent to 361.00 tons of alfalfa hay.

913 × 26.36 pounds per day × 30 days (one month) =
722,000.04 pounds of hay/ 2,000 pounds per ton = 361.00 tons

In completing adjustments for the different degrees of ranch balance, the appraiser must recognize that the different types of land do not

**Table 19.8** Ranch Balance Analysis

| Month | BLM Pasture | BLM Pasture | National Forest | Native Pasture | Crested Wheat | Hay Aftermath Pasture | Meadow Hay | Total AUMs | Diff. AUMs |
|---|---|---|---|---|---|---|---|---|---|
| January | – | – | – | – | – | – | 750.00 | 750.00 | – |
| February | – | – | – | – | – | – | 750.00 | 750.00 | – |
| March | – | – | – | – | – | – | 750.00 | 750.00 | – |
| April | – | – | – | 250.00 | 500.00 | – | 913.00 | 1,663.00 | 913.00 |
| May | 645.00 | – | – | 105.00 | – | – | – | 750.00 | – |
| June | 645.00 | – | – | 105.00 | – | – | – | 750.00 | – |
| July | – | – | 408.00 | 342.00 | – | – | – | 750.00 | – |
| August | – | – | 408.00 | 342.00 | – | – | – | 750.00 | – |
| September | – | – | 408.00 | 342.00 | – | – | – | 750.00 | – |
| October | – | 406.00 | – | 182.00 | – | 162.00 | – | 750.00 | – |
| November | – | 406.00 | – | 71.00 | – | 51.00 | 222.00 | 750.00 | – |
| December | – | – | – | – | – | – | 750.00 | 750.00 | – |
| Total | 1,290.00 | 812.00 | 1,224.00 | 1,739.00 | 500.00 | 213.00 | 4,135.00 | 9,913.00 | 913.00 |
| Summer pasture | | | | | | | | | – |
| Excess fall pasture | | | | | | | | | – |
| Excess hay AUMs | | | | | | | | | 913.00 |
| Excess hay in tons | | | | | | | | | 361.02 |

contribute the same value as they do in forage production. For example, consider that rangeland that produces 0.40 AUMs per acre sells for $125 per acre. An animal unit of rangeland of this productivity will require 30 acres or $3,750 per animal unit (30 × $125). Alfalfa hay land producing 8.85 AUMs per acre sells for $1,500 per acre, equal to $169.49 per AUM, or $2,033.88 per animal unit.

When a ranch is out of balance, a complete study of alternative uses should be conducted. For example, if a ranch is short on winter feed, is hay or supplemental pasture readily available? If hay is not available in the marketplace, are term cattle or rental cattle available to use the extra supply of summer or winter forage? An obvious conclusion is that the highest and best use of a ranch property is directly related to ranch balance.

## Cow Unit

The preceding sections of this chapter have discussed carrying capacity (analysis of available forage) and ranch balance (timing of available forage). The analysis of a property as a 750-animal unit, balanced ranch does not indicate 750 producing cows. Replacement heifers and breeding bulls must be included in the analysis of the carrying capacity of a ranch operation on a cow/calf basis.

A producing cow unit is defined as the carrying capacity necessary to sustain a mature cow for a year and her nursing calf for six months, together with the necessary percentage of replacement heifers and breeding bulls. Typical death loss, culling rate, normal pregnancy rates, and required cow-to-bull ratios must be considered. These factors all vary with the geographical area and the style of the ranch operation.

For example, consider the following factors:

- Cow death loss and culling rate 16 %
- Replacement heifer retention rate 80 %
- Cow-to-bull ratio in herd count 20:1

Because of a combination of death loss, culling for physical problems, and infertility, 16% of the mature cow herd must be replaced each year. Death loss and infertility factors also effect replacement heifers and the number being retained so that only 80% of those selected ever make it into the breeding herd. Therefore, 16 replacements divided by 80% retention indicates that 20 heifers must be exposed during the breeding period to be able to replace 16 mother cows. Many factors influence the pregnancy rate; however, a generally accepted management practice is to provide five bulls for each 100 cows, plus one bull for each 20 heifers. These numbers could represent bulls actually placed in the breeding pasture or could account for a percentage held in reserve.

Measured at the time of pregnancy testing and culling of the mature cow herd, the following calculations result in the required animal unit carrying capacity:

| Animal Type | AUs Per Head | Total AUs |
|---|---|---|
| 100 Calving cows (as defined) | 1 | 100 |
| 20 Replacement heifers (as calculated) | 0.85* | 17 |
| 6 Breeding bulls (as defined) | 1.25 | 7.5 |
| 126 Head count | | 124.5 |

* Heifers are equivalent to 0.7 AU at weaning time and 1.0 AU at time of pregnancy test, but they require extra feed for development, with some culled at an early date—summarized to 0.85 AU.

These calculations indicate a required 1.25 AU carrying capacity for every mature cow to calve in the herd. Thus, a 100-cow unit ranch requires 125 AUs of feed availability.

The assumptions employed in determining the number of head of mature cows, replacement heifers, and bulls should be tailored to the appropriate ranching area. There are at least five assumptions that must be considered.

1. Cow culling rate
2. Heifer replacement rate
3. Cow-to-bull ratio
4. Animal unit classification
5. Death loss

## Valuation Procedures

All three approaches to value are applicable to the valuation of livestock ranches. Livestock ranches are inherently land-intensive, with building improvements comprising a nominal portion of the overall value. Units of comparison for the appraisal of livestock ranches vary with locality. The animal unit (AU) is normally accepted and understood everywhere. The animal unit can be refined to reflect local units of comparison such as cow units, head, and yearlings yearlong. Various other marketing units can be used for appraisal analysis, including price per acre for deeded land and price per AUM for government leases.

### Cost Approach

The cost approach can be applied to a ranch in the same way it is applied to any other type of agricultural property. These basic procedures are covered in Chapter 13. Ranch real estate improvements, such as livestock fences, irrigation systems, corrals, and stock water facilities, are found on nearly all of the sales and will be included in the unit price. Care should be exercised, however, since specialized irrigation systems such as center pivots or wheel lines contribute a significant portion of the irrigated land value. These improvements are not found on most sales and can vary considerably, so they need to be addressed separately in the cost approach.

The cost approach must also address the building improvements. The utility of a ranch improvement provides significant contributory value. Ranch properties often include a wide assortment of buildings that appear to have some remaining physical life, but are no longer functional for a modern ranch operation. The appraiser needs to understand the purpose and utility of each building and how the market is recognizing its contribution to value. The presence or absence of a specific building may or may not influence the market value of the subject.

The theoretical base for the cost approach is reproduction cost, but replacement cost may also be used to determine the new cost of a structure or building. In estimating new costs, appraisers must ensure that their cost data sources are thorough and reliable. It is important that good field techniques are employed. All the buildings that have functional utility should be measured and detailed notes should be taken so that the new cost of each structure or building can be estimated. The base used in most cost service guides is direct construction costs, which includes the cost of material and labor as well as the contractor's profit required to construct the improvement on the effective appraisal date.

Several methods can be applied to estimate depreciation: the economic age-life method, the modified economic age-life method, the breakdown method, cost-to-cure estimates, sales comparison techniques, and income capitalization techniques. These procedures are discussed in detail in Chapter 13.

Land value must always be considered in terms of the highest and best use as though vacant. Many market analysis techniques, including direct sales comparison, paired sales, bracketing, and statistical analysis, can be used to estimate the value of land in the cost approach to value. The available market evidence will determine which technique is most applicable.

Applying the cost approach to the real estate portion of a ranch will typically include developing incremental values by land type (e.g., irrigated meadow, dry rangeland, subirrigated pasture, sand grass, hard grass) and by lease type (e.g., state AUM lease, BLM AUM lease, U.S. Forest Service AUM lease). When completing a cost approach appraisal of a property with various components, it is critical to support that the sum of the values of the parts is equal to the value of the whole.

The tables on the following pages illustrate the types of analysis undertaken in preparation for the cost approach.

### Land Valuation

A land value opinion is developed in conventional fashion using unimproved ranch sales. Both the fee-owned land and the leased acreage in the comparable sales must be allocated by land type (e.g., irrigated crop and dry cropland, permanent or improved pasture, native range). Table 19.9 identifies six sale properties to be used in the land valuation process. Like the subject property, all the sales are of native rangeland, so no additional allocation between the land classes is required.

**Table 19.9** Unimproved Sales Analysis

| Record No. | 4405 | 4404 | 4410 | 14356 | 14355 | 14357 |
|---|---|---|---|---|---|---|
| Sale no. | Sale U-1 | Sale U-2 | Sale U-3 | Sale U-4 | Sale U-5 | Sale U-6 |
| Grantee | Haley | Cromera | Consolidated | Alpha Cattle | Gordon | Ellison |
| Sale date | 11 Months | 11 Months | 6 Months | Current | Current | 11 Months |
| Size | 11,726.49 | 5,760.00 | 5,157.99 | 10,431.30 | 18,910.00 | 7,431.00 |
| Sale price | $1,289,914 | $460,000 | $464,150 | $885,000 | $1,850,344 | $800,000 |
| Acre price | $110.00 | $79.86 | $89.99 | $84.84 | $97.85 | $107.65 |
| AU price | $3,192.86 | $3,194.44 | $3,157.48 | $3,105.26 | $2,937.05 | $2,476.78 |
| AUs | 404 | 144 | 147 | 285 | 630 | 323 |
| Acres/AU | 29.0 | 40.0 | 35.0 | 36.6 | 30.0 | 23.0 |
| Land type | Panhandle | R Plains | R Plains | R Plains | Panhandle | R Plains |
| Improvements | None | None | None | None | None | None |
| Oil field activity | None | None | None | None | Moderate | Extensive |

Once the sale properties have been identified, researched, and inspected, they must be adjusted for time and financing. The time adjustments are extracted in Table 19.10. All the sales in this data set were cash or cash equivalent and required no adjustment for financing. Table 19.10 demonstrates the pairing of sales, which is used to extract the time adjustment indicated in the subject market area.

**Table 19.10** Time Adjustment Analysis

| Sale | U-1 | U-3 |
|---|---|---|
| Sale U-1 vs. U-3 | | |
| Date | 11 Months | 6 Months |
| Price per AU | $3,192.86 | $3,157.48 |
| Compound change/month | -0.22% | |
| **Sale** | **U-1** | **U-4** |
| Sale U-1 vs. U-4 | | |
| Date | 11 Months | Current |
| Price per AU | $3,192.86 | $3,105.26 |
| Compound change/month | -0.25% | |
| **Sale** | **U-2** | **U-3** |
| Sales U-2 vs. U-3 | | |
| Date | 11 Months | 6 Months |
| Price per AU | $3,194.44 | $3,157.48 |
| Compound change/month | -0.23% | |
| **Sale** | **U-2** | **U-4** |
| Sale U-2 vs. U-4 | | |
| Date | 11 Months | Current |
| Price per AU | $3,194.44 | $3,105.26 |
| Compound change/month | -0.26% | |
| **Sale** | **U-3** | **U-4** |
| Sale U-3 vs. U-4 | | |
| Date | 6 Months | Current |
| Price per AU | $3,157.48 | $3,105.26 |
| Compound change/month | -0.28% | |
| Conclude | -0.25% per month | |

Once the time and financing adjustments are estimated, the land valuation process can proceed as shown in Table 19.11. Please note that the subject is all native rangeland as are the sales. Therefore, the carrying capacities of the various range sites are used to allocate the subject component values. When multiple land classes are present, the valuation process should be patterned after the processes described in the land valuation portion of the cost approach chapter.

Table 19.12 shows the allocation of the value conclusion to the subject property.

## Building Residual and Depreciation Analysis

The next step in the cost approach is to analyze the improved sales to extract the depreciation rates prevailing in the market area. Analysis of the improved sales must include building residual analysis to estimate how much the improvements contribute to the value of each improved sale. Table 19.13 illustrates the analysis of the improved sales to allocate the value to the land and improvements. Each sale is allocated based on the most comparable unimproved sales, which have been adjusted for time and financing to the improved sale date. The resulting indicated

**Table 19.11** Time and Financing Adjustments to Unimproved Sales

| Record No. | 4405 | 4404 | 4410 | 14356 | 14355 | 14357 |
|---|---|---|---|---|---|---|
| Sale No. | Sale U-1 | Sale U-2 | Sale U-3 | Sale U-4 | Sale U-5 | Sale U-6 |
| Grantee | Haley | Cromera | Consolidated | Alpha Cattle | Gordon | Ellison |
| Sale date | 11 Months | 11 Months | 6 Months | Current | Current | 11 Months |
| Size | 11,726.49 | 5,760.00 | 5,157.99 | 10,431.30 | 18,910.00 | 7,431.00 |
| Sale price | $1,289,914 | $460,000 | $464,150 | $885,000 | $1,850,344 | $800,000 |
| Acre price | $110.00 | $79.86 | $89.99 | $84.84 | $97.85 | $107.65 |
| AU price | $3,192.86 | $3,194.44 | $3,157.48 | $3,105.26 | $2,937.05 | $2,476.78 |
| AUs | 404 | 144 | 147 | 285 | 630 | 323 |
| Acres/AU | 29.0 | 40.0 | 35.0 | 36.6 | 30.0 | 23.0 |
| Land type | Panhandle | R Plains | R Plains | R Plains | Panhandle | R Plains |
| Improvements | None | None | None | None | None | None |
| Oilfield activity | None | None | None | None | Moderate | Extensive |
| Financing | Cash | Cash | Cash | Cash | Cash | Cash |
| Sale price/AU | $3,192.86 | $3,194.44 | $3,157.48 | $3,105.26 | $2,937.05 | $2,476.78 |
| **Adjustments** | | | | | | |
| Financing | 0 | 0 | 0 | 0 | 0 | 0 |
| Price adj. for financing | $3,192.86 | $3,194.44 | $3,157.48 | $3,105.26 | $2,937.05 | $2,476.78 |
| Time adjustment | | | | | | |
| Months | 11 | 11 | 6 | 0 | 0 | 11 |
| % adjustment | -2.75% | -2.75% | -1.50% | 0.00% | 0.00% | -2.75% |
| $ adjustment | -$88 | -$88 | -$47 | $0 | $0 | -$68 |
| Price adj. for time and financing | $3,105 | $3,106 | $3,110 | $3,105 | $2,936 | $2,408 |
| Conclude $3,100 per animal unit | | | | | | |

**Table 19.12** Subject Land Valuation

| Range Site | Acres/AU | Acres | Imputed Price Per Acre | Value |
|---|---|---|---|---|
| Clay loam | 30 | 9,440 | $103.33 | $975,435 |
| Very shallow | 40 | 8,110 | $77.50 | $628,525 |
| Shallow clay | 37 | 7,600 | $83.78 | $636,728 |
| Sandy land | 40 | 2,780 | $77.50 | $215,450 |
| Gypland | 50 | 1,908 | $62.00 | $118,296 |
| Loamy bottomland | 20 | 441 | $155.00 | $68,355 |
| Total | | 30,279 | | $2,642,789 |
| Estimated subject land value via cost approach | | | | $87.28/acre |

land contribution for each sale is then deducted from the sale price resulting in the indicated value that the improvement contributes to the improved sale.

After the total building contribution of the sale is known, the rate of depreciation can be estimated on an overall basis or improvement by improvement. Table 19.14 details the process of estimating the annual depreciation based on the allocated building values developed by sales comparison.

## Table 19.13 Improvement Residual Analysis

**Improved Sale I-1 15 Months** — **Ranch Type: Panhandle**

| Improved Sale Price | Improved Sale Land | Improved Sale Acres | Improved Sale AUs | Unimproved Sale U-1 Contribution Acre or AU | Improved Sale Land Contribution | Improved Sale Improvement Contribution |
|---|---|---|---|---|---|---|
| | Range land | 33,000 | 1,320 | $3,222 | $4,253,040 | |
| | Imp. range | 0 | 0 | 0 | 0 | |
| | Cropland | 0 | 0 | 0 | 0 | |
| | Leased land | 0 | 0 | 0 | 0 | |
| $4,300,000 | Total | 33,000 | | | $4,253,040 | $46,960 |

**Time Adjustment of Unimproved Sale to Improved Sale's Date of Sale**

| | | | |
|---|---|---|---|
| Sale U-1 | 11 months | Time adjustment | |
| Sale I-1 | 15 months | -0.25% per month | |
| Difference | -4 months | Sale U-1 adjusted price | $3,222 |

**Improved Sale I-2 11 Months** — **Ranch Type: Rolling Plains**

| Improved Sale Price | Improved Sale Land | Improved Sale Acres | Improved Sale AUs | Unimproved Sale U-6 Contribution Acre or AU | Improved Sale Land Contribution | Improved Sale Improvement Contribution |
|---|---|---|---|---|---|---|
| | Range land | 8,548 | 389 | $2,476 | $963,164 | |
| | Imp. range | 0 | 0 | 0 | 0 | |
| | Cropland | 0 | 0 | 0 | 0 | |
| | Leased land | 0 | 0 | 0 | 0 | |
| $1,025,750 | Total | 8,548 | | | $963,164 | $62,586 |

**Time Adjustment of Unimproved Sale to Improved Sale's Date of Sale**

| | | | |
|---|---|---|---|
| Sale U-6 | 11 months | Time adjustment | |
| Sale I-2 | 11 months | -0.25% per month | |
| Difference | 0 months | Sale U-6 adjusted price | $2,476 |

**Improved Sale I-3 6 Months** — **Ranch Type: Panhandle**

| Improved Sale Price | Improved Sale Land | Improved Sale Acres | Improved Sale AUs | Unimproved Sale U-1 Contribution Acre or AU | Improved Sale Land Contribution | Improved Sale Improvement Contribution |
|---|---|---|---|---|---|---|
| | Range land | 43,585 | 1,362 | $3,150 | $4,290,300 | |
| | Imp. range | 0 | 0 | 0 | 0 | |
| | Cropland | 0 | 0 | 0 | 0 | |
| | Leased land | 0 | 0 | 0 | 0 | |
| $4,532,077 | Total | 43,585 | | | $4,290,300 | $241,777 |

**Time Adjustment of Unimproved Sale to Improved Sale's Date of Sale**

| | | | |
|---|---|---|---|
| Sale U-1 | 11 months | Time adjustment | |
| Sale I-3 | 6 months | -0.25% per month | |
| Difference | 5 months | Sale U-1 adjusted price | $3,150 |

**Improved Sale I-4 2 Months** — **Ranch Type: Rolling Plains**

| Improved Sale Price | Improved Sale Land | Improved Sale Acres | Improved Sale AUs | Unimproved Sale U-4 Contribution Acre or AU | Improved Sale Land Contribution | Improved Sale Improvement Contribution |
|---|---|---|---|---|---|---|
| | Range land | 8,916 | 637 | $3,121 | $1,988,077 | |
| | Imp. range | 0 | 0 | 0 | 0 | |
| | Cropland | 0 | 0 | 0 | 0 | |
| | Leased land | 0 | 0 | 0 | 0 | |
| $2,100,000 | Total | 8,916 | | | $1,988,077 | $111,923 |

**Time Adjustment of Unimproved Sale to Improved Sale's Date of Sale**

| | | | |
|---|---|---|---|
| Sale U-4 | 0 months | Time adjustment | |
| Sale I-4 | 2 months | -0.25% per month | |
| Difference | -2 months | Sale U-4 adjusted price | $3,121 |

Note that when adjusting the price of the land sale to the improved sale for time produces a negative adjustment, the adjusted price will be higher in a declining market.

**Table 19.14** Market Depreciation Extraction Analysis

| | Sale I-1 | Sale I-2 | Sale I-3 | Sale I-4 |
|---|---|---|---|---|
| Sale price | $4,300,000 | $1,025,750 | $4,532,077 | $2,100,000 |
| Land contribution | $4,253,040 | $963,164 | $4,290,300 | $1,988,077 |
| Building contribution | $46,960 | $62,586 | $241,777 | $111,923 |
| RC dwelling | $49,000 | | $298,800 | $135,000 |
| Depreciation | $28,000 | | $123,743 | $60,000 |
| Dwelling contribution | $21,000 | | $175,057 | $75,000 |
| % Depreciated | 57.1% | | 41.4% | 44.4% |
| Effective age | 25 | | 25 | 20 |
| % Dep./year | 2.3% | | 1.7% | 2.2% |
| RC tenant dwelling | | $52,500 | $103,600 | $52,500 |
| Depreciation | | $22,500 | $59,200 | $29,250 |
| Dwelling contribution | | $30,000 | $44,400 | $23,250 |
| % Depreciated | | 42.9% | 57.1% | 55.7% |
| Effective age | | 20 | 25 | 25 |
| % Dep./year | | 2.1% | 2.3% | 2.2% |
| RC bunk/cook | $64,000 | | | |
| Depreciation | $47,360 | | | |
| Bunk/cook contribution | $16,640 | | | |
| % Depreciated | 74.0% | | | |
| Effective age | 20 | | | |
| % Dep./year | 3.7% | | | |
| RC barn(s) | $30,000 | $136,000 | $72,000 | $48,000 |
| Depreciation | $20,680 | $103,414 | $50,400 | $34,327 |
| Barn(s) contribution | $9,320 | $32,586 | $21,600 | $13,673 |
| % Depreciated | 68.9% | 76.0% | 70.0% | 71.5% |
| Effective age | 20 | 25 | 20 | 25 |
| % Dep./year | 3.4% | 3.0% | 3.5% | 2.9% |
| RC shed | | | $2,700 | |
| Depreciation | | | $1,980 | |
| Shed contribution | | | $720 | |
| % Depreciated | | | 73.3% | |
| Effective age | | | 20 | |
| % Dep./year | | | 3.7% | |

The depreciation analysis shown in Table 19.14 produces overall annual depreciation rates for each improvement class. Based on these indications, an appropriate rate can be selected for application to the subject improvements. In this analysis depreciation has been treated on an overall basis. A complete breakdown allocating the various forms of depreciation to each class of improvement was not conducted. If market properties demonstrates significant functional or external obsolescence, the breakdown method as described in Chapter 13 is preferred. Whether an overall depreciation estimate or the breakdown method is used, the results of this analysis are ultimately applied to the subject improvements based on the appraiser's estimate of the effective age of the subject improvements. Table 19.15 shows the application of the estimated rates

**Table 19.15** Subject Cost Approach Estimate

| | Units | Unit Cost | Amount |
|---|---|---|---|
| RC—owner's home & garage | 2,136 | $75.50 | $161,268 |
| Less depreciation | | | |
| % Depreciation | 22.00% | | |
| Depreciation amount | | | – $35,479 |
| Improvement contribution | | | **$125,789** |
| RC—foreman's home & garage | 2,016 | $42.50 | $85,680 |
| Less depreciation | | | |
| % Depreciation | 26.40% | | |
| Depreciation amount | | | – $22,620 |
| Improvement contribution | | | **$63,060** |
| RC—cook shack | 1,216 | $32.00 | $38,912 |
| Less depreciation | | | |
| % Depreciation | 55.50% | | |
| Depreciation amount | | | – $21,596 |
| Improvement contribution | | | **$17,316** |
| RC—hunting lodge (mobile home) | 1,600 | $26.20 | $41,920 |
| Less depreciation | | | |
| % Depreciation | 40.00% | | |
| Depreciation amount | | | – $16,768 |
| Improvement contribution | | | **$25,152** |
| RC—utility barn | 3,600 | $11.50 | $41,400 |
| Less depreciation | | | |
| % Depreciation | 60.00% | | |
| Depreciation amount | | | – $24,840 |
| Improvement contribution | | | **$16,560** |
| RC—corrugated metal barn | 2,400 | $7.50 | $18,000 |
| Less depreciation | | | |
| % Depreciation | 52.50% | | |
| Depreciation amount | | | – $9,450 |
| Improvement contribution | | | **$8,550** |
| RC—hay barn | 3,600 | $4.50 | $16,200 |
| Less depreciation | | | |
| % Depreciation | 14.80% | | |
| Depreciation amount | | | – $2,398 |
| Improvement contribution | | | **$13,802** |
| RC—3-bay overhead cake bin | 60 | $160.00 | $9,600 |
| Less depreciation | | | |
| % Depreciation | 16.00% | | |
| Depreciation amount | | | – $1,536 |
| Improvement contribution | | | **$8,064** |
| Total improvement contribution | | | **$278,293** |
| Plus land value via unimproved sales | | | |
| 30,279.26 acres @ $87.30 per acre = | | | $2,643,379 |
| Estimated subject value via cost approach | | | $2,921,673 |
| Rounded to | | | $2,922,000 |

of depreciation to the subject property's improvements, resulting in the estimated value of the improvements. Land value is then added to arrive at the estimated value of the subject property via the cost approach.

### Sales Comparison Approach

Nearly all of the market analysis techniques covered in Chapter 14, including direct comparison, paired sales, and statistical analysis, can be used to estimate the value of a ranch using the sales comparison approach. The available market evidence will determine which technique is most applicable and which unit of comparison is most appropriate. Possible units of comparison for this approach include animal unit, cow unit, yearling unit, and acres.

Regardless of the technique is used, if the properties are compared based on animal units, the appraiser should note that the value per AU very often decreases as the percentage capacity on assured leases increases. The value per AU may also decrease as the percentage capacity on irrigated or cultivated land increases. The appraiser must fully understand the local market and complete all analyses on a market-proven basis. Adjustments may be required for the physical characteristics listed below. The list is not all-inclusive, but many of these factors are considered in most ranch appraisals.

- Land use, which could include assured leases
- Water, both livestock and irrigation
- Forage productivity
- Balance and convenience of operation
- Real estate improvements
- Building improvements
- Size
- Percentage of leased land
- Location/access
- Special amenities/recreation

It is important to develop a list of physical characteristics that are recognized by the buyers of ranch properties in the area of the property being appraised. As in all appraisals, the need for adjustments reflects the actual market.

In applying the sales comparison approach to ranch properties, the appraiser follows the standard methodology described in the Chapter 14. The Sales Comparison Approach. The specific adaptation of the methodology to ranch properties is demonstrated in Tables 19.16 through 19.19. As previously discussed, certain nuances must be considered in appraising hay-based ranches versus Great Plains ranches. Hay-based ranches derive a portion of their productive capability from the growing of hay in a cultured or semi-cultured environment and harvesting and storing the production to use for feeding later in the year; a typical Great Plains operation uses the livestock to harvest the forage produced on the property.

When analyzing hay-based ranches and using animal units or animal unit months as the unit of comparison, the carrying capacity of the properties must be converted to a consistent unit of comparison. This may be accomplished by converting the carrying capacity of the sales and subject to "grass equivalent" animal units or animal unit months. This technique recognizes that hay-based ranches usually consist of rangeland, irrigated crop/hay, and irrigated pasture. Irrigated lands are more management-intensive than rangeland and the costs associated with harvesting the feed, which include labor and machinery, must be considered in the analysis. To equalize the data, conversion is appropriate. The basis for conversion is the typical cropshare arrangement, in which the owner provides the land and another individual provides the labor and machinery for irrigation and harvesting. Depending on the area, the share split may vary slightly, but it is generally close to a 50-50 split. To recognize the contribution of the hay/crop land while keeping the unit of comparison constant, the number of animal units from the hay base is reduced by 50% in the case of a 50-50 split on both the sales and the subject property. The corresponding "grass equivalent" animal units, therefore, account for any difference in capacity derived from the hay-based lands.

As in any valuation, the unit of comparison selected for the property must be the unit recognized by the market in the subject area. In most ranching areas the preferred unit is the animal unit or animal unit month. However, in areas where the market is driven by nonagricultural uses such as recreational use, commercial hunting, fishing, or other uses, a per-acre unit of comparison may be indicated by the market. Table 19.16 summarizes the sales to be analyzed and adjusted in the sales comparison approach.

The improved sales must be analyzed to allocate the improvements' contribution via residual analysis properly, as described in Chapter 14.

**Table 19.16** Improved Sales

| **Record no.** | **4411** | **12809** | **4407** | **12808** |
|---|---|---|---|---|
| Sales no. | Sale I-1 | Sale I-2 | Sale I-3 | Sale I-4 |
| Grantee | Willard | Porter | TBP Group | Williams |
| Sale date | 15 months | 11 months | 6 months | 2 months |
| Size | 33,000.00 | 8,548.08 | 43,584.58 | 8,916.04 |
| Sale price | $4,300,000 | $1,025,750 | $4,750,000 | $2,100,000 |
| Acre price | $130.30 | $120.00 | $108.98 | $235.53 |
| AU price | $3,258 | $2,640 | $3,488 | $3,297 |
| AUs | 1320 | 389 | 1362 | 637 |
| Acres/AU (rounded) | 25 | 22 | 32 | 14 |
| Land type | Panhandle | Roll. plains | Panhandle | Roll. plains |
| Improvements | Y | Y | Y | Y |
| Oil field activity | None | Extensive | None | Moderate |

The underlying analysis of carrying capacity is critical to the process and must be completed before processing of the approach can commence. The building contributions can be allocated for application in the sales comparison approach as shown in Table 19.17.

The improvement and land adjustment process can proceed once the building allocation analysis is complete. This analysis allows direct comparison of the sale improvements with the subject's improvements. In both processes the appraiser must exercise informed judgment in making comparisons and deriving adjustments. Table 19.18 demon-

**Table 19.17 Building Allocation**

| Sale No. | Sale I-1 | | Sale I-2 | | Sale I-3 | | Sale I-4 | |
|---|---|---|---|---|---|---|---|---|
| Improvement | Util./Cond. | Contrb. $ | Util./Cond. | Contrb. $ | Util./Cond. | Contrb. $ | Util./Cond. | Contrb. $ |
| | Size* | $/Unit | Size | $/Unit | Size | $/Unit | Size | $/Unit |
| Main | F/F | $21,000 | | | G/G | $175,057 | F/F | $75,000 |
| Dwelling | 1,400 | $15.00 | | | 4,980 | $35.15 | 3000 | $25.00 |
| Tenant | | | A/F | $30,000 | F/F | $22,200 | F/F | $23,250 |
| Dwelling | | | 1500 | $20.00 | 1,480 | $15.00 | 1,550 | $15.00 |
| Tenant | | | | | F/F | $22,200 | | |
| Dwelling | | | | | 1,480 | $15.00 | | |
| Bunk/cook | F/F | $16,640 | | | | | | |
| House | 2,560 | $6.50 | | | | | | |
| Barn(s) | F/F | $9,320 | P/F | $32,586 | A/F | $21,600 | F/F | $13,673 |
| | 3,000 | $3.11 | 13600 | $2.40 | 7,200 | $3.00 | 4,800 | $2.85 |
| Shed(s) | | | | | A/F | $720 | | |
| | | | | | 360 | $2.00 | | |
| Total building contribution | | $46,960 | | $62,586 | | $241,777 | | $111,923 |

* Size in square feet

**Table 19.18 Building Comparison to Subject**

| Improvements | Size | Utility Cond. | $/Sq. Ft. | Contrib. Value |
|---|---|---|---|---|
| **Sale I-1** | | | | |
| Dwelling | | | | |
| Tenant dwel. | 1,400 | F/F | $15.00 | $21,000 |
| Tenant dwel. | | | | $0 |
| Tenant dwel. | | | | $0 |
| Mobile home | | | | $0 |
| Bunk/cook | 2560 | F/F | $6.50 | $16,640 |
| Barn(s) | 3000 | F/F | $3.11 | $9,330 |
| Shed(s) | | | | $0 |
| Bins/etc | | | | $0 |
| Total improvement value | | | | $46,970 |
| Price per animal unit @ 1,320 AUs | | | | $35.58 |

| Improvements | Size | Utility Cond. | $/Sq. Ft. | Contrib. Value |
|---|---|---|---|---|
| **Subject** | | | | |
| Owner's house | 2,136 | G/E | $60.00 | $128,160 |
| Foreman's hse. | 2016 | G/G | $30.00 | $60,480 |
| Tenant dwel. | 1220 | G/G | $20.00 | $24,400 |
| Hunting lodge | 1600 | F/A | $15.00 | $24,000 |
| Mobile home | 1310 | F/F | $10.00 | $13,100 |
| Cook shack | 1216 | F/G | $12.00 | $14,592 |
| Barns 2 | 6000 | F/F | $3.00 | $18,000 |
| Hay barn/shed | 3600 | G/G | $2.00 | $7,200 |
| Cake bins | 60 | G/G | $135.00 | $8,100 |
| Subject improvement value via Sale I-1 | | | | $298,032 |
| Price per animal unit @ 850 AUs | | | | $350.63 |
| Indicated adjustment | | | | $315.05 |

**Table 19.18** Building Comparison to Subject *(continued)*

| Improvements | Size | Utility Cond. | $/Sq. Ft. | Contrib. Value | Improvements | Size | Utility Cond. | $/Sq. Ft. | Contrib. Value |
|---|---|---|---|---|---|---|---|---|---|
| **Sale I-2** | | | | | **Subject** | | | | |
| Dwelling | | | | | Owner's house | 2136 | G/E | $60.00 | $128,160 |
| Tenant dwel. | 1500 | A/F | $20.00 | $30,000 | Foreman's hse | 2016 | G/G | $30.00 | $60,480 |
| Tenant dwel. | | | | $0 | Tenant dwel. | 1220 | G/G | $20.00 | $24,400 |
| Tenant dwel. | | | | $0 | Hunting lodge | 1600 | F/A | $15.00 | $24,000 |
| Mobile home | | | | $0 | Mobile home | 1310 | F/F | $10.00 | $13,100 |
| Bunk/cook | | | | $0 | Cook shack | 1216 | F/G | $12.00 | $14,592 |
| Barn(s) | 13600 | P/F | $2.40 | $32,640 | Barns 2 | 6000 | F/F | $3.00 | $18,000 |
| Shed(s) | | | | $0 | Hay barn/shed | 3600 | G/G | $2.00 | $7,200 |
| Bins/etc | | | | $0 | Cake bins | 60 | G/G | $135.00 | $8,100 |
| Total improvement value | | | | $62,640 | Subject improvement value via Sale I-2 | | | | $298,032 |
| Price per animal unit @ 389 AUs | | | | $161.03 | Price per animal unit @ 850 AUs | | | | $350.63 |
| | | | | | Indicated adjustment | | | $189.60 | |
| **Sale I-3** | | | | | **Subject** | | | | |
| Dwelling | 4980 | F/G | $35.15 | $175,047 | Owners house | 2136 | G/E | $60.00 | $128,160 |
| Tenant dwel. | | | | $0 | Foreman's hse | 2016 | G/G | $30.00 | $60,480 |
| Tenant dwel. | 1480 | F/F | $15.00 | $22,200 | Tenant dwel. | 1220 | G/G | $15.00 | $18,300 |
| Tenant dwel. | 1480 | F/F | $15.00 | $22,200 | Hunting lodge | 1600 | F/A | $15.00 | $24,000 |
| Mobile home | | | | $0 | Mobile home | 1310 | F/F | $10.00 | $13,100 |
| Bunk/cook | | | | $0 | Cook shack | 1216 | F/G | $12.00 | $14,592 |
| Barn(s) | 7200 | A/F | $3.00 | $21,600 | Barns 2 | 6000 | F/F | $3.00 | $18,000 |
| Shed(s) | 360 | A/F | $2.00 | $720 | Hay barn/shed | 3600 | G/G | $2.00 | $7,200 |
| Bins/etc | | | | $0 | Cake bins | 60 | G/G | $135.00 | $8,100 |
| Total improvement value | | | | $241,767 | Subject improvement value via Sale I-3 | | | | $291,932 |
| Price per animal unit @ 1,362 AUs | | | | $177.51 | Price per animal unit @ 850 AUs | | | | $343.45 |
| | | | | | Indicated adjustment | | | | $165.94 |
| **Sale I-4** | | | | | **Subject** | | | | |
| Dwelling | 3000 | F/F | $25.00 | $75,000 | Owners house | 2136 | G/E | $60.00 | $128,160 |
| Tenant dwel. | | | | $0 | Foreman's hse | 2016 | G/G | $30.00 | $60,480 |
| Tenant dwel. | | | | $0 | Tenant dwel. | 1220 | G/G | $15.00 | $18,300 |
| Tenant dwel. | 1550 | F/F | $15.00 | $23,250 | Hunting lodge | 1600 | F/A | $15.00 | $24,000 |
| Mobile home | | | | $0 | Mobile home | 1310 | F/F | $10.00 | $13,100 |
| Bunk/cook | | | | $0 | Cook shack | 1216 | F/G | $12.00 | $14,592 |
| Barn(s) | 4800 | F/F | $2.85 | $13,680 | Barns 2 | 6000 | F/F | $3.00 | $18,000 |
| Shed(s) | | | | $0 | Hay barn/shed | 3600 | G/G | $2.00 | $7,200 |
| Bins/etc | | | | $0 | Cake bins | 60 | G/G | $135.00 | $8,100 |
| Total improvement value | | | | $111,930 | Subject improvement value via Sale I-4 | | | | $291,932 |
| Price per animal unit @ 637 AUs | | | | $175.71 | Price per animal unit @ 850 AUs | | | | $343.45 |
| | | | | | Indicated adjustment | | | | $167.74 |

strates the development of the improvement adjustments, which will be utilized in the sales comparison adjustment grid.

The final step in the process is the direct comparison of the sale properties to the subject property. In this example only one type of land, native rangeland, was present. When multiple land types are present, a land adjustment must be developed based on the improved sales as detailed in the Chapter 14. Hay-based ranches will generally require development of a "grass equivalent" analysis as well as categories for land subject to various types of federal or state grazing permits or leases. Once the pairings and adjustment estimation process is complete, the sales adjustment grid can be constructed. Table 19.19 incorporates the adjustments developed above.

**Table 19.19 Improved Sales Adjustment Grid**

| Record No. | | 4411 | 12809 | 4407 | 12808 |
|---|---|---|---|---|---|
| Sale no. | Subject | Sale I-0 | Sale I-0 | Sale I-0 | Sale I-0 |
| Grantee | | Willard | Porter | TBP Group | Williams |
| Sale date | Current | 15 months | 11 months | 6 months | 2 months |
| Size | 30,279.00 | 33,000.00 | 8,548.08 | 43,584.58 | 8,916.04 |
| Sale price | | $4,300,000 | $1,025,750 | $4,750,000 | $2,100,000 |
| Acre price | | $130.30 | $120.00 | $108.98 | $235.53 |
| AU price | | $3,258 | $2,640 | $3,487 | $3,297 |
| AUs | 850 | 1320 | 389 | 1362 | 637 |
| Acres/AU (rounded) | 36 | 25 | 22 | 32 | 14 |
| Land type | Roll. plains | Panhandle | Roll. plains | Panhandle | Roll. plains |
| Improvements | Y | Y | Y | Y | Y |
| Oil field activity | Slight | None | Extensive | None | Moderate |
| **Financing adjustment** | | | | | |
| Financing | | Cash | Cash | Liberal | Cash |
| Sale price/acre | | $130.30 | $120.00 | $108.98 | $235.53 |
| Adjustments | | | | | |
| Financing adjustment | | $0.00 | $0.00 | -$5.00 | $0.00 |
| Financing-adjusted price/acre | | $130.30 | $120.00 | $103.98 | $235.53 |
| Financing-adjusted price/AU | | $3,258 | $2,640 | $3,327 | $3,297 |
| Financing-adjusted price | | $4,300,000 | $1,025,750 | $4,532,077 | $2,100,000 |
| **Time adjustments** | | | | | |
| Months | | 15 | 11 | 6 | 2 |
| Time adjustment | | -$120 | -$72 | -$50 | -$16 |
| Time-adjusted price/au | | $3,138 | $2,568 | $3,277 | $3,281 |
| **Other adjustments** | | | | | |
| Improvements | | $315 | $190 | $166 | $168 |
| Improvement adjusted price | | $3,453 | $2,758 | $3,443 | $3,449 |
| Oil field activity | | $0 | $695 | $0 | $0 |
| Oil field-adjusted price | | $3,453 | $3,453 | $3,443 | $3,449 |
| Location | | 0 | 0 | 0 | 0 |
| Indicated subject value/AU | | $3,453 | $3,453 | $3,443 | $3,449 |
| Estimated subject value via direct sales comparison approach | | | | 850 AUs @ $3,450 = $2,932,500 | |

The adjustment process shown in Table 19.19 required downward adjustments for time based on an average annual decline of 3%, which was derived from paired sales analysis. The time adjustments were developed based on compound change. Straight-line or compound-change calculations are certainly acceptable, but using straight-line change requires extra analysis if sales both before and after the date of valuation are analyzed. Assuming compound change eliminates this extra step and produces the same result.

The only other adjustments required, based on the sales analysis, were the improvement adjustments detailed in the building adjustment analysis and adjustments for oil field activity, which were estimated through paired sales analysis.

## Income Approach

Nearly all of the income and expense information and methods of capitalization discussed in Chapter 15. The income approach can be applied to ranch properties. The most common technique employed is direct capitalization, in which a single year's income estimate is converted into a indication of value in one step, either by dividing the income estimate by an appropriate rate or multiplying it by an appropriate factor.

Deriving capitalization rates from comparable sales is essential. Data on the sale price, income, expenses, and market condition of each property at the time of sale is needed. In addition, the appraiser must be certain that the net operating income of the subject property and each comparable sale is calculated and estimated in the same manner.

When analyzing the income and expenses of ranch properties, there are three generally accepted methods: cash lease, cropshare rental, or owner-operator. A ranch lease can be structured to fit any situation. Most ranch properties are leased for a flat rental rate, based on a per-animal-unit or animal-unit-month rate. In areas where hunting or recreation is the primary motivating factor in the market, income is often quoted on a per-acre basis. Many current leases tend to be short-term. The appraiser may treat these short-term leases with a vacancy loss, if applicable, depending on how vacancy is perceived in the market. Caution should be exercised in determining when a ranch property becomes vacant.

In the appraisal of a commercial office building, for example, vacancy occurs if a tenant leaves on Friday and the manager cannot find a new tenant before Monday. For ranch properties, if a tenant does not renew a lease in October, after the growing season and forage has been harvested, the land does not become vacant until the spring when the next growing season begins. Considering the physical characteristics of ranch properties and agricultural operations, the supply and demand relationship, and local economic conditions, a vacancy loss rate for a flat fee short-term lease is generally not required. The appraiser should survey the local market to determine if a vacancy estimate is necessary. As long as the vacancy rate is treated consistently (whether used or not used) in analyzing the comparable sales and the subject property, the appraisal will be market-supported.

There are many different types of leases written on ranch properties. Leases can be written with different language that means the same thing. For example, most ranches on the eastern plains rent on a per-acre basis, usually with limits on the carrying capacity. Ranches in the intermountain region tend to rent on a carrying-capacity basis per day, animal unit month, or animal units. Many different kinds of leases can be found in each area, and it is important to find the best method of lease analysis to correlate the different lease information into a single, common unit. For example, a $4.80 per acre rent for pasture in an area that requires 30 acres per animal unit is the same rental rate as $12.00 per animal unit month or $144.00 per animal unit.

Although information on intensive farming areas may be easy to find, it is often very difficult to find perfect rental information on ranch properties. Consider ranches that have a high percentage of their carrying capacity from assured lease land. Special arrangements must be made with the U.S. Forest Service and the BLM to allow nonowner livestock to be grazed. As the spread or percentage capacity from assured lease land changes, it causes a large variance in the capitalization rates. This makes it somewhat difficult to compare and analyze these rates.

When rental information is not adequate or consistent, an owner-operator statement is an alternative method of capitalization. Owner-operator analysis is a very detailed procedure, and the appraiser who applies it must be intimately familiar with the area and the ranching industry. Income and expense information is available from most agricultural colleges and local extension services. The owner-operator method requires that the appraiser research the comparable sales with the same detail, intensity, and consistency as the subject property. Care should be taken to assume typical management and practices, not necessarily the practices of the current operator.

Although it is sometimes considered subjective, the income approach can provide supportable market evidence for valuing a ranch property and can offer an indication of expected market returns on the investment, which is of use to many clients. It is not the appraiser's responsibility to determine the exact or perfect capitalization rate or to measure management. Rather, it is the appraiser's responsibility to be objective and consistent throughout the capitalization process.

The income approach is applied to ranch properties with the same procedures used for all other classes of property. The only difference is the unit of comparison used, which should be the unit prevalent in the subject property's market area. The units of comparison most commonly used are animal unit, animal unit month, and acre. Table 19.20 demonstrates the development of capitalization rate analysis based on the improved sales analyzed in the sales comparison approach.

**Table 19.20** Capitalization Rate Analysis

| Sale No. | Subject | Sale I-1 | Sale I-2 | Sale I-3 | Sale I-4 |
|---|---|---|---|---|---|
| Sale price | | $4,300,000 | $1,025,750 | $44,750,000 | $42,100,000 |
| Acres | 30,279.00 | 33,000.00 | 8,548.08 | 43,584.58 | 8,916.04 |
| Date | Current | 15 months | 11 months | 6 months | 2 months |
| Land type | Roll. plains | Panhandle | Roll. plains | Panhandle | Roll. plains |
| Acre price | | $130.30 | $120.00 | $108.98 | $235.53 |
| Price per AU | | $3,258 | $2,640 | $3,487 | $3,297 |
| Total AUs | 850 | 1,320 | 389 | 1,362 | 637 |
| Estimated gross income | | | | | |
| Range land | | $118,800 | $34,969 | $122,582 | $57,317 |
| Impvd. range | | $0 | $0 | $0 | $0 |
| Cropland | | $0 | $0 | $0 | $0 |
| Hunting (per acre @ $0.50) | | $15,675 | $0 | $20,703 | $4,235 |
| Total gross income | | $134,475 | $34,969 | $143,285 | $61,552 |
| Total gross income/acre | | $4.08 | $4.09 | $3.29 | $6.90 |
| Less operating expenses | | | | | |
| Taxes | | $16,500 | $3,590 | $21,792 | $3,745 |
| Maintenance | | $6,724 | $1,748 | $14,329 | $6,155 |
| Management | | $5,379 | $1,399 | $5,731 | $2,462 |
| Total operating expenses | | $28,603 | $6,737 | $41,852 | $12,362 |
| Net operating income | | $105,872 | $28,232 | $101,433 | $49,190 |
| Net operating income /acre | | $3.21 | $3.30 | $2.33 | $5.52 |
| Net operating income/AU | | $80.21 | $72.58 | $74.47 | $77.22 |
| Est. capitalization rate | | 2.46% | 2.75% | 2.14% | 2.34% |
| Estimated capitalization rate | | | | | 2.40% |

The analysis of income data on the sale properties indicated that the current market rent for grazing in the subject area was $80 to $100 per AU annually. Sale 3 was the only sale that was leased for grazing at the date of sale for $90 per AU. Therefore, $90 per AU was adopted as the prevailing market rent. The hunting income estimate was based on prevailing lease rates in the subject area. Sale 1 and Sale 3 were leased for hunting at $0.50 per acre based on 95% of the total acreage. The headquarters and improvements areas were excluded from the hunting lease area.

Taxes were based on actual taxes at the date of sale. Maintenance and management fees were based on typical rates prevailing in the market area for ranch properties.

The indicated capitalization rates developed from the sales range from 2.14% to 2.75%. In the final analysis, a rate of 2.4% was adopted as the overall rate for application to the subject.

The application of the income approach to the subject property is detailed in Table 19.21 based on a landlord/tenant arrangement.

The pro forma analysis in Table 19.21 uses the data derived from the analysis of sales in the market area. In this example no privilege or improvement income was included since all of the leases in the market

**Table 19.21** Subject Pro-forma Operating Statement

| Income Estimate | Units | Price/Unit | Total |
|---|---|---|---|
| Grazing (per AU) | 850 | $90.00 per AU | $76,500 |
| Hunting (acres) | 28,765 | $0.50 per acre | $14,383 |
| Total gross income | | | $90,883 |
| Less operating expenses | | | |
| Ad valorem taxes (land) | $0.42 | $0.42 per acre | $12,717 |
| Management (% of gross) | 4.00% | | $3,635 |
| Maintenance (% of gross) | 5.00% | | $4,544 |
| Total operating expense | | | – $20,896 |
| Net operating income | | | $69,987 |
| Capitalization | $69,987 | @2.4% | $2,916,125 |
| Estimated value via the income approach | | | $2,916,125 |
| Rounded to | | | $2,916,000 |

area included the use of improvements at the lease rate. In some jurisdictions separate line items for improvements may be applicable depending on local custom and market activity.

In conclusion, the appraiser should recognize that the appraisal of ranch properties is no different than the appraisal of any other type of property. There are three approaches to value and all three are applicable to ranches. Certainly, the appraiser must be competent to deal with ranch property, as required by the Uniform Standards of Professional Appraisal Practice, but appraisers of all other types of property are subject to the same competency requirements.

# *The Valuation of Permanent Plantings*

## Introduction

In 1998 the average American consumed 293.9 pounds of fruit: 47.4 pounds of apples, 125.6 pounds of citrus, and 120.9 pounds of other fruit. Figures for the 1998 season indicate that growers produced 17.5 million tons of citrus fruit valued at $2.6 billion, 14.7 million tons of non-citrus fruit valued at $6.9 billion, and 2.2 million tons of tree nuts valued at $1.9 billion, for a total value of $11.4 billion for U.S. fresh fruit and nut production.[1]

Fruit and nuts are classified according to the kind of trees, deciduous or evergreen, that bear them. Deciduous fruit plants lose their leaves, usually in the winter, and have a period of apparent dormancy. They produce:

- Pome fruit such as apples and pears
- Stone fruit such as peaches, plums, apricots, plumcots, and cherries
- Berries or small fruit such as grapes
- Nut crops such as almonds, chestnuts, pecan, pistachios, and walnuts
- Miscellaneous crops such as figs, kiwifruit, mulberries, and persimmons

Evergreen fruit-bearers retain their leaves throughout the year, even when dormant, and produce citrus fruit, including oranges, lemons, grapefruit, tangerines, and tangelos, plus other fruits such as olives and avocados.

---

1. United States Department of Agriculture, *October 1999 Fruit Situation and Outlook Yearbook.*

Orchards and vineyards vary in size from small family-owned and -operated enterprises to large agri-business of more than 50,000 acres. In 1998 the USDA reported that 3,900,700 acres in the United States were improved with permanent crops. Of these acres, 1,138,800 (29%) were planted to citrus fruit; 1,797,900 acres (46%) were planted to major deciduous fruit (including apples, peaches, pears, grapes, cherries, plums, prunes, and apricots): 245,400 acres (7%) were planted to minor fruit (including nectarines, figs, olives, avocados, and dates); and 718,600 acres (18%) were planted to nut trees.[2]

## Special Considerations for Appraisal

An orchard or vineyard is a permanent planting, a specialized property that presents a unique appraisal situation. The appraiser must focus on the specific horticultural crop produced by the permanent planting, which requires in-depth knowledge of the particular variety of tree, vine, or bush and its characteristics, production, and requirements. If the appraiser is not knowledgeable about the variety of produce being cultivated on the property, a horticultural consultant should be engaged.

Permanent plants require a number of years to reach maturity, and these plantings commit the land to that particular use for the economic life of the trees, vines, or bushes. Specialized properties such as orchards, vineyards, and groves usually have a higher value per acre than open land in the same locale.[3] The development and maintenance of permanent plantings requires specialized skills to ensure that the plants reach and maintain their genetic potential. Poor management can cause stress due to parasites or lack of water; the resulting damage may adversely affect the value of the property.

## Development Time and Income-Producing Life

Orchards, vineyards, and groves are unique agricultural properties. There is significant variation in permanent crops, even within small geographic regions such as the Napa and Sonoma Valleys of California, where more than 100 varieties of wine grapes are grown. Appraisers must be knowledgeable of the land's potential capacity to produce different varieties of permanent crops as well as all costs associated with their development and production. In addition to the normal ground preparation, planting, and fertilizing, it usually takes several years before permanent plantings generate positive cash flow. The time required to establish a permanent planting typically varies from three to 12 years, as compared to coastal vegetable land in California which is capable of producing two to three crops per year. The life cycle of permanent

2. Ibid.
3. There are some exceptions, such as highly specialized vegetable land in rich coastal valley areas of California.

plantings moves from infancy to the point of establishment, when annual income exceeds pre-harvest and harvest costs, through a period of mature production, and ultimately to decline. For example, one variety of apple orchard may take five to seven years to reach mature production, produce at a stable level for 30 years, and decline after 40 years. Certain peach varieties may reach mature production after seven years but begin to decline after 20 years. Some pear varieties do not reach mature production until 12 years, but only start to decline at 45 to 50 years. The life cycles of permanent crops vary in timing and duration, changing with different crops and even different varieties of the same crop. Figure 20.1 illustrates the projected life of a typical almond orchard in the Southern San Joaquin Valley of California, with establishment occurring in the sixth year and mature production occurring in Years 8 through 19.

**Figure 20.1** Cash Flows for an Almond Orchard

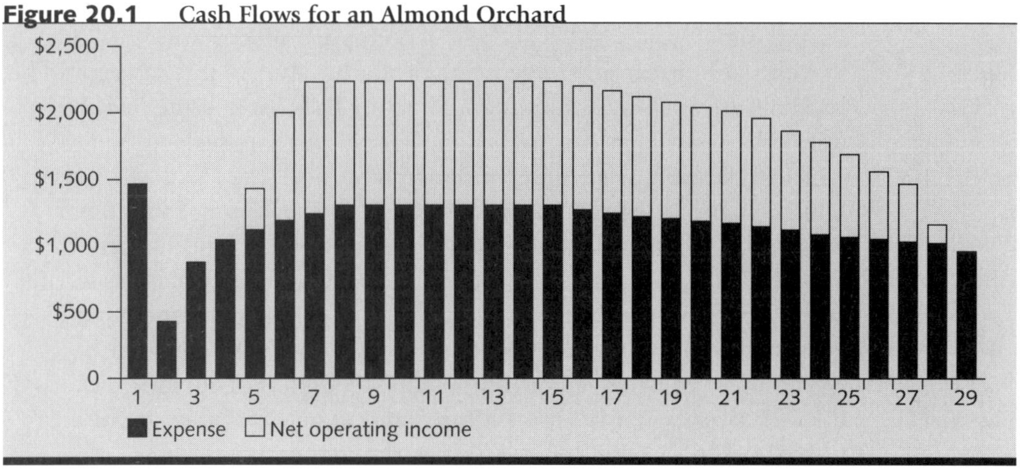


An understanding of bioengineering, genetics, and plant development has become increasingly important in recent years. Scientists can isolate genes that control plant development and growth, and laboratory techniques based on plant physiology and biochemistry may accelerate the process of creating, identifying, and incorporating new and useful plant traits. Table grapes offer one example. For many years, the Thompson seedless variety was the only seedless grape. In 1973 the Agricultural Research Station in Fresno, California, jointly funded by the USDA and University of California cooperative extension, introduced the Flame seedless variety, a flavorful red table grape which was well accepted by consumers. In the relatively brief period since 1973, several other red seedless table grape varieties have been developed, market tested, and accepted by consumers. A review of recently released varieties indicates that approximately 12 to 15 years are spent researching and developing improved varieties of trees and vines. So much research is going on and so many new varieties are being brought to market, however, that the market advantage of new successful variety may only last for a short time, e.g., five to seven years for some types of tree fruit. Appraisers of permanent crops should remember that trees, vines, and bushes are wasting, or depreciating, assets. Therefore, their age, condition, and expected life are crucial factors which must be understood and considered.

## Stages of Development

Permanent plantings are typically found in one of three stages: developing or immature, mature, or decline; each of these three stages requires a different appraisal emphasis.

The permanent planting must represent the highest and best use of land as improved. During an investigation of a developing or immature vineyard or orchard, it may be discovered that the property cannot be expected to develop into a successful orchard, and its value for alternative uses should be considered. The property value, in this case, could be less than vacant land value, after deducting the cost of removing the trees or vines and preparing the land for alternative uses.

The maximum value of a permanent planting is realized and sustained when: 1) the soil is appropriate for the species and varieties cultivated; 2) the locale, including climatic conditions and water availability and quality, are appropriate for this crop; 3) markets for the crop are well-established and accessible; and 4) the permanent planting is effectively managed. Effective management includes consistent replacement of damaged or dead plants; reduction of climatic hazards through irrigation, frost protection, and drainage facilities; and utilization of appropriate and cost-effective pest control measures.

## Soil Conditions

While it is generally accepted that superior soil quality is advantageous to field, row, and permanent crop production, the appraiser of permanent crops should know that poor quality soils are compatible with certain crops. For example, the presence of wine grape vineyards on the hillsides of California's Napa Valley have shown that Class IV and Class V rocky soils can, with good climatic conditions and proper management, produce good yields, superior quality fruit, and higher incomes than vineyards on Class I soils in other, less compatible, areas of California. Similarly, citrus groves on lesser quality soils located in more temperate zones may be more productive than groves on higher quality soils in colder locations. In these cases, climatic conditions are the dominant factor, not soil quality.

## Local Conditions

Climate not only limits what kinds of crops can be grown, but also determines the degree of success and the quality of the crop. If the climate is appropriate for a particular variety, the fruit can be produced more economically with less risk. Temperate-zone trees have a complex set of requirements. For example, winter temperatures must not kill or seriously damage the trees, but must be cold enough to satisfy the trees' dormancy requirements. The alternative is delayed foliation, which will reduce the crop production in that year. The area temperature and light must be adequate to mature the crop during the growing season for the particular variety, and commercial production is usually concentrated in appropriate climates. Climate also affects the time of harvest and the time window during which the crops mature. Harvest time has become crucial to producers, both domestic and foreign, who compete for the high prices commended by the first fruits of the season.

In most orchards, vineyards, and groves, irrigation is provided to ensure adequate water at crucial times. Insufficient water while the fruit is growing, for example, may result in fruit that is too small to be commercially marketable. The appraiser must explore the adequacy, quality, dependability, source, terms, water rights, and cost of the irrigation water supply. In the United States, new and expanding water issues are complicating the appraiser's research and understanding of water supply and quality problems in permanent crop production. The central issues are: 1) limited water supplies accompanied by increasing agricultural water demand; 2) decreasing groundwater quality; 3) increasing water costs; and 4) environmental and social attitudes that support reallocation of traditional water supplies for urban expansion, wildlife habitat, and fisheries. The costs to pump groundwater for irrigation have increased in recent years due to higher energy costs and declining groundwater levels.

## Established Markets

Fruit can be produced for either the fresh market or the processing market, which includes processing for juice, freezing, canning, and dehydrating. The demands of each market are specific and distinct. For example, the fresh market generally equates bigger with better. Therefore, a grower aiming for the fresh market may thin the fruit to increase its size. This kind of grower usually is more quality-conscious and has higher operating expenses than the grower who aims for the processing market. Fresh market prices are usually higher than those obtainable in the processing market. Harvest costs are usually lower for fruit grown exclusively for processing.

The fresh market often depends on cold storage or controlled-atmosphere storage facilities. Since field heat must be removed from the fruit as soon as possible to extend its shelf life, the availability of nearby storage facilities is important. Some growers have their own on-site storage facilities, and some fresh fruit operations are completely integrated with storage facilities, packing lines, and shipping facilities.

Controlled-atmosphere storage can extend the storage life of some varieties of fruit. The fruit can be harvested when ready and stored in such facilities without deterioration until demand warrants marketing it. Effective controlled-atmosphere storage necessitates an air-tight chamber in which the air has been replaced by an inert gas (usually nitrogen) refrigerated at a constant temperature. When a sealed room is opened, the entire contents should be marketed rapidly.

Fruit processing facilities should also be located relatively close to the producer. Soft fruits are particularly susceptible to damage from long and rough hauling, which causes them to lose grade. Some fruit producers operate orchards that produce both fresh and processed fruit. Production may be graded by quality or size for combination use. For example, a pear orchard may grow Bartlett pears for processing and Bosc and D'Anjou pears primarily for the fresh winter pear market. Bosc and D'Anjou pears that fail to reach market quality can usually be sold for processing.

A few growers in suburban locations operate successful "mom-and-pop" pick-your-own orchards. These are usually not considered commercial operations as they represent a minuscule portion of the permanent crop industry. Prices for the fruit still on the trees may be higher, lower, or approximately equal to contract prices. But, by avoiding harvest costs, a grower can usually sell fruit to be picked by the consumer for less than the contract price.

In appraising an orchard, the appraiser should note the size and quality of the fruit produced, the average production (preferably for the last five to 10 years), the method of marketing, and significant varieties. The desirable size is affected by the characteristics of the variety and the consumer's preferences. For example, California navel oranges are typically richly flavored, seedless, deeply colored, and large. Good-quality, medium-size navel oranges may bring the best prices, but all vineyards, orchards, and groves produce varied sizes of fruit, from small to large. Packing house reports usually supply information on grower returns and the size, quality, and pricing of fruit.

The appraiser should also understand and evaluate the marketing strategy for the subject and competing crops, which may vary from direct sales to the consumer to sales trough sophisticated marketing cooperatives such as the Almond Board of California. This board represented growers of 521.3 million pounds of almonds in 1992-1993, approximately 64% of the world's supply. Distant foreign events such as droughts or war, international treaties like GATT or NAFTA, and trade barriers may affect the value of a domestic permanent planting.

## Management

Even with a compatible combination of crop, soil, and climate, the profitability of a permanent planting is largely dependent on management decisions concerning pollination, spacing, pruning, replacement, frost protection, pest eradication, and energy sources.

Many varieties of trees cannot pollinate themselves. Even for self-fertile trees, better fruit is usually produced through cross-pollination. Since the pollen from one variety is used to fertilize another variety, the proper variety, number, and arrangement of pollinating trees is necessary for the success of the orchard. The sweet cherry, for example, can present particular difficulty because most varieties cannot fertilize themselves, and a high percentage of fertilized blooms is needed for a full crop of these small fruits. A typical solution is to have equal numbers of the pollinizer and the main variety, both good marketable cherry types, even though each type of cherry usually needs to be harvested and marketed separately. The pollen of some varieties is carried by the wind, while other pollen is transferred by domestic bees. The orchard should be arranged in a pattern that takes advantage of the wind currents and the normal flying behavior of the pollinating insects.

One of the objectives of an orchard is to develop maximum yields of high-quality fruit per acre, with a minimum of time and cost. The variety of rootstock planted, the size of the trees, and their pattern of arrangement directly affect this goal. The original planting pattern should be suitable for the vigor of the stock and the ultimate size of the trees. It should facilitate pest control, cultivation, harvesting, and efficient use of the land, whether dwarf trees, standard trees, or high-density plantings are used. In high-density plantings, the excess trees may eventually be removed or hedge-rowed to give the mature trees adequate room and sunlight. The economic soundness of this decision will be determined by future production.

Sometimes growers try to renew old orchards by interplanting young trees alternately with old trees. The plan is to establish the young trees, prune back the older ones gradually, and then remove the old ones entirely. This plan has disadvantages, however, because the young trees are often shaded by the older trees, choked by weeds, damaged by equipment, or not properly irrigated. Planting density should be determined by variety, rootstock, and soil fertility. The development of dwarf and semi-dwarf rootstocks, particularly of apples and pears, has resulted in very high density plantings, which can be trellised like vines with special training and pruning. Trellis, or plant support, systems may represent a significant part of the development costs for many permanent plantings and may be a relevant element of comparison. Many types of plantings are developed on sophisticated, costly support systems. In grape and kiwifruit vineyards, for example, the trellis system may be the single most costly item in the development process.

Most permanent plantings require pruning, which can be accomplished either by training or cutting young plants into a certain form or shape or by maintenance pruning. Some standard-sized fruit trees are trained to a certain form, such as modified leader, central leader, or open center. Dwarf apple or pear trees are often trained in hedgerow systems, forming a tree wall designed to provide a maximum number of trees per acre and to facilitate mechanical cultivation and harvesting. Maintenance pruning is done to maintain fruitfulness and to aid orchard management, harvesting, and pest control. To be successful, pruning should be adapted to the vigor of the particular rootstock and to the growth, flowering, and fruiting habits of the variety.

Prudent control of damaging vine and tree pests is an essential part of good management, and a wide range of options is available. For example, the traditional choice has been timely applications of sufficient commercial pesticides. In recent years, however, research and consumer agencies have focused attention on the possible risk of many of these chemicals to the consumer. In June 1993 a National Academy of Sciences (NAS) report focused on the possible effects of food-borne pesticide residues on the health of infants and children. It is likely that this report, which concluded that there is a "potential concern" that some children may be ingesting unsafe amounts of pesticides, will impact pesticide use and availability. Appraisers of permanent crops need to monitor government policies on the regulation and registration of pesticides and how they impact pest control and pre-harvest costs.

One outgrowth of consumer concerns about pesticides has been the development of "organic" farms. Since the mid-1970s, organic farming has evolved from a philosophy to part of mainstream agriculture, albeit on a small scale. An appraiser of organically grown permanent crops should be aware that cultivation costs may be higher due to the increased management, labor, and machinery modification required to compensate for the reduction or elimination of pesticides, herbicides, fungicides, and chemical fertilizers. Another economic consideration is the fact that most certifying agencies charge fees and require a transition period of three years before certification is granted. Organic plantings typically experience decreased yields and revenue for several years.

Another management choice that can affect the economic production of a permanent planting is the selection of electricity, natural gas, petroleum, or, in the future, solar power to operate equipment and machinery. The appraiser of permanent crops must have the experience to identify and understand the impact and subtle implications of each significant aspect of the planting, from layout and design, through physical planting, care, and production, to consumer recognition and marketing strategies. The use of this information in the sales comparison, cost, and income approaches to value is discussed in the following pages.

## Trend Analysis

Proper market trend analysis is most important in the valuation of permanent plantings. As previously stated, the development of an orchard or vineyard commits the land to the production of that crop for a finite economic life. The user of the land has thus lost the flexibility typical of open land. The owner-operator can no longer adjust to changes in the marketplace by switching crops in response to changes in the market; he or she is now committed to the production of the single crop chosen for development. The future viability of the market for the crop produced by that orchard or vineyard is, therefore, critical to the value of the asset.

Markets for most permanent crops are typically regulated or controlled in some manner by marketing orders, commissions, boards, or other commodity groups. It is incumbent on the appraiser to understand the structure of the market for the permanent crop under analysis. In many cases, these marketing groups are voluntary. Growers have the option of joining or complying with specific orders or regulations and in return they benefit from the group's marketing program. In other cases, participation is mandated by federal regulations, as in the case of the raisin industry. These entities may control both the production and sale of the crop, either through volume shipment controls or production diversion programs. The controlling entity may also establish size or quality standards for shipment of the product, adding another element of control to the market. Crop shipments into the market may be controlled by industry reserve requirements. Under this arrangement, predetermined portions of a crop are held in reserve for carryover into subsequent years or for shipment to export markets.

Payments for production controlled by such entities may also be nontraditional. Crop proceeds may be distributed by various methods which prolong payment over a significant period of time after the actual physical delivery of the crop. Returns to the grower, then, may not follow the typical pattern of cash payment upon delivery of the crop, which is assumed in most markets.

Marketing or regulatory commissions or entities may also be fruitful sources of data for market analyses. They may be able to offer the appraiser historical information on relevant factors such as bearing acres, nonbearing acres, past production, prices, and yields. Some groups track sales of nursery stock of trees or vines for plantings, which can allow the appraiser to monitor trends in recent, nonproductive plantings.

For many crops, varietal preferences can shift rapidly, as new varieties with superior physical characteristics are developed, introduced, and gain popularity over older, established varieties. A new variety of plum, for example, may have a deeper, brighter color, mature a few weeks earlier, or have a longer shelf life and thus be preferred over an older variety. The appraiser must keep up with market preferences and develop techniques to measure or track these trends. Varietal preferences may be quantified by tracking market share or by surveying packers or fruit brokers.

In the analysis of market trends, the appraiser should attempt to reduce yield and price information down to the net return per acre. Prices and yields may fluctuate dramatically in response to changes in supply and demand, so merely tracking unit prices may reflect an inaccurate picture of true conditions. For example; a severe frost or hailstorm in a particular year may greatly reduce yields, which could trigger an unusually high unit price for that crop. Reporting the high unit price for that one year would give an unrealistic view of the market conditions for that crop. A more accurate method of measuring such trends would be to track the total gross revenue per acre or develop a weighted average of the commodity price. Years of low yields and high prices, then, will reflect a stabilized return with the higher price offset by the lower yield: conversely, for years of low prices and high yields, the lower price will be offset by the greater level of production.

For example, to track the historical price of almonds, an appraiser could view the simple price per pound over the past five years, use the mathematical average these unit prices as an indicator of stabilized income, and consider the result a reflection of the commodity's recent market health. However, another appraiser could calculate the weighted average—i.e., the total value of all five years' crops divided by the same five years' total crop production—and arrive at a more realistic vision of the stabilized income for the particular almond orchard and overall conditions in the industry.

Table 20.1 presents data on California almond production from 1975 to 1996. The inordinately high prices in 1986 and 1995 reflect extremely short crops in those years resulting from inclement weather conditions. These prices tend to distort the data, elevating the average price per pound derived as a simple arithmetical average. Developing a weighted average, however, smooths the effect of these high prices by compensating for the very low volume of nuts sold at that price. The weighted average return per acre presents the clearest picture of the trend in this commodity. Again, merely viewing the high prices in 1986, 1993, and 1995 would provide a distorted view of market conditions. The 1986 price resulted in a very low return per acre, due to the diminished yields caused by bad weather, while the lower price in 1994, combined with that year's high yield, actually resulted in a higher return per acre.

Also, for many permanent crops, marketing entities control the volume of product released to the market. For example, the Raisin Administrative Committee recommends to the Secretary of Agriculture the total amount of the raisin crop to be designated as "free" tonnage.

**Table 20.1** California Almonds

| Crop Year | Bearing Acreage | Total Production (1,000 Pounds) | Yield (Pounds Per Acre ) | Average Grower Return/Lb. | Average Grower Return Per Acre |
|---|---|---|---|---|---|
| 1975 | 247,948 | 186,000 | 750 | $0.68 | $513 |
| 1976 | 257,854 | 283,747 | 1,100 | $0.65 | $722 |
| 1977 | 273,417 | 313,085 | 1,145 | $0.84 | $943 |
| 1978 | 303,592 | 181,026 | 596 | $1.45 | $864 |
| 1979 | 321,370 | 376,027 | 1,170 | $1.54 | $1,796 |
| 1980 | 324,878 | 321,796 | 991 | $1.47 | $1,457 |
| 1981 | 326,206 | 407,444 | 1,249 | $0.78 | $974 |
| 1982 | 334,258 | 346,729 | 1,037 | $0.94 | $976 |
| 1983 | 356,223 | 241,893 | 679 | $1.04 | $706 |
| 1984 | 379,565 | 586,909 | 1,546 | $0.77 | $1,190 |
| 1985 | 409,243 | 464,660 | 1,135 | $0.80 | $908 |
| 1986 | 412,674 | 251,597 | 610 | $1.92 | $1,164 |
| 1987 | 410,898 | 659,677 | 1,605 | $1.00 | $1,606 |
| 1988 | 407,141 | 590,007 | 1,449 | $1.05 | $1,521 |
| 1989 | 409,393 | 488,508 | 1,193 | $1.02 | $1,217 |
| 1990 | 411,294 | 656,189 | 1,595 | $0.93 | $1,483 |
| 1991 | 380,887 | 485,928 | 1,276 | $1.19 | $1,517 |
| 1992 | 390,240 | 545,902 | 1,399 | $1.30 | $1,833 |
| 1993 | 401,652 | 488,254 | 1,216 | $1.90 | $2,318 |
| 1994 | 403,226 | 730,000 | 1,810 | $1.25 | $2,263 |
| 1995 | 400,000 | 370,000 | 925 | $2.40 | $2,220 |
| 1996 | 401,000 | 510,000 | 1,272 | $2.00 | $2,544 |
| Overall averages: | | | 1,235 | $1.28 | $1,523 |
| 1988–96 averages | | | 1,348 | $1.45 | $1,880 |

The remainder of the crop is then designated as "reserve" tonnage. The Raisin Bargaining Association negotiates the "field price" for free tonnage, and member growers are paid based on this price.

In years of large crops and soft markets, free tonnage may be as low as 50% of the total crop. While the reported field price may still reflect higher values, the grower may be receiving this price on only 50% or 60% of the total crop. The reserve is held in storage, at the grower's expense, to be sold at a later date or in other markets. Typically, returns to the grower on this reserve portion of the crop are significantly lower. The appraiser must be aware of the complexities within the market for the specific crop in question and recognize the resulting impact of such peculiarities on the incomes generated by properties producing these crops.

## Sales Comparison Approach

In applying the sales comparison approach, the appraiser identifies market-recognized factors of comparability and their influence on the prices paid for a commodity. The appraiser usually compares each variety of produce on a per-acre basis. The elements of comparison can be different, but some appear to be universal, such as time, size, age, production, and profitability. This approach is considered to produce a good indication of the market value of permanent plantings when sufficient recent sales of comparable properties are available for verification and analysis. The appraiser of permanent plantings must have adequate knowledge of the specific crop to identify and quantify significant factors of difference, which vary for different crops. For instance, the appraiser should know the advantages and disadvantages of specific varieties, planted rootstocks, budwood, irrigation water requirements, water sources and costs, types of irrigation distribution systems, typical cultivation practices, and pre-harvest, harvest, and marketing costs.

Appraisers of permanent plantings throughout the United States use data from the cooperative extension services chartered under the land grant university system and funded by federal, state, and county governments. Wholesale nurseries can also provide the appraiser with essential information on genetic capability, patents, and consumer response. Patented varieties, which are developed for the sole use of a private breeder, sold by license, or developed for public use and distributed without royalties, are subject to the forces of supply and demand which can affect value significantly. The values of permanent crops, even patented varieties, are also affected by the principle of change, in that geneticists are constantly trying to develop strains with improved vigor, size, color, shelf life, productivity, and harvesting dates. While it may take 12 to 15 years to develop, test, patent, and market a new variety, its market advantage may last only five to seven years or until a newer variety is preferred by consumers.

Much information on sale properties can be obtained through interviews with sellers or buyers. The physical data on specific properties should be supplemented with

- Data from wholesale nurseries
- Information from local farm advisors
- Information from grower organizations and cooperatives
- Statistical data from county and state agencies
- Sale verification data, including property rights conveyed, buyer and seller motivation, financing terms, date of possession (Will the buyer take possession at close of escrow or after harvest?), cultivation costs (Were cultivation costs reimbursed as part of the sale or outside escrow?), holding period by seller, marketing period, and net operating income (for OAR and IRR analyses).

### Wholesale Nurseries

Wholesale nurseries are often a source of valuable information on new varieties that have not been market-tested or varieties for which data is not available due to the limited acreage planted. Data can include any special soil requirements, drainage, size of plants, preferred densities of plant population, expected yields, dates of harvest, and production during development of planting. The nursery may be a patent holder or have a license to sell a patented variety (if applicable) and may demand a royalty on the patented variety. Government-funded varieties are not patented.

## County Agents and Farm Advisors

County agents are part of the cooperative extension program mentioned above. They are a reliable source of information on production and management, including varietal selection, soils, water, irrigation efficiencies, fertilization problems, cultivation practices, test plantings, and disease control.

## Grower Organizations and Marketing Orders

Federal marketing orders are voluntary commodity programs initiated by producer groups. They are enacted only when approved by referendums in which all producers within the proposed production area are eligible to vote. They are self-governed under legislation established in the 1937 Agricultural Marketing Agreement Act and, for the most part, are self-financed. They are intended for specialty crops and are not permitted for livestock or basic field crops. A marketing order allows an agricultural crop industry to seek orderly marketing of its production. A federal marketing order sets up a mechanism for all producers of a crop in a given area to exercise control over selected aspects of the marketing of their crop and still be exempt from antitrust prosecution.

Each crop marketing order is developed by and for the particular needs of the commodity group seeking the marketing order. One or all of the following provisions may be included in a crop marketing order:

- Specification of grades, size, quality, or maturity
- Advertising, promotion, market development, and research
- Allotting the amount each processor may handle or purchase
- Establishing how much may be marketed during a set period
- Establishing methods of determining surpluses as well as their control and disposition
- Establishing a reserve product pool
- Inspection of the product
- Fixing the size, capacity, weight, dimensions, or packing of the containers used in marketing
- Prohibiting unfair competition and unfair trade practices
- Requiring processors to file their selling prices and not to sell product below the prices filed

Only those marketing tools included in the marketing order may be used by that commodity group. Any one tool or a combination of tools described above may be written into a marketing order.

To start a marketing order, an order proposal must be submitted to the U.S. Secretary of Agriculture along with a request for hearings on the order. If sufficient grower support is shown, the secretary will hold public hearings on the proposal. An opportunity for written comments follows the hearings. Then the secretary makes a decision whether or not to submit a proposed marketing order to a vote of all the growers. The marketing order is established if two-thirds of the voting growers vote in favor of the marketing order or if those growers representing two-thirds of the total production vote for it. Marketing orders are ended when more than half of the growers or those representing more than half the production vote against the order. An order may be amended through a procedure similar to that followed to establish the order.

A marketing order is administered by an elected board of growers and processors and a public member who is elected by the other board members. Board members, other than the public member, are elected by those they represent on a one person-one vote basis. The U.S. Secretary of Agriculture oversees board actions to make sure the board does not act beyond its authority as specified in the marketing order.

Under the marketing order, it is the processors who are regulated. Assessments for enacting the order are collected from processors or first handlers. However, they can pass these costs on to buyers or deduct them in making their purchases from growers. Imports are not regulated under a marketing order.

The advantages of a marketing order include industry self-control through use of selected marketing tools. A marketing order provides a means for all growers and processors to join together for various marketing activities. The disadvantage of a marketing order is that it is compulsory for all operators in the defined area. Marketing orders are attractive because they carry only modest administrative costs.

For example, the Almond Board of California administers Federal Marketing Order No. 91 over all handlers and/or marketers of almonds. This agency assesses growers per pound of production and provides research, quality control, promotional services, and marketing in both domestic and foreign markets. One of the oldest and largest farmer cooperatives in the world is Sunkist Growers, Inc., which represents 6,000 citrus growers in California and Arizona who farm more than 200,000 acres and produce 55% of the citrus grown in the United States. Sunkist primarily monitors quality control at the packing house level. While Sunkist does market some lower-quality fruit, only high-quality citrus is "stamped" with the Sunkist seal.

### Statistical Data From County and State Agencies

Statistical data on bearing and nonbearing planted acreage, average production, and average price is very useful to the appraiser of permanent plantings in identifying and projecting trends in supply, demand, and commodity prices. This information is compiled and distributed by various government agencies, including the USDA Economic Research Service, cooperative extension services, state agricultural statistics offices, county/parish agricultural commissioners, and some commodity organizations.

## Sales Verification

Appraisers need to verify more comprehensive data for sales of permanent crops than for sales of less complex agricultural properties. The verification sheet should include all the information needed to analyze the sale property in all three approaches to value. The holding period is used to estimate the typical period of ownership for particular varieties and, when derived from market evidence, can assist in developing discounted cash flow analyses and internal rates of return. Appraisers also need to confirm the marketing periods of sale properties to project estimated marketing periods for the subject property. The date of possession is a significant issue as "high input" permanent plantings, such as almonds and table and wine grapes, can accrue significant cultivation costs prior to harvest, and sales often postpone possession until after harvest. Also, additional monies are often paid for reimbursement of the cultivation costs incurred to date or for the actual crop planted. The net operating income *(NOI)* of sale properties will be derived from actual historical yields, commodity prices, and pre-harvest and harvest expenses. This figure is used in the sales comparison, income, and cost approaches. An example of a typical permanent crop sales verification sheet is shown in Table 20.2.

The sales comparison approach employs paired sales analysis in which elements of comparison are isolated and differences between the sale properties and the subject are analyzed. Adjustments approximating the degree of comparability or quantifiable difference between properties are applied to the prices of the sale properties to reflect the physical characteristics of the subject property. Paired sales analyses can be used to quantify elements of comparison but, in many markets, the data is limited and the appraiser may not be able to isolate or quantify adjustments. Appropriate elements of comparison may be varied and differ subtly. In many permanent planting markets, e.g., citrus, olives, pistachios, and some varieties of vines, plant age may not be critical, as plants tend to be quite long-lived. Conversely, plantings in other markets may experience quite short economic lives. Varietal preference may be a moving target, with today's jewel tomorrow's candidate for removal. Time can be a difficult adjustment, as the semi-permanence of these plantings makes them especially sensitive to cyclical change.

Climate is a critical element for some types of plantings, and the appraiser must draw on the market knowledge available to operators, buyers, and sellers. In most climate-sensitive industries, historical, site-specific temperature readings may be available for review and comparison. Water is a ubiquitous factor affecting the value of almost all permanent plantings. Value may depend on the quantity, quality, and

**Table 20.2** Data Collection

Verification sheet—Sale No. ______
APN(s): ______
Principal crop(s): ______ Sec: ______
Twsp: ______
Rnge: ______
Total assessed acreage: ______ Planted acreage: ______ Zoning: ______
Deed/escrow date: ______ County: ______ Recorder's book: ______
Recording date: ______ Recorder's page: ______
Document number: ______
Grantor: ______ Address: ______
Grantee: ______ Address: ______
Sale verified by: (buyer/seller/broker) ______ With: (appraiser) ______
Purchase price: ______ Cash to seller: ______
1st T/D amount: $ ______ Term: ______ Property rights conveyed: ______
Yrs: ______ Price adjustment: ______
Rate: % ______
Amortization type: ______ Cash equivalent price: ______ Motivation: ______
Price adjustment: ______
Holding period: ______ Marketing period: ______
Cultivation costs/crop included in sale: ______ Contributory value: ______
Permanent plantings: ______ Age/variety: ______
Spacing: ______ Yield: ______
Other land uses: ______ Contributory value: ______
Topography: ______ Soil: ______
Date of possession: ______ *NOI:* ______
Water supply/district: ______ Well: ______ Blended cost/af: ______
Water cost/district: ______ GPMs: ______ Cost: ______
Water requirement/acre: ______ Distribution system: ______
Comments: ______
Sale price/gross acreage: $ ______
Sale price/acre (permanent plantings): $ ______

**Table 20.3** Unadjusted Sales

| Sale No. | Sale Date | Area/ Acres | Production Lbs./Acre | Water $/Acre | Tree Age | Sale Price/Acre |
|---|---|---|---|---|---|---|
| 1 | 1/1/90 | 320 | 1,800 | $250 | 12 | $4,153 |
| 2 | 1/1/91 | 240 | 1,500 | $100 | 15 | $3,832 |
| 3 | 8/15/91 | 480 | 2,000 | $250 | 10 | $5,261 |
| 4 | 3/15/92 | 80 | 2,000 | $250 | 18 | $4,133 |
| 5 | 5/15/92 | 640 | 2,000 | $250 | 18 | $4,154 |
| 6 | 7/15/92 | 300 | 1,800 | $100 | 10 | $5,674 |
| 7 | 12/15/92 | 160 | 1,200 | $250 | 18 | $1,825 |
| 8 | 1/15/93 | 120 | 1,800 | $350 | 10 | $4,235 |
| 9 | 4/15/93 | 320 | 1,800 | $250 | 12 | $4,564 |
| 10 | 8/1/93 | 240 | 1,500 | $100 | 15 | $4,150 |
| Subject | 9/1/93 | 320 | 1,900 | $225 | 14 | |

cost of available water or, conversely, on the ability to drain, dispose, or control excess water or normal runoff.

In Tables 20.3 and 20.4, 10 sales of comparable almond orchards are considered, analyzed, and adjusted to develop specific indicators of value relevant to the subject property. Adjustment were derived for time, production, water cost, and tree age. Water cost and tree age were observed to be negative adjustments.

Time adjustments at an annualized rate of 3% were developed, as indicated by paired analyses of Sales 1 and 9 and Sales 2 and 10, which are equal in all respects except time. The two sale pairings indicate annualized rates of 2.9% and 3.1%, so a 3% rate is applied.

**Table 20.4** Adjusted Sales Grid

| Sale No. | Time | Production Lbs./ Acre | Water Cost/ Acre | Tree Age | Total Adjustments | Indicated Value/ Acre |
|---|---|---|---|---|---|---|
| 1 | $476 | $300 | $150 | -$300 | $626 | $4,779 |
| 2 | $315 | $1,200 | -$750 | $150 | $915 | $4,747 |
| 3 | $329 | -$300 | $150 | -$600 | -$421 | $4,840 |
| 4 | $183 | -$300 | $150 | $600 | $633 | $4,766 |
| 5 | $163 | -$300 | $150 | $600 | $613 | $4,767 |
| 6 | $193 | $300 | -$750 | -$600 | -$857 | $4,817 |
| 7 | $39 | $2,100 | $150 | $600 | $2,889 | $4,714 |
| 8 | $80 | $300 | $750 | -$600 | $530 | $4,765 |
| 9 | $52 | $300 | $150 | -$300 | $202 | $4,766 |
| 10 | $11 | $1,200 | -$750 | $150 | $611 | $4,761 |

| Sale No. | 1 | 1 > 9 | 9 | 2 | 2 > 10 | 10 |
|---|---|---|---|---|---|---|
| Date | 1/1/90 | 3.2877 | 4/15/93 | 1/1/91 | 2.5836 | 8/1/93 |
| Acres | 320 | = | 320 | 240 | = | 240 |
| Lbs./acre | 1,800 | = | 1,800 | 1,500 | = | 1,500 |
| $ /Acre | $250 | = | $250 | $100 | = | $100 |
| Age | 12 | = | 12 | 15 | = | 15 |
| /Acre | $4,153 | $411 | $4,564 | $3,832 | $318 | $4,150 |
| Annualized rate | | | 2.9% | | | 3.1% |

Once this time adjustment is made, Sale 4 can be adjusted for time at the indicated rate of 3% and Sales 4 and 5 can be compared for size.

| Sale No. | 4 | 5 |
|---|---|---|
| Date | 3/15/92 | 5/15/92 |
| $ / Acre | $4,133 | $4,154 |
| Time (years) | 0.1671 | |
| Adjustment | $20.47 | |
| Adjusted | $4,154 | $4,154 |
| Acres | 80 | 640 |
| Lbs./acre | 2,000 | 2,000 |
| $ /Acre | $250 | $250 |
| Age | 18 | 18 |
| Indicated size adjustment | | $0 |

Similar analyses indicated production adjustments of $3.00 per pound of production, with higher production meriting higher unit prices per acre. Water cost adjustments were established at $6.00 per dollar of cost, with higher costs equating to lower orchard values. Tree age adjustments of $150 per year of age were developed, with older orchards receiving lower unit prices per acre. The final adjusted values indicate a relatively narrow range of per-acre values for the subject orchard.

The sales comparison approach is a useful tool, but it requires attention to detail and comprehensive analysis of the general parameters of the subject market and the physical characteristics of the comparable sales and the subject property. The appraiser must learn to recognize which factors impact value in the market for the specific property type and develop relevant adjustments for differences in these elements of comparison that lead to meaningful value conclusions.

## Cost Approach

The cost approach to value is based on the principle of substitution, which affirms that no prudent investor would pay more for a property than the cost to acquire a comparable site and establish permanent crops of equal desirability and utility, without undue delay. The reproduction cost of a permanent planting, plus site value, provides a measure against which prices for similarly improved properties can be judged.

### Site Value

Contributory land value is estimated by sales comparison, using sales of land similar to the land included in the subject property. These sales may be of developed row crop farmland, native land, or land previously used for other permanent plantings.

The land investment required to develop permanent plantings can be substantial, and the typical investor or grower will research water costs, soil adaptability, and climate conditions before making a commitment to develop a vineyard, orchard, or grove. Good-quality farmland is adaptable to a wide variety of field, vegetable, and permanent crops. The market recognizes the ease of adapting farmland use from one year to the next, depending on the economic outlook for alternative crops such as cotton, alfalfa, grain, or vegetables. However, once land is planted to trees or vines, it becomes an orchard, grove, or vineyard and no longer has the economic versatility of open land. For example, in times of drought, irrigated farm land can be left idle or lie fallow, awaiting the return of a viable water source. If the land has been developed into a permanent planting, however, it must be maintained and the owner or operator must acquire water, which is often costly in such times. Thus, the capitalization rate, which reflects the risk of ownership, changes from the relatively low market capitalization rate for farmland, or the even lower rate for native pasture, to the higher rates associated with permanent plantings.

The contribution of the land, as if vacant and available for development to its highest and best use, is derived from the market by direct comparison with unimproved sales that are typically adjusted only for financing and time. From verified sales data the appraiser estimates contributory land value, which can then be incorporated into the cost approach to value.

## Contributory Value of Improvements

### Cost To Establish

The cost to establish a permanent planting includes four types of expenses: 1) capital costs, 2) pre-harvest expenses, 3) harvest expenses, and 4) the cost of capital and entrepreneurial profit. Capital costs typically include the cost of acquiring land and preparing it for planting, purchasing and planting nursery stock, and purchasing and installing the irrigation system. As noted earlier in this chapter, the appraiser relies on various public and private sources for significant cost information relating to the specific planting being appraised. The available data on land value, irrigation improvements, water cost, and management must be evaluated as to its reliability and applicability to the subject property. Sample costs to establish an almond orchard are shown in Table 20.5.

The format shown in Table 20.5 can be modified to develop the estimated cost to establish any permanent crop. This format presents detailed items of expense and offsetting income through the development period. The cost of capital should reflect the risk rates associated with the permanent crop being appraised as well as the cost of borrowing capital from private or institutional lenders. Typically, permanent plantings are perceived as high-risk enterprises, and the yield and direct capitalization rates observed in the

**Table 20.5** Cost to Establish Almond Orchard

| Year | 1 | 2 | 3 | 4 | 5 |
|---|---|---|---|---|---|
| Capital costs | | | | | |
| Land | 2,700 | | | | |
| Irrigation system | 1,000 | | | | |
| Trees | 279 | 6 | 3 | | |
| Land preparation | 368 | | | | |
| Layout, plant, prune | 77 | 6 | 3 | | |
| Total capital | 4,424 | 12 | 6 | 0 | 0 |
| Pre-harvest costs | | | | | |
| Prune and train | | 18 | 29 | 33 | 33 |
| Brush disposal | | 9 | 13 | 18 | 18 |
| Disc 3X | 19 | 19 | | | |
| Mow middles | | | 17 | 23 | 23 |
| Contact strip weed control | 8 | 8 | 8 | 8 | 8 |
| Pre-emergence strip control | | 27 | 27 | 18 | 18 |
| Misc. irrigation expense | 34 | 48 | | | |
| Irrigation labor | 12 | 12 | 12 | 3 | 3 |
| Irrigation water | 100 | 100 | 150 | 200 | 200 |
| Fertilizer and application | 10 | 18 | 33 | 52 | 66 |
| Pest control | 61 | 44 | 65 | 103 | 103 |
| Dormant shaking | | | 54 | 54 | 54 |
| Pollination service | | | 25 | 50 | 50 |
| Misc. expenses | 40 | 40 | 40 | 40 | 40 |
| Equipment cost | 75 | 75 | 75 | 75 | 75 |
| Office, business, supervision | 50 | 50 | 50 | 50 | 50 |
| Taxes | 30 | 30 | 40 | 40 | 50 |
| Total pre-harvest costs | 439 | 498 | 638 | 767 | 791 |
| Total capital and pre-harvest costs | 4,863 | 510 | 644 | 767 | 791 |
| Harvesting costs | 0 | 0 | 126 | 186 | 223 |
| Total annual costs | 4,863 | 510 | 770 | 953 | 1,014 |
| Less income @ $1.10/meat lb. | 0 | 0 | 275 | 550 | 1,100 |
| Net annual cost | 4,863 | 510 | 495 | 403 | -86 |
| Accumulated cost | 4,863 | 5,859 | 6,940 | 8,037 | |
| Cost of capital, @ 10% | 486 | 586 | 694 | 804 | |
| Total development cost | 5,349 | 6,445 | 7,634 | 8,841 | |

marketplace reflect this higher risk. Entrepreneurial profit is often observed in the expansion phase of planting cycles, as demand for the product outpaces supply, and growers find they can sell young, immature plantings at a profit. However, few growers actually develop new plantings with the intent of selling them at a quick profit. More often, the grower plans on holding the planting over a long term and extracting profit from the ongoing operation of the development.

A permanent planting is considered "established" or "developed" at the end of the last year of negative cash flow—i.e., in the year prior to the first year in which projected annual revenues exceed total annual expenses. While the planting may continue to mature for some years past this point in time, in Table 20.5 the cost to establish, or RC, of the almond orchard is projected as $8,841 at the end of Year 4.

### Depreciation

Permanent plantings, like most other improvements, tend to suffer loss in value, or depreciation, from physical, functional, and external causes. When sufficient verified sale data is available, the most reliable indicator of depreciation is the actual loss of value observed in the market, which may be divided into physical, functional, and external depreciation. If not enough information is available to allow the appraiser to extract market-derived depreciation, the appraiser may use a traditional age-life method, isolating the respective components of the property and applying appropriate levels of depreciation to each. Many properties may include short-lived items such as drip irrigation or trellis systems, which may experience accelerated depreciation. When such items are present, depreciation analysis should be performed. The various components of depreciation may be either curable or incurable. The appraiser should also attempt to determine when the living improvements actually begin to accrue physical depreciation. If the permanent planting is truly the highest and best use of the land, the living planting may not experience physical depreciation until it reaches maturity, although the ancillary infrastructure (irrigation and trellis systems) may begin to lose value from the moment of installation.

Table 20.6 illustrates the analysis of physical depreciation based on the effective ages of the permanent plantings. As the table shows, the market appears to recognize economic lives of 25 to 30 years for these types of orchards. Physical depreciation rates should be extracted from sales of plantings with ages similar to the subject because rates may vary with plant age—i.e., higher or lower rates may be observed in sales of younger or older trees. As noted previously, many types of permanent plantings are relatively long-lived, and the markets for these types of properties may exhibit little, if any, physical depreciation. Short-lived items such as drip irrigation systems often experience accelerated depreciation.

Functional obsolescence may be observed when the planting is inconsistent with contemporary market preferences in terms of the varieties planted, plant densities, or trellising systems. Advances in technology, and consequent cost savings or production increases, may make certain types of plantings obsolete. Table 20.7 illustrates the functional obsolescence observed in a comparable sale with less-than-optimum production of 1,800 pounds per acre. This obsolescence is derived from the market as shown in the table.

**Table 20.6** Depreciation Rates Derived From Sales Data

| Sale No. | 1 | 2 | 3 | 4 |
|---|---|---|---|---|
| Sale date | 1/1/90 | 1/1/91 | 8/15/91 | 3/15/92 |
| Sales price /acre | $7,000 | $7,200 | $7,150 | $6,800 |
| Less land value/acre | – $4,000 | – $4,400 | – $4,650 | – $4,000 |
| Contributory value of trees | $3,000 | $2,800 | $2,500 | $2,800 |
| RC of orchard (less land) | $6,141 | $6,423 | $6,150 | $6,320 |
| Less contributory value | – $3,000 | – $2,800 | – $2,500 | – $2,800 |
| Observed depreciation | $3,141 | $3,623 | $3,650 | $3,520 |
| Effective age of trees (years) | 14 | 17 | 15 | 14 |
| Annual depreciation ($) | $224 | $213 | $243 | $251 |
| Annual depreciation (%) | 3.65% | 3.32% | 3.96% | 3.98% |
| Indicated economic life | 27.4 | 30.1 | 25.3 | 25.1 |

Table 20.8 shows an analysis of the functional obsolescence suffered by a Chenin Blanc (white wine grape) vineyard during a period when consumer preference strongly favored red wines.

**Table 20.7** Functional Obsolescence Estimate

| | |
|---|---|
| Contributory orchard value, per acre | |
| Sales price per acre | $5,400 |
| Less land value per acre | – $3,000 |
| Contributory value of orchard (less land) | $2,400 |
| Total depreciation | |
| RC of orchard, per acre | $6,300 |
| Less contributory value | – $2,400 |
| Total observed depreciation | $3,900 |
| Physical depreciation, long-lived | |
| Total observed depreciation | $3,900 |
| Less curable physical (irrigation system) | – $150 |
| Less incurable physical, short-lived | – $500 |
| Physical depreciation, long-lived | $3,250 |
| RC of orchard less curable physical and incurable short-lived depreciation ($6,300 less $150 and $500) | $5,650 |
| Effective age of trees | 15 years |
| Estimated annual physical depreciation rate (from sales) | 3.3% |
| Estimated physical depreciation | |
| 3.3% per year × 15 years = 49.5% × $5,650 | – $2,797 |
| Physically depreciated RC | $2,853 |
| Additional depreciation from functional obsolescence—low production | |
| Total observed, less physical, depreciation ($3,250 less $2,797) | $453 |
| Functional obsolescence as percentage of physically depreciated RC ($453 ÷ $2,853) | 15.9% |

The appraiser is faced with a question as to whether the functional obsolescence derived in Table 20.8 is curable. The obsolescence may be physically cured by grafting the existing vines to one of the red varieties that are currently popular, thus allowing the vineyard to command higher grape prices. The cost to graft the vines and to maintain the grafted vines until full production is restored must then be measured against the resulting value of the fully productive, new variety. Solving this problem requires income approach techniques.

External obsolescence can be observed in permanent crops when conditions outside of the subject property affect value. For example,

**Table 20.8** Functional Obsolescence Due to Change in Preference

| | | |
|---|---|---|
| Contributory vineyard value | | |
| Sale price per acre | $5,700 | |
| Less land value per acre | – $3,000 | |
| Contributory vineyard value | $2,700 | |
| Total depreciation | | |
| RC of vineyard (per acre) | $7,900 | |
| Less contributory value | – $2,700 | |
| Total observed depreciation | | $5,200 |
| Effective age of vines | 15 years | |
| Estimated annual physical depreciation | 2.0% | |
| Estimated physical depreciation | | |
| 2.0% per year × 15 years = 30.0% | | |
| $7,900 × 30.0% = physical depreciation | | $2,370 |
| Functional obsolescence = total dep., less physical | | $2,830 |
| Physically depreciated RC (PDRC) | $5,530 | |
| Functional obsolescence as a percentage of PDRC | | 51.2% |

supply and demand imbalances caused by natural disaster, changes in consumer preference, new varieties, or fluctuations in competitive bearing acreage can impact the market, lowering the commodity prices and values of those permanent crops. The modern market is increasingly a global market. The modern appraiser must be cognizant of the true market for the product of the permanent plantings and the influence of global competition

## Income Approach

The income approach measures the present worth of future benefits. The application of direct capitalization may be appropriate in the case of long-lived permanent plantings such as some types of grape vineyards. The Thompson seedless grape vine, for example, can be extremely long-lived, continuing to produce at economic levels for many decades. In the valuation of such properties, the direct capitalization approach can be a valid indicator of value. The appraiser must, however, have sufficient knowledge of the plantings being appraised and the market's treatment of those plantings to determine the appropriate method of valuation.

Use of the income approach in the valuation of many permanent plantings, however, is complicated by the permanence or semipermanence of these developments. Traditional direct capitalization using an overall rate may not be appropriate in many cases. In the valuation of young or immature plantings, or of older plantings past their years of peak production, application of this methodology may be misleading or inaccurate. In such cases, discounted cash flow is the proper method of analysis. The appraiser must attempt to measure the income stream generated by the subject property over an appropriate period of time.

Young, immature plantings require the use of discounted cash flow analysis. The appraiser must measure the income stream generated by the planting during the early years of development and the years of maturing production as well as the value of the development as a mature, productive asset.

In the case of older plantings, the holding period will be defined by the remaining economic life of the planting. The economic life of any permanent planting can be considered to end when the planting fails to generate sufficient revenue to offset production costs. Thus, an orchard has reached the end of its economic life when total gross revenues no longer exceed total costs of production. This determination will, of course, be affected by the price received for the commodity produced. A decline in the unit price for the particular commodity will hasten the end of the development's economic life. As stated previously, market trend analysis is critical in the determination of probable crop prices and the resultant economic lives of the permanent plantings that produce these crops.

At the end of the planting's economic life, the residual value of the underlying land, less any costs to be incurred in restoring the land to a useful state, must be estimated as the final component of the stream of income. Cash flows for a sample permanent planting development are illustrated graphically in Figure 20.2. (This graph appeared earlier in the chapter.)

**Figure 20.2** Cash Flows for an Almond Orchard

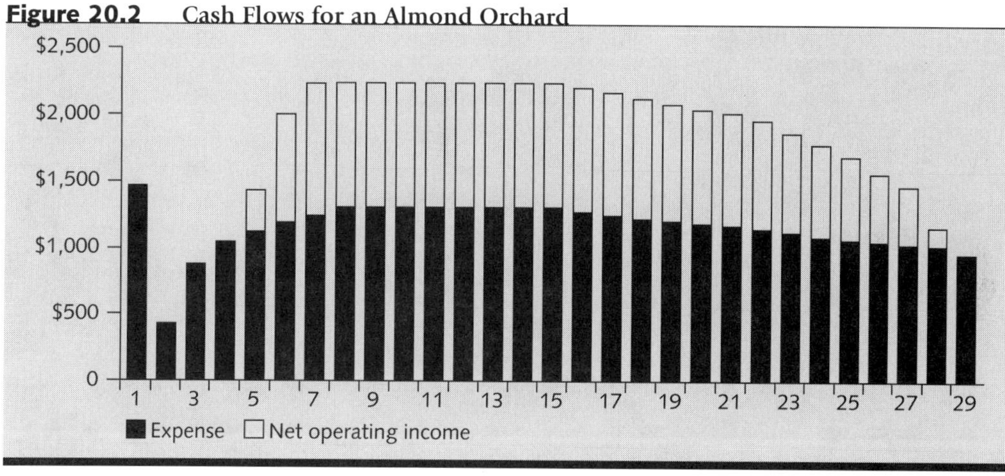


The first years of development reflect the initial costs of establishing the planting, including the cost of the raw land in the first year. Maturity is reached some years later, as the value of production overtakes cultivation costs. Net operating income continues to increase for several more years, as yields continue to expand, and a period of stabilized income is achieved during the mature years of production. Net income begins to decline after the years of peak production as the planting enters the final phase of its life. The economic life of the planting ends when net operating income fails to offset cultivation and harvest costs.

The application of discounted cash flow analysis to this cycle measures the income over the total life of the investment and the rate of return earned by the capital invested in the initial development. In Table

20.9, the internal rate of return for the same investment is measured as 15.37%. The reversionary value in the final year of the holding period is measured as the sum of the annual net operating income, (-$29), plus the market value of the underlying land, ($2,400), minus the cost to remove the obsolete trees ($150), or $2,221. This analysis serves to graphically illustrate the time value of money. While this enterprise generates a total income stream of $16,257, the present worth of this income stream is, by definition, $0 at the indicated internal rate of return of 15.37%.

**Table 20.9** Discounted Cash Flow Analysis

| Crop Year | Costs (Development + Cultivation) | Yield (Pounds Per Acre) | Price (Dollars Per Pound) | Total Gross Revenue | Net Operating Income | Net Cash Flow (Plus Reversion) | Present Value Factor | Present Value |
|---|---|---|---|---|---|---|---|---|
| 1 | 3,819 | | | | -3,819 | -3,819 | 0.866798 | -3,310.30 |
| 2 | 441 | | | | -441 | -441 | 0.751339 | -331.34 |
| 3 | 737 | 400 | $1.07 | 428 | -309 | -309 | 0.651259 | -201.24 |
| 4 | 880 | 800 | $1.07 | 856 | -24 | -24 | 0.564510 | -13.55 |
| 5 | 880 | 1,500 | $1.07 | 1,605 | 725 | 725 | 0.489317 | 354.75 |
| 6 | 897 | 1,800 | $1.07 | 1,926 | 1,029 | 1,029 | 0.424139 | 436.44 |
| 7 | 913 | 2,000 | $1.07 | 2,140 | 1,227 | 1,227 | 0.367643 | 451.10 |
| 8 | 1,058 | 2,200 | $1.07 | 2,354 | 1,296 | 1,296 | 0.318672 | 413.00 |
| 9 | 1,186 | 2,400 | $1.07 | 2,568 | 1,382 | 1,382 | 0.276224 | 381.74 |
| 10 | 1,186 | 2,400 | $1.07 | 2,568 | 1,382 | 1,382 | 0.239431 | 330.89 |
| 11 | 1,186 | 2,400 | $1.07 | 2,568 | 1,382 | 1,382 | 0.207538 | 286.82 |
| 12 | 1,186 | 2,400 | $1.07 | 2,568 | 1,382 | 1,382 | 0.179894 | 248.61 |
| 13 | 1,186 | 2,400 | $1.07 | 2,568 | 1,382 | 1,382 | 0.155932 | 215.50 |
| 14 | 1,186 | 2,400 | $1.07 | 2,568 | 1,382 | 1,382 | 0.135161 | 186.79 |
| 15 | 1,159 | 2,208 | $1.07 | 2,363 | 1,203 | 1,203 | 0.117157 | 140.99 |
| 16 | 1,134 | 2,031 | $1.07 | 2,174 | 1,039 | 1,039 | 0.101552 | 105.53 |
| 17 | 1,112 | 1,869 | $1.07 | 2,000 | 888 | 888 | 0.088025 | 78.17 |
| 18 | 1,091 | 1,719 | $1.07 | 1,840 | 749 | 749 | 0.076300 | 57.15 |
| 19 | 1,071 | 1,582 | $1.07 | 1,693 | 621 | 621 | 0.066137 | 41.08 |
| 20 | 1,054 | 1,455 | $1.07 | 1,557 | 503 | 503 | 0.057327 | 28.86 |
| 21 | 1,037 | 1,339 | $1.07 | 1,433 | 395 | 395 | 0.049691 | 19.63 |
| 22 | 1,022 | 1,232 | $1.07 | 1,318 | 296 | 296 | 0.043072 | 12.73 |
| 23 | 1,009 | 1,134 | $1.07 | 1,213 | 204 | 204 | 0.037335 | 7.61 |
| 24 | 996 | 1,043 | $1.07 | 1,116 | 120 | 120 | 0.032362 | 3.87 |
| 25 | 984 | 959 | $1.07 | 1,026 | 42 | 42 | 0.028051 | 1.18 |
| 26 | 974 | 882 | $1.07 | 944 | -29 | 2,221* | 0.024315 | 53.99 |
| Internal rate of return 15.37% | | | | | | 16,257 | | 0.00 |

* Includes reversionary value of land, less cost to remove trees

The critical components of the income approach are revenue (income), expenses (costs), and rate.

## Revenue

The typical means for measuring revenue in the income approach is to measure the rental income potential of the subject property. This approach is frustrated, in many cases, because most permanent plantings are owner-operated, not rented or held by investors for rental income. When plantings are rented, it is often the result of an involuntary acquisition by a lender, who then offers the property for rent rather than incur the significant costs of management and production. Typically, these rentals are for a short term, while the owner seeks to market the property. When permanent plantings are held for investment purposes, they are commonly leased on a cropshare basis or for a combination of a fixed minimum cash rent, plus owner participation in a share of revenues above a certain level. The appraiser is obligated to survey the market and determine the rental terms typical in the specific market under examination. Given the paucity of rentals of permanent plantings, some appraisers erroneously mix cash rents and cropshare rents in appraisals of permanent plantings. Typically, cropshare rents will exhibit a higher rate of return on investment because the investor shares in the risk of the crop and thus demands a higher rate of return. Consequently, if both types of rents must be used, the appraiser should measure the increment of return attributable to the risk factor and adjust accordingly.

Proper analysis of the revenue generated by a permanent planting may require the appraiser to explore a variety of factors. In most cases, the historical production of the planting is the key to this estimate, but historical information must be viewed in light of many aspects of production and the value of that production. Production in any specific crop year may have been impacted by climatic conditions such as a frost at bloom time. The appraiser must determine if the recurrence of such events is a reasonable probability to determine if that year's production should be included, or excluded, in calculating average yields. A particular property may be located in an area susceptible to frosts or other negative events. The appraisal of that property should properly measure the impact of the adverse location on the property's income. Conversely, another property may enjoy a location that it is not significantly impacted by events that affect other properties in the general area.

The value of the crop may be determined by the size, maturity, color, or market timing of the crop produced. Many permanent planting crops are highly volatile, and a particular box of fruit may command one price in June and a markedly lower or higher price two weeks later in July. The appraiser must also isolate factors attributable to the specific management or marketing skills of the grower. Helpful sources of information include commodity groups, marketing commissions, farm advisors, university extension offices, county agricultural commissioners, county crop reports, and federal and state market news services.

When rental data is available, the gross returns to the land may be estimated by multiplying the total gross revenue generated by the property by the appropriate cropshare rental factor. A farm generating total gross revenues of $2,000 per acre in a market where a 30% crop share is typical would yield a gross return to the land of $600 per acre ($2,000 × 30%).

## Expenses

Expenses must be deducted from gross revenues to reflect the typical terms found in the specific market. In some markets, rental rates are net, net, net, with the tenant (lessee) assuming all costs associated with the land, including property taxes and water district or special assessments. In other markets, the landowner (lessor) may be responsible for property taxes and other expenses. In the leasing of permanent plantings in some markets, it may be typical for the lessor to share in costs associated with maintenance of the plantings—e.g., in a vineyard lease, the lessor may agree to share in the cost of trellis replacement, with the lessor responsible for the direct cash costs of materials and the lessee agreeing to provide the necessary labor to retrofit those materials. Net operating income may then be established by simply deducting appropriate expenses from gross revenue.

## Rate

Direct, overall rates can be calculated from comparable sales using traditional methods. The appraiser selects an appropriate overall rate and applies this rate directly to the net operating income developed for the subject property. In the absence of adequate market data to extract overall rates, rates may be developed as they are in the appraisal of other types of properties. Again, the appraiser should take care to select rates from homogenous data, i.e., from cash rents or cropshare rents but not from both. Typically, cropshare rents will yield higher overall rates as the lessor demands a higher return for the increased level of risk associated with cropshare leases.

Table 20.10 shows data on a raisin vineyard and four comparable sales. It illustrates a typical application of direct capitalization.

The four sales are leased on a cropshare basis. The appraiser has developed average returns for raisins over five years prior to the sale dates and the effective date of appraisal for the subject property. Revenues, expenses, and overall rates were then calculated as shown. The lease on Sale 3 calls for the tenant to pay taxes, so no expenses were estimated for this sale. Professional farm management is not typical for small parcels such as these in this market, thus no management expenses were included. The analysis indicates a range of overall capitalization rates from 7.08% to 7.74%, with the highest and lowest rates indicated by the least similar sales, Sales 2 and 3. The most similar sales, Sales 1 and 4, indicate rates of 7.55% and 7.47%, so a rate of 7.5% can be selected for the subject. The analysis indicates an estimated net operating income of $27,199 for the subject property. The indicated overall rate of 7.5% is then applied to the subject property's net income, yielding an estimate of market value by the direct capitalization of income of $362,653, or $7,555 per acre.

**Table 20.10** Sales and Subject Income Analysis

| Sale | 1 | 2 | 3 | 4 | Subject |
|---|---|---|---|---|---|
| Price | $327,600 | 372,500 | 365,760 | 347,300 | |
| Size (acres) | 42 | 50 | 48 | 46 | 48 |
| Price/acre | $7,800 | 7,450 | 7,620 | 7,550 | |
| **Revenue** | | | | | |
| Yield(tons/acre) | 2.8 | 2.4 | 2.65 | 2.5 | 2.6 |
| Price/ton | $890 | $890 | $890 | $890 | $890 |
| Gross/acre | $2,492 | 2,136 | 2,359 | 2,225 | $2,314 |
| Share | 25.00% | 26.00% | 25.00% | 27.00% | 26.00% |
| Share rent/acre | $623.0 | 555.36 | 589.62 | 600.75 | $601.64 |
| Total revenue | $26,166 | 27,768 | 28,302 | 27,635 | $28,879 |
| **Expenses** | | | | | |
| Expenses/acre | $34 | 28 | 0 | 37 | $35 |
| Total expenses | $1,428 | 1,400 | 0 | 1,702 | $1,680 |
| *NOI* | $24,738 | 26,368 | 28,302 | 25,933 | $27,199 |
| Overall rate | 7.55% | 7.08% | 7.74% | 7.47% | |

Discounted cash flow (DCF) analyses may be performed using similar methods to develop estimates of revenues, expenses, and net operating income. In the DCF analysis, however, the appraiser must also develop an estimate of the reversionary value of the property at the end of the holding period. This may be accomplished by direct capitalization of the next year's anticipated income or by another relevant method. Using the data presented in Table 20.8, a DCF analysis is performed for the vineyard the end of the first year of development (see Table 20.11). The land was acquired one year ago at a cost of $3,400 per acre and development began immediately. The vineyard is considered established at the end of Year 3, when revenue first exceeds expenses, even though grape yields will continue through Year 5, when the vineyard reaches full mature production. The reversionary value of the vineyard at the end of Year 5 is estimated by direct capitalization of the anticipated mature net income of $3,724, capitalized at a terminal rate of 20%. At a discount rate of 20%, the present worth of the annual cash flows is estimated as shown in Table 20.11. The value of the property, then, is estimated as the sum of the present values of each annual cash flow, or $9,093.

**Table 20.11** Discounted Cash Flow Analysis

| Year of Development | 2 | 3 | 4 | 5 |
|---|---|---|---|---|
| Yield (tons/acre) | | 3 | 6 | 9 |
| Price ($/ton) | | $600 | $600 | $600 |
| Revenue | $0 | $1,800 | $3,600 | $5,400 |
| Expenses | $3,920 | $1,247 | $1,526 | $1,676 |
| *NOI* | -$3,920 | $553 | $2,074 | $3,724 |
| Reversion | | | | $18,620 |
| Net cash flow | -$3,920 | $553 | $2,074 | $22,344 |
| PVF @ 20.0% | 0.833333 | 0.694444 | 0.578704 | 0.482253 |
| Present value | -$3,267 | $384 | $1,200 | $10,775 |
| Sum of present values: | $9,093 | | | |

The income approach can be a valid, accurate, and relevant technique for the valuation of permanent plantings if the appraiser is competent and accomplished in the basic techniques of direct capitalization and discounted cash flow analysis.

APPENDIX
• A •

# *Professional Practice*

*Webster's Ninth Collegiate Dictionary* defines a *craft* as an occupation or trade requiring manual dexterity or artistic skill, a *trade* as the business or work in which one engages in regularly, and a *service* as useful labor that does not produce a tangible commodity. A *profession* is defined as a calling requiring specialized knowledge and often long and intensive academic preparation. Accordingly, a *professional* is characterized as conforming to the technical or ethical standards of a profession. The profession must have high standards of competence, scholarship that is continually supplemented by its members, and a code of ethics and standards of practice accepted by members who are willingly regulated by their peers.

The appraisal of real estate is not an exact science. Relevant data must be available in sufficient quantity for an appraiser to reach a supportable conclusion. Moreover, the appraiser must be skilled in analyzing the data and apply judgment, experience, and education to solve appraisal problems. Appraisers must be objective and unbiased and form their conclusions without considering preconceived values that favor their own interests or their client's interests. Professional appraisers have the necessary education and skills to reach sound conclusions objectively and to convey this information in a manner that is clear and not misleading.

The basic objective of the American Society of Farm Managers and Rural Appraisers (ASFMRA) is to create and maintain a professionally trained group of accredited farm managers and rural appraisers capable of providing expert guidance and assistance to farmland owners, farmers, and other groups responsible for farmlands and rural properties. Rural appraisers serve landowners and lenders by providing property valuations on which lenders, agriculturists, and investors can base lending or investment decisions. They keep informed about today's ever-changing land values to provide expert estimates and analysis of the market value of rural real estate. Appraisal clients include such diverse groups as owners, operators, bankers, lawyers, government agencies, and insurance and mortgage agencies.

The ASFMRA has adopted the Uniform Standards of Professional Appraisal Practices (USPAP) promulgated by the Appraisal Standards Board of The Appraisal Foundation and has its own Code of Ethics. The authority for the ASFMRA Code of Ethics and Standards of Professional Practice lies in the Society's Bylaws.

## ASFMRA Code of Ethics

The 15 articles in the Code of Ethics are summarized below.

### Article 1

A member shall cooperate with the Society and expose within the Society's ranks any corrupt or dishonest conduct on the part of any member. A member shall uphold the laws of the land and not knowingly submit false or misleading information to an officer or committee of the Society. A member shall not injure or attempt to injure, falsely or maliciously, the professional reputation of another member.

### Article 2

A member shall conduct the operation of his profession in accordance with this Code of Ethics and the Standards of Professional Practice.

### Article 3

A member shall render services without fear, favor, partiality, or the accommodations of his/her own interests.

### Article 4

A member may accept only those assignments for which the member has all necessary capabilities to adequately complete the assignment. Should technical ability be lacking, a member may associate with a person who has that ability upon acceptance of the assignment.

### Article 5

A member shall not violate the confidential nature of the appraiser-client relationship.

### Article 6

A member must not represent conflicting interests without written consent of all parties involved.

### Article 7

A member must disclose all previous and present relationships with the property, case, or purpose of employment.

### Article 8

All advertising of an appraisal practice must not be false nor misleading, and it must be dignified and not demeaning to the Society or its members.

### Article 9

A member must not operate a business in violation of state and federal laws.

### Article 10

A member may not claim professional qualifications which are not possessed or which may be subject to erroneous interpretation.

### Article 11

A member shall not co-mingle the funds of his client(s) and his own funds.

### Article 12

Service fees shall not be contingent upon appraised values, and a member shall accept no bonuses, favors or special inducements.

### Article 13

A member shall not reveal information about a property or a client's business which is not of public record unless the member has written permission from the client, or is directed by a court of law.

### Article 14

A member who testifies in court in a case in which another member testifies for the opposing party is required to file a case report with the Society within 30 days.

### Article 15

Appraisal reports and supporting documents must be submitted to the Society within 30 days when the Society is investigating a review of complaint.

## Uniform Standards of Professional Appraisal Practice (USPAP)

The Uniform Standards of Professional Appraisal Practice form the core of a truly professional appraisal practice. The Appraisal Standards Board (ASB) of The Appraisal Foundation develops, publishes, interprets, and amends USPAP on behalf of appraisers and users of appraisal services.

The ASB publishes USPAP in a bound edition that includes all prior changes, a history of changes, Statements on Appraisal Standards, Advisory Opinions, and an index. The Appraisal Standards Board (ASB) develops and amends appraisal standards through a dialogue among its members, other appraisers, and the users of appraisal services.

## Origin and History

The present standards are based on the original USPAP document developed in 1986-1987 by the Ad Hoc Committee on Uniform Standards and copyrighted in 1987 by The Appraisal Foundation. Before the ASB was formed in 1989, USPAP had been adopted by major appraisal organizations in North America and many state real estate boards and commissions. USPAP was rapidly recognized throughout the United States as the generally accepted standards of appraisal practice.

On January 30, 1989, the ASB was organized and unanimously approved and adopted the original USPAP as the appraisal standards promulgated by the ASB. Important dates in the history of USPAP are listed as follows:

- Effective date of original standards 4/27/87
- Ethics Rule added 12/4/89
- Preamble through Standard 2 revised 4/20/90
- Standards 3, 4, and 5 revised 6/5/90
- Standard 6 revised 9/10/90
- Standards 7 and 8 revised 3/3/92
- Standards 9 and 10 revised 9/16/92
- Glossary added 5/23/95

The ASB continues to review and revise USPAP and to issue Statements on Appraisal Standards and Advisory Opinions.

## Summary by Format

### Introduction

### Preamble

An overview of the need for USPAP and its organization and requirements is offered in this section.

### Ethics Rule

The ethics rule is divided into four sections that have broad implications and includes one very specific requirement. The topics addressed are conduct, management, confidentiality, and record keeping, which includes specific requirements for the retention of records.

**Competency Rule**

This rule states that an appraiser must have the knowledge and experience to complete the assignment completely.

**Departure Rule**

Under this rule, limited exceptions to USPAP are allowed, but the result must not be confusing or misleading. Some federal agencies prohibit use of the departure rule.

**Jurisdiction Exception**

This section states that if any part of USPAP is contrary to the law or public policy of any jurisdiction, then that part is void in that jurisdiction.

**Supplemental Standards**

This section holds that USPAP represents minimum standards that may be supplemented by public agencies and certain client groups.

**Definitions**

Terms germane to USPAP are defined here.

**Standard 1**

In developing a real property appraisal, an appraiser must identify the problem to be solved and the scope of the work necessary to solve the problem, and correctly complete research and analysis necessary to produce a credible appraisal.

**Standard 2**

In reporting the results of a real property appraisal, an appraiser must communicate each analysis, opinion, and conclusion in a manner that is not misleading.

**Standard 3**

In performing an appraisal review assignment involving a real property or personal property appraisal, an appraiser acting as a reviewer must develop and report a credible opinion as to the quality of another appraiser's work and must clearly disclose the scope of the work performed in the assignment.

**Standard 4**

In performing real estate or real property consulting services, an appraiser must be aware of, understand, and correctly employ those recognized methods and techniques that are necessary to produce a credible result.

**Standard 5**

In reporting the results of a real estate or real property consulting service, an appraiser must communicate each analysis, opinion, and conclusion in a manner that is not misleading.

**Standard 6**

In developing a mass appraisal, an appraiser must be aware of, understand, and correctly employ those generally recognized methods and techniques necessary to produce and communicate credible appraisals.

**Standard 7**

In developing a personal property appraisal, an appraiser must identify the problem to be solved and the scope of the work necessary to solve the problem, and correctly complete research and analysis necessary to produce a credible appraisal.

**Standard 8**

In reporting the results of a personal property appraisal, an appraiser must communicate each analysis, opinion, and conclusion in a manner that is not misleading.

**Standard 9**

In developing a business or intangible asset appraisal, an appraiser must identify the problem to be solved and the scope of the work necessary to solve the problem, and correctly complete the research and analysis steps necessary to produce a credible appraisal.

**Standard 10**

In reporting the results of a business or intangible asset appraisal, an appraiser must communicate each analysis, opinion, and conclusion in a manner that is not misleading.

## Legislation Pertaining to Appraisals

### Federal Legislation

The Federal Interagency Real Estate Property Appraisal Committee **(FIRPAC)** is headed by the Office of Management and Budget **(OMB)** and made up of various federal agencies that use appraisal services. It includes many agencies not regulated by the Financial Institutions Reform, Recovery, and Enforcement Act of 1989 (FIRREA).

**Circular A -129 of 1988** affects appraisals made for any federal credit agency regulated by the OMB, except land acquisition agencies. It requires the use of certified appraisers.

**Banking Circular 225,** "Guidelines for Real Estate Appraisal Policies and Review Procedures" and "Clarification of Guidelines for Real Estate Appraisal Policies and Procedures," regulates appraisals made for commercial banks and specified for the Federal Reserve Board, Federal Deposit Insurance Corporation (FDIC), and the Office of Comptroller of the Currency (OCC).

**12 CRF 7.3025,** "Other Real Estate Owned," also regulates appraisals made for commercial banks and is specific to the OCC.

**Internal Revenue Service Regulation** concerns charitable contribution appraisal reports for income tax purposes and is, of course, specific to the Internal Revenue Service (IRS).

**Executive Order 12630** was issued in 1988 and requires all federal agencies to determine whether an action will result in a taking.

**Uniform Relocation Assistance and Real Properties Acquisition Act of 1970** was amended in 1987 and provides for just compensation for owners and tenants who are displaced by land acquisition for federally funded projects. This specifically affects the Department of Transportation (DOT), Department of Agriculture, Department of Interior - National Park Service, and 11 other agencies.

**Competitive Equality Banking Act of 1987 (CEBA)** ensures that appraisals made for savings and loan institutions conform to the same requirements as those for commercial banks. It is specific to the five Federal Financial Institutions Regulatory Agencies (FFIRA). Those agencies are:

1. Office of Thrift Supervision (OTS)
2. Office of Comptroller of the Currency (OCC)
3. Federal Reserve Board
4. Federal Deposit Insurance Corporation (FDIC)
5. National Credit Union Administration (NCUA)

CEBA is the foundation for FIRREA

**Financial Institutions Reform, Recovery, and Enforcement Act of 1989 (FIRREA)** is also known as the "S&L bailout act." Title XI of this act required each of the five FFIRAs plus the Resolution Trust Corporation (RTC) to draft rules to give substance to this regulation. It affects appraisals made for the FFIRAs and RTC as well as all federal S&Ls and banking institutions.

These agencies developed the following circulars in response to this act:

- OCC - 12 CFR Part 34, Subpart C
- Federal Reserve Board - 12 CFR Parts 208 and 225
- FDIC - 12 CFR Part 323
- RTC - 12 CFR Part 1608
- OTS - 12 CFR Parts 506, 545, 563, 564, 571
- NCUA - 12 CFR Parts 701, 722, 741

## Environmental Legislation

Most environmental legislation does not regulate appraisal policies and procedures, but these laws are vitally important to appraisers.

**Comprehensive Environmental Response, Compensation, and Liability Act (CERCLA),** which passed in 1980, set up the superfund to clean up contaminated sites. While CERCLA does define the methods of setting risks and liability, a procedure known as **environmental property assessment (EPRA)** is usually required on commercial and industrial sites to identify environmental concerns before a property is transferred.

**Superfund Amendments and Reauthorization Act (SARA) of 1986** provides certain exemptions for third parties who acquired contaminated property provided they had appropriately inquired into the previous ownership and uses of the property consistent with proper stewardship of the property in an effort to minimize liability.

Professional appraisers, and rural appraisers in particular, should be familiar with other legislation including

- **RCRA, Resource Conservation Act** - controls hazardous waste and underground tanks
- **CWA, Clean Water Act** - regulates discharge of pollutants and authorizes permits to dredge and fill
- **ESA, Endangered Species Act** - habitat regulation
- **TSCA, Toxic Substance Control Act** - regulates polychlorinated biphenyl (PCB) and asbestos
- **CAA, Clean Air Act**- establishes air quality standards
- **IFRAF, Federal Insecticide, Fungicide, and Rodenticide Act** - provides for the registration, storage, transportation, and disposal of these harmful chemicals
- **SDWA, Safe Drinking Water Act** - sets limits of lead in drinking water
- **NFIA, National Flood Insurance Act** - set up the National Flood Insurance Program
- **FEMA, Federal Emergency Management Act** - issues directives on where to build in coastal areas and floodplains. Appraisers must report whether a property is in a flood hazard area.

### State Legislation

With the enactment of the Financial Institution Reform, Recovery, and Enforcement Act (FIRREA) in 1989, Congress recognized the need to improve appraisal practices. Title XI of FIRREA addresses real estate appraisals and encourages states to form state appraiser and licensing agencies.

All 50 states have responded and enacted legislation that establishes an organization to certify and license appraisers with the Appraiser Qualification Board (AQB) criteria as the minimum qualifications.

APPENDIX

• B •

# Appraisal Review

Appraisal reports are the basis for decisions involving large sums of money or valuable property. Decision makers for governmental agencies or large corporations, who are responsible for property acquisitions or exchanges, and decision makers for institutions that loan money or take mortgages on real estate base their decisions on appraisals.

Appraisal reports have been reviewed for years. Many government programs require appraisal reviews. It is a long-standing government rule that appraisals for federally supported loans be reviewed. Appraisals for federal real estate purchases, exchanges, and condemnations have been reviewed since the 1930s. Banks, savings and loan institutions, large corporations, and even some private clients, such as relocation firms, recognize the value of appraisal reviews conducted by qualified review appraisers.

The appraisal review function has become an important and necessary adjunct to the appraisal process. Appraisal review is now a specialized branch within the real estate appraisal profession. The Uniform Standards of Professional Appraisal Practice has a separate section (Standard 3) for appraisal review.

## Technical Review versus Administrative Review

There are two categories of appraisal review: technical review by a qualified review appraiser and administrative review by an administrator or manager. The class of review required for a particular assignment is most often determined by the client.

An administrative appraisal review involves appraisal management—i.e., managing the procurement and use of appraisals to meet the goals of the institution or agency. Decisions such as whether to loan money and how much to pay for a property are administrative decisions.

Administrative reviews are typically performed by a nonappraiser who may not have the appraisal education or data resources necessary to perform an appraisal. The objective of the administrative reviewer is not to pass judgment on the technical details in an appraisal report, but rather to:

- Determine the acceptability of the report and its compliance with the purpose for which it was prepared
- Assess the report's conformity with institutional or agency requirements
- Evaluate whether the appraisal serves a specific purpose within the framework of making a business decision.

The extent of an administrative review may vary with each institution or agency. However, it must, at a minimum, address key acceptability standards established by federal regulations and/or user-agency policy. Administrative reviews are generally desk reviews and are often prepared using a checklist format.

A technical appraisal review, on the other hand, is conducted to determine if the appraisal report is sound, conforms to accepted, contemporary guidelines, and supports the appraised value.

Whether the value opinion meets the client's needs (whether it is high enough or low enough) is irrelevant to the technical appraisal review. How much the institution should loan or how much the agency should pay for a property are administrative decisions. They have nothing to do with the adequacy or inadequacy of an appraisal report. The purpose of an appraisal report is to develop an opinion of value, generally market value. Consequently, management's or the client's decision about how much to pay or loan **must not** be considerations of the review appraiser. Again, the technical review must be devoted to ascertaining the adequacy of the appraisal report.

The requirements described in the Uniform Standards of Professional Appraisal Practice, Standard 3, and the discussion below relate solely to technical appraisal review.

## The Need for (Technical) Appraisal Review

The increased complexity of the appraisal process, which includes more technical and sophisticated methodology, has resulted in an increased need for appraisal review by competent review appraisers. Most administrators, especially those working for public agencies, large corporations, and financial institutions, want or need a double-check. They want an objective, unbiased analysis of their appraisal reports by an appraisal expert who can tell them whether:

- The appraisal is technically correct
- The opinion of value is accurate as of the valuation date

- The appraisal report represents the product they need and are paying for

Both the public and private sectors realize that appraisal review is simply good business practice. As in many other fields of endeavor, the product prepared (the appraisal report) is subject to examination. Because it is also good business practice to obtain a second opinion about a product, two appraisals of the same property are often obtained. Many times, the value opinions are quite different. Management (i.e., the agency head or the individual responsible for the acquisition, loan, or other decision for which the appraisal is made) needs to know which of the appraisals, if either, is correct. An appraisal review by an expert appraiser is needed to resolve the differences in the appraisals and determine the value opinion that management will use as the basis for its actions.

Appraisal review is needed to assure management (the decision makers) that the possibilities for error or fraud (or merely the appearance of fraud) are minimized. In the case of government agencies, appraisal review by competent, unbiased appraisers is needed to assure the government administrators and the public that the government is not paying too much or too little to the landowners involved.

Appraisal review is one of the checks and balances used to reassure the public that administrators and the appraisers are working under ethical controls. Review of appraisals by competent appraisers will give the public and users of appraisal services confidence in the accuracy of appraisals. Appraisal review gives agencies, institutions, and clients confidence that they are not paying or loaning too much, that the owner is receiving proper value, and that they will not be accused of having inadequate checks and balances.

Appraisal review is also a means of policing appraisers. This is emphasized by the savings and loan scandals of the mid-1980s, when appraisers and the lack of a competent review process for appraisals were blamed for huge monetary losses.

## Purpose of Appraisal Review

Appraisal review is a quality control function. Since an appraisal is often the basis for large expenditures of money—e.g., for loans, purchases or exchanges of real estate, investments—and appraising is not an exact science, the main purpose of the appraisal review is to satisfy the client that the chance for error is minimized. Thus, the appraisal review is necessary to ascertain whether the opinion of value accurately represents the market value of the property appraised.

Other purposes of the appraisal review are:

- To provide a test of reasonableness and appropriateness for the user of the report.

- To check the appraisal for technical adequacy, to ascertain whether the appraisal procedures have been properly and accurately developed, to determine whether the data has been thoroughly and logically analyzed, and to ascertain whether or not conclusions are consistent with the data presented.
- To correct errors and omissions.
- To give the client (management)an enhanced degree of confidence. The review assures management that the proper property and property rights were valued and that the appraiser has examined all relevant indicators and elements of value and applied appropriate appraisal techniques. Finally, the review ensures that the value opinion is supported by the data and the analysis.
- To identify divergencies (i.e., significant differences in conclusions) in appraisals of the same or similar properties.
- To raise appraisers' level of professionalism. Appraisal review should ensure a more uniform, higher-quality product. This will aid the appraiser in professional advancement.
- In some agencies and institutions, appraisal review establishes the basis for accepting or rejecting an individual as an "approved appraiser." In many regulatory agencies, banks, and private corporations, appraisal review is a part of the ongoing maintenance of the client's "approved appraiser" list.

## Appraisal Review versus Appraisal

The definition of an *appraisal* is, "An unbiased estimate of the nature, quality, value, or utility of... identified real estate." Appraisal review has been defined as "an analysis of a completed appraisal to determine if it conforms to specific requirements and guidelines, and to ensure that the appraisal report is mathematically correct and reasonably consistent with data cited. More simply stated, the appraisal review is an *evaluation of an appraisal.*

The similarities between the appraisal and the appraisal review are that they both require objective analysis and reporting by competent appraisers. Just as an appraisal is an objective opinion of a property's value, so the appraisal review is an objective analysis of the appraiser's work.

The appraisal review, in short, is a check to ascertain if the appraisal problem has been correctly solved and if the opinion of value is accurate. Appraising is a complex art, not a science, and it is founded largely on judgment and opinion (based on technical analysis and observation). Therefore, the appraisal review must be consistent with the appraisal problem. Merely reading an appraisal report to verify factual data or mathematical accuracy does not constitute an appraisal review.

The appraisal review is more. It involves an analysis of the property appraised, the appraisal problem, the factual data, and the information supplied. The appraisal review is an analysis of the appraiser's thought process; it involves interpretation of the appraiser's analysis and explanations and evaluation of the data and analyses used to develop the valuation approaches.

The appraisal review is not an appraisal of the property; rather, it is an analysis of the appraisal report. The result of the appraisal review is a second professional opinion about the appraisal problem.

## Qualifications of the Review Appraiser

A review appraiser examines the reports of other appraisers to determine whether their conclusions are consistent with the data reported and with other generally known information.[1] Several, less formal definitions characterize the duties of a review appraiser. The "technical" review appraiser is an appraiser who reviews and analyzes the relevant facts assembled by the appraiser and, using reason and judgment, forms an opinion or conclusion with respect to the real estate problem.

To be qualified as a review appraiser, an appraiser must first be a competent appraiser—educated, trained, and experienced. The reviewer should have qualifications equal to or better than the appraisers whose work is being reviewed. These qualification criteria include:

- *Certification by the state as a general appraiser.* A qualified review appraiser is one who is academically qualified in all applicable appraisal techniques. State-general certification implies a minimum level of education consisting of broad-based approved professional valuation courses and a minimum level of experience.
- *Specialized experience.* Several years of specialized experience in the reviewer's field of employment is important to prepare the reviewer for the complexities of current appraisal assignments for a variety of clients. If the review appraiser is involved in condemnation appraisals, he or she must be well versed in the applicable federal and state condemnation law and in the distinction between compensable and noncompensable items. Court experience is helpful if the reviewer is expected to participate in judicial proceedings as an expert witness. Administrative or consulting duties in the reviewer's field of employment help round out the full range of valuation opportunities to which the reviewer may be exposed.
- *Advanced appraisal training.* The review appraiser should keep abreast of the profession by completing courses or seminars on topics such as condemnation appraising, investment analysis,

1. *The Dictionary of Real Estate Appraisal,* 3d ed. (Chicago: Appraisal Institute, 1993.)

subdivision analysis, mineral property valuation, advanced case studies of a variety of property types, discounted cash flow analysis and appraisal, regression analysis, and highest and best use analysis. The review appraiser should complete suitable course work related to the types of appraisals he or she may be asked to evaluate. The review appraiser must be competent in the contemporary appraisal methods applicable to the type of property being appraised.

- *Personal traits.* One of the more important traits of a good review appraiser is keen analytical ability. From the facts, analyses, and conclusions found in the appraisal report, and his or her own knowledge and investigations, the review appraiser must be able to make logical and sound decisions. Reviewers must be able to analyze a report and "separate the wheat from the chaff," identifying which items about an appraisal are significant and which are not.

  A review appraiser should also have good verbal and written communication skills and the ability to convince, train, and persuade others. The reviewer exhibits mature judgment in reviewing the work of other professionals by not nitpicking reports nor, conversely, rubber-stamping them. The reviewer should have the ability to read between the lines, combining logical, analytical methods of evaluation with good intuition. Above all, the review appraiser must have the confidence to review another professional's work authoritatively and provide constructive criticism in a tactful, responsible manner.

  Finally, the review appraiser must be able to cope with controversy because controversy is part of the review process. If this were not so, there would be no need for the review appraiser.

- *Objectivity and fairness.* In addition to being educated, trained, and experienced, the review appraiser must demonstrate competence, objectivity, fairness, and integrity as a professional. The reviewing appraiser is a valuation expert hired to render objective opinions concerning real estate values and appraisal reports. The reviewer subscribes to a code of ethics and to the Uniform Standards of Professional Appraisal Practice.

  To ensure objectivity and independence in the review process and to preclude the appearance of conflicts of interest or wrongdoing, review appraisers should not:

  1. Be responsible for negotiating the sale, purchase, or loan approval of the appraised property
  2. Review an appraisal prepared by the reviewer's supervisor
  3. Review an appraisal of a property personally and recently appraised by the reviewer

4. Review an appraisal prepared by someone who has recently reviewed the reviewer's own appraisal work
5. Participate in any appraisal or review activities that could give the appearance of conflicting with the Ethics Provision of the Uniform Standards of Professional Appraisal Practice.

Like other appraisers, the reviewer must not become an advocate. One of the reviewing appraiser's greatest tasks is to remain objective and impartial and evaluate the technical aspects of the appraisal, not the political aspects. The review appraiser is often subject to severe pressures, both internal and external, and some pressures may be of his or her own making. Although expenses and deadlines are affected by the reviewer's actions, the reviewer cannot be concerned whether the loan officer makes the loan, whether the government buys the property or has to condemn it, that the "owner will not sell at that price," or that the boss might not meet acquisition or loan targets and get a good performance rating. Reviewers who become concerned with any of these issues risk becoming advocates. Their decisions might be vulnerable to criticism, create controversy, or damage the reputation of the agency, organization, or client for whom they work.

There are often many personalities involved. Subtle pressures might be applied by people within or outside the organization who, for political, monetary, or other reasons, want higher or lower values or want appraisals approved or disapproved. Often a reviewing appraiser must critique the work of highly reputable appraisers. The reviewing appraiser must not allow the wishes or whims of others to influence his or her decision concerning the accuracy or acceptability of the appraisal report. Despite outside influences, the professional reviewing appraiser focuses on determining whether the appraisal report is technically correct and appropriately supported. The reviewer can, however, work with management, clients, and appraisers to secure proper appraisals or make the necessary revisions and improvements.

- *Membership in a professional appraiser organization.* Although not required, the reviewer's skills may be enhanced by membership in one of the professional appraiser organizations that sponsor The Appraisal Foundation. These organizations have minimum qualifications requirements, fair testing, and a code of professional ethics and conduct.

## Responsibilities of the Review Appraiser

### Objectives

The main objective of the review appraiser is to secure *approvable* appraisals. To present professional valuation advice to the client, e.g., management, loan officers, attorneys, the review appraiser must first

- Identify the property to ensure that the correct property was appraised
- Identify the property rights to be valued
- Define the appraiser's assignment.

The appraisal reviewer does not have to find something wrong with each appraisal. Each appraisal should be evaluated in the context of what is reasonable. Thus, the review appraiser ensures that the appraisal has no fatal flaws and includes proper support for the value opinion.

### Primary Responsibilities

The main responsibilities of the review appraiser are to:

- Ensure that the appraisal conforms to contractual requirements (that is, the appraiser did the job he or she was hired to do).
- Ensure that the appraisal conforms to recognized, contemporary appraisal principles, practices, and techniques
- Ensure that the appraisal is based on applicable law and departmental, agency, or institutional regulations and policy. (For example, if the appraisal is prepared for a federal agency, did the appraiser follow applicable federal law and appraisal rules?)
- Ensure that the appraisal approaches are properly developed and technically correct; that the data has been analyzed in a logical fashion and the analysis conclusions are consistent with the data in the report; and that the quantity and quality of the data are suitable for the problem at hand.
- Ensure that the value opinions are supported by the data cited or other information available.
- Clarify contract specifications, law, rules, policy, and regulations with regard to the property appraised. The review appraiser is responsible for ensuring that the appraiser valued the correct property and property rights and that the value conclusion represents the value sought (e.g., market value of the whole as a single unit, value of the sum of the parts).
- Correct significant technical deficiencies. While insignificant deficiencies that have no measurable effect on value can be disregarded, the review appraiser must ensure that significant deficiencies are corrected.

- Ensure that he or she is competent or associate with a reviewer who is competent to evaluate specialist reports regarding resource appraisals (such as timber or minerals) or other peculiarities of the valuation assignment.
- Conduct personal investigations when there is reason to believe the cited facts are not correct or are incomplete.
- Resolve valuation divergencies. When two appraisals for the same property have significantly different value opinions, it is the review appraiser's responsibility to ascertain which, if either, is the most accurate and best opinions of the property's value.
- Inform the client of all defects in the appraisal report.
- Alert the client to any limiting conditions affecting the valuation.

The review appraiser must ultimately convince management that the opinion of value represents the value sought. If the objective is market value, as it normally is, the review appraiser must convince management that the value opinion represents the price for which the property, with all its limitations and potential, would sell for after reasonable exposure time on the open market under current market conditions.

The review appraiser also has responsibilities to the appraiser who prepared the report. The review appraiser should read the entire appraisal report before making any judgments. No section should be taken out of context. If the review appraiser disagrees with the appraiser, the appraiser is entitled to know the reasons why and examine the evidence or analysis supporting the reviewer's viewpoint.

## Limitations, Authorities, and Obligations

In carrying out their responsibilities, working with management, and working with appraisers, review appraisers must understand the limitations on their actions as well as their obligations and authorities (or prerogatives).

### Limitations

The reviewing appraiser **cannot**

- Change an appraisal report (except for minor, clerical or typographical errors, and then only with the appraiser's concurrence)
- Direct an appraiser to conclude a certain value opinion
- Try to influence the appraiser's judgment
- Substitute his or her judgment for the appraiser's
- Dismiss careful appraisal judgment solely because it cannot be conclusively supported by sales data
- Direct an appraiser to rely on a particular appraisal approach or method

- Impose his or her personal theories, rules, or procedures on the appraiser. (The review appraiser can only insist that the appraiser use procedures, theories, and techniques recognized by the appraisal profession, federal or state case law, or applicable regulations.)

### Authorities (or Prerogatives)

Although the review appraiser cannot impose his or her judgment, opinions, and findings on the appraiser, the reviewer **can**

- Question the appraiser's judgment when it is illogical or not reasonably inferred from data cited
- Consider all pertinent data available (e.g., other appraisals, sales data)
- Clarify information with engineers, other appraisers, real estate people, other landowners, and zoning authorities
- Discuss the appraisal and its problems with the appraiser to resolve the factual, legal, or conceptual differences that are the basis of the problems.
- Request from the appraiser clarification, supplemental data, and further analysis.
- Reconfirm factual data such as sales and zoning regulations
- Require the appraiser to consider all pertinent facts and information
- Require the appraiser to support the appraisal conclusions with reasonable facts and/or analysis
- Make a field review

### Obligations

In addition to things that the reviewing appraiser can and cannot do, there are certain things that he or she must do. The reviewing appraiser **must**

- Determine if the alleged facts cited are correct, if the assumptions are valid, if the analyses and appraisal approaches are properly processed, and if the appraiser did an adequately thorough job.
- Determine whether the analyses and conclusions are consistent with the data cited in the report.
- Get all the facts before making the final decision on an appraisal.
- Correlate the value opinions against all available, pertinent information. Other appraisals of the same, nearby, or similar properties may be compared to the appraised property to ascertain the reasonableness of the appraisal. Other data on hand, such as other relevant sales, time-trend studies, and economic analysis may be considered.

- Be unwilling to accept anything less than a proper appraisal report.
- Evaluate the appraisal report in the context of the data and market conditions in effect at the time the appraisal was conducted. The reviewer must not refute the appraisal report using events subsequent to the appraisal date unless those events were assured of occurring at the appraisal date.
- Demonstrate reasonable judgment. Reasonable limits are considered for each appraisal, based on data available. The reviewer should not evaluate the appraisal in the context of right versus wrong, but rather in terms of what is reasonable versus what is unreasonable.
- Be unwilling to approve (or recommend approval) until all significant appraisal problems and divergences have been resolved. In this regard, the review appraiser must distinguish between significant deficiencies and those that might be perceived as significant, but are actually interpretive.
- Clearly understand the consequences of approving or disapproving an appraisal.
- Approve or recommend for approval the opinions in compliance with institutional or agency policy.
- Write a convincing and defensible review report, describing the reasons why the appraisal report is approvable or not approvable.
- Share, with the appraiser who prepared the appraisal report, responsibility for the value opinion by signing the review report. The review appraiser must be willing to convince management or the client of the accuracy of the appraisal and the value opinion. The reviewer must be prepared to defend the appraiser and to testify in court as to the value of the property, if necessary.
- Document all actions to get the appraisal report approvable. All contacts with the appraiser concerning the appraisal report must be documented, including responses by the appraiser. Also, the review of other material, inquiries, and confirmation of appraisal data should be documented.
- Render an independent, objective, and impartial decision. Again, the review appraiser must not become an advocate.

## Reviewing Appraiser's Options and Actions

The reviewing appraiser's primary objective is to secure an approvable appraisal. At the conclusion of the appraisal review, the review appraiser has several options. The most common actions a review appraiser may take are: 1) approve the report, 2) accept the report without approval, 3) request or order another appraisal 4) disapprove, and 5) perform another appraisal (i.e., the reviewer becomes the appraiser). At the outset of an assignment, the review appraiser must clearly understand the consequences of approving or disapproving an appraisal.

### Approval

The review appraiser should approve the appraisal report or recommend its approval if the report conforms to existing standards, follows current corporate or agency policy, is based on the proper premises, adequately supports the value opinion, is consistent with the review appraiser's personal knowledge, and is in compliance with Standard 2 of the Uniform Standards of Professional Appraisal Practice.

The review appraiser must not approve an appraisal with speculative or unreasonable limiting conditions and assumptions or a report that is substandard. Significant weaknesses, differences, and divergences noted during the current review, or previous reviews, of the appraisal must be resolved. The review appraiser must document that previously noted deficiencies, weaknesses, or questions have been corrected or clarified.

The review appraiser must not approve an inadequate report or an inadequately supported value opinion. However, in attempting to make an appraisal approvable, the reviewer may support an otherwise weak appraisal by adding, through the review report, additional vital data and personal analyses. The value opinion may then be approved as supported by the reviewer, provided the appraiser's value opinion is not changed. (The reviewer should first, however, attempt to have the appraiser improve the appraisal report to the required standard.)

Each appraisal report should be sufficiently complete and thorough to support itself. While this is not always possible, the record must show that the value opinion is supported.

### Acceptance Without Approval

An appraisal may be technically acceptable even though the value opinion is in doubt or unconvincing. In such cases, the review appraiser "accepts the appraisal without approval." Such acceptance must be followed by, or made in conjunction with, an action to approve a better-supported appraisal report or to obtain another appraisal report to substantiate, support, or replace the original report. Approving one appraisal report over another "acceptable" report, where the divergence in value opinions is within acceptable parameters, constitutes acceptance without approval of the unapproved report. The reviewer must clearly state and support the reasons for approving one report over another.

### Recommend A Second Appraisal

If an appraisal is not approvable or does not provide strong support for the value opinion, or if the reviewer questions the value estimate without having defensible reasons to disapprove the report, the reviewer may request a second appraisal to refute or support the first appraisal. In such a case, the second appraiser should be instructed to appraise the property without knowledge of the results of the first.

When the second appraisal report is received, the reviewer should compare the reports, resolve any differences, and approve the best report. The unapproved report may be simply "accepted," as explained above. On the other hand, both reports may be approved if they are essentially compatible. If one report is not acceptable, it may be disapproved, but it should not be disapproved simply because the other report is better. There must be sufficient weaknesses to warrant disapproval.

### Disapproval

If the report does not meet acceptable standards, after diligent efforts have been made to get the needed support and improve it, and the value opinion is still not adequately supported, the review appraiser should disapprove the appraisal report with full documentation regarding the basis for disapproval.

Some states require state-certified and state-licensed appraisers to report (often under statutory immunity) any real estate appraiser who has acted in a manner that does not meet the generally accepted standards of professional appraisal practice. In addition, some states require that real estate appraisers who have performed appraisal services beyond their level of competency be reported to the applicable licensing authority. Reviews that disapprove an appraisal report should be written with state reporting requirements in mind so that they may be provided to the state licensing authority.

### Perform Another Appraisal

In certain situations, *but only as a last resort,* the review appraiser may recommend (not approve) a value opinion or opinion of just compensation different from that cited in the appraisal report under review. The reviewer may recommend such a value when all of the following apply:

1. Despite diligent efforts to improve the report, the problem still cannot be resolved satisfactorily with appraiser of record;
2. The review appraiser still cannot concur with the appraiser's opinion; and
3. Obtaining another appraisal is not feasible or economical.

If all these conditions are met, the review appraiser may offer a different opinion of value. That value opinion then becomes a second appraisal. The new opinion is the sole responsibility of the original reviewer; who is now the appraiser of record. The original reviewer may cite and support the new opinion in the appraisal review report, using the original appraisal report as the foundation. Or the reviewer may prepare a separate, self-supporting appraisal report. In either event, the documentation must comply with Standard 2 of the Uniform Standards of Professional Appraisal Practice and include inspection of the appraised property and the comparable sales cited to support the value opinion.

Recommending a new opinion of value does not constitute changing an appraisal report. The review appraiser's new opinion of value must be reviewed at the next higher level.

## Review Procedures and Methods

### Categories of Appraisal Review

Generally, there are two categories of appraisal review: the *terminal review* and the *cumulative review.*

In the terminal process, the appraisal report is reviewed in full by a single reviewer or a group of reviewers located near the project. The terminal reviewer makes the ultimate decision concerning the appraisal report.

In the cumulative review process, which is often called the *agency process,* the appraisal report is reviewed in whole or in part at several levels, with final approval decided at a level and a location away from the project. The first level of review is the most detailed, usually involving a field review. The highest level is often a desk review, which is explained below.

## Review Procedure

The type of appraisal review may vary, depending on the scope or nature of the assignment, the client or policy requirement, and nature of the property appraised. The review procedure generally includes five steps:

1. Project familiarity
2. Office review or desk review
3. Field review
4. Final office review
5. Written review report

### Project Familiarity

The review appraiser must first become familiar with the assignment, the appraisal report, and the appropriate market area. Familiarity is best achieved by reviewing the data presented and examining the appraised property and comparable sales. Often this is accomplished in the field review, but sometimes it is done preceding the appraisal. This is the normal procedure for a project in which several appraisals from the same area will be submitted. In such cases, pre-appraisal inspection of the area and the project is necessary. Often, the reviewer makes this inspection with the appraisers.

The review appraiser learns about the local economics, land types, land uses, and different property types in the area. A personal understanding of the local economics that affect the real estate market is particularly useful.

## Office Review or Desk Review

Typically, especially in second-level reviews, the desk review is made without a field inspection other than the preceding project familiarity step. In other cases, such as in the one-level (terminal) review process, the office review is preliminary to the field review.

The office review involves reading the report in its entirety and noting any inconsistencies, contradictions, items that will require further evaluation or special mention in the review report, and items to be checked in a field review. In the office review, data and analyses supporting the appraisal conclusions are checked as are all mathematical computations. The legal description and estate appraised are checked against the deeds or options, reported sizes, maps, and construction plans provided to ensure that the correct property and property rights are appraised. Limiting conditions are reviewed. The office review should identify any legal matters to be resolved.

The appraisal methodology and techniques employed in the appraisal are reviewed for their appropriateness and to ensure that they are properly and accurately developed. Data cited in the report may or may not be confirmed by the desk review. If not, data that need additional confirmation are highlighted to be checked in the field review. In the desk review, a review checklist is often used, although experienced reviewers have the checklist memorized.

The appraiser's qualifications are also checked during the office review to ensure that the appraiser is qualified to do the job at hand. Desk reviews are usually limited to the data presented in the appraisal report, although pertinent data in the reviewer's files may also be used.

## Field Review

A field review, conducted with the appraisal report in hand, is a mental reappraisal of the property by the reviewer through the eyes of the appraiser. Data contained in the appraisal report are confirmed in the field review. The appraised property and the comparable sales are inspected on the ground. Field reviews include at least an exterior examination of any improvements. Other data, such as zoning and sales listings, may also be checked. Additional competitive data may be identified. Sometimes the field review is made with the appraiser.

Often, only a partial field review is necessary. This involves inspection of only the subject property or inspections of the subject property and the most important sales cited in the appraisal.

The main purpose of a field review is to assure the reviewer of the thoroughness of the appraiser's investigations into trends and conditions and the accuracy of the reporting. The depth and extent of the field review investigation will depend in large measure on the reviewer's evaluation of the appraiser's reliability, the nature of the property appraised, and the requirements for exercising necessary quality control.

Some agencies and institutions require a field review for every appraisal review. Others only require a designated percentage of appraisals be field reviewed. In other cases, only problem appraisals are field reviewed. A field review is normally conducted by government agencies, lending institutions, and others as a check on the appraisal process. A field review may be used by a professional appraisal organization to verify a candidate's experience.

### Final Office Review

In the final office review, the reviewer correlates all the data in the appraisal report, in the reviewer's files, gathered during the field review, and obtained by additional investigation and inquiry. Then, the review appraiser makes the critical decision: Is the appraisal report good enough to approve? Is it bad enough to disapprove? Is additional data and analysis needed from the appraiser? Should another appraisal be obtained?

The final review is made during the office review when a field review is not needed. Only by evaluating the appraiser's qualifications and studying the information, data, and analysis presented in the report can the reviewer determine if the report conforms to applicable law, regulations, institutional or agency standards, and the Uniform Standards of Professional Appraisal Practice.

### Written Review Report

After completing the evaluation of an appraisal, review appraisers must prepare a written review in compliance with Standard 3 of the Uniform Standards of Professional Appraisal Practice to document their findings and support their conclusions. The types and content of review reports are described below.

## Appraisal Review Reporting Methods

The two recognized formats for preparing an appraisal review report are the narrative review report and the form review report. These formats are described below. Each format can be prepared using one of two general methods: the detailed report or the exceptional report. When the detailed reporting method is used, all actions taken by the appraiser and the reviewer are recorded. The reviewer comments on all parts of the appraisal. Conversely, using the exceptional method, the reviewer records and comments **only** on exceptional points of the appraisal.

Review reports using the exceptional method must still comply with Standards Rule 3-2 of the Uniform Standards of Professional Appraisal

Practice regarding content, albeit in an abbreviated form. The exceptional system is often used for higher-level reviews within an organization when the first-level review is a narrative review report.

Generally, an appraisal review report need not duplicate the work that was adequately performed in the appraisal report. The review report merely needs to summarize the appraisal so an individual can read the review report and have a good idea what the appraisal report contains and how well it complies with the standards under which it was prepared.

The level and type of review report should be predefined. There should be no misunderstanding regarding the type of review report requested by the client. The scope of the review should be clearly understood by both the reviewer and the client at the outset.

Whether the reviewer is approving the appraisal report, returning it for improvement, or disapproving the appraisal report, the review report must support the reviewer's conclusion. The report must demonstrate why the appraisal report is approvable, or why it is not approvable.

## Narrative Review Report

The narrative appraisal review report provides a detailed explanation of all facets of the review process, from recitation of the assignment and summarization of key components of the property being appraised to itemized comments describing deficiencies or items worthy of commendation. The narrative review report summarizes the important elements of the appraisal, usually commenting on each part of the appraisal, and records the review appraiser's actions, analysis, and conclusions. The length and depth of the review report depend on the client's (or management's) needs and the complexity of the appraisal problem.

The narrative review report may be detailed or exceptional. It generally includes the following elements:

- What the appraisal report is.
  This summary should include the appraiser's name, date of the appraisal, a brief description of the property ( i.e., size, location, estate appraised), and the value opinion.
- What the appraiser did.
  The review report should summarize how the appraiser solved the appraisal problem, the approaches used, and the analyses and adjustments made.
- What the reviewer did.
  The review report should summarize the scope of the review—i.e., what the reviewer did in the process of the review. For example, the review report should state whether a field review was made, or whether the review is limited to a desk review. The review report should cite the actions the reviewer took to check and reconfirm the data and analysis.

- What the reviewer thinks of the appraisal report.
  The reviewer should comment on the strengths and weaknesses of the appraisal report. The reviewer should point out why minor deficiencies do not detract from the credibility and accuracy of the report.
- Conclusion.
  The review report should state the review appraiser's conclusion—approval, disapproval, or second appraisal—and support that conclusion.

An example of a narrative review report format follows.

**Introduction**

Property appraised

Appraiser, co-appraiser

**Appraisal Summary** *(Identify property appraised per Standard 3-1(b))*

Owner of record

Location of property

Size of property

Estate appraised

Improvements

Highest and best use

Date of value opinion

Opinion of value

Date of the review

**Scope of the Review** *(Standard 3-1(c), 3-2(c))*

Desk review

Field review

Discussions with appraiser

**Property Data Summary** *(Property rights appraised per Standard 3-1(b))*

Briefly describe the property—location, size, shape, land types, predominant characteristics contributing to value.

**Estate Appraised** *(Property rights appraised per Standard 3-1(b))*

Identify the estate appraised. This is a narrative discussion of the estate appraised, describing the effect that any outstanding rights or other limitations on the property title have on use of the property and how the value of the property is affected.

**Valuation**

Summarize the appraisal approaches and techniques used, how they were developed, and the value conclusions derived from each approach.

**Comments** *(Opinion as to appropriateness per Standard 3-1(f))*

The reviewer should comment on the appropriateness of the appraisal approaches applied, their development, and the analyses leading to the appraiser's conclusions. Point out the strengths of the appraisal report. Describe significant weaknesses in the appraisal and point out the effect on the valuation and overall quality of the report. Describe why minor weaknesses are insignificant.

**Action**

Approve, disapprove, accept, conditional approval, or order a second appraisal.

**Reviewer's Certification**

Reviewer's name, title, signature, certification, and list of limiting conditions.

## Form Review Report

In a form review report, the documentation associated with reviewing an appraisal report is abbreviated, but the review process is not. The full review process is necessary regardless of the review report format. To prepare a form review report, the reviewer summarizes some very basic information about the property being appraised and then records a positive or negative response for each item printed on the form. The form report normally provides an opportunity to elaborate on items requiring special mention or clarification.

Form review reports are typically developed for specific institutional or agency needs. The Standard Appraisal Review Report (SARR) form is a common format for rural appraisal reviews and is available through ASFMRA.

## Review Checklist

A useful tool for either review report format is an appraisal review checklist. A checklist helps the reviewer systematize the review process to ensure that no section is inadvertently overlooked and that the appraisal complies with the client's requirements. Strict adherence to a checklist can, however, cause the reviewer to shortchange qualitative aspects of the appraisal report. The checklist does not determine the acceptability or unacceptability of the appraisal. It is merely a tool to help the reviewer systematize the review process. Currently, no one review checklist is universally accepted. Most checklists are tailored to the specific criteria of the institution or agency that uses them.

## Divergency in Appraisal Reports

A reviewer encounters a major problem when two competent appraisers develop widely different values for the same property, or where competent appraisers derive incompatible values for different, but nearly similar properties. One of the greatest tests of the review appraiser's skill and judgment is in resolving divergence.

*Divergency* is defined as a significant difference between the conclusions of two appraisers or between one appraiser and the reviewing appraiser. Even the term "significant" is subject to interpretation. For years it has been conceded that if the value opinions prepared by two appraisers (appraising the same property as of approximately the same valuation date with the same data) vary by 10% or less, the value opinions are compatible. This range, however, applies to a noncomplex property. The value spread between "compatible" appraisals can be even larger if the properties are complex and appropriate data are scarce.

## Causes of Divergency

**Divergency** usually can be attributed to one of the following factors:

1. *Factual, i.e., different understandings of the facts.* The appraisals might be based on different facts. For example, the appraisers might be using a different legal description or have a different understanding of the property rights to be appraised. Other factual differences might be noted in property size, encumbrances, zoning, or sales dates or prices.
2. *Inaccurate assignment.* The divergency might be caused simply by erroneous or confusing instructions pertaining to the law (e.g., "larger parcel," rights taken, interpretation of the effect of easements) or a misunderstanding regarding the application and measurement of damages, benefits, and compensable or noncompensable items. The appraiser may not have been given the most current, correct legal description or the appraiser might have misunderstood the instructions. Often appraisers do not understand whether an appraisal is to be made following federal or state law. Sometimes a property is a split estate but is not identified as such at the beginning of the project. Many times the condemnor identifies the property to be taken, but does not identify the "larger parcel." Poor communication is one of the greatest causes of appraisal divergency. It is imperative that the appraiser be given precise instructions, and that he or she understands them.
3. *Judgment.* The divergency might be based simply on different judgments concerning the property's highest and best use, the depreciation of buildings, the amenities of the property, the comparability of comparable sales, and other appraisal elements. Appraisers have varying degrees of knowledge, skill, and judgment and may arrive at vastly different value opinions for the same property. Differences in skill and experience significantly affect judgment.
4. *Type of property appraised.* Often it is the property itself that is the cause of the divergency. Unique or difficult-to-appraise properties will usually result in divergent value opinions, especially when good, comparable data are scarce. As stated above, highly competent appraisers, using the same data and appraisal instructions, may conclude values that vary by much more than the standard 10%. When appraising complex properties, there must be some latitude for the application of different premises and judgments. The definition of "significant difference" must be expanded for these appraisals.

5. *Motivation.* Another cause of divergency can be attributed to the human desire to please one's employer. The landowner's appraiser believes it is best to err on the side of the landowner, while the condemnor's appraiser, who believes that the government wants to save public money, is inclined to be "conservative." A lack of complete independence on the part of one appraiser, or both, often is the cause of divergent opinions of value.
6. *Poor appraisals.* Sometimes the divergency can be attributed simply to one or more poor appraisals. Poor appraisals result from inexperience, incompetence, conflicting motivation, or, sometimes, dishonesty.

## Resolving Divergency

The first step in resolving divergency is to review both appraisals completely and thoroughly and ascertain the source of the differences. Different approaches are used to resolve divergencies resulting from different factors.

1. *Factual differences.* Factual differences can be verified as can differences in the appraisers' instructions. The appraisers should be asked to clarify or confirm the facts in question and rethink the analysis based on any changes.
2. *Inaccurate assignment.* If the divergency is a result of an inaccurate assignment or interpretation of the assignment, furnish both appraisers with correct assignment instructions—i.e., legal description, estate to be appraised, legal instructions, proper appraisal method or procedure to use—and have the reports revised according to the correct instructions.
3. *Judgment.* Frequently a divergency is an honest result of differing ideas about the property. While factual differences or appraiser instructions can be verified, resolving divergency resulting from differences in appraisers' judgment is the greatest test of the reviewing appraiser's skill and judgment. The reviewer must analyze the basis, reasoning, and analytical process used by each appraiser and make judgments as to their validity. He or she may call for further detail, supporting data, or analysis from one or both appraisers.
4. *Type of property appraised.* Often a value difference is due to the nature of the property. Data to solve appraisal problems involving some unique property types might be scarce or unavailable. Consequently, reputable appraisers might differ in their evaluation and conclusions concerning the inferred valuation of the appraised property. This type of difference will appear in appraisal reports as conceptual or judgmental differences and, in

these cases, more latitude must be given for honest differences. The review appraiser faced with this situation must carefully analyze both appraisal reports and select the appraisal report that, in the reviewer's opinion, provides the strongest support for the appraiser's conclusions and best represents the value of the appraised property under current conditions.

5. *Motivation.* Divergency can result when a value opinion is influenced by the appraiser's motivation. Again, this will usually appear in an appraisal report as a conceptual or judgmental difference. The appraiser may select an inappropriate highest and best use which supports a higher value or make unsupported and/or extreme adjustments to the comparable sale prices. Thus, divergency resulting from a lack of objectivity on the part of the appraiser is usually resolved as a judgmental divergency. If the reviewer cannot persuade the appraiser to make a truly objective appraisal, the appraisal must be disapproved. Moreover, the appraiser should not be considered for future appraisal work.
6. *Poor appraisals.* Divergency resulting from inadequate appraisals may be resolved by simply requiring the appraiser to do an adequate job. Often a poor appraisal report will be prepared by a hired appraiser who was assumed to be competent or by an assistant who does not have the necessary qualifications to perform the appraisal. In this case the contracted appraiser should be required to perform the appraisal and furnish an appraisal report that meets the client's requirements. Sometimes a poor appraisal results because the appraiser is actually not competent to solve the appraisal problem. In this case the appraiser must, according to USPAP, employ or associate himself or herself with a competent appraiser. If the appraiser declines to do this, the inadequate appraisal report must be disapproved.

## Final Decision

Finally, the reviewing appraiser must attempt to get each appraisal improved to the quality needed and document all steps taken. In contacts with the appraiser, the reviewing appraiser should not try to influence the appraiser's opinion of value. The reviewer should be merely trying to resolve questions and find support for the value opinion. If the problem is one of judgment, the reviewer should discuss the need for additional information, analysis, and support for the judgments and value opinion presented.

Ultimately the reviewing appraiser must determine which appraisal report is the most correct and which value opinion is the best supported and most accurate. If the reviewing appraiser does not substantially agree with either of the appraisals and the divergency and any other problems cannot be satisfactorily resolved, the reviewer may obtain another appraisal or prepare his or her own appraisal.

The review report should summarize all the appraisals. If one report is not approved, the reviewer should cite the weaknesses of that report and demonstrate why one appraisal is better than the other. The review report must support the reviewer's conclusion. In addition, the review report should summarize the attempts to bring all reports up to an acceptable level.

## Obtaining Good Appraisals

Often the reviewing appraiser is responsible for procuring appraisals. In these situations, the review appraiser has a good opportunity to obtain adequate appraisals by pre-appraisal work. A good review starts before the appraisal is made, even before it is assigned.

First, the review appraiser must become familiar with the project. Then he or she can determine the type and extent of appraisal required, the qualifications of the appraiser, and the information the appraiser will need. The appraisal is more likely to meet the requirements if the review appraiser follows these rules:

1. Visit the project (appraisal) site. Identify the appraisal needs, legal matters, trends, and other relevant factors. Evaluate the complexity of the appraisal problem.
2. Write clear instructions and specifications. Explain fully what is to be appraised, when, and the type of report needed. Explain the extent to which valuation approaches must be developed and conclusions supported.
3. Evaluate the appraiser qualifications needed in terms of the specific appraisal assignment. Selection of the appraiser should depend on the complexity of the problem. For example, an appraiser with outstanding credentials for appraising high-rise apartment or commercial buildings is not needed for the appraisal of a single-family residence. Conversely, an appraisal for condemnation of part of a farm or ranch should not be assigned to an appraiser with only commercial (downtown) appraisal experience. Partial acquisition appraisal training and experience would be desirable, if not mandatory, for such an assignment.
4. Select appraisers who have the qualifications to solve the appraisal problems involved and to write an acceptable appraisal report, who consistently provide a satisfactory work product, who can accurately apply accepted and appropriate appraisal methods and techniques, and who have demonstrated good judgment. Experience on similar projects is probably necessary if the project is difficult or unusual.

5. Furnish the appraiser what he or she needs. The chance for a good appraisal is enhanced when the review appraiser knows what the appraiser needs and ensures that the appraiser is given adequate information.

   For typical appraisals, the review appraiser should provide the appraiser with:

   a. An accurate legal description
   b. Information on the estate to be appraised, which identifies
      (1) reservations or outstanding rights
      (2) access rights
      (3) water rights
      (4) mineral rights
      (5) other restrictions, such as deed restrictions, restrictive covenants, easements, or other encumbrances.
   c. Title information
   d. Adequate maps and plats

      For condemnation appraisals or partial acquisition appraisals, the review appraiser will also need to furnish:
   e. Rules of law—federal versus state law or interpretation of such laws
   f. Details of the project—what it is, when it began, what it will do
   g. Property rights to be taken or acquired
   h. Interpretation of rights remaining with the landowner
   i. Physical details of the project—construction or other changes involved, what will be done as part of the project, and what will be left for the landowner to cure
   j. Physical effect of construction, if any
   k. Proper identification and size of the "larger parcel," and the "remainder"

6. Require pre-work discussions. If the project is large or complicated, pre-work discussions are helpful. At the pre-work conference and as the appraisal assignment progresses, the review appraiser is responsible for clarifying the assignment. Topics to be discussed should include:

   a. Definition of the property and the appraisal problem
   b. A clear understanding of who (the client or the appraiser) will furnish
      (1) Land surveys
      (2) Timber cruises

(3) Mineral surveys

(4) Mineral appraisals

(5) Engineering studies

(6) Soil studies, seepage tests

(7) Environmental impact studies/reports

(8) Hazardous materials surveys, disposal

c. Type and number of appraisal reports needed

(1) Narrative or form report

(2) Documentation required (e.g. five pages or 200)

d. Which approaches must be fully developed. For some complex properties, a subdivision approach might be desired. For mining properties, a full discounted cash flow appraisal might be needed. On the other hand, on large, unimproved land parcels that currently produce no income related to their highest and best use, an income approach might be superfluous. For example, in federal condemnation cases where there are ample sales to develop the value opinion by the sales comparison approach, the income approach is not needed. These decisions must be conveyed to the appraiser prior to the appraisal assignment so that he or she understands the full scope of the assignment. In partial acquisitions the appraiser must understand the applicable appraisal method to use—a full before-and-after appraisal, a modified before-and-after appraisal (part taken as part of the whole plus changes in value of the remainder), or the part taken as a separate unit method.

e. Legal instructions

(1) Federal or state law applicable?

(2) Consider enhancement or diminution from the project?

(3) Only special benefits offset or are all benefits offset? Are benefits offset against the whole award or only against damages?

(4) Definition of the project and date of project inception.

f. Review of the appraisal report. It is essential that the appraiser realize that the appraisal will be reviewed by a competent review appraiser and that questions will probably arise and must be resolved.

## Working with Appraisers

It is important that review appraisers learn how to communicate with appraisers. All questions should be expressed in a positive tone so the appraisal review can be a learning experience for the appraiser. All written communications should be grammatically correct, well documented, and tactfully presented. The reviewer's reasoning should be explained and supported with factual data and analysis. Review appraisers should avoid a patronizing or condescending tone and should not be unduly critical. Review appraisers must be careful not to make personal references in any written communications to or about the appraiser. These are irrelevant to the report and the appraisal review process.

The reviewer should persuade, rather than demand, that the appraiser improve or better support his or her appraisal report. The reviewer can ask the appraiser to "reconsider," "re-analyze," or "take another look at" the appraisal conclusion in light of data cited in the appraisal report or new data provided by the review appraiser. A review appraiser will be much more successful in dealing with an appraiser if their interaction is professional.

## Review Appraisers are Subject to the Standards Rules

Standards Rules 2-5, 3-1, and 3-2 of the Uniform Standards of Professional Appraisal Practice govern the activities of review appraisers.

Standards Rule 3-1 outlines the steps the review appraiser must take in conducting an appraisal review. The requirements cited above conform to this standards rule.

Standards Rule 3-2 requires that, after forming an opinion as to the appropriateness of the appraisal report, the review appraiser must prepare a separate review report outlining the nature and the results of the review process. All appraisal review reports, regardless of the format used, must contain a "certification statement." The review reports discussed above fulfill the requirements of Standards Rule 3-2.

Standards Rule 2-5 requires that if a review appraiser signs an appraisal report prepared by another, the review appraiser must accept full responsibility for the contents of the report. Normally, the review appraiser signs the review report, but not the appraisal report.

## Advice to Reviewing Appraisers

The following guidelines are suggested to reviewers:

1. Avoid collaboration with the appraiser. The reviewer's task is to conduct an independent, objective analysis. The reviewer may assist the appraiser in selecting the correct analysis or appraisal

technique, if asked, but must avoid getting so involved in the appraisal process that the appraisal is actually a result of the reviewer's own work and analysis.

2. Don't direct the appraiser to a value or a valuation premise. These must be the appraiser's decisions.
3. Don't require a mathematical formula for every appraisal process.
4. Don't require that everything be documented and supported beyond any doubt.
5. Require facts and analysis that lead to the appraiser's conclusions, but be open-minded.
6. Don't get so involved in the project that objectivity is lost. Some reviewers get so involved in the project for which the appraisal is made that they lose their objectivity and approve substandard appraisals rather than impede the progress of the project.
7. Don't let the needs of management or the client regarding deadlines and priorities influence the review decisions.
8. Don't let management or the client pressure you (the review appraiser) into recommending approval of an unacceptable appraisal report. There are other options.
9. Treat the appraiser as a professional. In discussions concerning the need to improve an appraisal report, leave the appraiser a way out. Avoid making the appraiser lose face.
10. Don't accuse the appraiser of wrongdoing, carelessness, collusion, or trying to come up with a certain value.
11. Don't exaggerate your role.

## Conclusion

The reviewing appraiser's job is one of quality control. He or she is responsible for the quality of appraisal reports prepared for his or her employer or client. The employer or client must rely on these appraisal reports in making decisions about real estate transactions. In the end, the reviewing appraiser is responsible for approving or recommending a value opinion that represents the market value of a particular property.

Appraisal review is an analysis and evaluation of the valuation process employed by an appraiser. It is an analysis of the factual information cited in the appraisal report as well as other data available to the reviewer. Furthermore, it is an analysis and an evaluation of the validity and accuracy of the appraiser's work and the resulting value opinion. Like the appraisal, the appraisal review is an independent opinion by a professional appraiser of the valuation problem and its solution.

The reviewing appraiser is not only a highly skilled, well-trained, and experienced appraiser, he or she is a "valuation analyst" skilled in analysis and reviewing techniques. The review appraiser's objective is to secure acceptable appraisals. He or she must be able to read and comprehend an appraisal report and, with an attitude of fairness and objectivity, find within the report sufficient support for the value opinion.

The reviewing appraiser seeks to ensure that the appraiser supports his or her opinions and judgments with actual data and analysis and employs appropriate valuation approaches and techniques that result in an accurate and supported value opinion. The reviewing appraiser should not be concerned with trying to second-guess the appraiser's judgment or direct the appraiser to a preconceived value opinion or premise.

The reviewer's task is not to get a perfect appraisal, but to assure that the appraisal is adequately supported and reasonable. The review appraiser must make his or her decision in fairness and objectivity, based on facts, market evidence, and proper analysis.

"Reasonable" is a term on which both the appraiser and the reviewing appraiser must rely. What might be reasonable to one appraiser might not be reasonable to another. There will usually be differences—either between appraisers or between appraisers and review appraisers. If valid and pertinent data and analyses are included in appraisal reports, the divergencies and differences will be reduced.

In conducting an appraisal review, the reviewing appraiser has certain prerogatives, limitations, and authorities. Review appraisers, management, and appraisers are best served when review appraisers understand and utilize their prerogatives, limitations, and authorities wisely.

If the review appraiser cannot approve or recommend approval of an appraisal report, he or she must tactfully and diplomatically solicit the necessary improvements to the appraisal from the appraiser. If the reviewer approves or recommends approval of the appraisal, he or she must share with the appraiser responsibility for the value opinion. If the review appraiser makes a diligent effort to get the appraisal up to the required standard but the appraisal still cannot be approved, the review appraiser may obtain a second appraisal or prepare his or her own appraisal.

The final result of an appraisal review is a written appraisal review report that briefly describes the property appraised, outlines the appraisal process employed, and presents a decision as to the appropriateness of the appraisal report. The reasoning and support for the review appraiser's conclusions are also included. Reviewing, like appraising, is not an exact science. It requires judgment based on study, interpretation, and an analysis of facts.

The reviewing appraiser's job is highly sensitive and often controversial. Reviewers must review, analyze, and critique the work of qualified and often highly competent professionals. Because the review appraiser challenges or criticizes the work of reputable appraisers, the review appraiser must be a person of judgment, tack, and initiative as well as appraisal and analytical skill. In rendering an impartial, professional decision, he or she cannot shirk responsibility no matter how heavy the internal pressures, how severe the criticism, or how experienced and eminently qualified the appraiser might be. The reviewing appraiser should be bound by a code of ethics and strictly avoid conflicts of interest and the appearance of conflicts. Professional reviewing appraisers who are qualified, capable, and tactful work cooperatively and respectfully with competent appraisers to obtain sound, high-quality appraisals that meet the needs of employers and clients.

APPENDIX

• C •

# *Undivided Interests and Partitions*

In the appraisal of undivided or partial interests in real estate, the appraiser normally spends a great deal of time and effort valuing the fee simple title to the property, but ends up using subjective reasoning in arriving at a value for the undivided interest. Two court cases, *The Northern Trust Company*[1] and *Estate of Sullivan*,[2] have stated that opinions of value based on "gut feeling" are no longer acceptable and actual data from the marketplace should be used as the main criteria to estimate the value of a fractional or undivided interest. The judge in each case ruled that the analysis performed by the appraiser to establish the discount was not specific and, therefore, the appraiser's testimony was given little credence. In fact, the Tax Reform Act of 1984 requires that appraisers be more thorough in their analyses or they could be barred from practice under Sections 6700 and 6701 of the IRS code. Lack of analysis could result in an overevaluation or underevaluation of the property.

Before discussing the rights of property owners and the valuation of undivided interests, it might be appropriate to discuss and review the rights of real property and the bundle of rights theory.

## Real Property and the Bundle of Rights Theory

The term *real property* is a complicated legal concept that refers not to the physical parcel of land, but rather to the benefits and rights inherent in the ownership of the physical real estate. In other words, the property consists not of real estate, but of rights to real estate. As one court has held, "the term *property* may be defined to be the interest which can be acquired in external objects or things." The "things" themselves are not,

1. *The Northern Trust Co.*, 87 T.C. 348 (1986).
2. *Estate of Elizabeth O'Herron Sullivan*, 45 T.C.M. 1199 (1983).

in a true sense, property, although they constitute its foundation and material. The idea of property stems from the connection, or control, of interests which, according to law, may be acquired in or over objects. This interest may be absolute when the object in question is lawfully appropriated by an individual for his or her own use to the exclusion of all others. It is limited or qualified when the control acquired falls short of the absolute.[3]

The property rights inherent in real estate include the right to occupy, sell, lease, or bequeath it and any other rights or benefits that can be derived from the ownership of the land. These various benefits of ownership are sometimes known as the *bundle of rights.* When a property owner has all of these rights, he or she has a complete ownership interest in the property, which is known as *fee simple title.* To best illustrate the bundle of rights, imagine a group of sticks. Each one of these sticks represents a right inherent in the ownership of property. The total number of sticks represents fee simple ownership. A lease, an easement, or a deed restriction takes one of the sticks away from the bundle of rights, leaving the property with an encumbrance.

Fee simple title, the complete bundle of rights, can be divided in many ways: 1) by a lease (leased fee or leasehold estate), 2) through an easement (for a highway, power lines, pipelines, etc.), 3) by deed restrictions, 4) through a life estate, or 5) via concurrent or co-ownership of property. The first four examples differ from the fifth in that one of the bundle of rights is removed from fee ownership if the owner grants a lease or an easement. Under concurrent ownership or co-ownership of property, however, one of the sticks is not really removed from the bundle. That right or co-ownership is inherent within all the sticks in the bundle of rights. The main characteristic of an undivided interest or co-ownership in property is that each co-owner has the same rights as the other co-owners.

There are three types of concurrent or co-ownership of property: joint tenancy, tenancy by the entirety, and tenancy in common.

*Joint tenancy* may be created by two or more owners and generally includes a right of survivorship. This means that the survivors, rather than the heirs, will receive the interest once owned by a deceased joint tenant. A joint tenant cannot transfer his or her interest to heirs through a will, but a joint tenant can convey the proportional joint tenancy interest by "inter vivos" (lifetime deed without the consent of the other co-owners).

*Tenancy by the entirety* generally can be held only by a husband and wife. Each has an undivided interest in the property and the right of survivorship. The main difference between this type of ownership and joint tenancy is that neither spouse can convey his or her interest without the consent of the other. Both spouses usually must sign a deed before any interest can be conveyed.

3. *Griffith The. Charlotte* et al, 23 S.C. 25, 38 (1884).

*Tenancy in common* is an estate held by two or more persons, each of whom has an undivided interest. Tenancy in common will be discussed in the remainder of this chapter.

The distinguishing characteristics of a tenancy in common are the unity of the co-tenants and their right to possession of the property.

- All co-tenants have equal rights and may deal with their separate estates as they please. Each can convey, lease, mortgage, or otherwise encumber the interest without having to consult with the other co-tenants.
- No one co-tenant has the exclusive right to occupy any specific part of the co-owned property. Each co-tenant enjoys his or her own right to occupy the property. For example, assume a 500-acre ranch improved with a home is owned by three co-owners with equal, undivided interests. The home cannot be occupied by one co-owner to the exclusion to the others without some type of agreement or compensation to the other two co-owners.
- If a co-tenant occupies a portion of the property, this particular co-tenant is liable to the other co-tenants for the value of the possession. For example, if one co-tenant decides to graze livestock on the property, he or she is liable to the remaining co-tenants for the rental value of a grazing lease.
- A co-tenant may be reimbursed for any expenditures made to maintain and preserve the property. In other words, each undivided owner is responsible for a fair share of any expenditures or expenses incurred.

The distinguishing characteristics of co-ownership can be complicated and the arrangement is subject to specific problems. If a co-tenant is dissatisfied with the ownership interest, there are several remedies available to dispose of the undivided interest. The co-tenant may 1) sell the undivided interest, 2) seek voluntary partition of the property, or 3) demand partition by the courts.

Selling an undivided interest in a property may be difficult because most buyers prefer fee simple title. When a person purchases an undivided interest, a possible partition suit or problems with the other co-tenants is part of the deal. An individual would prefer to own 200 acres of property in fee simple rather than a one-third undivided interest in 600 acres unless the undivided interest in the 600 acres can be bought at such an attractive discount that the possible ordeal of disputes or a partition is justified. It is almost impossible to obtain financing on an undivided interest in property because of the problems of co-ownership and increasingly strict banking regulations. Any potential buyer would have to agree on the day-to-day operation and management of the property with the other co-owners, and rarely do all parties have the same economic goals and management capabilities. In general, the ownership of an undivided interest is not considered a good investment due to lack of marketability, poor liquidity, and lack of available financing. Moreover, the amenities of ownership are limited because of the co-owner's lack of control.

If a dissatisfied co-tenant cannot sell his or her undivided interest, the next step would be to seek a voluntary partition of the property with the other co-tenants or co-owners. The many costs incurred in executing a voluntary partition include legal fees, survey costs, appraisal fees, and title fees. It is not uncommon for a partition of an undivided interest in property to be unfriendly. In fact, most partitions are unfriendly due to the co-owners' different economic goals, management styles, or sentimental attachments to the property. An unfriendly partition will likely be settled in court, so the co-tenant will incur more attorney fees, court costs, and other expenses. Each party must be represented either as a plaintiff or a defendant if a partition is remanded to the courts even if the individual is agreeable to the division. If the court decides that a property cannot be divided equally among the co-owners, it will order the property to be sold and the proceeds to be divided among the co-owners after all expenses have been deducted. Some properties may be impossible to partition because of their physical features or improvements, or because the property cannot be divided into economically viable units.

## Valuation of an Undivided Interest

Undivided interests are appraised for a number of different reasons, including estate settlements, lawsuits brought by incompatible owners, a need or desire to sell, estate planning, and divorce. The valuation of an undivided interest begins like any other appraisal. The first step is to define the problem, the objective, and the highest and best use. After the preliminaries are ascertained, the market value of the whole property must be estimated. If possible, the three approaches to value, the sales comparison, cost, and income approaches, are applied. After the value of the whole property is estimated, the next step is to identify the undivided interest being appraised. The size of the interest is very important. A majority interest owner has the advantage of control over a minority interest owner. The type of property being appraised is also important. For example, a row-crop farm may be more easily divisible than a ranch.

A property may be too small to be divided if it cannot function as a viable economic unit after partition. Physical characteristics of the property should be considered: road access, improvements, river frontage, and productivity. It should be noted that the appraiser is looking at the property as if the final goal or outcome for the co-owners would be to divide the property in an equitable fashion. Market conditions at the time of valuation should also be considered. In a strong real estate market, the demand for an undivided interest may be greater and it may be easier to sell than it would be in a slow real estate market. If the marketing of fee simple properties is slow, an undivided interest is likely to sell slower.

In valuing an undivided interest, it is very difficult to assess the personalities of the other co-owners. The appraiser has to judge whether the other co-owners are compatible or will agree to a partition and whether the co-owners can operate the property together. Perhaps the most important aspect of valuing an undivided interest is determining whether the highest and best use of the property would remain the same after a partition.

In conclusion, all possible liabilities associated with the ownership of the undivided interest must be measured. In most cases, discounting the value of the undivided interest is justified; in a very few cases, a premium value may be applied to the undivided interest. The Internal Revenue Service has no regulations as to the discounts applicable to fractional interests in appraising property for estate tax purposes. However, the *IRS Training Manual for Estate Tax Examiners,* Section 6.05, appears to authorize discounts by audit personnel. Discounts for undivided interests in real estate have been approved by the courts as early as the court case known as *Andrews,* 13 B.T.A. 651 (1928), Aff'd 38 F. 2D 55 (C.A. - 2 1930). It must be remembered that the courts want facts to establish a discount or premium, not just an unsupported opinion.

While there are a great many reasons to support a discount for a fractional interest, sometimes an argument can be made that an undivided interest would sell at a premium. For example, if a co-owner wishes to acquire control or fee simple title ownership in a family-owned property, the acquired interest could bring full value or even a premium if the co-owner will gain a majority interest. Similarly, a small outstanding undivided interest may command a premium in a sale to a majority owner who wishes to reassemble the property to fee simple status. In this situation there are no potential problems with the undivided interest, such as the cost of partition, legal fees, and appraisals. The premium is actually paid for the control of the property and does not conform to the definition of market value because the probable price is not paid in a competitive, open market and it is affected by undue stimulus. Transactions between family members are not considered to be valid comparable sales.

An appropriate discount for an undivided interest in property can be ascertained in several ways—using market analysis, cost analysis, and an income analysis.

Cost analysis is often used by valuers, but it is the most speculative method and the most difficult to develop due to the many assumptions required. In applying a cost analysis, the appraiser estimates all of the "hard costs" and "soft costs" that would be incurred by the undivided interest owner to partition the property. The appraiser assumes the worst-case scenario that the property will be partitioned in court. The hard costs for the partition would include legal fees and court costs, appraisal fees, surveying costs, and title insurance. The soft costs to be considered are the time required to accomplish the partition and the attendant time value of money relative to the capital investment and operational opportunity costs, which are usually frozen during the partition process. The estimation of these costs requires extensive research as does the development of proper assumptions concerning the time and expense required for partition. Clearly, cost analysis should only be used when market sales are not available.

A market, or sales comparison, analysis is always the best indicator of whether or not a discount is appropriate. However, few sales of undivided interests exist and many of these sales occur among the co-owners; only rarely are sales to outside parties consummated. A collection of verified undivided interest sales transacted over time is by far the best tool an appraiser can have when appraising these types of properties. If undivided interest sales are found, a matched pair analysis can be made between these sales and similar fee simple sales to arrive at a discount. Table C.1 illustrates how sales are compared to estimate the discount for an undivided interest.

**Table C.1 Undivided Interest Sales**

| Sale No. | Total Acreage | Interest | Purchase Price Per Acre | Market Value Per Acre | Indicated Discount |
|---|---|---|---|---|---|
| 1 | 900 | ⅓ | 340 | 400 | 15% |
| 2 | 1,200 | ¼ | 300 | 380 | 21% |
| 3 | 1,600 | ¼ | 300 | 400 | 25% |
| 4 | 800 | ⅛ | 210 | 350 | 40% |
| 5 | 1,000 | ½ | 350 | 400 | 12.5% |
| 6 | 1,000 | ¾ | 375 | 400 | 6.25% |

Sales 2 and 3 indicate that the discount for a ¼, or 25%, interest is between 21% and 25%. The remaining sales indicate that the discount increases as the size of the undivided interest decreases.

Sometimes a discount can be estimated based on the loss of income to the property during partitioning. This is demonstrated in the case of the *Estate of Della Walder Van Lober Seis.*[4] The decedent had an undivided interest in timberland totaling 79,705 acres in 11 properties ranging in size from 26 acres to 26,045 acres. The undivided interests held in the properties ranged from about 2% to 25%. The number of co-owners in each of these properties ranged from 3 to 30. There were other lands in the estate that were not at issue in the court.

The estate filed a tax return claiming a total value of $1,589,750. At trial the IRS claimed the value of the estate to be $9,414,572. In estate tax court, the burden of proof is on the estate to prove the value except on any new matter, i.e., increases or deficiencies after the case has arrived in court. One of the more interesting aspects of this case was that

4. *Estate of Della Walder Van Lober Seis,* cite as II 86,501 P-H Memo T.C.

one of the factors considered in justifying a discount to the fee value was the loss of income to the property during partition. With fluctuations in the price of timber and the lengthy process of partition, a general moratorium on contracting the timber was needed during the partition process. This resulted in a sharp decrease in income during the partition. A 60% discount was eventually established in the case based on evidence of a sale reflecting this discount.

In determining the appropriateness of a discount and the size of the discount, it is important to correlate the data and see how it compares to the subject. The following criteria should be considered when examining the individual characteristics of the property and deciding how much of a discount to apply:

- Size of the property. Will the property still be an economically viable unit after partition? This question could arise in regard to a property in an area where the stocking rate for livestock is such that a divided property will not be economically viable when there is no other alternative use. The appraiser should consider whether the size of the property will change its highest and best use.
- Type of property. Sometimes commercial properties that are co-owned are not as complicated as other undivided interests because outside management can be obtained to operate the income-producing property and avoid management conflicts. For rural properties, however, the production of income is not always the primary motivation of ownership. Recreational benefits and other amenities are sometimes significant to owners.
- Physical characteristics. Can the property be fairly partitioned with respect to physical features? Will each co-owner have access, water, and similar productivity?
- Market conditions. When market conditions are poor, even fee simple properties are hard to sell. A less marketable property like an undivided interest will be harder to market.
- Personalities of co-owners. Co-owners who have different motivations or economic goals and are unwilling to cooperate can affect the marketability and value of a property adversely.
- Highest and best use. Will the highest and best use of the property remain the same before and after partition? If not, will the new highest and best use have a negative or positive effect on value?
- Equity requirements. Can each co-owner be financially responsible for maintenance of the property? Financing for undivided interests is extremely difficult, if not impossible, to obtain.
- Benefits or liabilities accruing from the ownership of the fractional interest. Few benefits can be found in owning a fractional

interest. One benefit may be that a co-owner is receiving a favorable income from a property that requires few management decisions and little control. The many disadvantages of co-ownership have been discussed earlier, but they can be summed up as: lack of control, lack of marketability, lack of value as collateral, and few amenities of ownership.

If, at the end of the analysis, the appraiser decides to discount the undivided interest, the procedure shown in Table C.2 can be applied.

**Table C.2** Discounting Procedure for Undivided Interest

| | |
|---|---|
| Market value of fee simple property:<br>1000 acres @ $400/ac. | $400,000 |
| Pro rata value: 25% of fee simple value | $100,000 |
| Less discount for undivided interest:<br>25% of $100,000 | $25,000 |
| Final estimate of value for 25% undivided<br>interest in subject property | $75,000 |

## Partitions

An appraiser may be asked to provide counsel in the partition of a property. The appraiser should focus on dividing the property equally in terms of value, not acreage. This is important because the tract or tracts will not have the same characteristics or per-acre value throughout the whole property. The appraiser should also strive to identify the method of division that would be the most practical and least expensive. The divided shares should have at least the same utility after the division as before the division. Partition lines should be drawn with consideration of physical features, productivity, and reducing the amount of new fencing to be constructed.

When dividing a property into equal shares, the first requirement is to appraise the fee simple title to the property or properties as one would in any other assignment. After the market value of the whole property is determined, division of the property can occur. The following example illustrates the procedure for partitioning an estate in which four separate properties are to be divided into two equal shares. The properties are described as follows.

- Property A is a 15,755.54-acre ranch. The property is a long, narrow tract in native brush with highway frontage along the east boundary. The east portion of the ranch is more productive than the west portion. The highest and best use of the property is as an agricultural and recreational property.
- Property B consists of 930.28 acres of native brush across the road from the east part of Property A. The highest and best use is the same as Property A, but the value per acre is higher due to the property's smaller size.
- Property C consists of 465.32 acres with an unusual and specific use. The property is held under a long-term lease, which is secure

and specifies an above-market rent for the area. The property is located adjacent to and on the east side of Property A.

- Property D consists of 835.10 acres of land, mostly in native brush with some cultivated land that can be irrigated from a river. A major U.S. highway divides the property into two tracts. Its current use is agricultural and recreational, but it is located away from the other three properties and in fairly close proximity to a metropolitan area. The location and physical features of the property make it feasible to divide the property into smaller tracts for resale.

The estimated market value of the four subject properties is shown in Table C.3. The appraiser hopes to divide the property into two shares so equal that the undivided interest holders will have difficulty choosing one share over the other.

In this illustration, three properties (Properties A, B, and C) are nearly contiguous. Analysis of the neighborhood indicates that the largest tract, Property A, has road frontage and superior physical features on the east portion (7,664.24 acres) as compared to the west portion (8,091.30 acres). It would be most practical to keep the east part of Property A with Property B and Property C. The features on the east part were estimated to be 7% better, indicating an upward adjustment of 7% per acre for the east part of the divided ranch, for an adjusted value of $267.50 per acre. The value of the west portion remains at $250 per acre.

The calculations for the proposed division are shown in Table C.4. This division keeps the properties contiguous in Share 1. The larger portion of Property A is given to Share 2, because of this portion's poorer physical features and access. Property D is also included in Share 2 because it is ready for development into small acreage tracts. Property B, an income-producing property, is included in Share 1. Both shares have the potential for an income-producing use as well as agricultural and recreational uses, and both shares are approximately equal in value.

**Table C.3** Market Value of Separate Properties

| | | |
|---|---|---|
| Property A | $250/ac × 15,755.54 acres | $3,938,885 |
| Property B | $310/ac × 930.28 acres | $288,387 |
| Property C | $658.47/ac × 465.32 acres | $306,400 |
| Property D | $745/ac × 835.10 acres | $622,150 |
| Total value | | $5,155,822 |

**Table C.4** Division of Property into Two Shares

| | | |
|---|---|---|
| | **Share 1** | |
| Property B | 930.28 acres | $288,387 |
| Property C | 465.37 acres | $306,400 |
| Property A (East) | 7664.24 acres | $2,050,184 |
| Total value Share 1 | | $2,644,971 |
| | **Share 2** | |
| Property D | 835.1 acres | $622,150 |
| Property A (West) | 8091.30 acres | $2,022,825 |
| Total value Share 2 | | $2,644,975 |

## Conclusion

The valuation of undivided interests and consultation

on partitions can be a very challenging. More detailed research is needed than in a typical appraisal. The appraiser must have vision to consider the property before and after division and make the most equitable division possible.

APPENDIX
• D •

# *Leasehold and Leased Fee Valuation*

Many appraisals require the valuation of an interest in a property that is less than the fee simple interest. The fee simple interest in a property consists of all rights inherent in ownership, limited only by the governmental powers of taxation, eminent domain, police power, and escheat. Technically, the fee simple title ceases to exist the moment two different parties hold any rights in a given property. Easements, life estates, remainder interests, undivided interests, reservations of mineral or air rights, and transfer of development, agricultural, recreation, or any other rights limit the fee simple estate; therefore, the fee simple estate is almost unattainable.

The first and most important step in valuing a leased fee or leasehold interest is to accurately identify the rights available to the owner and/or tenant. A lease agreement transfers certain property rights. The lessor retains certain rights, such as the repossession of the property at the end of the lease term (typically known as a reversion) and the right to collect rent during the term of the lease. Certain property rights are transferred to the lessee during the term of the lease.

Throughout much of the United States and Canada, many farms and ranches are based entirely on deeded land held in fee simple title. Regardless of geographical location, a farm or ranch may include both deeded and leased land. Ranching operations in the 17 western states are often made up of deeded land as a base unit to provide forage production or base water, together with land leased from federal or state governments. Farms, ranches, grain storage facilities, agricultural processing facilities, and similar properties throughout the United States often include land leased from other entities such as municipalities, utilities, transportation companies, timber companies, and private owners. The lease term will vary from annual to multiyear. Proper analysis of deeded or leased lands requires research and understanding of all considerations of the real estate used in the agricultural operation.

## State and Federal Leases and Permits

In order to stimulate development of the unsettled portions of the country west and south of the original 13 colonies, the United States Government ceded land to many state governments when the rectangular survey system was instituted. States usually received specific sections within each township, typically Sections 16 and 36. Although some states have sold this land to private individuals, many still retain ownership. Most state-owned lands are leased to private enterprise, often for agricultural purposes.

State ownership usually involves a wide variety of land use types, including dry rangeland, subirrigated pasture or meadow, forest, or cropland. Some states allow multiyear leases whereby the state-owned land becomes an integral part of the agricultural operation. Over time, this state-leased land may become extremely important to the feasibility of the operation of the associated deeded land. Changing social and economic factors may heavily influence the utility and financial feasibility of state leases.

The U.S. Government owns and leases many millions of acres suitable for livestock ranching, particularly in the western states. Because various divisions of the federal government may administer the land, a variety of lease/permit restrictions or opportunities may be involved.

As an example, the professional appraiser should recognize that the lands within the National Forest system are administered by the U.S. Forest Service of the Department of Agriculture. Lands under the stewardship of the Department of the Interior are administered by the Bureau of Land Management (BLM). Both entities have a variety of leases or permits, with terms varying from annual to 10 years. Both entities apply strong controls as to the use of the leased land. Leases and permits are specific as to types of uses, number and species of livestock, and allowable grazing season.

Livestock grazing of either Forest Service or BLM lands is considered a privilege. Although the privilege is held by a private individual and is associated with specific deeded land, permits and leases are not automatically conveyed with the sale of a ranch. Permitees of Forest Service or BLM lands must possess commensurate deeded land or water base. The permitee must have the ability to maintain the livestock on deeded land during portions of the year that the leases/permits cannot be used.

Upon the sale of the deeded land, the permitee must request a transfer of the grazing privileges to the buyer of the deeded land. Although permits typically remain with the base property to which it has been attached for many years, the permit technically reverts to the Forest Service or BLM and does not have to be granted to the new owner. A second alternative for the transfer of Forest Service and BLM permits is by transfer with owned cattle. A permitee may sell cattle to a new owner and request that the federal permit be transferred with the cattle. Again, the new owner/permitee must possess commensurate deeded land.

Grazing permits are closely supervised by both the Forest Service and BLM. These government administrators have the right to set up strict pasture rotations, salting patterns, and periodic counting of the livestock. It is always a sound appraisal practice to verify all details of a government grazing permit with the appropriate Forest Service or BLM personnel. Various other governmental entities, including the U.S. Armed Forces and the U.S. Fish and Wildlife Service, may allow grazing leases to private operators on lands controlled by those entities.

In addition to state and federal grazing leases, many ranches today incorporate additional lands through private leases. The private leased land may be owned by entities including individuals, families, grazing associations, investment companies, railroads, or municipalities. Railroad ownership in some western states is checkered with private or state ownership of adjoining sections. As with land subject to federal grazing permits, land encumbered by private leases may become an integral part of a ranching operation. Regardless of private, state, or federal ownership, feasibility and value of leased land within livestock ranching operations may be extremely variable due to the requirements of the specific administrator.

## Types of Leases

There are five general types of leases:

1. A *cash rental lease* has a specific level of rental payment throughout the term of the lease and is a common lease type associated with agricultural properties. These leases may be on a cash basis or on a crop share basis. On a cash basis, the rental payment is stated. On a share rent basis, the annual rental varies depending upon production levels and commodity prices, causing the annual rental from crop share leases to fluctuate.
2. A *bushel* (or other unit—pound, crate, bag, barrel, etc.) *per acre lease* guarantees the landlord a specific level of production per acre.
3. A *cash plus percentage of production lease* establishes a base level of rental and then additional percentage rental based on reaching a specified production level.
4. An *indexed lease* is more common in commercial properties, but it occasionally may be found in rural and agricultural operations and is generally associated with processing facilities or grain storage operations. These leases generally adopt one or more

economic indices, such as the Consumer Price Index, that are used to adjust the rental rate on an annual or other basis. These leases are unlikely to be associated with crop or pasture land.

5. Other types include *step-down, step-up, and annual reevaluation leases.* Typical of these lease types are the ranch leases found in the western states whereby all or a portion of the rental is tied to livestock prices.

In addition to the foregoing types of leases related to crop and pasture land, there are many grain elevators and related facilities constructed on land owned by others, principally railroad companies or port authorities, in which the land is leased from the entity (in a ground lease) to construct a processing or storage facility that has easy access to transportation modes for movement of products to domestic or foreign markets.

Typically, a lease sets out rights of the lessor and lessee, the terms or price paid for the rights leased, the length of the term of lease, and the condition of the property at the termination of the lease. Generally, the owner has the right to receive the lease payments and the right to the reversion at the end of the lease. Rental payments may be received in advance or at the end of each specified payment period (monthly, quarterly, semi-annually, etc.).

## Lease Interests

When an ownership interest in a property is encumbered by a lease, separate estates in the property are created. A leased fee estate is the property interest held by the landlord with the right of use and occupancy conveyed by lease to a tenant. A leasehold estate is the property interest held by the lessee who has the right of use for a stated period.

A positive leasehold estate is created when the *contract rent* (the rent stated in the lease agreement) is less than market. A negative leasehold estate is created when contract rent exceeds market rent. No leasehold value exists when contract rent and market rent are equal. Figure D.1 illustrates the various leasehold/leased fee positions and their positive or negative positions.

**Figure D.1**

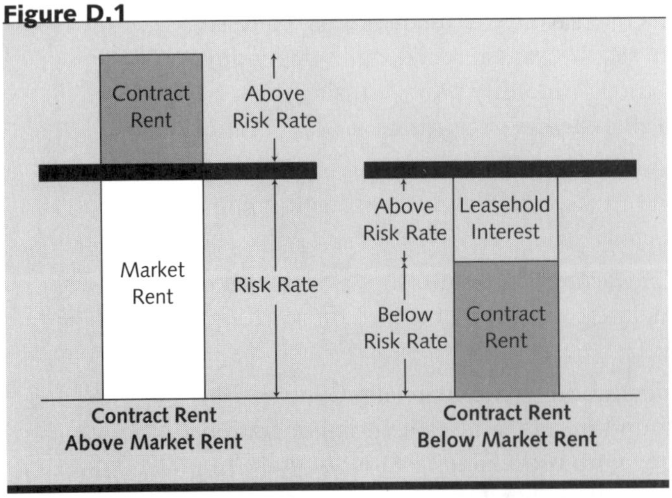


In theory, these estates exist on any term lease; however, in practice they become a factor in long-term leases. For example, on a year-to-year lease, Joe Jones leases 1,000 acres of grazing land to Ron Brown at $1 per acre per year. Market rents in the area are $5 per acre per year. Brown receives a $4 per acre annual advantage on the lease, but it is a year-to-year lease and Jones can cancel the lease at the beginning of any year. The $4,000 annual savings is a leasehold interest; however, it is insignificant because it will occur for only one year. The owner has given up $4 per acre for one year, but he has not given up any significant property rights. Hence, in typical appraisal practice, leases of one year or less are not considered to have any significant value and therefore no positive leasehold exists.

The ability of an appraiser to recognize and deal with the existence of a leasehold/leased fee estate is critical to properly developing an opinion of value of a property that is subject to a lease. If a client has specifically instructed the appraiser to value the fee simple interest for a property under a long-term lease, a hypothetical condition now exists in the appraisal. In this situation, USPAP requires that the appraiser disclose the hypothetical condition and indicate its impact upon value.

## Example

The following example of the treatment of a leasehold/leased fee estate is often typical of situations found in many agricultural property appraisals. The subject property is a vegetable farm owned by the Bradley Estate subject to a lease to John Rogers. Jess Bradley had been good friends with a young farmer in the area for a number of years. Just before he died, Bradley asked his lawyer to draw up a 25-year lease on his 300-acre vegetable farm with Rogers for $75 per acre per year, with Rogers paying the taxes on the property, which were $25 per acre per year at the time of the lease. Bradley felt by securing a lease at market rates for a long term that he was assuring his wife of a steady income stream from the property should anything happen to him. The lease was signed five years ago, and both Bradley and his wife had died three years ago with the property going to their two children in an estate. Over the five years since the execution of the lease, rental rates have increased to $150 per acre per year, with the landowner paying annual taxes of $30 per acre.

Similar land in the area is selling for $2,000 per acre for the fee simple interest. What are the values of the leasehold and leased fee estates?

From the data given above, an appraiser can ascertain that Rogers, the tenant, has a $45 per acre rent advantage ($75 per acre plus taxes of $30 per acre versus $150 per acre). The overall capitalization rate indicated by the data is 6% ([$150 – $30]/$2,000). It would seem obvious that the values of the leasehold position (the tenant, Rogers) would be $750 per acre ($45/0.06) and the leased fee position (Bradley Estate) would be $1,250 per acre ($75/0.06). This may well be the answer, but additional market analysis is required before that conclusion can be drawn.

The first step in analyzing the leasehold and leased fee estates is to examine the security of the two positions. The primary motivating factor in any income-producing investment is the security of the investment, either the original corpus of the investment or the income stream to it. The soundness of these two factors incorporate to make up the risk inherent in the investment recognized by the market.

In the example under consideration, which income stream would be the most attractive to an investor? Which income stream holds the least amount of risk? From the landowner's standpoint, the $75 net income is a secure income stream until cash lease rates drop by more than $45 per acre. Obviously, at a 6% rate of capitalization and without additional information, the landowner's position is more secure. This analysis should lead to the conclusion that it would be inappropriate to merely capitalize the two income streams at 6% and conclude that as the appropriate value for the two estates. Additionally, the capitalization rate for the leased fee is most likely below 6%, and the capitalization rate for the leasehold is most likely 6% or greater.

Additional market analysis is required in order to solve for the value of the leased fee and the leasehold estate. It should be apparent that the leased fee and the leasehold values are not direct relationships and the sum of the two parts (leased fee value plus leasehold value) may not equal the whole (fee simple value).

Obviously, the ideal situation is to obtain sales of the two positions in the market and apply indications drawn from those sales to the subject property. The following analysis will assume an active market for the two interests.

## Leasehold Valuation

The value of the Rogers leasehold interest in the Bradley property is estimated as follows: The tenant has a contract rental advantage of $45 per acre with a remaining term on the lease of 20 years. Contract rent is $75 per acre net to the landowner. The following sales of leasehold estates with long term remaining on the lease were found in the market area:

| | |
|---|---|
| Sale 1: | $35 per acre rental advantage sold for $350 per acre |
| Sale 2: | $25 per acre rental advantage sold for $250 per acre |
| Sale 3: | $50 per acre rental advantage sold for $500 per acre |

All three sales indicate a 10% rate of capitalization, so the value of the subject leasehold interest is $450 per acre ($45/0.10).

## Leased Fee Valuation

Similarly, market research uncovered three sales of farms encumbered by leases with 20 years remaining on term and capitalization rate analysis as follows:

| | |
|---|---|
| Sale 4: | $100 per acre lease sold for $1,800 per acre |
| Sale 5: | $60 per acre lease sold for $1,350 per acre |
| Sale 6: | $70 per acre lease sold for $1,460 per acre. |

Assume that all sales had market rents of $120 per acre net to the landowner and the tenants pay the taxes on all sales. Capitalization rate analysis is as follows:

| Sale | Net Income | Sale Price | $R_O$ |
|---|---|---|---|
| Sale 4 | $100 per acre | $1,800 per acre | 0.0556 |
| Sale 5 | $60 per acre | $1,350 per acre | 0.0444 |
| Sale 6 | $70 per acre | $1,460 per acre | 0.0479 |

From this data the appraiser can conclude that the appropriate rate of capitalization for the subject leased fee estate is near 4.79% because Sale 6 has conditions most similar to the subject. Additionally, it would be appropriate to assume the rate to be somewhat above 4.79% due to the risk factor that is obviously present in this market. Sale 4 indicates a capitalization rate of 5.56% with a $100 rental rate nearest the market rental rate of $120. From this data set, it appears risk increases in correlation to rent increases.

Graphic analysis of the data (Figure D.2) indicates close to a straight-line, direct relationship between net income to the property and overall cap rate ($R_O$). Therefore, the appropriate rate of 4.8% to 4.9%—say, 4.85%—can be concluded for the subject. This results in an indicated value for the subject leased fee estate of $1,546 per acre.

**Figure D.2**

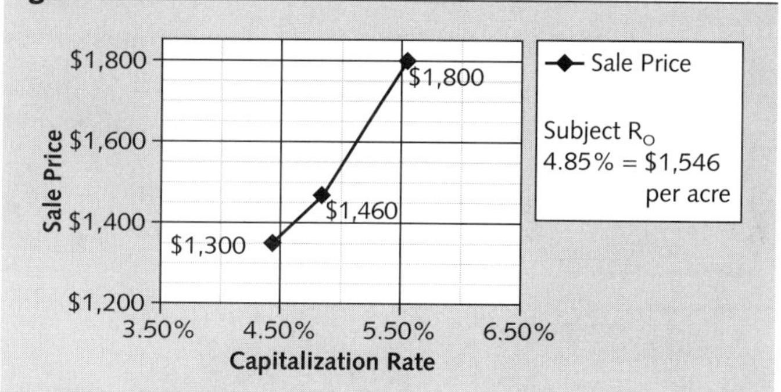


In this example, the leased fee value ($1,546 per acre) and the leasehold value ($450) added together essentially equal the fee value of the subject property ($2,000 per acre). Thus, a value of $600,000 is indicated.

## Income Approach (Annuity Capitalization)

Often there are insufficient sales of leased fee properties or leasehold interest upon which to develop reliable value conclusions using direct capitalization techniques. In the absence of adequate sales of directly comparable properties, analysis of leasehold/leased fee interests in related agricultural properties can be used to develop discount rates for application to the subject property.

The valuation of the leased fee consists of two components:

- Present worth of the contract rent payments
- Present worth of the reversion

The valuation of the leasehold interest consists of the present worth of the rental advantage. In a leasehold there is no reversion.

Discount rate derivation is critical to properly assessing both the leasehold and the leased fee positions in a property. The most effective and appropriate technique for estimating these discount rates is by extracting them from sales of properties with leased fee and leasehold interests. Rate derivation should be based on an analysis of sales of both leased fee and leasehold properties.

The following sale of a leased fee estate can be used to extract a discount rate for the leased fee interest:

| | |
|---|---|
| Contract rent: | $10,000 annually |
| Remaining term: | 10 years |
| Estimated value of fee simple interest at end of lease: | $225,000 |
| Sale price of leased fee interest: | $170,000 |

### Discount Rate (Y) derivation

(Using HP-12c)

| | |
|---|---|
| $170,000 | [CHS] [PV] |
| $225,000 | [FV] |
| 10 | [n] |
| $10,000 | [PMT] |
| | [i] = 8.104529—say, 8.1% |

### Overall Rate ($R_O$) indication

$10,000/$225,000 = 0.0444444 or 4.44%

Note that the $R_O$ from the sale of the leased fee is significantly lower than the discount rate.

The subject leased fee estate has a contract rent of $12,000 and a remaining term of nine years. The estimated value of the fee simple interest at the end of the lease is $240,000.

| | | |
|---|---|---|
| $12,000 | [PMT] | (contract rent) |
| $240,000 | [FV] | (estimated value of the reversion 10 years hence) |
| 9 | [n] | (remaining term of lease) |
| 8.10 | [i] | (indicated discount rate from sales analysis) |
| | [PV] = $193,716—say, $194,000 | |

The following sale of a leasehold estate can be used to extract a market discount rate for leasehold positions:

| | |
|---|---|
| Contract rent: | $10,000 annually |
| Market rent: | $17,000 annually |
| Rental advantage: | $7,000 annually |
| Remaining term: | 10 years |
| Sale price: | $50,000 |

### Discount Rate (Y) derivation

(Using HP-12c)

| | |
|---|---|
| $50,000 | [CHS] [PV] |
| 10 | [n] |
| $7,000 | [PMT] |
| | [i] = 6.637326—say, 6.64% |

### Overall Rate ($R_O$) derivation

$7,000/$50,000 = 0.14 or 14%

Note that the $R_O$ from the sale of the leasehold is significantly higher than the discount rate.

The subject leasehold has a $9,000 rental advantage with nine years remaining on the lease. Using the discount rate derived from the above sale, the value of the leasehold interest can be estimated as follows:

| | |
|---|---|
| $9,000 | [PMT] |
| 9 | [n] |
| 6.64 | [i] |
| | [PV] = $59,546 |

If data is not available for proper abstraction of market-derived discount rates, a far less desirable method is via the "built-up rate." The components of a built-up rate in leased fee valuation generally consist of

- Safe rate
- Risk rate
- Non-liquidity
  - Tax benefits
  - Enjoyment of ownership
  - Appreciation (hedge against inflation)

The components of a built-up rate in leasehold valuation generally consist of

- Safe rate
- Risk rate—may be 0% if significant rental advantage exists

Different rates are typically used in the valuation of leased fee estates compared with leasehold estates. The different components of discount rates used in valuing the two positions of future income may reflect levels of perceived risk. Each rate reflects the different levels of risks—real or perceived—by the landlord and tenant.

## Sandwich Leases and Multiple Leasehold Values

The rights and obligations of the parties to a lease agreement must be considered. Without a prohibition clause in the lease, the tenant has the right to sublease the property to a third party, as long as the obligations under the lease are met. Assuming a positive leasehold estate in the original lease of a property, risk analysis similar to that discussed previously can be applied to a sandwich leasehold analysis.

In a sandwich situation, the landowner is always one piece of bread and the current user of the property is the other piece with the other lessees being sandwiched in the middle as the sandwich ingredients.

Assume that a landowner leases a farm to Farmer A for a 25-year lease. After operating the tract for five years, Farmer A leases to Farmer B at a higher rate than the original contract rent. This places A in the sandwich position. Assuming that market lease rates continue to rise and after a second five-year period, the two leasehold positions and the leased fee must be appraised. Risk analysis dictates that the lowest discount rate or rate of capitalization would be applicable to the leased fee position, a higher rate is applicable to Farmer A's position, and the highest rate of the three is applicable to Farmer B's leasehold estate, all predicated on the relative risk associated with the three positions.

The following analysis is used to illustrate the valuation of the various positions in a property that was leased five years ago for a total of 20 years. The farm includes 500 acres that were leased for $75 per acre with the landowner paying taxes of $10 per acre. Current market rent is $150 per acre. Earlier this year, the tenant subleased the farm for $100 per acre. Market value for farms that are not subject to long-term leases is $1,000 per acre.

| | |
|---|---|
| **Sale No. 1** | Leased fee estate sale |
| | 600 acres leased for 15 years at $125 per acre sold for $800 per acre |
| | Property taxes: $10 per acre |
| **Sale No. 2** | Sandwich position sale |
| | 500 acres with 15 years remaining on the lease |
| | Original lease: $75 per acre |
| | Second lease: $115 per acre |
| | Sale price: $200 per acre |
| **Sale No. 3** | Leasehold estate sale |
| | 700 acres leased for $80 per acre with 15 years remaining |
| | Sale price: $400 per acre |
| | Market rent: $150 per acre |

### Analysis of Sale No. 1

| | |
|---|---|
| Overall rate leased fee estate: | $125 – $10 = $115/$800 = 14.375% |
| Discount rate: | $800 [CHS] [PV] |
| | $115 [PMT] |
| | $1,000 [FV] |
| | 15 [n] |
| | [i] = 14.90% |

### Analysis of Sale No.2

| | |
|---|---|
| Income to sandwich position: | $115 – $75 = $40 |
| Discount rate: | $200 [CHS] [PV] |
| | $40 [PMT] |
| | 15 [n] |
| | [i] = 18.42% |

### Analysis of Sale No. 3

| | |
|---|---|
| Income to leasehold position: | $150 – $80 = $70 |
| Discount rate: | $400 [CHS] [PV] |
| | $70 [PMT] |
| | 15 [n] |
| | [i] = 15.48% |

### Analysis of Subject

| | |
|---|---|
| Value of leased fee estate | |
| Net income: | $75 – $10 = $65 |
| | $65 [PMT] |
| | 15 [n] |
| | 14.90% [i] |
| | $1,000 [FV] |
| | [PV] = $506 |

### Value of Leasehold Estate

| | |
|---|---|
| Income to leasehold position: | $150 – $75 = $75 |
| | $75 [PMT] |
| | 15 [n] |
| | 15.48% [i] |
| | [PV] = $429 |

### Value of Sandwich Position

| | |
|---|---|
| Income to sandwich position: | $100 – $75 = $25 |
| | $25 [PMT] |
| | 15 [n] |
| | 18.42% [i] |
| | [PV] = $125 |

In this analysis the total of all of the values of the individual estates is $1,060 per acre, which is $60 per acre greater than the fee simple value. This difference occurs because the discount rates are not taken from sales that are identical to the subject property in terms of size, quality, and lessee's financial strength. It is considered that the value of the individual estates still approximates the value of the fee simple unencumbered.

## Summary

In conclusion, the appraisal of leased fee or leasehold interests is relatively straightforward but requires employment of the techniques discussed herein if the opinion of value developed is to be reliable. Market-derived rates must be used whenever possible to properly value leased fee and leasehold interests.

APPENDIX

• E •

# Agricultural Capital Markets and the Farm Credit System

## Overview

Rural property appraisers need to be aware of how the availability of capital influences the market for agricultural property and how the U.S. agricultural industry's capital requirements are met. Capital is the agent of production that brings land, labor, and coordination (the other three agents of production) into use. Capital investment in farm real estate in the United States represented over $807 billion by the end of 1995.[1]

Agricultural property attracts capital investment when the return on that investment is competitive with other investment opportunities. In today's market, investment returns are measured in terms of economic, social, governmental, and/or environmental benefits.

Agricultural property represents an investment of capital that is in some ways similar and in other ways significantly different from investments in the urban sector of the nation's economy. As with commercial property in the urban sector, agriculture requires capital investment to support operations and acquire the equipment, fixtures, and real estate used in production and processing. The capital consumed by agriculture takes the form of equity and debt, and it may represent a short-term (for one crop or season), intermediate-term (for equipment or livestock), or long-term (real estate) investment. In contrast to the urban sector, agricultural asset investments are usually in a property that requires skilled management, are frequently dependent on the economics of a single commodity, and often involve a residential use along with the production agriculture use.

Income to provide a return on and recapture of the capital invested in an agricultural property is highly dependent on weather, commodity

1. *Farm Real Estate Values,* Stock Number 86010 (Washington, D.C.: USDA, Economic Research Service, updated 4/96).

price/production cost relationships, and, particularly in grain and fiber production, government programs. To a far greater degree than in the urban sector, the rural property appraiser must consider the viability, stability, and durability of the agricultural enterprise on the property in the course of completing an appraisal.

Reliable rural property appraisals are important as contributors to the stability and future of the agricultural industry. Appraisals are used by owners and investment managers in often complex business transactions and capital investment (buy, sell, or hold) decisions. Agricultural property lenders look to the appraisal for essential information about a property's income-generating capacity under typical ownership and management, and the value of the property as collateral when used to secure a loan. Lenders also look to the appraisal to provide important information about the physical and economic character of the property and the nature of the market for the property. Long-term lenders are also interested in the stability of the market for rural properties and the influence social, governmental, or environmental forces have on the property's market.

## Background

The sources of capital and means of financing agribusiness enterprises changed dramatically over the course of the twentieth century. At the beginning of the twentieth century, most farmers and ranchers relied on trade credit for their operating capital. Long-term financing for agricultural property from public sources was not widely available. Recognizing the need for more stability and competition in financing for the agricultural industry and the increasing need for intermediate and long-term debt capital, the federal government facilitated development of a cooperatively owned credit delivery system, which is today known as the Farm Credit System. The Farmers Home Administration (FmHA) and a variety of commodity and conservation programs were created after the boom/crash cycle in agriculture following World War I, as the federal government also saw the need for both government-funded or government-directed loan programs to aid the farmer and rancher. A multitude of banking law changes, at the state and federal level, also brought new financing sources to agricultural property owners.

After World War II and throughout the 1950s, farmers and ranchers adopted new technology and agribusiness management practices. During the 1960s and 1970s, while the technology continued to evolve, new ways to manage capital and risk were adopted. Beginning in the 1970s environmental awareness and public resource policy concerns also began to affect the agricultural industry. Another change following World War II was the growing importance of farm commodities in foreign trade and international diplomacy. Particularly during the past two decades, the federal government's commodity programs have influenced conditions in the agricultural industry.

A combination of economic, political, and global climatic conditions during the 1970s resulted in strong expansion of U.S. farm industry and farm land investment. The industry expansion was primarily driven by heavily reliance on growth in export markets and continued high inflation. Much of the expansion, for both land and equipment, was funded with debt rather than equity.

Early in the 1980s borrowers and lenders in all sectors of the agricultural industry became acutely aware of the need to better recognize commodity and enterprise risk in their financial decisions. Farm loan defaults and property foreclosures increased dramatically when farm exports and commodity prices declined, significantly reducing the ability of borrowers to service the debts they accumulated during the 1970s.

Financial stress in the agricultural industry during the first half of the 1980s was addressed in a combination of actions taken by borrowers, lenders, and the federal government. Borrowers cut back on the use of production inputs, deferred equipment replacement, and curtailed expansion of their operations. Lenders began to apply more rigorous financial tests before advancing loan funds and gradually worked through the inventory of foreclosed property. The federal government used commodity programs and a variety of direct loan programs to help stabilize the farm economy.

By the mid-1980s the agricultural industry entered a recovery phase as farm exports increased, inflation subsided and the debt service burden began to decline. As the industry entered the 1990s, borrowers had reduced their long- and short-term debt by 30% from their 1980s peaks, and they adopted a more cautious approach to the use of debt to finance expansion or new equipment. Lenders were also more cautious. Debt-service ability and stability of the agricultural enterprise and borrower returned as primary considerations for making sound agricultural loans. Loan support and security in the form of real estate collateral returned to a subordinate role in credit decisions. While credit at competitive prices was available, demand for loans was restrained because borrowers and lenders both had a new appreciation for the risk associated with making long-term capital investments in assets exposed to strong external economic, social, governmental, and environmental forces.

## Current Era

During the 1980s forces were set in motion that substantially changed the methods of analyzing financial strengths and providing credit to production agriculture. From 1985 through 1990 the nation's finance industry went through a process of examination and new regulation that has affected both agricultural and urban lending. Every sector of the financial community saw new regulation or significant reregulation in this period. For example, the Farm Credit System's 1971 Farm Credit Act was amended three times in as many years (1985, 1986, and 1987). These amendments significantly affected the System's structure and operations.

Through most of the decade the farm financial industry was in turmoil, a situation caused by (among other factors) insufficient or underused methods for analyzing the true profitability of various farm enterprises. This environment created an increased interest in farm financial education and sophisticated financial analysis techniques. However, the lack of standardization in farm financial analysis caused problems in understanding and using data for decisions, and was often cited as a substantial barrier to the accessibility of funds from capital markets. The need for development of standardization in farm financial analysis became more evident with the passage of the Agricultural Credit Act of 1987 because it contained provisions for the government to sponsor establishment of a secondary market for agricultural real estate loans (e.g., Farmer Mac, which is discussed below).

While Congress was working on the 1987 Agricultural Credit Act, the National Commission on Agricultural Finance was examining the farm financial industry. This commission cited a need for standardization of agricultural credit analysis and farm financial statements. In response to this need, an industry-wide Farm Financial Standards Task Force (FFSTF) was formed. The FFSTF, which consists of experts from all facets of the farm financial industry, is an independent organization sponsored by the Farm Foundation. Membership in FFSTF includes representatives of all types of lenders—commercial bankers, Farm Credit System loan officers, insurance industry lenders, Farmers Home Administration lenders, and other, non-institutional lenders—as well as regulators of financial institutions, academicians, farm financial research specialists, representatives of the U.S. Department of Agriculture (USDA) Extension Service and the Economic Research Service, accountants, representatives from farm groups and farm software firms, and other industry specialists.

The strategic goal of FFSTF is to develop and establish universally acceptable farm financial guidelines for U.S. production agriculture. Specific goals are as follows:

1. Identify certain ratios common to all areas of the country
2. Identify standard methods of calculating these ratios
3. Draw up standardized farm financial statements formats that may be used by all farm lenders
4. Review and recommend content guidelines and/or calculation guidelines for farm financial software

The FFSTF recommends its material to the American Institute of Certified Public Accountants (AICPA), which in turn publishes statements on generally accepted accounting principles (GAAP). The Financial Accounting Standards Board (FASB) uses GAAP as a reference in FASB's oversight of accounting practices used in the financial and securities industries.

The FFSTF published suggested guidelines in 1991. The recommendations were arranged into three subject areas:

1. *Universal Financial Reports.* This material contained suggested procedures and principles for constructing farm financial statements for financial analysis purposes (i.e., the balance sheet, the income statement, the statement of cash flows, and the statement of owner equity).
2. *Universal Financial Criteria and Measures.* This material addressed computations, interpretations, and limitations of some of the most widely used farm financial ratios.
3. *Universal Information Management.* This material contained suggestions for collecting and using standardized farm financial data for the benefit of agricultural producers and those who serve them.

The work and impact of FFSTF continues as the agricultural industry and accounting profession study, assimilate, and adjust to the changing financial analysis needs of U.S. agriculture.

During the 1980s changes underway in the urban sectors of the finance industry ultimately affected the agricultural finance industry, also. The use of junk bonds to fund leveraged buyouts eventually brought turmoil and then tighter control to the securities and commercial banking industries. Likewise, the combination of dramatic shifts in federal tax policies and problems in the thrift industry caused severe stress in the urban real estate markets, particularly for lenders with portfolio concentrations in development properties. Stress in the commercial banking and thrift industries was addressed by a variety of measures, including the Financial Institutions Reform, Recovery and Enforcement Act of 1989 (FIRREA).

Title XI of FIRREA elevated The Appraisal Foundation to a status similar to the Financial Accounting Standards Board. The Appraisal Foundation was created in 1987 by a group of organizations representing professional appraisers. The Foundation's Appraisal Standards Board (ASB) and Appraiser Qualifications Board (AQB) promulgate minimum standards that establish how appraisals may be done and qualifications for appraisers. Title XI of FIRREA also created the Appraisal Subcommittee of the Federal Financial Institutions Examination Council (FFIEC) and referenced The Appraisal Foundation's ASB and AQB standards as a set of basic criteria certain federally regulated financial institutions must follow. The Appraisal Subcommittee also required those institutions to develop supplemental standards to address the character of their particular lending market.

FIRREA covered financial institutions regulated by the Federal Reserve System, Office of the Comptroller of the Currency, Office of Thrift Supervision, National Credit Union Administration, and Federal Deposit Insurance Corporation, and the act affected a broad range of financial institution management, capitalization, and operating areas. Title XI of FIRREA had direct impact on how an institution obtained and used appraisals in its lending activities. State laws implementing real property appraiser licensing and certification programs in response to Title XI of FIRREA were in place across the country by the end of 1992.

Federal agencies making direct real estate loans or responsible for federal real estate resources management, while exempt from appraiser licensing and certification programs under state law, were obligated to address the appraisal requirements of Title XI by Circular A-129 issued by the Office of Management and Budget. In early 1993, the Farm Credit System's regulator (the Farm Credit Administration) issued its collateral regulation to closely parallel Title XI of FIRREA.

As the last decade of the twentieth century began, rural property appraisers faced a new era. In the 1990s and in the first years of the new millennium, the professional appraiser must consider and analyze the market for agricultural property under a new set of professional and market requirements. While basic agricultural economics remains the primary force influencing the market for rural property, the influence of social, governmental, and environmental forces must be recognized. In this new era, agriculture must compete for capital and market share on a global stage. In addition, farmers and ranchers must contend with powerful urban interests for federal and state resources (such as water and use of public lands) and the benefits of federally funded programs as well as for capital.

## Capital Markets

Agriculture requires the investment of capital to support operations and provide the assets used in the production and processing of agricultural goods. The profit and other benefits from this use of capital must be sufficient to reward the investor at a rate at least equal to that offered by alternative investments. Today's agricultural property owners and managers, while having access to a wide variety of financing sources, also find a broader group of competitors for capital. National, regional, and state economies, multinational as well as national corporations, and the urban commercial and residential markets are competitors. Private individuals and companies, community banks and thrifts, as well as insurance companies and money center banks all have the opportunity to channel investment funds to maximize yields through the securities markets, syndications, pooling, and a host of other financing mechanisms. Agricultural lending must also compete with these investment alternatives for capital.

Investment of capital in agricultural property consists of equity, debt, or a combination of the two. Agricultural property owners and managers, particularly after the experience of the 1970–1980 business cycle, evaluate the risk in their operations before committing the future of their livelihood to debt service. Lenders likewise evaluate the risk inherent in an agricultural property in the course of making a loan commitment.

Lenders typically evaluate the risk in any credit decision by considering five factors:

1. Character
2. Credit
3. Capacity
4. Collateral
5. Conditions

Character focuses on the applicant's experience, personal history, and management abilities, particularly with the type of agricultural enterprise and business entity associated with the property. The character evaluation, together with the applicant's credit history, provides an indication of the applicant's willingness and ability to repay the debt. Capacity deals with the applicant's previous financial performance and well-supported business and enterprise cash flow projections. The evaluation of the cash flow provides an indication of the applicant's ability to repay the debt from business operations or other sources. Collateral focuses on the amount of security the property to be taken as collateral represents. When the character and capacity evaluations result in a need to take collateral as security, the collateral taken must reasonably protect the lender while allowing the applicant to manage the business constructively. The conditions factor deals with loan terms (advance rate, amortization period, type[s] of security, specific performance requirements, etc.) and covenants. Appraisals completed for lenders do not usually involve the credit and conditions factors.

Although the work of appraisers in connection with lending must focus on the collateral category, consideration to the capacity category is also needed in the course of the appraisal work. The capacity of a property to produce an income flow is a function of physical property characteristics and management, subject to commodity price level fluctuations. The commodity associated with the property establishes a reference point when considering the quantity and stability of income a property can be expected to generate.

When developing an opinion of market value, the appraiser must focus on the level of management that is "typical" for the subject property. This is because, for example, in an income analysis capitalizing an income stream that results from superior management will tend to overstate the value of the real property. Conversely, capitalizing an income stream that results from inferior management will tend to penalize the property. In other words, when developing an opinion of market value (value-in-exchange) for real property, the appraiser must focus on what is typical for the property rather than what superior or inferior management achieves.

If the appraiser is developing an opinion of investment value or some other value specific to a particular ownership (but not value-in-exchange), consideration may or may not be appropriately given to a specific management level other than what is typical.

The commodity price and climatic cycles of the past two decades have also renewed an appreciation for the need to develop "stabilized" estimates of production and income for use in market analysis and credit decisions. The borrower or lender who makes a decision to increase leverage, particularly with long-term debt, in an agricultural property investment after two or three years in an up cycle of commodity prices invites a refresher course in basic agricultural economics.

## Financing Rural Properties

Rural properties include several primary property types and uses, each with its particular capital needs and sources. The primary categories addressed in this text are

- Agribusiness
- Production agriculture
- Timberland
- Rural residential

Rural properties are also used for a variety of other purposes, including recreation, open space or ecological preservation, and mineral extraction. At times, rural property serves as a vehicle for speculative investment. In such cases, the property takes on a divided market characteristic. The physical highest and best use of the property continues (usually under a landlord/tenant arrangement) while the investor seeks a profit by either working to change the physical use of the property in some future time period or expects profit to come from the interim rental income and eventual resale of the property to another investor or to an owner-user.

### Agribusiness

Agribusiness properties typically are vertically and/or horizontally integrated units, with at least a portion of the property devoted to packing, processing, or shipping agricultural commodities. Although exceptions occur, most such properties are single-purpose by design and have limited market appeal as operating units without the associated business entity and its intangible property (franchise rights, trade name, label, contracts, etc.). When such properties are sold as operating units, the sale most often involves a business entity transfer with the price for the tangible real and personal property and intangible personal property components allocated on the basis of negotiations involving accountants and tax attorneys. When such properties are sold as vacant and/or non-operating units, the buyers are typically paying only the price they view as reasonable for the property's productive capacity, net of their cost to reconfigure or rehabilitate the property to suit their needs.

Agribusiness properties are commonly included in corporate finance lending as part of a general security instrument taken by the lender. The development of a stand-alone market value for the real and/or personal property portions of such properties, as components in an operating unit, is difficult. Most often, the value developed represents the contributory value of the property appraised, based on its historical or designed throughput capacity, to the going concern that makes up the related agribusiness entity. When the property appraised is part of a viable business operation, the function of the appraisal is most often to aid the lender in developing a reasonable allocation of the loan principal between short-, intermediate-, and long-term amortization schedules.

When the property appraised is not operating due to financial failure of the entity in ownership, the function of the appraisal is most often to aid the owner, the lender, or a court in determining the amount the property will bring when liquidated to convert the capital investment to money that is then used to retire debt or invest elsewhere.

## Production Agriculture

Production agriculture includes a vast array of crop, livestock, and other rural properties. The capital needs of production agriculture property owners vary with the type of enterprise associated with the property and its ownership. Some properties, such as those held in ownership by one family for generations, have very low or zero long-term debt and/or have no need for borrowed operating funds, while other operations involve complex debt structures including a full range of loan types and terms.

The enterprise located on the property can involve a single commodity or a broad range of commodities. Integrated units involving several commodities are not unusual. Some properties are the sites of very specialized and intensive operations. Examples are seen in stand-alone livestock or livestock commodity production units, such as a farrow-to-finish swine facility, dry-lot dairy, poultry (broiler or egg) unit, aquaculture unit, and the like. These single-purpose properties, while perhaps not involving vertical integration, take on the character of an agribusiness property in the farm real estate market and in the type of financing they use.

Appraisals of production agricultural properties for lenders must address the highest and best use and market value of the property under typical ownership and management, and, in particular, these appraisals must develop a sound basis for estimates of the property's productive capacity and earnings ability.

When an appraisal is in connection with a long-term loan, the specific characteristics of the market for the property need to be described fully. The lender and borrower are in the process of making a long-term capital investment decision and need to be aware of the stability and durability of the market for the property well beyond the current production cycle. Although appraisers are not expected to forecast the future, professional rural property appraisers have a unique perspective on the market for such properties, and lenders as well as owners need the benefit of that perspective when long-term capital investment decisions are made.

## Timberland

Financing timberland (exclusive of farm woodlots) is most often handled by insurance companies or Farm Credit System institutions. Large corporations finance their timberland acquisitions internally or through such methods as private placement bonds. Where private individuals require timberland financing, the loan amount is generally dependent on the merchantable timber value, and the lender includes loan covenants requiring partial reduction of the loan as the timber is harvested.

As environmental concerns, particularly over endangered species, have grown in the past two decades, the need for professional appraisers to be familiar with the ecology and local forest practice regulations as well as with the economics of the timber industry has become apparent. Timberland financing can extend well beyond the 20- to 30-year loan term seen with other production agriculture property, and future harvest rights can be remarkably different than in the present. This type of financing is highly specialized, and any appraiser practicing in this market needs to be appropriately qualified and well aware of the market's characteristics.

## Rural Residential

The financing of rural residential properties is carried out by a variety of institutions, including those serving production agriculture and urban property markets. Commercial banks and thrifts, Farm Credit System institutions, credit unions, and mortgage brokerage firms are active in financing rural residential properties, and Federal Housing Administration (FHA), Veterans Administration (VA), and Rural Development (RD) loans can be obtained in rural areas. USDA Rural Development also makes loans available for community development in rural areas of a specified size.

The appraisal of rural residential properties other than farm properties generally relies heavily on the use of the market comparison approach to value. In rural communities, residential developments as well as commercial developments are equivalent to urban use of land in rural settings. In such settings the appraiser must apply the same considerations as are required in a more populated setting, but the appraiser also needs to evaluate how an entirely new type of property in the community will be received by the market.

Appraisals of rural residential properties have much in common with residential properties in an urban setting, but they also have important differences. Such elements as access, water and sanitation, power and telephone services, and distances and convenience to community services (schools, medical facilities, churches, markets, fire protection, etc.) can cause significant differences in the market for a rural residential property. Other important factors include the property setting, adjacent property uses, linkage to employment centers, community income levels, and local land use or land subdivision restrictions. The latter element can be highly significant when the rural residential property involves a parcel that is larger than is needed for a single dwelling.

## Types of Agricultural Property Financing

The unique needs of the agricultural industry have given rise to the expansion of credit sources to meet its increased credit demands. These demands include loans to service short-, intermediate-term, and long-term needs.

### Short-term Credit

Short-term credit, generally called *production credit,* provides working capital for annual and seasonal operational costs. The purchase of production inputs such as seed, feed for animals, and fertilizer, as well as other essential operating expenses, are often financed. Loans or advances are repaid from the proceeds from the harvest of annual or seasonal crops, which often serve as the collateral for this type of credit. The customary maximum duration of this type of loan is the crop season or livestock production cycle, i.e., generally one year.

### Intermediate-term Credit

Intermediate-term credit is extended for inputs that are used for more than one season, such as equipment, breeding livestock, and dairy cattle, and that are financed for three to ten years. Intermediate-term credit may also be used to refinance short-term debts during adverse periods in the farm economy. However, on the basis of the 1970–1980 cycle experience, such use of intermediate-term credit has become less prevalent.

Personal property is ordinarily used as collateral for this type of loan, but lenders may require mortgages on real estate as additional collateral for short- and intermediate-term loans when borrowers have only marginal security otherwise. However, this use of real estate as additional security for a loan is a result of short- and intermediate-term lending rather than real estate lending.

The major sources of short- and intermediate-term credit are commercial banks, Farm Credit System institutions (including Banks for Cooperatives), the Farm Service Agency (FSA), merchants, dealers, suppliers, and processors.

### Long-term Credit

Long-term credit is used primarily for the purchase of real estate or the consolidation of short-term or intermediate-term debts. Homes in rural areas, as well as farm and ranch land and buildings, are financed through long-term loans. Soil and irrigation improvements may also be financed by long-term credit. In all cases, the loan is secured by a mortgage or deed of trust on the real estate itself. The multiple purposes of long-term credit make a variety of terms necessary, but most real estate loans are made for periods ranging from 10 to 30 years.

The major sources of long-term loans for rural properties are individuals, commercial banks, life insurance companies, Farm Credit System institutions, and the Farm Service Agency. Processors have increased their long-term financing activity, particularly to help producers establish or expand permanent plantings or make other production unit improvements.

## Credit Sources

Agricultural property owners use a variety of sources for credit. Institutional lenders, such as banks, insurance companies, Farm Credit System institutions, and certain government agencies—mainly the Farm Service Agency, which was previously the Farmers Home Administration—provide the major share of debt financing. Private sources, including individuals, trade creditors, and processors, represent a significant share as well. The percentage of debt financing provided by these various sources changes over time. Table E.1 shows the relative proportions of real estate debt provided by these sources (private sources are all included in the Individual category).

Table E.2, showing the various FmHA loans and grants by program, was developed from information provided by the Farmers Home

**Table E.1** Market Share of Total Farm Debt

| | 1980 | 1985 | 1990 | 1995 | 1996 | 1997 | 1998 |
|---|---|---|---|---|---|---|---|
| Farm Credit System | 31.8% | 31.6% | 25.9% | 24.8% | 25.4% | 25.5% | 25.8% |
| Commercial banks | 22.6% | 25.0% | 34.5% | 39.8% | 39.5% | 39.7% | 41.0% |
| Farm Service Agency | 10.4% | 13.8% | 12.3% | 6.7% | 6.1% | 5.4% | 4.8% |
| Life insurance cos. | 7.2% | 6.3% | 7.0% | 6.0% | 6.1% | 6.1% | 5.8% |
| Individuals and others | 28.0% | 23.3% | 20.3% | 22.7% | 22.9% | 23.3% | 22.6% |
| Total | 100.0% | 100.0% | 100.0% | 100.0% | 100.0% | 100.0% | 100.0% |

Administration National Office, Information Resources Management Division, prior to the division of FmHA into two agencies, the Farm Service Agency and Rural Development.

**Table E.2** Lending Activity of FmHA Before Division into FSA and RD

Statistics on the types and number of loans made under FmHA programs in operation through September 30, 1992.

| | Number of Loans or Grants | Dollar Amount (in thousands) |
|---|---|---|
| **Agricultural Credit Insurance Fund** | | |
| Direct farm ownership loans | 412,448 | 12,268,496 |
| Guaranteed farm ownership loans | 17,246 | 2,543,682 |
| Direct farm operating loans | 2,679,219 | 26,055,176 |
| Guaranteed farm operating loans | 94,209 | 8,332,355 |
| Direct emergency disaster loans | 1,246,706 | 23,270,898 |
| Direct soil and water loans | 45,246 | 520,933 |
| Guaranteed soil and water loans | 84 | 4,399 |
| Indian land acquisition loans | 97 | 101,518 |
| Resource conservation and development loans | – | – |
| Interest buydown program | 941 | 31,166 |
| State mediation grant | 69 | 11,483 |
| Credit sales of acquired property | 351 | 26,182 |
| **Total ACIF** | **4,496,616** | **73,166,288** |
| **Rural Housing Insurance Fund** | | |
| Sec. 502 Direct single-family housing | 2,091,520 | 43,835,487 |
| Sec. 502 Guaranteed single-family housing | 5,729 | 290,960 |
| Sec. 515 Direct multifamily housing | 24,653 | 12,663,440 |
| Sec. 504 Housing repair loans | 63,180 | 175,447 |
| Sec. 514 Farm labor housing loans | 2,214 | 278,638 |
| Sec. 524 RH site development loans | 138 | 18,701 |
| Sec. 521 Rental assistance | 251,839 | 4,278,971 |
| Sec. 502 Rental assistance | 295 | 8,246 |
| **Total RHIF** | **2,439,568** | **61,549,890** |
| **Rural Housing Support Programs** | | |
| Sec. 516 Farm labor housing grants | 350 | 2,249,034 |
| Sec. 504 Housing repair grants | 80,814 | 228,616 |
| Sec. 523 Mutual S/H housing grants | 764 | 145,736 |
| Sec. 509 Construction defects | 842 | 2,841 |
| Sec. 523 Self-housing land development loans | 53 | 10,652 |
| Sec. 533 Housing preservation grants | 1,141 | 141,700 |
| Credit sales of acquired property | 5,443 | 183,802 |
| **Total Housing Support** | **89,407** | **2,962,381** |
| **Rural Development Insurance Fund** | | |
| Direct water and waste loans | 33,740 | 11,576,008 |
| Guaranteed water and waste loans | 20 | 5,275 |
| Direct community facility loans | 6,514 | 2,875,171 |
| Guaranteed community facility loans | 73 | 63,599 |
| Guaranteed B&I loans | 7,581 | 6,245,207 |
| Guaranteed B&I drought/disaster acts | 73 | 47,122 |
| Resource construction and development loans | 292 | 27,465 |
| Watershed and flood prevention loans | 493 | 179,224 |
| **Total RDIF** | **48,786** | **21,019,071** |

**Table E.2** Lending Activity of FmHA Before Division into FSA and RD *(continued)*

| | Number of Loans or Grants | Dollar Amount (in thousands) |
|---|---|---|
| **Rural Development Loan Fund** | | |
| Intermediary relending loans | 100 | 111,667 |
| **Rural Development Grants** | | |
| Water and waste grants | 16,576 | 4,129,965 |
| Rural development grants | 1,857 | 157,455 |
| Solid waste management grants | 48 | 4,500 |
| Emergency community water assistance grants | 88 | 20,000 |
| **Total RD Grant Programs** | **18,669** | **4,423,587** |
| **Grants other than FmHA** | **1,219** | **281,643** |
| **Programs Not Currently Funded** | | |
| Direct B&I loans | 48 | 60,281 |
| Guaranteed emergency disaster loans | 3,727 | 101,369 |
| Economic emergency loans | 126,770 | 7,227,952 |
| Guaranteed emergency livestock loans | 7,738 | 1,045,660 |
| Bureau of Reclamation-soil and water loans | 283 | 4,936 |
| Water facilities loans (individual) | 18,296 | 29,695 |
| Recreation loans (individual) | 463 | 20,528 |
| Economic opportunity loans (individual) | 64,660 | 109,395 |
| Economic opportunity loans (association) | 1,476 | 21,107 |
| State corporation operating loans | 185,904 | 158,964 |
| Irrigation and drainage loans | 610 | 32,240 |
| Grazing loans to associations | 605 | 129,420 |
| Land conservation and development loans | 226 | 268 |
| Rural rehabilitation loans | 3,031,331 | 1,015,826 |
| Weatherization loans | 7,089 | 7,080 |
| Housing supervisory assistance grants | 57 | 5,000 |
| Department of Army/Navy grants | 24 | 5,237 |
| Office surface mining projects grants | 2 | 728 |
| Regional development grants | 693 | 54,453 |
| Economic development administration grants | 43 | 20,416 |
| Emergency job grants | 14 | 2,871 |
| Water and sewer planning grants | 1,834 | 22,293 |
| Finance corporation grants | 3 | 14,264 |
| Guaranteed finance corporation loans | 12 | 19,140 |
| Recreation association loans | 1,035 | 110,510 |
| Rural renewal loans | 56 | 6,550 |
| Special impact grants | 2 | 2,700 |
| Rural development planning grants (ADA) | 605 | 19,000 |
| Energy grants—growth management | 158 | 6,299 |
| Energy grant—site acquisition | 277 | 52,448 |
| State corporate farm ownership loans | *(# not available)* | *(# not available)* |
| **Total of Programs Not Currently Funded** | **3,454,041** | **10,306,630** |
| **Grand Totals** | **10,548,306** | **173,709,490** |

Source: Farmers Home Administration National Office, Information Resources Management Division

## Commercial Banks

Commercial banks[2] are private corporations chartered by either federal or state governments. For the purpose of this discussion, the category "banks" includes thrifts (savings and loan companies), although thrifts may have significantly different business structures.

Banks derive their loan funds from money deposited in savings and demand accounts and are required by law to keep a certain percentage of these deposits in reserve. As depository institutions, banks are subject to federal regulation by the Federal Deposit Insurance Corporation (FDIC). Depending on its charter, a bank may also be regulated by state bodies and/or federal agencies, such as the Office of the Comptroller of the Currency (OCC) or Office of Thrift Supervision (OTS). State banks that are members of the Federal Reserve System and bank holding companies or non-bank subsidiaries of bank holding companies are regulated by the Federal Reserve Board (FRB).

Commercial banks were not historically a primary source of funds for financing agricultural real estate. Banks historically concentrated on commercial lending to agriculture with short- and intermediate-term loans for production financing and for equipment, fixtures, or property improvements. Real estate mortgages or deeds of trust were taken by banks as additional security on some commercial loans. However, in response to the modern farmer's increased credit needs after World War II, and particularly since the 1970s, banks expanded their long-term as well as commercial agricultural lending activities. By creating branch offices and by participating with urban banks and other production credit lenders through correspondent banking relationships, banks have made more funds available to farmers and ranchers.

In 1998, commercial banks accounted for about 41% of the outstanding farm debt (see Table E.1). The interest rate charged by commercial banks is subject to any usury laws in effect, and the loans themselves must be fully amortized. Commercial banks make conventional loans (loans whose security is based solely on the property used as collateral) and loans that are insured or guaranteed by agencies of the federal government, such as FHA, VA, and FSA.

Commercial banks figure most significantly in the extension of short- and intermediate-term credit to eligible farmers and ranchers, and those institutions are the most active source of loan funds in this area.

## Life Insurance Companies

Life insurance companies are governed by both the state in which they are incorporated and the states in which they operate. Life insurance companies are of two principal types—mutual and stock. Mutual life

2. Readers are encouraged to review Chapter 6 of *The Appraisal of Real Estate,* 11th ed. (Chicago: Appraisal Institute, 1996), 97–132, for additional detail on bank and thrift sources of credit.

insurance companies are owned by policyholders, who share in net earnings by receiving dividends that can be used to reduce their premium expenses. The profits of stock life insurance companies belong to shareholders, who may or may not be policyholders.

For many years, life insurance companies were the largest institutional lenders in farm real estate loans. As recently as the mid-1960s, their share equaled that of the Farm Credit System (Federal Land Bank). In 1981, life insurance companies were the second most active institutional lenders, holding 14% of outstanding farm real estate loans. However, in more recent years their share of farm real estate loans has decreased. The Farm Credit System and commercial banks have largely absorbed this lost share by dramatically increasing their market share since 1976.

Life insurance companies have traditionally invested heavily in real estate. Their activities include both mortgage lending and property ownership. Life insurance companies may acquire full ownership of real estate for their own investment accounts or for the separate investment accounts that they manage for pension trusts. Although real estate ownership may amount to only 3% to 5% of a company's assets, a large dollar investment is involved because major life companies may have billions of dollars in assets.

Insurance companies are also substantial mortgage lenders for their own accounts, under contracts with pension managers, and as managers of separate accounts. Many portfolio managers, especially those of the larger companies, view agricultural mortgages and equities as safe and risk-rewarding investments for a diversified portfolio.

Historically insurance companies have usually acquired real estate positions that are long term and that related well to their regular business in which policy premiums are collected over extended periods. Through normal insurance sales and operations, life insurance companies accumulate large amounts of funds that they place in diverse investments.

The recent development and popularity of "universal" life insurance, whereby the policyholder, not the company, directs the investment of policy reserves, impacts the dependability of the availability of these funds for extended long-term periods. Additional major sources of funds to develop recently include guaranteed investment contracts (GICs), annuity considerations, and other types of contracts with pension fund managers having various term lengths. For example, according to recent statistics compiled by the American Council of Life Insurers, in 1970 life insurance premiums made up 44% of total income, and annuities made up 8%. In 1980 the proportions were 31% and 17%, and in 1991 were 19% and 30%, respectively. Annuity considerations may, and usually do, have contractual obligations as to minimum returns and thereby impact the terms that may be offered a borrower.

These recent changes, together with a narrowing of spreads and margins, have created a need to match assets more precisely to liabilities, thereby altering the investment strategies of the life companies to more closely match the available cash flows required by their liabilities.

The mortgage investments of life insurance companies cover the full range of realty types—e.g., residences, apartments, offices, shopping malls, hotels, and industrial properties. In addition are all types of agricultural real estate such as row crop, permanent plantings, timber, livestock ranches, and other agribusiness type real estate. Although their principal rural investment activity is in loans for farm real estate, they also invest in a limited amount of loans to agricultural support businesses such as feedlots, dairy, poultry facilities, hog facilities, and grain elevators. Because many companies have great financial resources, they have been important in mortgaging large income-producing properties.

Life insurance companies have consistently invested in safe, high-yield enterprises. Although a large percentage of their funds is invested in commercial and industrial real estate loans, life insurance companies also invest in agricultural properties that present little risk and offer reliable profit. Much of their agricultural lending activity originally took place in specific geographic areas such as the north central United States and in the southern Plains states. Large commercial farms located in these areas, where soil and weather conditions are predictably good, attracted life insurance company loans. More recently, because of investor requirements for higher yields, there has been an increase in agricultural investments in the western and southeastern states, where invested funds obtain a higher return.

Their underwriting regulations allow life insurance companies to lend on improved or income-producing property. Loans on unimproved, nonproductive land have different characteristics and require different underwriting requirements. Life companies have a limited appetite for such loans.

As with other lenders serving agriculture, the financial crisis in the 1980s has influenced the life companies to review their activities and focus of operations. This has resulted in fewer players, larger average loan sizes, and shifts in the geographic distribution of loans.

The events of the 1980s led to concentration of farm mortgage assets within the industry into fewer companies. The number of life insurance companies making new loans in the conventional farm mortgage market dwindled from 12 in 1980 to six at the beginning of 1992. The five largest companies active in 1992 dominated farm lending among insurance companies prior to 1980. At the end of 1992 these five companies held over 86% of the industry's farm mortgage assets, up from 74% in 1980.

More recently, as the life insurance industry focused more on commercial real estate, farm mortgages as a percentage of total mortgages declined from 9.9% in 1980 to 3.9% in 1991. In contrast, commercial real estate increased from 61.6% to 80.8%. The overall trend—to diversify their portfolios—has been underway for decades. For example, in 1920 farm mortgages made up half of all the life insurance industry's mortgage assets and 17.3% of its total assets. Farm mortgages at the end of 1992 made up less than 1% of total industry assets.

While the six remaining companies have different investment strategies, a policy common to most companies is to favor larger loans. At least three have stated minimum loan sizes of $500,000 and one company is at $1 million. For all practical purposes, life insurance companies are no longer players in the market for loans less than $150,000 and relatively minor players under $500,000. For Corn Belt farmers, this implies that most insurance companies are not in the market for mortgages on less than 160 acres. This trend to larger loans has also resulted in more emphasis on loans to agribusiness and commercial timber lending.

Geographically, farm mortgage holdings by insurance companies continued to shift from the Corn Belt to the Pacific Coast and the Southeast during the 1980s. The Corn Belt's share of outstanding industry mortgage volume declined from 23.5% to 16.4%, and the share captured by the Pacific Coast region increased to 33.7% from 19.3%. California has the largest concentration of life insurance farm mortgage loans. In the Northeast, Lake States, northern Plains, and even the Corn Belt, insurance companies are becoming an inconsequential factor in farm mortgage lending. The industry's market share of total farm real estate debt is below 10% (except 16.4% in the Corn Belt) in these regions and fell in all but three USDA regions during the 1980s, a trend predating 1960. A few of the life insurance companies are establishing correspondent networks to originate the loans. The loans are then purchased in pools in a single transaction. This methodology serves to incorporate some of the economics of a larger single loan transaction to smaller individual mortgages.

The farm financial problems of the 1980s had a profound impact on life insurance company lending to agriculture. Defaults among life insurance company mortgages were among the highest of any farm lender. Farm mortgages in 1992 represent an inconspicuous share of the life insurance industry's total asset base. The six remaining companies accounted for 86% of the industry's farm mortgages and generally have both high total assets and large farm mortgage portfolios.

Life insurance companies are not limited in the amount they can lend on a single property, and, except for the Farm Credit System, insurance companies tend to make larger loans than other lenders. They primarily base amounts on a given percentage (from 66 ⅔% to 75%) of the market value of the property and the repayment ability of the enterprise. Like commercial banks, these lenders must charge interest rates that are consistent with usury ceilings in states where these are in effect. Depending on interest rate relationships, this has caused some limitation in their activity in these states. Life insurance companies are not required to adhere to full amortization schedules and have been fairly creative in their loan repayment plans. Depending on the size and term of the loan, it may be fully amortized, or the borrower may pay interest only during the loan term and make a balloon payment of the principal amount upon expiration of the term.

The term of the loan granted by a life insurance company may be longer than other institutional lenders. Because of this, they may incorporate additional loan covenants such as due on encumbrance or due on sale clauses. This practice allows the lender to reevaluate the soundness of the loan on the basis of the new farm operator.

## Seller Financing

Seller financing has consistently been an important source of credit for rural properties. This source of credit remains strong. The importance of seller financing may be seen as an effect of general economic trends in the United States as well as trends that affect agricultural property value and the nature of the industry. As recent experience has clearly illustrated, the value of agricultural real estate can decline as well as increase. Lenders are controlled in the amounts they may lend and the interest rates they may charge, whereas private owner-lenders are not under these constraints (other than by usury laws). When market conditions cause lenders to reduce the supply of funds available for real estate loans, the importance of seller financing tends to increase.

Another trend in agriculture that influences the use of seller financing is the need for funds to expand productive capacity, particularly when such expansion involves unproven technology or a new commodity to the borrower's operation. With such extensive cash outflow and related increased risk, the borrower seeks ways to expand capacity with as little cash investment as is both wise and feasible. Seller financing often provides such an alternative because the risk can exceed an institutional lender's limits.

Farm and ranch land transfers often occur between relatives. Parents who wish their children to take over the management and production of their operation often begin the process of transfer long before they expect it to be fully accomplished. In this manner, the younger farmer's or rancher's credit credentials become established, and the final transfer of ownership is facilitated. Appraisers are often involved in estate appraisals even before the estate is created, for the purpose of aiding in the transfer of ownership.

The primary instrument used when a farm or ranch owner finances the sale of property is the land contract for sale (also known as the *contract for deed*, the *land contract*, and the *contract for sale*). In financing by land contract, the buyer pays a relatively low down payment for the land, although the sale price for the property may be higher than if it were financed through an institutional lender. In exchange for providing the buyer with the benefit of a lower initial capital outlay, the seller receives the prospect of higher future gains. The seller holds the deed until the terms of the contract are fulfilled or until the buyer has enough equity in the property to renegotiate for a conventional mortgage. In such a transaction, the buyer takes possession of the property and pays all taxes, assessments, and insurance. The advantage to the seller, in addition to the higher sale price, is that payment of taxes on the sale may be spread out over a number of years.

A second instrument that may be used in seller financing is the purchase-money mortgage, through which a seller conveys the property to the buyer by means of a conventional deed but takes back a mortgage on the property as security for the unpaid balance of the purchase price. This method is used when the purchaser has a down payment and the seller wants to delay paying income taxes on the capital gain or in order to receive the interest that may be higher than savings rates.

### The Farm Credit System

The Farm Credit System (FCS) came into being gradually between 1916 and 1933. Some of its parts were mandated by Congress, others by executive order. In its efforts to help farmers become self-sufficient owner-operators with greater economic strength, the U.S. Government developed a system for evaluating the credit needs of the American farmer, providing a stable source of funds for these needs, and creating the means through which credit needs could be met throughout the normal favorable/unfavorable/favorable production and economic cycles that affect the agricultural industry.

In 1992 FCS provided about 33% of the real estate credit and 16% of the non–real estate credit used by America's farmers, ranchers, and cooperatives, in addition to serving the credit needs of rural homeowners, electric and telephone cooperatives, and rural water systems.

The Farm Credit System is a nationwide network of borrower-owned lending institutions and specialized service organizations. Components of the system include

- Farm Credit Administration (FCA)
- Farm Credit Banks (FCBs)
- Federal Farm Credit Banks Funding Corporation (the Funding Corporation)
- Farm Credit System Insurance Corporation (FCS Insurance Corporation)
- System associations (ACAs, PCAs, FLCAs, FLBAs)
- Farm Credit Leasing Services Corporation (FCL)
- Farm Credit Council (FCC)

The Farm Credit System is an instrumentality of the federal government, and it is independently regulated by the Farm Credit Administration (FCA). The FCA was created by executive order in 1933 as the supervising and coordinating agency for the entire Farm Credit System. The role of FCA was changed to that of an arm's-length regulator by the 1985 and 1987 amendments to the Farm Credit Act of 1971. Located in McLean, Va., FCA is directed by a three-member board nominated by the President and confirmed by the Senate. The FCA is responsible for examining and ensuring the safety and soundness of all FCS institutions.

The Federal Farm Credit Banks Funding Corporation manages the sale of FCS securities in the nation's money and capital markets through a network of investment dealers and dealer banks in New York. Farm Credit Consolidated Systemwide Debt Obligations are the primary source of loan funds for FCS and are the joint and several liability of system banks. The Funding Corporation provides the investment community and the System with an efficient method of funding the loans made to the nation's farmers, ranchers, and cooperatives. Loans are priced on the basis of the current market, with the borrower's rate reflecting the nature of the credit, its risk and term, and the cost of providing the credit services.

The Funding Corporation also provides financial advisory services and supports FCS banks in the management of interest rate risk. The Funding Corporation is owned by FCS banks and is governed by a 10-member board. Seven elected members appoint two members from outside the system. The Funding Corporation President serves as a non-voting board member.

The Farm Credit System Insurance Corporation was established by the 1987 amendments to the Farm Credit Act of 1971, primarily to ensure timely payment of principal and interest on consolidated or system-wide debt securities. The FCS Insurance Corporation is managed by a board of directors consisting of the Farm Credit Administration board. Each Farm Credit Bank is required to contribute to the Insurance Fund through a current annual assessment of 0.15% on accruing loans and 0.25% on non-accrual loans.

Another FCS entity created by the 1987 amendments to the Farm Credit Act of 1971 was the Farm Credit System Financial Assistance Corporation (FAC). The FAC was created to provide capital to help financially troubled FCS institutions. The funds for assistance were raised through the sale of federally guaranteed securities to private investors. The FAC was governed by the same board of directors as the Funding Corporation. Of the $4.0 billion in total assistance authorized by the 1987 legislation, the Farm Credit System used only $1.26 billion and has repaid all of the assistance. The FAC dissolved on December 31, 1992. The FCS Insurance Corporation was designated the legal successor to agreements involving the FCS Assistance Board.

### FCS Banks and Associations

The Farm Credit System, through its banks and associations, provides agricultural credit to farmers, ranchers, and cooperatives. All FCS banks and associations are governed by boards of directors elected by member borrowers/stockholders. The institution's bylaws provide for at least one member of the board to be elected from outside the system by other directors. System lending and leasing institutions are primarily capitalized by a combination of stock borrower's purchase in the course of obtaining a loan or lease and retained earnings.

Typical services include operating and real estate loans with fixed, adjustable, or variable rates; rural home mortgages; credit life insurance and crop insurance; and various financially related services such as farm record keeping and financial planning. Unlike commercial banks, thrifts, or credit unions, FCS banks and associations do not take deposits.

The Farm Credit System is divided into districts covering the United States and Puerto Rico. The FCS districts are

- CoBank
- AgFirst Farm Credit Bank
- AgriBank, FCB
- AgAmerica, FCB
- FCB of Wichita
- FCB of Texas
- Western FCB

As is commonplace in the financial services industry, there is ongoing potential for change (through mergers and affiliation) in the current FCS district and association alignments.

Following the Agricultural Credit Act of 1987 amendments to the Farm Credit Act of 1971, various district Banks for Cooperatives formed a national bank for cooperatives, under the name CoBank. CoBank is headquartered in Denver and has offices throughout the United States. CoBank also finances exports of U.S. agricultural products and provides international banking services for farmer-owned cooperatives.

The Farm Credit Banks (FCBs) resulted from the 1988 merger of the Federal Land Bank and Federal Intermediate Credit Bank in each district. Six FCBs provide a source of funds as well as supervision and support services to 39 Federal Land Bank Associations (FLBAs), 63 Production Credit Associations (PCAs), 54 Agricultural Credit Associations (ACAs), and 33 Federal Land Credit Associations (FLCAs), for a total (as of February 24, 1999) of 189 locally owned Farm Credit associations serving rural America. FCBs provide a risk-pooling mechanism on a district basis for the associations in the district. In some districts, this facility spreads risk across the district and serves as a first line of defense against financial stress resulting from cycles of adverse weather or commodity price/production cost relationships in one area of the district.

### FCS Associations

The various FCS institutions make loans to farmers and ranchers through local associations. With the exception of the FLBAs, the local associations are direct lenders that make loans using funds generated through retained earnings and a credit line from the FCB. FCBs make long-term mortgage loans through the FLBAs that are agents of the FCB. PCAs make short- and intermediate-term loans. ACAs, created by the merger of one or more FLBAs and PCAs, have received a transfer of authority from their FCB to make and participate in long-term real estate mortgage loans as well as make direct short- and intermediate-term loans. FLCAs are associations (formerly Federal Land Bank Associations) to which direct long-term real estate lending authority has been transferred from their FCB. In some districts, FLBAs, PCAs, ACAs, and FLCAs operate under joint management and are commonly referred to as Farm Credit Services.

The Farm Credit Leasing Services Corporation (FCL), headquartered in Minneapolis, is owned by FCS banks and has sales offices located throughout the nation. FCL provides leasing services to eligible borrowers, including agricultural producers, cooperatives, and rural utilities.

The Farm Credit Council (FCC) is the national trade association representing the interests of the Farm Credit System. There are two offices: Washington, D.C., where the council represents the legislative and regulatory concerns of FCS, and Denver, where the council provides business-related services to its system membership.

The council provides an important mechanism for grassroots involvement in the development of FCS legislative positions and policies while also providing its membership with a wide array of system-related services. Membership in the FCC is made up of Farm Credit District Councils and CoBank. These member councils, in turn, draw their membership from Farm Credit Banks and associations, as well as other cooperative farm lenders and associations.

## USDA and Other Government Sources

### The Farm Service Agency

The Farmers Home Administration (FmHA) was conceived and organized during the 1930s to provide credit to individuals who were unable to obtain adequate credit from commercial lenders. The FmHA has never been in competition with commercial lenders; rather, it has tried to supplement credit that is available from traditional lending sources. Congress dissolved the agency in 1996. The responsibility for the farm loan programs of FmHA was given to the Farm Service Agency (FSA). To qualify for an FSA loan, the applicant must meet both income and geographic requirements and must be unable to qualify for all the needed credit at reasonable rates and terms from conventional lenders or other loan sources.

Through FSA's program, borrowers receive loans and supervision designed to aid them in improving their financial situations. The loans are intended to be intermediate in nature, the duration only until the borrower's financial condition warrants transfer to other sources of credit.

Two categories of loans are made either by or through FSA: direct loans and guaranteed loans. Funds for loans made directly by or guaranteed by FSA are derived from annual appropriations by Congress. Since 1988, emphasis has shifted from FSA making direct loans to the Guaranteed Loan Program. Commercial lenders supply funds for FSA guaranteed loans. Under this program, up to 90% of the loan may be guaranteed by FSA. Unlike direct loans that are originated, processed, and serviced by FSA personnel, the commercial lender performs these functions under the Guaranteed Loan Program.

Farm ownership loans are extended to qualified family farmers who obtain a substantial portion of their income from farming and are considered owner-operators even if part of their income is derived from off-the-farm labor. The loans are extended for purchasing land, refinancing debts, constructing or repairing farm buildings or improving farmland, developing water facilities, and establishing farm-based business enterprises to supplement farming income.

Under the category of farm operating loans, the FSA extends short- and intermediate-term credit to eligible operators of family farms. Such loans may be made for purchasing livestock or equipment and supplies needed for farm production, for refinancing debts, for making minor real estate improvements, and for establishing non-farm enterprises to supplement farm income.

Farm emergency loans that cover losses from natural disaster that are not otherwise compensated are sometimes made available in cases of natural disasters.

### Rural Development

The Rural Housing and Community Program type loans of the old Farmers Home Administration (FmHA) were folded into a new USDA agency named Rural Development (RD). The RD agency is involved in an extensive variety of loan types. Rural Development directs its major lending activities toward rural communities, businesses, and low-income rural families and individuals who do not currently own safe and adequate housing. Qualified veterans are given preference in the processing for RD loans, but they must meet the same qualifications as nonveterans. An RD loan can be made on up to 100% of the property's market value, but there are maximum amount limits that vary according to the loan category. Interest rates on some direct loans by RD may be subsidized based on the borrower's income. Rates on guaranteed loans are set by the lending institution that carries the loan.

Rural housing loans are made for short- and long-term needs of low- to moderate-income families to build, purchase, or improve homes in rural areas. Towns whose populations are below specific levels qualify as rural areas for RD loans. Families with above moderate income living in such areas are eligible for RD rural housing loans under the Guaranteed Loan Program. Short-term loans are made for repairs and rehabilitation of substandard rural properties.

Several rental housing loan programs available through RD provide rental units for low- to moderate-income rural residents and farm laborers. Loans are made to eligible individuals, trusts, associations, partnerships, limited partnerships, state or local public agencies, consumer cooperatives, and profit and nonprofit corporations. The maximum repayment period for most of these multifamily housing loans is 50 years.

Three categories of RD community programs aid in the development of rural America through direct or guaranteed loans, grants, or a combination of loans and grants. Water and waste disposal loans and grants are made to public entities to assist rural communities of less than 10,000 people to develop water and waste disposal systems. Community facilities loans are made to public entities to construct, enlarge, extend, or otherwise improve community facilities in rural areas and towns of less than 10,000 people. Hospitals, fire stations, libraries, and city halls are commonly developed under this program. The Guaranteed Business and Industry Loan Program is available in communities of 50,000 people or less. The program provides guarantees on loans made by eligible lenders to improve, develop, or finance business, industry, and employment and to improve the economic and environmental climate in rural areas.

### The Small Business Administration

The Small Business Administration (SBA) was created by Congress in 1953 as a permanent and independent government agency. Its primary purpose is to protect the interests and serve the needs of small business concerns in the United States, particularly by providing loan funds in cases where such funds are necessary and appropriate. In 1976 legislation was passed to include farms among small business concerns that SBA may serve. However, because FSA is still the principal government agency serving the credit needs of the farm industry, SBA's role in this area is secondary to that of FSA.

The SBA basically guarantees loans funded by lending institutions with SBA agreeing to cover up to 90% of the loss in the case of default. Very few direct loans are made from SBA funds, and there is an established $150,000 maximum limit. The lender's interest rate on guaranteed loans is established by lending institutions but is subject to maximum limitations set by SBA. Interest rates also vary according to the type and purpose of the loan.

Although SBA does make agribusiness loans to farmers, the principal responsibility for such loans now rests with RD or FSA. Most SBA loans are made to farmers under the administration's regular guaranteed business loan program. Under its regular business loan programs, SBA can guarantee or make loans for both short- and long-term needs of rural business ventures.

The loan maturity varies according to the purpose of the loan, usually five to seven years for working capital and up to 25 years for real estate or equipment. However, SBA expects the borrower to repay the loan as quickly as possible, and maximum loan terms are not automatically available in every case.

### The Federal Agricultural Mortgage Corporation (Farmer Mac)

The Federal Agricultural Mortgage Corporation (Farmer Mac) is a federally chartered instrumentality of the United States established by the Agricultural Credit Act of 1987, which amended the Farm Credit Act of 1971. Farmer Mac was established primarily to attract new capital for the financing of agricultural real estate and rural housing loans and to provide liquidity to agricultural real estate and rural housing lenders. Farmer Mac is intended to aid the development of a secondary market for agricultural real estate and rural housing loans (referred to as *qualified loans*) made by participating *originators,* pooled by eligible agricultural mortgage marketing facilities (referred to as *poolers*) and secured by first liens on agricultural real estate, including rural housing, by guaranteeing the timely payment of principal and interest on obligations (Farmer Mac I securities) backed by such loans and securities representing interests in such loans (the Farmer Mac I program).

Farmer Mac also issues certificates (Farmer Mac II securities) representing interests in pools of portions (i.e., guaranteed portions) of farm ownership, operating loans, and certain other loans guaranteed by the United States acting through the Farm Service Agency (the Farmer Mac II program).

The Agricultural Credit Act authorizes Farmer Mac to borrow up to $1.5 billion from the Secretary of the Treasury, subject to certain conditions, to enable Farmer Mac to fulfill its guarantee obligations. The debt created by such borrowing will bear interest at a rate determined by the Secretary of the Treasury, taking into consideration the average rate on outstanding marketable obligations of the United States as of the last day of the calendar month ending before the date of the purchase of such obligations.

Currently both Farmer Mac I and Farmer Mac II programs have numerous agricultural loan products for lenders to use in servicing the needs of their customers. Overall, the agricultural real estate and rural housing secondary markets are in their early developmental stages, but interest appears to be strong and growing. If successful, Farmer Mac could have a significant impact on the way agricultural real estate and rural housing are financed in the future. As documentation and analysis of these types of loans improves, so should the availability of products and their pricing.

## Summary

Capital invested in agriculture includes short-term investments to support enterprise operations, intermediate-term investments to provide equipment and fixtures, and long-term investments in the real estate used to produce food and fiber for the nation and for export throughout the world. Agricultural investment capital takes the form of equity or debt.

Equity capital typically comes from accumulated profits and savings, but it may involve other sources such as joint ventures, syndications, grants from government bodies, and the like. Debt capital may be obtained from a wide variety of institutions, including commercial banks, insurance companies, government agencies, and the Farm Credit System, as well as from private individuals, trade creditors, and processors. Lease financing is available through commercial banks, the Farm Credit System's Leasing Services Corporation, equipment suppliers, and government agencies. Long-term credit, as well as joint-venture or syndication equity investment capital, is available from insurance companies and private individuals.

Agribusiness financing is available through commercial banks, Farm Credit System banks or associations, and the Farm Credit System's Bank for Cooperatives (CoBank). Some cooperatives and large scale agribusiness entities are internally financed or issue private debt instruments for their financing needs.

As a result of the creation of the Federal Agricultural Mortgage Corporation (Farmer Mac), long-term agricultural lenders have available a mechanism to obtain loan funds through securities markets in a way similar to urban property lenders and the Farm Credit System. The investor's risk is reduced through the pooling of loans, with each pool having a diversity of commodities and property types from different geographic regions of the nation.

Rural property appraisers need to recognize the role their professional opinions play in agricultural capital markets. With a broader competitive arena, the continued availability of both equity and debt capital for agriculture is not assured unless risk is properly identified and understood by agricultural property investors and lenders. Well-supported value opinions that reflect the particular characteristics of the property and its market, favorable or unfavorable, are needed at every level of the industry. The experience of the most recent boom and bust cycle should not be quickly lost. When internal or external forces cause changes in the market for agricultural property, the appraisal should identify the changes and factually report the impact so that those who rely on the perspective of the appraiser will be well informed in their business and financing decisions.

APPENDIX

• F •

# Narrative Appraisal Report Outline

Although not all appraisers organize their reports in the same manner, most reports contain certain common features. The required common features are specified in the Uniform Standards of Professional Appraisal Practice (USPAP). Appraisals that comply with all the requirements of USPAP are acceptable appraisals from the standpoint of appraisal content, and the following outline provides information required by USPAP. Some appraisal organizations require members writing demonstration reports to follow a more specific outline than is required by USPAP. The following outline is recommended to members of the American Society of Farm Managers and Rural Appraisers writing demonstration reports.

The *Handbook of Technical Writing,* 4th edition, by Charles T. Brusaw, Gerald J. Alred, and Walter E. Oliu (New York: St. Martin's Press, 1993) is an excellent reference book and guide that contains numerous examples of writing various portions of technical reports. Throughout the following outline references are made to appropriate topics in Brusaw, and readers are encouraged to use Brusaw as a technical writing reference when preparing appraisals. Appraisals are divided into four parts:

1. Introduction
2. Factual data
3. Analysis and conclusions
4. Addenda

## Introduction

The introduction states the purpose (market value, investment value, insurance value, etc.) and function (use of report, e.g., mortgage, estate, etc.) of the appraisal, identifies and describes the subject property, and presents the value conclusion. The introduction also includes any assumptions and conditions that limit the report.

Two alternative organizations are appropriate for appraisal introductions. The traditional organization is title page, letter of transmittal, table of contents, summary of salient facts and conclusions, and assumptions and limiting conditions. An acceptable alternative eliminates the letter of transmittal and renames the summary of salient facts and conclusions as the executive summary. The resulting organization is title page, table of contents, executive summary, and assumptions and limiting conditions. The major components of the introduction are as follows:

## I. Title Page

The title page includes the following information:

- Property name and identification
- Address or location of the property being appraised
- Name and address of the client
- Name and address of the appraiser
- Effective date of the valuation
- Date of report preparation[1]

## II. Letter of Transmittal (traditional organization only)

The letter of transmittal formally presents the appraisal report to the client. Letters of transmittal are drafted in proper business style, being as brief as the character and nature of the assignment permits. The letter of transmittal is part of the appraisal report and is not to be confused with the *cover letter* used to convey the report to a client. The letter of transmittal is used to provide a permanent record for identifying the writer and intended reader(s). A suitable letter of transmittal includes

- Property name or identification
- Effective date of the value estimate
- Date of report preparation
- Type of value estimated
- Statement that the property inspection and all necessary investigation and analysis were made by the appraiser
- Final estimate of value
- Any extraordinary assumptions and limiting conditions
- A statement providing location of the property and a brief description, if necessary

1. Charles T. Brusaw, Gerald J. Alred, and Walter E. Oliu, *Handbook of Technical Writing*, 4th ed. (New York: St. Martin's Press, 1993), 259-260.

- Signature of the appraiser[2]

In most cases the letter of transmittal is the first item readers will read. Use special care to reflect your professionalism.

## III. Table of Contents

The table of contents shows the page numbers of the various sections and subsections of the report. The table of contents is arranged to help readers locate specific items in the report. When the appraisal includes more than five tables, a "list of tables" containing table numbers, titles, and page numbers follows the table of contents. When there are more than five figures (drawings, photographs, maps, charts, or graphs), a "list of figures" containing figure numbers, titles, and page numbers follows the list of tables.[3]

## IV. Summary of Salient Facts and Conclusions or Executive Summary

The summary of salient facts and conclusions provides readers with a summary of the appraisal problem, factors considered, analyses performed, and conclusions reached. Often the summary of salient facts and conclusions is presented in the form of a bulleted list with brief statements made about each topic covered. Little or no effort is made to tie the points together.[4]

Executive summaries are usually prepared in narrative form to provide busy readers an opportunity to be informed without reading the complete report. Executive summaries should fully inform readers of the problem studied and conclusions reached. Executive summaries are usually proportional in length to the appraisal document. Complex appraisal problems require more detailed summaries to adequately inform readers. Normally, executive summaries should have the same organization as the appraisal report.

At a minimum, both the Summary of Salient Facts and Conclusions and the Executive Summary cover the following points:

A. Brief property identification
B. Effective date of the valuation and date of report preparation
C. Property ownership and property rights appraised
D. Property location
E. Acreage breakdown and brief list of improvements
F. Zoning
G. Highest and best use
H. Brief summary of subject leases, if any

---

2. Ibid., 701.
3. Ibid., 261, 674–675.
4. Ibid., 232–237.

I. Value indicated by the cost approach

J. Value indicated by the income capitalization approach

K. Value indicated by the sales comparison approach

L. Final opinion of value

### V. Assumptions and Limiting Conditions

These statements inform the client and other readers of the report of the limiting conditions that are outside the scope of the assignment or not within the responsibilities of the appraiser. The appropriate conditions should be stated clearly and be relevant to the current appraisal problem. The best location for the assumptions and limiting conditions section is early in the report to alert readers to the premises underlying the analysis and used to prepare the appraisal. The number and type of assumptions and limiting conditions vary depending upon the nature of the appraisal assignment. The letter of transmittal should contain a summary of extraordinary assumptions or limiting conditions. Special notice is given to any assumptions and/or limiting conditions that have a direct impact on value, and it is imperative that appraisal readers understand those important assumptions and limiting conditions. The assumptions and conditions must be reasonable and justifiable. It is important that the client agree to the limiting conditions at the time of assumption. Otherwise, you may be assuming away a condition that the client wants the appraisal to address.

*The Appraisal of Real Estate,* 11th edition (Chicago: Appraisal Institute, 1996) and *Communicating the Appraisal: The Narrative Report* (Chicago: Appraisal Institute, 1988) are recommended reading. Both books contain numerous examples of specific assumptions and limiting conditions that you can incorporate into appraisal reports. Available report writing courses, such as those offered by the American Society of Farm Managers and Rural Appraisers, also cover and provide examples of assumptions and limiting conditions.

## Factual Data

The factual data section of a report contains facts affecting the appraisal. In general, all information used to prepare the appraisal (except comparable sales information) is included as factual data, and it is not appropriate to introduce additional information in the analysis that is not presented as factual data. Information included depends on the appraisal assignment. However, the following items are generally required:

### I. Property Identification

The property location is presented clearly along with distances to the nearest population centers. The size and overall type of the property should be mentioned—e.g., "a 759-acre farm 17 miles northwest of Boland, Iowa."

## II. Legal Description

The legal description of the estate appraised is provided. If the description is longer than one or two paragraphs, the description should be placed in the addenda and readers referred to its location.

## III. Appraisal Objective

The appraisal objective informs readers of the purpose and function of the report. The appraiser, perhaps in consultation with the client, determines the appraisal's purpose while the client indicates the intended use or function of the report.

### A. Purpose of the Appraisal

The purpose of the appraisal is answering the value question the client wants to answer. The purpose may be to estimate market value, insurable value, or some other type of value.

### B. Function of the Appraisal

The client's expected use of the appraisal should be clearly stated. Readers should be told whether the appraisal is to be used for loan purposes, estate settlement, bankruptcy, or some other use. The name of the person authorizing the appraisal report should be included.

## IV. Definition of Value

A precise definition of the value sought should be provided. After reading the definition, readers should not question the type of value being determined. When estimating market value, the appraisal should state whether the estimate is expressed in terms of cash, terms equivalent to cash, or other precisely defined terms. If there are not clear guidelines as to what the definition of value ought to be for a particular appraisal problem, the definition of value provided by the Uniform Standards of Professional Appraisal Practice should be used.

## V. Scope

The process of collecting, confirming, and reporting data is described. Readers are informed about the extent of the investigation undertaken. This section aids readers who use the report to make decisions.

## VI. Property Rights Appraised

The particular rights or interests being valued are identified. This is particularly important in assignments that involve partial interests such as limited rights (surface or mineral rights only), fee simple estates subject to long-term leases, or leasehold interests. Also, other encumbrances such as easements, mortgages, or special occupancy or use requirements impacting the property should be listed. These encumbrances should be discussed relative to the value being estimated.

## VII. Effective Date of the Appraisal

The effective date of the value estimate must be stated along with the date the report is prepared. Appraisal assignments may call for estimating current value or value at some other point in time. Stating the date of valuation precisely is essential for proper use and interpretation of the report. The effective date of the appraisal establishes the context for the value estimate. The date of the report indicates the perspective from which the report is prepared.

## VIII. Ownership History of the Property

At a minimum, the appraisal should include a summary of any agreement of sale, option, or listing of the property being appraised, all sales of one- to four-family residential properties within the last year, and sales of all other properties within the last three years. If a sales history is not included, there should be an explanation of efforts undertaken to obtain the information. A preferred ownership history for the subject property is five years. The sale price, terms, and other pertinent information should be included in the history.

## IX. Area-Regional and Neighborhood Analysis

Prepare the area-regional and the neighborhood analysis to provide a foundation that will lead readers to recognize and accept your conclusions of the property's highest and best use(s). Include only information pertinent to the appraisal problem. Discuss both positive and negative aspects of the area. If only positive or only negative factors are included, the report is likely to be misleading. If significant supporting statistical data is needed, incorporate the information into tables or include a brief summary in the body of the report that refers to more lengthy material in the addenda.

Having information on file is no reason to include it in the appraisal report. For each item included, you should ask

- Is the data relevant?
- Does the information add to the understanding of the physical, social, political, economic, or market setting of the subject property?

One good test is continuity. Does the material flow logically from section to section? The type of appraisal problem and the nature of the subject property dictate, to a large extent, the information that you should include in this section. The relevant area for a specific farm may, in some cases, have a five- to ten-mile radius, while in other cases, several counties or entire states are included. The number and type of agriculture enterprises in the area determines the size of the area-regional and neighborhood analysis. Always address physical, social, economic, and governmental considerations having an effect on real estate values within the general area and neighborhood. Use the following partial list of topics as a guide for preparing the area-regional and neighborhood analysis:

- Physical considerations
  - Location and the boundary of the area defined
  - Access and transportation within the area defined
  - Land use, tenure, and expected changes in land use
  - Terrain, soils, and climatic considerations within the general area
  - Adequacy of utilities
- Social considerations
  - Community facilities and population
  - Schools, church, and recreation available
  - Heritage of community and prestige of area
  - Conformity of development
- Economic considerations
  - Proximity and adequacy of commodity markets
  - Types of farms and rent levels
  - Crop production levels
  - Typical management practices
  - Livestock carrying capacities
  - Local economy and possibilities for off-farm employment
  - Urban growth or population changes that may indicate trends in property values
  - Detriments and hazards
- Governmental considerations
  - Discussion of taxes and their trends
  - Zoning or other governmental restrictions
  - Special easements or assessments that are common to the area
  - Government policies as they impact returns to agricultural land

Conclude the discussion of the area-regional and neighborhood analysis with a summary of trends relevant to the subject property. Keep in mind the area and neighborhood is fundamental to the highest and best use of the property. Always interpret the relevant data in terms of how it affects the subject property. Without this interpretation and summary, the area and neighborhood analysis lacks significance.

Avoid stereotypical or biased assumptions relating to race, age, color, religion, gender, or national origin. Do not presume racial, ethnic, or religious homogeneity is necessary to maximize value in the neighborhood.

Either conclude each portion (physical, social, economic, and governmental) with a few sentences or add a conclusion for the entire section summarizing the points influencing the value of the property being appraised.

## X. Property Description

Describe the property so the reader develops an accurate mental image. Use sufficient detail to answer all important questions that the client or reader might pose as to the physical characteristics of the property, its economic contribution, and any specific benefits or constraints attributed to it. From the following list, select topics that are relevant for the property being appraised and include any additional items that are essential:

- Size and general shape of the property
- Location and access, including distance to markets and major trade centers
- Land use (e.g., acreage in cultivation, acreage irrigated, woodland acreage, minerals, etc.)
- Crop history, including yields, carrying capacity, rotation, fertilizer program management methods, etc.
- Farm/ranch operational and lease information
- ASCS crop program information
- Terrain, drainage, and other physical features
- Soil information
- Vegetation
- Water rights, irrigation systems, stock water, well information, etc.
- Utilities
- Taxes and assessments
- Zoning
- Easements, encroachments, and deed restrictions
- Descriptions of land improvements, including fences, corrals, levees, roads, ditches, etc.
- Description of building improvements, including condition, quality, size, amenities, and utility of each
- Disadvantages of the property in relation to its market area
- Advantages of the property in relation to its market area

If helpful, include photographs of major components of the subject property. Also, include good-quality maps of the property if they help the reader. Photographs and maps can either be placed in the text or in the addenda. All items placed in the addenda should be referred to in the body of the report. Some appraisers place all photographs and maps in one location. This helps concentrate the reader's attention on the report rather than having them repeatedly breaking concentration to look at photographs and maps. Where the photographs and maps are located is a matter of style.

As appropriate for the appraisal problem, include soil maps, topographical maps, site characteristics maps, building plats, and the like. Agricultural properties are diverse and some require significantly more data than others. For example, if the majority of the overall property value consists of land, include fewer improvement photos. By contrast, if a significant portion of the total property value is attributable to the improvements, include fewer land photos. Assist readers in understanding maps, photographs, and other exhibits by adequately describing what is shown.

## Data Analysis and Conclusions

Use the data analysis and conclusions portion of the report to lead readers through the logical sequence of steps you are completing to determine value. Base your analysis and valuation conclusions on the data presented in preceding sections and be sure nothing in the analysis or valuation process conflicts with information you provide earlier in the report.

### I. Highest and Best Use

Relate the highest and best use conclusions to information presented in preceding and succeeding sections of the appraisal report. Always provide a definition of highest and best use and convey the following ideas to your readers:

- Highest and best use conclusions are the foundation of the appraisal and the appraisal report.
- Highest and best use conclusions must be consistent with the area-regional data and subject property data. You must understand market behavior to develop the concept of highest and best use.
- Highest and best use conclusions form the foundation for applying the three approaches to value, performing the reconciliation, and selecting the final value.

Determine the highest and best use for the property "as if vacant" using the following four criteria. If the property is improved, also determine the highest and best use of the property "as improved" using the same four criteria:

1. Physically possible. The use must be physically possible considering the size, shape, soil, and terrain of the property.

2. Legally permissible. The use must be legally permissible considering both public and private restrictions. Also, discuss any possibility and/or likelihood of change in what is legally permissible (change in zoning).
3. Financially feasible. The use must be financially feasible by producing returns greater than operating expenses. Probable uses expected to produce a positive return are regarded as financially feasible. Provide a list of the financially feasible alternatives.
4. Maximally productive. The maximally productive use is the financially feasible use producing the highest price or value consistent with the rate of return warranted by the market.

Report the highest and best use of vacant tracts assuming they will be developed to their highest and best use. Decide the "as if vacant" highest and best use of an improved property by using the four criteria to evaluate how the property would be used if the existing improvements are removed. Decide the "as improved" highest and best use by using the four criteria to identify how the property would be used if the existing improvements are not removed.

Conclude both the "as if vacant" and "as improved" sections with summaries. In each summary discuss the four criteria for determining highest and best use. Finally, conclude the highest and best use section with your conclusion about the highest and best use of the property being appraised.

## II. Appraisal or Valuation Process

Use this section to present the mechanics and theory of the appraisal process, particularly the rationale underlying the three approaches to value. There are alternative organizations for this section. You can include a section on valuation methodology after the highest and best use section and before the sections discussing the approaches to value. Title the section "Appraisal Process" or "Valuation Theory." Discuss the steps you are following in the current appraisal assignment and indicate why certain approaches to value are being used while others are excluded. When using this alternative, discuss the mechanics of each approach as an introduction to the section on that approach.

An alternative organization of the appraisal or valuation process is to completely discuss the methodology and mechanics of each approach in this section. When using this approach, begin the discussion of the three approach sections with little or no introduction to the methodology you are using.

## III. Cost Approach

If you have completely discussed the cost approach methodology in the above section, proceed to discussing land values. If you have not discussed the cost approach, summarize the theoretical basis, the principle of contribution on which the cost approach is based, the methodology, and the procedures you are using to determine the cost approach estimate to value. Readers should have a clear understanding of how you are performing the cost approach steps after reading this introduction. Use terms that will be understood by readers and, if necessary, define technical terms.

### A. Land Value

Present market data and other information pertaining to land values along with data analysis and reasoning leading to the land value conclusion. In a clear and precise manner, identify and discuss all factors affecting land values on the property being appraised.

For most rural appraisals, the cost approach begins with valuing each of the various classes or types of land found on the subject property. Common land types include dry cropland, irrigated cropland, pasture, woodland, or other land uses. Ranch land types include rangeland, irrigated, meadow, subirrigated, and dry cropland. Use whatever types are appropriate for the appraisal problem. Information for placing values on each land type separately comes from comparable sales having no improvements (whenever possible). Describe each comparable sale's location, size, zoning, utilities, soils, and other physical characteristics. When determining the land value using the cost approach, the best comparable sales have no improvements and one land type. The most desirable situation is having adequate comparable sales that only require adjusting for time and financing. For each land type in the available comparable sales, choose estimates of value for the land types on the property being appraised. After the values per unit for each land type are determined, calculate the total value of the land component.

### B. Cost of Improvements

Estimate either the replacement or reproduction cost of the subject property's improvements. For a particular appraisal problem, use either replacement or reproduction cost, but not both. For major improvements, present cost data on an item-by-item basis. Indicate the source of the cost estimates (contractor's bid, equipment supplier, or evaluation service such as Marshall & Swift, Dodge, Boeckh, etc.).

Many technical words are usually included in this portion of the appraisal. Include definitions for all technical terms such as physical deterioration, functional obsolescence, external obsolescence, replacement cost new, reproduction cost new, etc. *The Dictionary of Real Estate Appraisal,* 3d edition (Chicago: Appraisal Institute, 1993) and *The Appraisal of Real Estate,* 11th edition, contain excellent definitions of commonly used technical terms.

### C. Depreciation

Estimate depreciation from all sources including physical deterioration (curable and incurable), functional obsolescence (curable or incurable, excess or deficiency), and external obsolescence. Prepare the depreciation analysis from relevant market data. Present all computations to help readers follow and understand your measurement of depreciation. Analyze all transactions as of the date of sale.

Explain the procedure you are using to estimate accrued depreciation. Normally, you will be using one of the following methods: physical age-life, economic age-life, modified age-life, breakdown method, or a sales comparison approach. Carefully define all technical terms and explain methods of analysis within the report. If possible, use market extraction of depreciation. At a minimum, extract an overall rate of depreciation using the improvement's effective age.

### D. Conclusion

Summarize your conclusions at the end of the cost approach presentation. Add the depreciated cost of improvements to the land value estimate to arrive at a cost approach estimate of the property's value. Explain to readers the applicability of the cost approach and point out any limitations that the cost approach may have for this particular problem.

## IV. Income Capitalization Approach

Unless discussed as part of the valuation process, introduce this section with a summary of the income approach and its foundation, including the principle of anticipation. Include a discussion of the theoretical basis and the relationship of the income approach to the highest and best use of the property. Point out that income and expense estimates are consistent with the highest and best use conclusion and that any renovation or conversion of improvements suggested in the highest and best use section is reflected within this approach.

### A. Subject Property Operating Statement

Prepare a properly constructed operating statement for the subject property; indicate whether it is based on cash rental, share rental, or owner/operator information; and explain the basis for the estimated potential gross income. Include comparable rents, yields, and terms documenting your income estimates.

### B. Income Adjustments and Expenses

Explain and justify estimated mortality losses (livestock), vacancies, or other typical and annualized losses—especially for share and owner/operator analyses. Explain in detail the basis for estimating fixed and operating expenses. Identify your sources, whether they are historical data (subject and sales), suppliers, USDA, Cooperative Extension Service, etc.

### C. Capitalization Methodology

Explain in detail the capitalization methodology applied. Frequently, this section of the report is confusing to readers, so strive for clarity and simplicity. Always use definitions and explain your rationale for the methodology used.

### D. The Capitalization Rate

Provide calculations showing the development of the capitalization rates. Again, rates are calculated as of the date of the sale with rents, yields, and expenses applicable at that time. If possible, estimate capitalization rates using direct comparisons based on comparable sales information. Present the sale price and net income figures for comparable properties and calculate the indicated rates. If this analysis is lengthy, present a summary in this section, probably using a table, and include a more detailed analysis in the addenda. However the rate is determined, readers should have a clear understanding.

### E. Value Calculation

Show the calculations for capitalizing income into value. If more than one capitalization method is used, provide a reconciliation of the value indications into a single value for the income approach.

### F. Conclusion

Summarize your conclusions at the end of the income approach presentation. Explain to the reader the applicability of the income approach and point out any limitations that the income approach may have for this particular problem.

## V. Sales Comparison Approach

Unless previously discussed as part of the valuation process section, provide a brief summary of the methodology used in the sales comparison approach, emphasizing that the approach is based on the principle of substitution. Summarize how proper execution of the sales comparison approach is tied to the conclusions reached within the area-regional, neighborhood, property description, and highest and best use sections.

### A. Units of Comparison

Establish and discuss units of comparison. Depending on the property type and location, these might be price per bushel, price per box, price per animal unit, GIM/GRM, or other units, rather than, or in addition to, the "overall price per acre."

### B. Selection of Comparable Sales

Choose comparable sales that have a highest and best use similar to the property being appraised and that have an area-regional situation similar to the property being appraised. Discuss the rationale for selecting the properties and discuss similarities and differences between the comparable sales and the property being appraised.

## C. Comparable Sales Description and Analysis

Place individual comparable sales descriptions and analyses either in the body of the report or in the addenda. Place lengthy descriptions ("courthouse" data, such as book, page, type of instrument, dates, legal description, grantor, grantee, etc.; or physical information and economic information related to the sale, including size, soils, land type, USDA base acreage, yields, etc.) in the addenda. You may wish to present a condensed version of the sales information in the body of the report either in paragraph form or as a summary table. The goal is providing readers clear and accurate information comparing the sales to the subject property while not reducing the "readability" of the report with unnecessary data.[5]

When you execute the sales comparison approach, you should support your value adjustments using paired sales. The potential for "pairings" exists in most markets if you have all available data and sufficient knowledge of the circumstances surrounding each sale. In certain instances, pairings may result in isolating the effects of two or three variables. You may then use other sales data to further refine the indications into meaningful adjustment estimates. The lack of definitive market support does not eliminate the necessity to present a complete discussion of your adjustment reasoning for each category—e.g., location, size, time, productivity, etc. Demonstrating that market data is conflicting or inconclusive is an important part of the process of providing a defensible data analysis for the sales comparison approach.

Prepare a comparable sales map showing the location of the comparable sales relative to the subject property and include the map either in this section or in the addenda. Label maps clearly and completely to help readers understand the information provided.

## D. Sales Comparison Grid

Prepare a market data grid showing adjustments to sales and indicating a subject property value estimate based on each comparable sale. Follow the grid with a paragraph discussing the adjustment factors used. All adjustments entered in the grid should be directly supported using market evidence developed in the sales analysis section. Any adjustments that cannot be supported by market extraction should be clearly supported

5. Where you place the detailed sales analysis is your choice. You can place most material in the addenda with a summary in the body, or you might place all material in the body as part of the sales comparison approach. A third alternative is preparing a separate appraisal report section titled Comparable Sales Analysis and included as part of the factual data portion of the report. If you choose this approach, place all sales data and related sales analysis for all three approaches in this section. Along with the normal sales analysis for the sales comparison approach, the sale price for each comparable sale is separated into components to provide information for the cost approach, and the rate of return expected on each comparable sale is estimated for use in the income approach. As each approach to value is discussed, refer to the appropriate part of the sales analysis section.

using solid market logic. The sales comparison approach becomes harder to defend as the number of subjective adjustments increases.

### E. Conclusion

The sales comparison grid provides an indication of the subject property's value based on each comparable sale. Summarize the strengths and weaknesses of the value indicated by each sale used in the grid and discuss how you are reconciling the several indications of value into a single sales comparison approach estimate of value. Finally, discuss the strengths and weaknesses of the sales comparison approach value estimate relative to the sales data available for this appraisal problem.

## VI. Reconciliation and Final Value Estimate

Use the reconciliation to lead readers to your final opinion of value. Discuss the interdependence of the three approaches and the limitations of each approach for the appraisal problem. Restate the value conclusions derived from the three approaches and provide the reasoning for selecting your final opinion of value. Base your final value on the data and analysis presented. In effect, your discussion represents an appraisal of the appraisal. Discuss the quantity, reliability, and sensitivity of the data used in each approach, the reliability of the value indication for each approach, and what makes one of the approaches provide a better value estimate than one or both of the other approaches.

State the final opinion of value as a single figure, a range, or a single figure within a range. The appropriate choice depends on the nature and complexity of the appraisal problem.

## VII. Certification

The purpose of the certification is to certify that you are preparing the appraisal in a professional, objective, and ethical manner. Indicate that you are abiding by the code of ethics of an appraisal organization and state whether your are or are not state-licensed or certified. Indicate whether state and/or appraisal organization continuing education requirements have or have not been met. Sign the certification page showing the effective date of the appraisal report.

The certification should be similar in content to the following statement taken from USPAP Standards Rule 2-3:

> I certify that, to the best of my knowledge and belief:
>
> - the statements of fact contained in this report are true and correct.
> - the reported analyses, opinions, and conclusions are limited only by the reported assumptions and limiting conditions, and are my personal, impartial, and unbiased professional analyses, opinions, and conclusions.
> - I have no (or the specified) present or prospective interest in the property that is the subject of this report, and no (or the specified) personal interest with respect to the parties involved.

- I have no bias with respect to the property that is the subject of this report or to the parties involved with this assignment.
- my engagement in this assignment was not contingent upon developing or reporting predetermined results.
- my compensation for completing this assignment is not contingent upon the development or reporting of a predetermined value or direction in value that favors the cause of the client, the amount of the value opinion, the attainment of a stipulated result, or the occurrence of a subsequent event directly related to the intended use of the appraisal.
- my analyses, opinions, and conclusions were developed, and this report has been prepared, in conformity with the *Uniform Standards of Professional Appraisal Practice.*
- I have (or have not) made a personal inspection of the property that is the subject of this report. (If more than one person signs the report, this certification must clearly specify which individuals did and which individuals did not make a personal inspection of the appraised property.)
- no one provided significant professional assistance to the person signing this report. (If there are exceptions, the name of each individual providing significant professional assistance must be stated.)

If you are an Accredited or Professional member of the American Society of Farm Managers and Rural Appraisers who has met the continuing education requirements, you may include the following statement:

> The American Society of Farm Managers and Rural Appraisers conducts a mandatory program of continuing education. I am current with the requirements of the program.

## Addenda

Include necessary supplemental exhibits in the addenda. Examples of items that may be included are leases, lengthy legal descriptions, lengthy soil descriptions, supplemental maps of the subject property or comparable sales, additional photographs of the subject or comparable sales, more detailed comparable sales information, and copies of governmental restrictions placed on the subject. You may place the Qualifications of the Appraiser section either in the body of the report after the certification or in the addenda. Provide qualification information pertinent to your education, technical training, type of work, and years of experience.

## General Suggestions

The written narrative appraisal report reflects your professionalism as an appraiser. Organize the report in a logical manner clearly stating the facts, assumptions, and conclusions. The importance of correct spelling, grammar, punctuation, and sentence structure cannot be overemphasized.

The following suggestions may also be helpful:

- Have the report proofread carefully, preferably by someone not involved in writing the report.

- Rarely are first drafts acceptable as final products. Edit the report carefully to make sure the appraisal is being presented in a clear, concise manner.
- Consider what factors affecting the subject property in one approach should also be considered in the other approaches to value.
- Explain any departure from normal appraisal practice and state the effect of any adverse conditions on the subject property.
- Place North directional arrows on maps, plats, and aerial photos.
- Always include original copies of materials in the report unless available copies are clear and legible.
- Refrain from hand-lettering maps, plats, or charts.
- Either type the report or have it professionally printed.
- Provide sources of definitions in an acceptable footnote format.
- Double-spacing or spacing at one-and-one-half lines is not essential, although it is usually desirable.
- Separate sections of the report to enhance the appraisal's appearance.
- Never conclude that present use is the highest and best use simply because it conforms to other properties in the area or to zoning.
- Express estimates to the degree of precision compatible with good judgment.
- Always include a table of contents and number all pages.
- Avoid using technical terms prior to defining those terms.
- Double-check all calculations.
- Use attractive, business-like covers. Whenever possible, use vinyl or professional weight cover stock covers, cut to slightly larger than standard 8½-by-11-in. paper. A professional binding service is recommended.
- Date and describe photographs of the subject property, sales, and rent comparables, if required. Use good-quality photographs and affix them so they do not fall from the report. Place photos in a logical sequence.
- Include original or high-quality copies of ownership, soil, topographical, and aerial maps of the subject property and comparable sales. Draw floor plans to scale and show dimensions. Reductions of actual surveys or blueprints may also be used.
- Be careful of copyright laws because some items may be copyrighted by the author.

- Title all exhibits included in the addenda of the report. Use professional-quality exhibits and refer to them by title in the body of the report. Copies of original documents should be clear and legible. When feasible, present critical data such as sales data adjustment charts, adjustment analysis, and income information in chart or table form to enhance reader understanding.
- Use graphics to help readers understand the analysis.

APPENDIX
• G •

# Water Rights and Mineral Rights

## Water Rights

The legal right to water is as important to the value of a property as the physical source of the water. A *water right* is a right to a definite or conditional flow (quantity) of water, generally for use at stated times and in stated quantities, for irrigation, livestock consumption, real estate developments (subdivisions), hydroelectric power development, etc. Water rights vary greatly throughout the United States and are affected by numerous compromises, exceptions, and changes. Most water rights, however, are controlled by state law. Many disputes occur because of the fluid nature of water and the fact that laws were passed without scientific knowledge of the interdependencies of the water cycle. Water rights are discussed under two major headings: *surface water doctrines* and *ground water doctrines.*

### Surface Water Doctrine

Laws regarding surface water rights in the United States are complex primarily because of differences in the availability of water. The *riparian doctrine* is applied in states where water has historically been plentiful. In the most arid states, water rights are based on the doctrine of *prior appropriation;* some states employ both doctrines with the appropriative doctrine dominant. The major differences between the riparian and appropriative doctrines of surface water are summarized in Table G.1.

The riparian doctrine of surface water is based on English common law and gives the owner of the land bordering a lake or stream the right to use the water on the contiguous land. The riparian right exists solely because the land is located next to a natural water supply, and the right to the surface water resides in the ownership of the land.

The riparian owners up and down the stream have equal usufructuary rights. They are entitled to share the water for reasonable use without

**Table G.1** Summary of Riparian and Appropriative Rights to Surface Water

| Riparian Right | Appropriative Right |
|---|---|
| Prevails in Eastern U.S. | Prevails in Western U.S. |
| Is based on English common law | Assumes public ownership of water |
| Is linked to land ownership and physical contiguity | Is acquired by diverting water for beneficial use and meeting specific local requirements |
| Entitles user to reasonable use, without materially diminishing quality of watercourse; "right to simple usufructurary while it passes along" | Entitles user to physical consumption of a certain amount of water for specified land |
| Entitles user to equal sharing | Entitles user to consumption on a seniority basis; "first in time is first in right" |
| Depends on ownership of land in physical proximity to water and transfers automatically with the land | Is a real property right, separate from right to land and separately transferable |

seriously diminishing the quantity or quality of the water used by other owners. The owners' rights are equal, regardless of location along the stream or the time when each property was purchased. The riparian doctrine does not define "reasonable use" or guarantee any particular quantity of water. The resulting uncertainty can lead to legal conflict among the owners along a water course. Reasonable use is determined in a particular case only when one party brings suit and the court makes a decision. As long as the practical application of riparian rights is confined to areas where the supply of water generally exceeds the demand, all users are able to share the available water and there is no need to specify quantity or reasonable use.

A riparian right can be lost by selling the land next to the water, leaving the remaining land cut off from the supply of water. Similarly, the right can be lost if the stream shifts or moves away from the riparian land. A riparian right can also be lost by condemnation or by failing to prevent someone else from diverting water for a certain number of years, as set by the statute of limitations (prescription).

Under the doctrine of prior appropriation, water is owned by the state and may be granted for beneficial use through appropriation. Appropriative surface water rights are acquired by use and the performance of certain legal requirements, as directed by the state, the county, or an irrigation district. These rights apply to a specific quantity of water and often to particular times of use, places of use, and methods of diversion. Unlike riparian surface rights, which depend on ownership of land in physical proximity to water, appropriative surface rights can entitle the owner of the water right to surface water that is not contiguous to the land on which it is used. In addition, the owners of appropriate surface rights are entitled to consume or use up the water, provided the water is for beneficial use. The oldest water right has precedence over more recently acquired water rights. Simply stated, first in time means first in right. This means that in times of scarcity, the holder of the oldest right is entitled to a full quantity of water before any is given to later holders. This principle of seniority clearly differs from the equal sharing of water in the riparian doctrine. Many appropriative water rights can also be transferred independently of the land.

Although the details vary with state regulations, the usual pattern for establishing appropriate water rights is:

1. The landowner files an application to use a certain quantity of water on specifically designated land, taking the water from a designated diversion point.
2. The landowner proves beneficial use for this water.
3. The appropriate authority in that state grants a decree or adjudication.

The water rights may be evidenced by a contract with the Water and Power Resources Service (formerly the U.S. Bureau of Reclamation), with the Secretary of the Interior, or with a public utility water distributor. Water rights may also be given by an individual state certificate or decree, by shares of stock in an irrigation company, or by location in an organized irrigation district. These agencies differ widely in administration and the terms on which water is granted. The appraiser must therefore be aware of the different regulations, local considerations, and any changes in the regulations. If water is obtained from public agencies, the laws governing use can be changed. Such a change can put limitations on the number of acres for which federal water can be used. For example, the Reclamation Act of 1902 specified that federal water can be provided for only 160 acres per owner (320 acres for a farm couple). The Reclamation Reform Act of 1982 increased the acreage limitations to 960 acres for a family farm. The act also affects the use of federal water on corporate-owned lands, methods for calculating the cost of water used on excess acres, and sale of lands in excess of the acreage limitations.

Private agencies can also change the cost of water, the terms of delivery, or the dependability of delivery. For this reason, the appraiser should be aware of the management practices of the agency, the condition of indebtedness, type of payments or assessments, and any shares or voting rights accorded the owners of water rights. The members of a private irrigation company could, for instance, vote to raise the individual assessments on water rights, with a corresponding increase in the expenses of the irrigator dependent on this water. These organizations usually keep records of the amounts, timing, and costs of water used by irrigators, which can be employed by the appraiser in estimating the amount, sufficiency, dependability, and expense of the water used for a particular tract of land.

To maintain an appropriative right, the owner must demonstrate "beneficial use." An appropriative right can be lost by condemnation, prescription, abandonment, or nonuse for three to five years, depending on state law. With regard to water, the rule is: "Use it or lose it."

## Ground Water Doctrines

Most state ground water laws were developed after the surface water laws. Sometimes, ground water doctrines follow or combine the surface water laws of a state.

Resembling the riparian doctrine for surface water, the *absolute ownership doctrine* states that a landowner can take as much water as desired from beneath the owned land. The water is considered part of the landowner's soil. The *reasonable use doctrine* is a modification of the previous principle, since use of ground water can affect other landowners' water supplies. This modification limits each landowner to reasonable use, recognizing equal rights to ground water resources. A further modification of the reasonable use doctrine is the *correlative rights doctrine,* which states that the landowner's use must be correlated with the use of others, especially during times of water shortage. When water is in short supply, each landowner is entitled to ground water in proportion to the percentage of land he or she owns in relation to the other lands overlying that ground water supply.

Under the doctrine of prior appropriation, ground water, like surface water, belongs to the people of the state and is subject to appropriation for beneficial use through a priority system similar to that used for surface water. In 1959, Hawaii passed a ground water use act providing for regulation of withdrawals of ground water from designated areas by a state commission.

## Other Water Rights

A neighbor may gain *prescriptive rights* by openly diverting water to which others have prior rights. If the diverted water is used without interruption for the number of consecutive years prescribed by state law and other legal requirements are met, a prescriptive right against the original owner is granted. A prescriptive right may be lost in the same way—i.e., it may be prescribed by someone else openly diverting the water for the specified period of time. A *vested right* is a clause in appropriation law that protects the rights of persons who appropriated water for beneficial use before the passing of appropriation laws. A vested right can also be lost by lack of use.

The water rights in a particular location can affect the value of property in different ways. In an appraisal of irrigated property, the appraiser must know whether the water rights are appurtenant to the land or transferable separately from the land. If they are separate, the legal source, conditions, terms, and value of the water rights should be known and noted by the appraiser. If water rights are not transferred with the land, the property's value may decline significantly and its highest and best use may be changed.

## Mineral Rights Valuation

Mineral rights are valued for a variety of reasons, including changes of ownership, tax purposes, mortgages or collateral for a stock offering, condemnation, acquisitions, mergers, and equipment additions. The appraisal of mineral-producing properties does not differ in many respects from the appraisal of more traditional real estate. With some exceptions, the same valuation techniques and methods can be applied. However, care should be exercised to ensure consistent analysis of the available data and re-application of the extracted indications to the property interest appraised. This concept, called *the purity of re-application,* should always be followed. Simply stated, the various indications, multipliers, relationships, etc., derived from sale information should be reapplied to the subject at the same income level, production point, etc., from which they were extracted from the sales.

Mineral rights consist of the right to extract all minerals contained in or below the surface of a property. Mineral rights may be granted with surface rights or without surface entry because the "mineral estate" is the dominant tenant in most states. Where mineral rights are severed from the surface, the owner of the mineral interest has a right to occupy the amount of surface area that is reasonably necessary for extraction purposes. In such cases, however, the mineral rights holder must pay the market value of the surface acres occupied and any damages. In some areas, the surface damage income is greater than the owner's share of the mineral income from royalties.

A mining right for a specific tract of land is a right to enter and occupy the land for the purpose of extraction, either by underground or surface methods. In some instances, this right may also include use of vegetative resources, such as timber, on the surface for the extraction process.

### Types of Ownership

Substantial mining and quarry properties are owner-operated. The owner may have purchased the property from another private party or by patenting a mining claim. The development or transfer of the mineral rights in privately owned (fee title) land is accomplished within the framework of state laws relating to property titles, sales and conveyances, leases, licenses, and contracts. Private rights are typically subject to federal or state statutes relating to conservation, wildlife and wetlands mitigation, taxation, safety, and other matters.

When fee title is not owned, leases provide the vehicle for mining lands owned by private parties or federal and state agencies for specified types of minerals. Leases between mining operators and private owners have these characteristics:

- They may include a purchase option.
- Lease payment may take the form of a royalty based on production.

- The rights may be subject to abandonment or forfeiture.
- Written leases may cover subletting;

Leases between mining operators and public owners have these elements:

- The state "land commission" or a similar agency has jurisdiction over all mineral operations on state-owned lands or on lands that contain minerals reserved by the state.
- Mining operations on public lands are subject to assessment as possessory interests just as possessory rights for other purposes are assessed for tax purposes.
- A permit or lease to prospect for minerals, other than for oil and gas, may be required by the state.
- Leases include a production agreement, which will likely call for reclamation bonds and possible escrow for road work, utilities, etc.
- The right to extract minerals on federal lands may be obtainable by lease or permit from the Secretary of the Interior, as opposed to filing a claim or patenting.
- Federal mineral deposits subject to "leasing" include coal, petroleum, natural gas, phosphates, oil shale, sodium salts, and potash.
- Federal "permits" are available for pumice, clay, sand and gravel, and other "common varieties."

A royalty originally was the rent or tax paid for the privilege of mining. In the United States, a royalty is a reservation to the lessor or seller of a portion of the minerals, or proceeds from their sale, at no cost. There have been many sales of mineral properties in which the seller retained a royalty interest even though title changed hands.

The royalty is actually part of the production costs (gross royalties are typically deducted from net sales). It should be included in valuing the leasehold estate when it can be accurately determined and is not contingent on unknown future commodity prices, profit levels, or other intangible measures.

Valuations of minerals may entail analysis of:

- A patented claim
- An unpatented claim
- A mill site
- In-place reserves, i.e., proven and probable reserves
- Structures only (leasehold or improvements held in fee)
- An operating mine

Patented mining claims are legally equivalent to a fee simple title acquired from the United States under the 1872 mining law. A *patent* is the conveyance by the United States of title to a citizen, where the property was in the public domain and is being conveyed into private ownership for the first time. A patent is the equivalent of a quit claim deed from the sovereign. All valuable mineral deposits (i.e., gold, silver, iron, lead, copper, and other metals plus many nonmetallic items like gypsum, limestone, and rare earths) on or in public domain lands, not withdrawn from mineral entry, are open to exploration, occupation, and patent. Patented claims are subject to taxation since they are, in effect, fee property. They can be mortgaged or sold like any other fee interest. Lands not open to mining claims include federal lands not in the public domain, e.g., acquired lands; public domain lands withdrawn from mineral entry; privately held lands; patented townsites; and land deeded to the state for unpaid taxes.

Unpatented claims are, in effect, the right to explore and produce locatable minerals under the 1872 mining law. No application for patent is necessary to operate a mine on public domain land. In most cases, a legal opinion is required as backup to determine the disposition of unpatented claims. Unpatented claims include millsite claims, lode claims, and placer claims. Lode claims must be based on the discovery of a valuable mineral deposit. Lode claims may include extra lateral rights, if they include the top or apex of the vein or lode. Each lode claim is limited to about 20 acres (600 ft. x 1,500 ft, or slightly more than 20 acres) regardless of the number of locators. A placer claim may be located on all other types of valuable and identifiable mineral deposits. The placer location must also be based on a discovery, but a discovery not in a vein or lode formation. The maximum size for a placer claim is 20 acres per person. However, an association placer may contain 20 acres per associate, up to a maximum of eight associates and 160 acres in a single claim.

Millsite claims may be located on a maximum of five acres of nonmineral land in conjunction with either lode or placer claims for the purpose of milling or processing the mineral. If the mining claim is forfeited or abandoned, the right to the mill site may also be lost, unless it is used for custom milling.

## Mineral Categories

Minerals include, but are not limited to, several broad categories including sand and gravel, precious metals, quarry types such as building stone, clay deposits, gem stones, and oil and gas. Essentially, *minerals* is a term used to refer to any subterranean commodity that has or may have economic value. Specifically, minerals are solid, inorganic elements or elemental compounds with a definite atomic structure and chemical composition.

Rocks are the essential components of the planet. They are classified into three major groups, depending on where they were formed:

igneous, metamorphic, and sedimentary. Rocks are aggregates of many different grains, which are fused, cemented, or bound together.[1] Petroleum is defined as an oily, flammable, liquid solution of hydrocarbons, yellowish-green to black in color, occurring naturally in the rock strata of certain geological formations. When fractionally distilled, petroleum yields paraffin, fuel oil, kerosene, gasoline, etc.[2]

Valuations of minerals fall into two broad categories—*operating units* with delineated reserves and *prospects*, which includes properties in all stages of development, including inactive mines, wells, and deposits. Analysis of sales of similar properties is the most desirable method for valuing small mineral properties, but it is often difficult to perform because small, profitable operations are rarely sold.

Appraisers typically estimate leased fee value or the contribution to the surface estate if the minerals are leased, being developed, or in production. For nonproducing rights, the question is usually contribution value only. When appraising an operating entity, standard methods of valuation are used and all three approaches of value may be applied. Units of comparison applied could include price per barrel, price per ton, price per cubic foot, price per ounce, or a gross income multiplier based on operations, royalty, percent of retail, etc. The income approach may employ direct capitalization or discounted cash flow analysis using a market discount rate or property yield rate. In counseling assignments, yield capitalization and/or discounted cash flow analysis can be applied with an equity yield rate based on certain investment criteria. The cost approach may be the strongest approach when buildings and structures are to be included in the analysis. However, component values may be estimated from sales for in-place reserves, surface estates, or portions with "possible" reserves.

## Mineral Reserves

Reserves are the volume of accessible mineral material of acceptable quality that will produce a return, after operating expenses, under present economic conditions. "The distinguishing point between 'ore' and 'metal or mineral bearing material' is the grade above which extraction of the end product is profitable under a given set of cost-price circumstances. Economic factors are key to this consideration. Without change of grade (quality) or location, what is 'mineral bearing material' at one time may become 'ore' at another through development of cheaper mining or extraction processes or because of changes in one or more of the factors entering into cost of production vs. sale of product."[3] After a sudden, sharp decline in the price of uranium, one mining company executive said, "We have a lot of uranium but no reserves."

---

1. Pellant, 1992.
2. *Webster's New World Dictionary*, 3d ed. (Simon & Schuster, Inc., 1986), 1011.
3. Baster and Parks, *Examination and Valuation of Mineral Property*, 4th ed. (Cambridge, Mass.: Addison-Wesley Press, Inc.), 2.

### "Proven," "Probable," and "Possible" Reserves

Various classifications of reserves have been proposed over the years by industry, governmental agencies, university engineers, and geologists. These classifications are qualitative, and most refer to metallic ores, although, by extension, these systems could be used for any mineral, including oil and gas. Generally, three sets of terms have been used or proposed over the years (see Table G.2).

Over time, the industry will likely gravitate to one set of terms. Whichever term, or set of terms, are used, understanding the distinctions is the key.

**Table G.2 Reserve Terminology**

| Current Use | Long-Time Use | C.K. Leith | USGS & US Bureau of Mines |
|---|---|---|---|
| Proven | Proven | Assured | Measured |
| Probable | Probable | Prospective | Indicated |
| Resources | Possible | Possible | Inferred |

- *Proven or measured reserves* are measured ore for which tonnage is computed from dimensions revealed in outcrops, trenches, workings, and drill holes and for which the grade is computed from the results of detailed sampling. The sites for inspection, sampling, and measurement are so closely spaced and the geological character is so well-defined that the size, shape, and mineral content are well established. The computed tonnage and grade are judged to be accurate within stated limits, and no such limit is judged to differ from the computed tonnage or grade by more than 20%.
- *Probable reserves* include ore for which tonnage and grade are computed partly from specific measurements, samples, or production data and partly from projection for a reasonable distance on geologic evidence. The sites available for inspection, measurement, and sampling are too widely or otherwise inappropriately spaced to outline the ore completely or to establish grade throughout.
- *Resources or inferred resources* include mineralization for which quantitative estimates are based largely on broad knowledge of the geologic character of the deposit and for which there are few, if any, samples or measurements. The estimates are based on an assumed continuity or repetition for which there is geologic evidence; this evidence may include comparison with deposits of a similar type. Bodies that are completely concealed may be included if there is specific geologic evidence of their presence. Estimates of inferred or possible ore should include a statement of the special limits within which the inferred ore may lie.

The disparity is evident when government agencies will accept a 20% margin of error in the "proven" category alone. Industrial rocks and minerals have less susceptibility for error because 1) their lateral boundaries are often observable at the surface, 2) an individual operator commonly establishes the lower boundary of, say, a sand and gravel deposit without much difficulty, and 3) such deposits are commonly homogeneous and are not often characterized by the barren streaks typically found in metallic ore deposits. In short, water-laid sediments such as sand and gravel are typically far more extensive and uniform in character; thus, they have more easily defined boundaries than metallic ores.[4]

Resources are generally regarded within the industry as too speculative to be included in the value estimate derived by the income method, but they can be used when the buyers and sellers have incorporated value for nonreserve mineralization in their mutually agreed-upon sale price.

## Valuation and Appraisal Process

Two basic items are needed to begin the process of valuing mineral properties. First, an estimate of the amount and quality of the reserves is required. Few, if any, real estate appraisers have the background and knowledge to make such an estimate; thus, a geologist or engineer familiar with the type of reserve should be consulted. Second, the appraiser should have some experience in mineral valuation and possess technical training in data analysis, i.e., rate extractions from market sales, experience with discounted cash flows, etc. With an accurate estimate and the necessary experience, the appraisal of mineral rights does not differ significantly from any other type of real property valuation.

The mineral contribution of a "known" mineral deposit has a different value than the mineral contribution of an "unknown or unmeasured" mineral right. In the mining of limestone, precious metals, and other quarried minerals, the surface is substantially altered or changed and most industrial purchasers place little value on lost surface rights. By contrast, buyers of "non-producing" mines allocate values to the implied reserve by attributing a portion of the sale price to the excess land and the remainder to the mineral deposit. Therefore, it is reasonable to assume that a property with a known reserve performs at a higher market level than a property with "non-producing" or "unknown" deposits.

### Definitions

Before beginning the appraisal process, definitions applicable to mineral valuation assignments should be set forth. *Market value* and *fair market value* are the most common terms used in most types of mineral valuations, and they are often misunderstood. *Market value* is defined as

4. *Valuation of Mines & Quarries*, Assessment Standards Division of Property Tax Department, California State Board of Equalization, pp. 62–63.

> the most probable price, as of a specified date, in cash, or in terms equivalent to cash, or in other precisely revealed terms, for which the specified property rights should sell after reasonable exposure in a competitive market under all conditions requisite to fair sale, with the buyer and seller each acting prudently, knowledgeably, and for self-interest, and assuming that neither is under undue duress.[5]

Fundamental assumptions and conditions presumed in this definition are:

1. Buyer and seller are motivated by self-interest.
2. Buyer and seller are well informed and are acting prudently.
3. The property is exposed for a reasonable time on the open market.
4. Payment is made in cash, its equivalent, or in specified financing terms.
5. Specified financing, if any, may be the financing actually in place or on terms generally available for the property type in its locale on the effective appraisal date.
6. The effect, if any, on the amount of market value of atypical financing, services, or fees shall be clearly and precisely revealed in the appraisal report.

This definition has been modified by federal agencies, but the "willing seller and willing buyer" concept is fundamental. The term *fair market value* has historically been used by the courts and applied by government agencies.

The 1992 Interagency Land Acquisition Conference, which produced the Uniform Appraisal Standards for Federal Land Acquisitions, has defined *fair market value* as:

> The amount in cash, or on terms reasonably equivalent to cash, for which in all probability the property would be sold by a knowledgeable owner willing but not obligated to sell to a knowledgeable purchaser who desired but is not obligated to buy.

This definition specifically implies the "value in-exchange" concept or the value to a disinterested third party.

By contrast, investment value actually targets a specific type of buyer or set of criteria. *Investment value* is "the specific value of an investment to a particular investor or class of investors based on individual investment requirements; as distinguished from market value, which is impersonal and detached".[6]

The key to this discussion is the difference between *investment value* and *market value.* Investment value reflects the subjective relationship between a particular investor and a given investment. Investment value

---

5. Appraisal Institute, *The Appraisal of Real Estate,* 11th ed. (Chicago: Appraisal Institute, 1996), 20.
6. Appraisal Institute, *The Dictionary of Real Estate Appraisal,* 3d ed. (Chicago: Appraisal Institute, 1993), 190.

is the price a specific or individual investor would pay for an investment in light of its perceived capacity to satisfy his or her individual desires, needs, or investment goals. To estimate investment value, specific investment criteria must be known.

### Cost Approach

The cost approach is usually one of the foundations of the appraisal process, but in mineral valuation it must be applied with caution. While values for various components, such as the surface estate (if any), mineral deposit, etc., may be identified, the approach is most appropriate when applied to properties with improvements and more than one other component value.

Typically, there is a relationship between cost and value. However, in mineral valuation, the cost of development rarely sets the minimum value of the deposit. In fact, the contribution of development is typically less due to the extensive costs of "proving" reserves. Obviously, the value of a deposit must materially exceed development costs if a mine is to be profitable. The value of the deposit can only be recognized, however, if additional expenditures are made for plant and equipment. The question then becomes, "Is the value of the deposit more of a business or enterprise value, or is it actually inherent in the deposit?" From a market standpoint, the deposit probably has substantially less value as an independent component part; thus, it is enhanced by the expenditure necessary for the processing facilities and/or operation.

### Income Approach

The income approach can be a strong methodology when valuing mineral-producing properties. Direct capitalization or discounted cash flow analysis using either "operations" or "royalty" income are commonplace. The income method is less applicable to "non-producing" deposits.

The values of common minerals such as sand, gravel, and building stone are tied to construction cycles and interest rates. Road construction and highway funding are also relevant factors. Sand and gravel tend to be less susceptible to value changes due to more consistent levels of demand. In fact, property yield rates, internal rates of return, and market discount rates extracted from sand and gravel deposits have frequently been viewed as threshold rates because most other types of minerals are viewed as more risky, more costly to develop, or more susceptible to change. Thus, the threshold rate may be used to set a lower limit for the discount rate applicable to a deposit of another, noncomparable mineral. Comparable sales are rarely available, so the appraiser's ability to extract a market threshold rate assists in the decision making and valuation process.

### Risk vs. Uncertainty

Many people equate risk with uncertainty but there is a substantial difference between the two. *Risk* is defined as the opportunity for loss. *Uncertainty* applies to factors where the outcome is not certain, but the opportunity for loss is not as apparent. Statistics show that monetary loss rather than gain is the norm for most mineral properties. In evaluating an investment, however, there are often many uncertainties regarding legislation, competition, etc., which must be considered. The most important single variable in the evaluation of mineral investments is risk. Risk can be quantified to some degree by placing the components into an equation:

Measured recoverable reserve + economics + risk = development decision

The ultimate objective is to quantify ignorance, or bias, or both. The proven reserve estimates quantify the first component of the equation to some degree, at least as far as reasonable. The remaining two factors—economics and risk—become the cornerstone around which the discount rate or decision to develop must be assessed by the appraiser.[7]

If discount rates, costs, projection periods, and other indicators are derived from market sales and reapplied to the subject at the same income and expense levels, the resulting value would most likely be a reflection of market value. Nevertheless, the appraiser must ask, "Can we duplicate the specific circumstances and characteristics (and resulting value opinion) in the market for other, somewhat similar properties with similar levels of risk?" Market value is a "value-in-exchange" concept, reflecting that amount of money for which the appraised estate or rights can be openly exchanged in the market. It is not the value to a specific investor based on set criteria. Market value is detached, impersonal, and objective.

For example, stating that the discount rate is 10% and reducing the cash flows via a discounted cash flow analysis to net present value is most likely to produce an estimate of investment value. The key is the assumption that the 10% rate is a "market rate." If the rates applied meet specific investor objectives or are narrowly interpreted, the resulting value is not market value. The question is, "Where did the 10% rate come from?" Is it an equity yield rate, a property yield rate, or an overall capitalization rate? If it is a property yield rate, was it computed using constant dollars or changing dollars? More importantly, what level of expense was applied, so the same levels of expense can be used to appraise the subject. Additionally, was it an "after-tax" or "before-tax" rate? Was it derived from an operation with $100,000 worth of equipment and start-up costs, or one requiring $150,000,000—150 times more investment? Although investment value is markedly different than market value, the terms are frequently, and erroneously, used interchangeably.

7. R.E. Megill, *Exploration Economics*, 2d ed. (Tulsa, Okla.: PennWell Books), 110–111.

The chief appraiser for a federal agency recently said this about market value: "An income stream can only be converted to a present worth, using a discount rate derived from sales with similar risks and quality of income. The use of 'built-up' rates based upon alternative investments is often proposed as an alternative to market derived discount rates. These synthetic rates might be appropriate for feasibility studies or risk assessments, but failing to be a product of market participants who are in the competitive arena for properties similar in nature to the property appraised, these estimates do not represent market value. Discount rates are often quoted from financial and business publications, within company stockholder reports, or internal financial analysis of a particular company's performance goals to use in the appraisal process, but these rates are not appropriate for the same reasons 'built-up' rates are inappropriate. As such, any resulting estimates of value are only 'numbers,' and do not reflect the market."

Equity yield discount rates have been used in discounted cash flow models, but they have generally been inappropriately applied. "Anticipated rates of return must not be confused with realized rates. Financial journals often publish information on realized rates of return on the book value of equity and on total investment, and these may sometimes be at low levels, even lower than interest rates. The published rates are low partly because they are 'after-tax' rates; however, they may also reveal that present profits are providing a low return on investments. What they do not reveal is that the present values of the properties involved may be greatly depressed. In short, realized rates of return on present owners' investments may diverge sharply from the investors' rates of return that might be derived if those same properties were sold."[8] Rates of return (property yield rates) on a sale property are always higher than equity rates, and markedly higher if a "before-tax" analysis is applied.

In the case of an operating field, mine, or pit, a valuation is an estimation of the future earnings discounted to reflect present value. A DCF analysis is appropriate when the subject is an operating property and the value of the "going concern" is sought. DCF analysis is inherent in the present worth calculations commonly used by mineral companies to determine whether an investment in the property is justified. Therefore, it typically represents an estimate derived using their investment criteria—not value to a third party. An appraisal of an operating entity hinges on what a buyer will pay to obtain the forecasted income stream. Common sense indicates that a buyer will not pay one dollar for each dollar of anticipated future profits, as there is risk inherent in the security of that future income. Commodity prices could decline, market share could be lost to competitors, etc. Risk, therefore, effectively

8. *Valuation of Mines & Quarries*, Assessment Standards Division of Property Tax Department, California State Board of Equalization, pp. 92–93.

reduces the market value of the income stream. In addition, the buyer seeks to profit from the money and time invested in the project. Hence, the anticipated rates of return used internally or by equity investors must not be confused with realized rates from the entire property.

The contrast between investment value and market value is illustrated by a sale in Alaska. In this case, 15 patented claims to the property were valued at $140,000 by a well-known Alaska mining engineer and placer miner who used traditional engineering discounted cash flow methods. The property sold to the highest bidder at auction for $27,600 in cash. While some may call this liquidation value because the property sold at auction, there were several bidders and it was advertised for several months. The buyer indicated "the property had a lot of potential," but sold for all that it was worth. The investment parameters applied by the mining engineer dictated the value of the property based on his assumptions. By contrast, the market indicated that those assumptions were not correct given the level of risk, location, etc. "This is not an approach that should be utilized by an appraiser who is not thoroughly experienced in appraising mineral properties. Even when used by an appraiser experienced in this field, however, this is a highly speculative appraisal method and great care should be exercised in its use. As one court stated: "Great care should be taken, or such valuations can reach wonderland proportions."[9]

For mineral properties, market value is much more subjective, and income forecasts rarely mirror actual performance. A knowledgeable real property appraiser will know if the discounted cash flow analysis (DCF) is incorrect at the outset. However, the appraiser's objective is to test the elements used in the cash flow analysis and arrive at a supported conclusion. The conclusion reached in the DCF analysis is not a value estimate. The DCF is only a model used to test the assumptions incorporated in the forecast. Income projections are typically only used for feasibility studies. Market tests of similar property should be used to "define" the market or market parameters. The appraiser should not arbitrarily assign a risk factor, estimate a dollar amount for unknown risk and other factors, and analyze the cash flow with a nonmarket, intercompany analysis rate or financial equity or stock rates. Generally, there are too many unknowns that cannot be quantified without direct market evidence.

The income approach is most commonly used and referred to for the appraisal of mineral properties. However, it is frequently defined incorrectly or involves mixing rates, expenses, and levels of income to produce an inaccurate estimate of value. The strongest application of the income approach is direct capitalization. Simply put, there is no better way to discount future income than to measure the relationship be-

---

9. *Uniform Appraisal Standards For Federal Land Acquisitions*, Interagency Land Acquisition Conference (Washington, DC: U.S. Government Printing Office, 1992), 23–24.

tween the net income and sale price of similar property. Nevertheless, alternatives techniques may be necessary when data is limited. The royalty income method may be the most appropriate when the value of the mineral "in place" is being sought because it reflects the landowner's interest or the value of the realty as opposed to the value of the business.

### Some Pitfalls of Discounted Cash Flow Analysis

The valuation of assets, which may include an existing or prospective income stream, is commonly approached with an analytic financial tool known as a discounted cash flow (DCF), net present value (NPV) technique. Discounted cash flow techniques mix historical cost and risk experience with projections of future income in an attempt to predict how time will impact the investment. The technique has been used for so long and abused so much by so many, that it is rarely called to account for its basic shortcoming.

Typically, a discussion of DCF-NPV valuation tends to focus on the components of the analysis—i.e., the basic assumptions—rather than the relevance of the analysis itself. The principal shortcoming of a DCF-NPV analysis is that the underlying assumptions constitute a forecast of operating and market conditions. Some scenarios are inherently lower in risk than others, but no forecast is ever 100% accurate and analyses are rarely risk-qualified. The farther the forecast extends into the future, the more likely it is that the forecast conditions will change. The second shortcoming of the technique is that, although it was developed to compare diverse investments, where various risks and problems tend to balance out or at least reach a comparable level of uncertainty, it is often applied to a single investment. In this application, no risk minimization or balancing can occur and the uncertainty inherent in the forecast is unmeasured.

Basic mathematical conventions, applied without thinking, can have an extraordinary impact on the DCF-NPV analysis. The typical conventions used in any forecast include simplification, rounding, and averaging.

Discounted cash flow techniques were developed before the advent of modern computers. Therefore, to complete the analysis in a timely manner using manual calculations, simplification was necessary. Simplification effects arise from common practices such as projecting uniform conditions into the future, holding discount rates constant, and lumping all investments into Year 0 and/or Year 0, Year 5, and Year 10 frameworks. Discount factor tables assume that all operating cash inflows and outflows take place at either the beginning of the year or at the end of the year when, in fact, cash flows ordinarily occur irregularly throughout the year. Compound interest tables, which more accurately reflect reality, are available but seldom used because the irregular nature of cash flows tend not to be readily predictable.

Financial theoreticians disagree on definitions for the cost of capital over and above "interest expense" on borrowed money, particularly if the asset is held free and clear. Choosing an appropriate discount rate under these conditions is problematical.

Considerable controversy and disagreement arises among experts concerning the selection of an appropriate discount factor. Part of the problem stems from a failure to understand what a discount factor is supposed to provide. The concept of interest suggests that the use of money has a value (or cost). Furthermore, the continuous use of money over time, rather than as a series of instantaneous transactions, is implied in DCF analysis. When experts confuse discount rates and interest rates, significant errors can occur. Components of the discount rate include: opportunity cost; transaction cost; and estimates for risk, uncertainty, and inflation. As a practical matter, when DCF techniques are applied, the expert typically goes to the level of detail that time and budget will allow, given the expert's field and applied knowledge, and then often throws caution to the winds in selecting a discount rate. The discount rate, then, becomes a sort of "Fibber McGee's Closet" where all unknowns are kept.

There are at least four ways to develop discount rates: 1) rates can be developed from the market (e.g., the appraiser's capitalization rate, 2) "built-up" rates are created by estimating the components mentioned above, 3) "industry practice" rates are generalizations that are probably not relevant to any individual project, and 4)the "cost of capital" rate is obtained by estimating the corporation's overall cost of capital and equating it with the discount. Most experts have a distinct preference, although they may not understand or be able to articulate the basis for their choice. It can be useful to distinguish between internal project risk and external business risk in choosing a discount rate.

Aside from use of inadequately supported assumptions ("garbage in, garbage out"), extreme reliance on DCF-NPV analysis as a substitute for experience and good judgment can lead to financial disaster.

### Sales Comparison Approach

In the sales comparison approach, value is estimated by comparing properties similar to the subject which have been sold. The approach is highly recommended and is held by the courts to be the strongest indication of value. However, appraisers are seldom blessed with comparable sales. At best, the appraiser has information on sales with inherent differences in location, size, and the quality of each transaction. The appraiser may have to analyze a variety of alternative property types for comparison purposes and to set valuation parameters. A key element in applying this approach is to reduce the sales information into units of comparison across several mineral types to provide a common basis for comparison.

**Market Effects of the Federal $100/Mining Claim Fee**

By 1993 the considerable impact of setting rental fees at $100 per claim was evident in the marketplace for unpatented mining properties. The fees reflect the actual ownership situation of title to unpatented claims. The United States is the permit holder and the claim holder is in the position of a lessee or permittee. In the past, the title owner was required to perform certain direct and indirect financial activities with regard to claims. Diligent development, including annual assessment work, was required along with routine filings, work descriptions, plans of operation, environmental studies, etc. The new fees, however, supplement the work requirements and are payable directly to the title owner, the United States. The immediate effect of the new fees was a nationwide reduction in the number of recorded claims. As of August 31, 1993, for example, 51% of all unpatented claims in Nevada were dropped. Logically, these claims had a value to the owner lower than the cost of maintaining them, which was $100 per claim per year. Claims in this category had previously sold, at times, for a few hundred to a thousand dollars each.

In Alaska, in particular, where the small miner's exemption was a factor, claim holders dropped claims until they retained only nine claims to qualify for the fee exemptions. In many cases, the excess claims were not dropped but sold or traded for small amounts of cash and/or equivalents to business associates and/or family members so that some control was maintained over the original group. Transactions of this nature were for $5.00 to $500 per claim, cash or barter. These transactions tended to be structured so that the purchaser could also qualify for the exemption.

Larger claim groups with corporate exploration targets, in Nevada and elsewhere, also saw values shift. In prior years, claim holders typically sought to lease or option their claims to an exploration company. Remuneration involved a signing bonus and periodic rental or option payments. Some transactions included work commitments, buyout provisions, and royalty payments if exploration was successful. In 1993 many of these claim holders were subjected to renegotiation. The exploration companies sought to propose that they take over the payment of fees to the United States and reduce or eliminate the rental or option payments to the property holder. Incentives offered to the claim holders included small, one-time bonus payments; the promise of future royalty interest; or simply the threat of withdrawal from the exploration agreement and cancellation of work. Subsequently, many of the professionals, mostly geologists, involved in this end of the business were forced to lower their payment demands and/or sought to replace these high-cost properties with lower-cost alternatives in Mexico or South America. The market for claims with delineated proven and probable reserves was unaffected as the incremental cost of holding was small compared with the potential or actual profits from mining activities.

The segment of the claim market most profoundly affected saw large declines in market value. Claims that formerly sold for $1,000 per claim or more suddenly brought only a few hundred dollars or less. Large claim groups containing exploration targets were reduced from annual payments of $20,000 to $40,000 (and up) to token payments plus a possible royalty interest. Corporate exchanges of exploration properties are not yet numerous enough to quantify losses in market value.

Other factors influencing the market for unpatented claims include increased environmental costs, longer lead times due to permit-related issues, increased taxation at the state level, and uncertainty about land tenure if the mining law of 1872 is changed or discarded. Considerable skepticism about the overall business climate and the current administration policies have also had an effect. These factors, along with the $100 fee, have served to redirect exploration budgets overseas.

Presently, mining venture capital finds a raw prospect in the jungle of Argentina, for example, more attractive than a well-defined target in Nevada. Some of the most active, small exploration companies are thinly capitalized and the $100 fee reduces their budgets so much that there is insufficient money left to perform meaningful exploration work. In addition, many Canadian exploration companies were also adversely affected by the decline of the Canadian dollar relative to the U. S. dollar, making the effective fee $130 to $140 Canadian per claim. The outcome is fewer discoveries and even longer lead times as the $100 fee is deducted from a fixed work budget. This has, in turn, resulted in a general lack of interest in projects in the Western United States. The overall cost of acquiring property overseas tends to compensate for the $100 fee when travel, foreign exchange, taxation, and logistics are taken into account.

More importantly, the political issues overshadowing the U. S. mining industry have been instrumental in diverting capital. In the United States, lead time for production is approaching five to 10 years when environmental impact analyses (EIAs) are challenged. U. S. government policies do not encourage, but rather discourage and delay, opening new mines. In Latin America, governments now seek new mining ventures, offering preferable tax treatment and requiring less rigorous review prior to issuing permits. Thus, lead times are shortened, positive cash flows are achieved earlier, overall mining costs are lower, and ore grades are often higher. All these factors combine to enhance the investor's return.

From the standpoint of stock market valuation, North American gold stocks have never fallen sharply because of increased fees and royalties in the United States. Consequently, stock prices did not rise when mining legislation failed in Congress. Some analysts view the net effects of the fee changes as positive because many companies have found better opportunities overseas. Mining has been an international industry and companies can change the focus areas for major investments very quickly and successfully.

Should new mining legislation pass Congress, a royalty in excess of 3% of net revenues is considered "onerous" by at least one major gold analyst. In the face of such legislation, property values could not be expected to rise over the short term without a compensating boost in commodity prices. A substantial increase in prices for most metals does not usually occur without the development of worldwide shortages and/or substantial increases in demand. The impact of the current political climate on a U.S. mine owner is to depress selling prices, sometimes to the extent that the market becomes inactive. Valuation of mine properties becomes problematic during periods of market inactivity. Caution should be exercised before reaching back into historic databases insofar as past transactions may not be relevant. Domestic political policies have little worldwide effect on metal prices when deposits in less developed countries are available on attractive terms.

In summary, the market for mine properties can be extremely volatile. Property values cannot always be statistically related to, or explained by, changes in the prices of the retail commodity. Detailed examination of the past and future market buyer motivation and other factors are critical to the analysis of sales data and subsequent valuation of a specific property.

## Oil and Gas Valuation

Oil and gas properties are, in general, quite similar to other mineral properties in terms of the valuation methods applied. As with other minerals, oil and gas are depleting resources. The value of oil and gas properties is a function of reserves (usually in thousands of barrels), of oil (in millions of cubic feet), and of gas quality (gravity of the oil or the BTU content per cubic foot of gas). Other relevant factors include the time required to produce the reserves, the costs of production and transportation to market, and the associated risks of oil and gas operations. The term *reserves* indicates that these resources can be extracted economically at the time the reserve estimate was made.

Oil and gas properties tend to be more complicated than other mineral entities because of the wide range of ownership interests that may be encountered. Ownership can be broken down into several categories. It is necessary to know what type of ownership is being evaluated in the appraisal process.

### Types of Oil and Gas Interests

Various ownership interests are described below.

| | |
|---|---|
| Mineral owner | The person or entity possessing title to the mineral property. The mineral owner may or may not hold title to the associated surface. |

| | |
|---|---|
| | In this case, the mineral estate is considered severed from the surface ownership. |
| Leasehold owner | The person or entity holding a lease on the mineral estate with the right to extract the oil and gas reserves present. |
| Royalty interest owner | That portion of the produced oil and gas retained by the mineral owner as partial compensation for leasing to the leasehold owner. This is normally a percentage of the gross production and is subject only to severance and income taxes. |
| Overriding royalty interest owner | That portion of the produced oil and gas retained by a third party or parties generally as compensation for services or subleasing of a particular mineral property. This is also normally a percentage of the gross production and is subject only to severance and income taxes. This interest is created from or carved out of the leasehold interest. |
| Working interest owner | The party who pays 100% of the costs of production. Working interest owners are usually the leasehold owners, but not always. |
| Reversionary interest | Sometimes called a *back-in,* this is an interest which will revert normally to the leasehold owner at a pre-designated time, such as pay out of the drilling and completion cost of an oil or gas well. This interest is usually an option to continue with a given overriding royalty, or to convert that interest to a larger working interest. |
| Net revenue interest | The net revenue to the working interest owners after all production costs, royalty, and overriding royalty interests have been paid. |

To further complicate many oil and gas properties, vertically segregated mineral ownership may exist. One company may have a leasehold interest in a given mineral property down to a certain depth below ground level, e.g., 8,000 feet below the surface, while a second company may have the leasehold for all zones below 8,000 feet.

It should be noted that royalty interest and overriding royalty interests are basically "no risk" positions since little or no capital is generally at risk. The potential incomes from these interests are subject to the capabilities of the operators of producing properties. The more efficient the operator, the longer the property will continue to be economic and generate income to the various parties.

## Determination of Oil and Gas Reserves

As with other mineral properties, the reserves present on a given property is of primary concern to a buyer. Several methods can be used by the geologist or petroleum engineer to determine the reserves for a given property. Since oil and gas are fluids rather than hard rock minerals, some differences do exist in the way that reserves are measured. Four basic methodologies have been used in oil and gas reserve determination: analogy, volumetrics, material balance, and decline curve analysis.

Analogy is a simple comparison of the production history of wells in an area and a projection of what is expected for the subject property in terms of rates and ultimate recoveries. This may work fairly well in areas of multiple wells and consistent operations.

Volumetrics is the calculation of oil in place using reservoir characteristics determined from well bore data and geological and geophysical mapping. Recovery factors for the types of oil or gas in question are then used to determine the amount of recoverable reserves to be expected. This is quite similar to measuring reserves of hard rock minerals based on ore body mapping using drill hole and geophysical data.

Material balance estimates use engineering formulas to determine remaining recoverable fluid volumes based on the amount of fluid already produced and the change in reservoir energy that has occurred with the removal of those fluid volumes.

Decline curve analysis is the plotting of production versus time. With increasing time, characteristic trends can be seen in these graphical presentations. Projections can then be made as to future production rates and cumulative volumes down to an economic producing limit. It is desirable to have at least two to three years of production history for a given well or field to improve the reliability of this method. It should be noted that decline curve analysis is probably the most common method for determining remaining producible reserves where production exists.

## Reserve Classification

Reserves of oil and gas are classified in several ways. Proven and producing reserves are usually considered to be 90% to 100% sure. Proven and developed, but not producing, reserves tend to be in the 80% to 90% sure range. Proven, but not developed, reserves may be risked in the 60% to 80% sure range.

Some companies carry estimates for probable reserves and sometimes possible reserves. Obviously the risk of these classifications is very high. Oil and gas property buyers will normally not attribute much value to such reserve estimates. The determination of proven producing reserves is best obtained by the material balance or decline curve methods. Volumetrics and analogy are more likely used in reserve estimates for proven but undeveloped reserves. In any event, what a buyer pays for an oil or gas property is determined by what the buyer felt the reserve numbers were. This may be different than the numbers determined by the seller. In some cases, a qualified third party (geologist or oil and gas expert) may be asked to make a determination of reserves for a property.

A further classification of oil reserves involves the potential for secondary and/or tertiary recovery of known in-place resources. Oil can only be produced from a reservoir as long as sufficient energy exists in the reservoir to move fluid to the well bore for extraction. Almost all reservoirs have enough energy present initially to allow oil to be produced either by flowing to the surface or by lifting of the oil to the surface by mechanical means. Most reservoirs have only enough energy to move a small percentage of the in-place oil to the well bore. Only about 10% to 20% of the oil in place in most reservoirs can be recovered using this natural reservoir energy. This is called *primary production.*

In many reservoirs, the primary production is the only economically recoverable oil to be extracted There are situations where artificial energy can be introduced into the reservoir to aid in moving more of the in-place oil to the well bore for recovery. This constitutes *secondary production,* which in some cases can result in the recovery of more than 50% of the original oil in place. The introduction of chemicals and surfactants into a reservoir may also enhance recovery. This is sometimes referred to as *tertiary production.* From an appraisal point of view, it is necessary to know what the property's total potential for producible reserves is. This information must also be known for the comparable sales used in analyzing the subject property.

Natural gas production is almost entirely dependent on natural reservoir energy. There is little or no way of economically introducing additional energy into a predominantly gas reservoir. However, the recovery of in-place gas resources is much greater than for oil, because of the expansive nature of this fluid.

## Commodity Value (Quality)

As mentioned earlier, the value of oil and gas reserves is related to the quality of the fluids produced. Oil type has a significant effect on the value in dollars per barrel received by producers. Crude oils are of two types: paraffin-based or asphalt-based. In addition, the gravity of the crude is also important. This is a measure of the viscosity of the oil and relates to the various refined products which can be obtained from a given type of crude oil. Higher-gravity oils tend to be more valuable than low- or very low-gravity oils. The difference in price per barrel can be several dollars, which is quite significant when the average price per barrel is in the $15 to $26 per barrel range. Crude oil is also classified as sweet or sour. Sour oil contains hydrogen sulfide, a nasty contaminate that must be handled and refined carefully. Sour crude oils are less valuable than similar-gravity sweet crude oils.

Natural gas also can vary in value. The primary determinant of natural gas value is its BTU content per cubic foot and the gas liquids that may be stripped from the gas. The standard normally set for natural gas is 1,000 BTU per cubic foot. Gas containing fewer BTUs per cubic foot than the standard is less valuable. Hydrogen sulfide can also be a problem in gas production and cause an associated loss in value.

The market value for crude oil has been quite unstable for the last several years. Seasonal swings in demand and international instability have caused significant changes in crude oil prices over fairly short periods of time. However, even with such fluctuations in price, the average benchmark prices have remained in the \$16 to \$18 per barrel range for the last few years. Most companies and their banking institutions have been using present values per barrel with very little price escalation into the future. Although many industry people are projecting price increases in the 5% per year range, recent history does not support even these modest price increases for economic analysis.

## Production Life

As with other types of minerals, the rate of production and the ultimate life of economic production are important in determining the value of oil and gas reserves. This is reflected in a discounted cash flow approach to the value of a given oil or gas property. However, regardless of the method used by a buyer to determine what price is to be paid for a property, the economic life and production rate are critical. Oil and gas production is different from other minerals in that the characteristics of the reservoir control the economic production rates. By contrast, mining production can often be sized to match the market and maximize extraction efficiencies.

The determination of production rates and the economic life for a given oil or gas property is usually based on the decline curve analysis for currently producing properties. Yearly averages for production can be extracted from the decline curve projections and used in estimating annual cash flow potential and the ultimate economic life of a given property. When proven, but not developed, reserves are being analyzed, it is necessary to use the more ambiguous method of analogy to obtain these estimates.

## U.S. Market for Oil and Gas

In the case of crude oil production within the continental United States, market demand is not a problem. All of the crude oil that can be produced has a market, although the price per barrel may vary due to international supply. Gas markets are more unreliable. There is and has been for several years a gas surplus in this country. Logistics and regional purchasers have a significant influence on the value of gas reserves. In many cases, existing production is limited by low demand. In some situations, there is no current market for known gas reserves.

# Index